Unmanned Aircraft Systems in the Cyber Domain

UNMANNED AIRCRAFT SYSTEMS IN THE CYBER DOMAIN

PROTECTING USA'S ADVANCED AIR ASSETS

Second Edition

R.K. Nichols, J.J.C.H. Ryan, H.C. Mumm, W.D. Lonstein, C. Carter, and J.P. Hood

NEW prairie PRESS
open access scholarly publishing

Second Edition

Cover design by Kira Miller
Cover image created by the Defense Advanced Research Projects Agency (DARPA)
and is available at https://www.darpa.mil/news-events/2016-03-31

Published by R.K. Nichols, Salina, Kansas

ISBN 978-1-944548-23-0

The first edition, published in 2018, was supported in part by Kansas State University Libraries' Center for the Advancement of Digital Scholarship under their Open/Alternative Textbook Initiative, grant approved by KSU Panel, January 2018.

Dedications

From: Professor Randall K. Nichols, DTM

I dedicate this book to three groups: **All USA serving and retired military personnel**, USA Coast Guard and federal and state law enforcement for keeping our country safe; to my Angel wife of 35 years, Montine, and children Robin, Kent, Phillip (USA Army), Diana (USA Army), and Michelle who have lived with a Dragon and survived; and finally, to all my students (over 50 years) who are securing our blessed United States from terrorism.

From: Dr. Hans C. Mumm

I dedicate this work to my students and colleagues and all those innovators; those dreamers that race against time as they create a future that is ever changing and evolving in ways that we cannot even imagine today. Your dedication to the field of autonomous systems will bring about positive change to the world landscape and humankind.

From: Wayne D. Lonstein

I dedicate this work to my wife and best friend Julie, my sons Ethan, Ari and Sam as well as my extended family and co-workers and my co-authors from whom I have learned so much. To all those brave souls who have made the ultimate sacrifice serving this nation, as well as those who have, are or will serve in our armed forces, police, fire and other emergency functions and their families who silently sacrifice. May our work in some way help you perform your duties more effectively and safely and through your service may the world becomes a more peaceful and harmonious place for all.

From: Dr. Julie J. C. H. Ryan

I dedicate this work to my husband Dan and to my students, who have taught me so very, very much.

From: Candice Carter:

I dedicate this work to an exceptional leader, mentor, and master of Bushido; Professor Randall Nichols. His commitment to training dragons to be successful in asymmetric warfare and in life is unprecedented. I am honored to be a lifetime dragoness trained by the master of Nito Ichi Ryu Ni To.

From: CPT John-Paul Hood:

I dedicate this work to my loving and supportive wife Katie, my two daughters Evelyn and Gwendelyn as well as my extended family whom continue to support me through this journey. Thank you for your love, encouragement and presence in my life.

Contents

Section VII. Technology Updates

Disclaimer

Information contained in this work has been obtained by the authors from sources believed to be accurate and reliable. However, neither New Prairie Press, R. K. Nichols (publisher), the U.S Army, the Department of Defense, Kansas State University, nor any of its authors guarantees the accuracy or completeness of the information published herein and neither any of the above mentioned parties nor its authors shall be responsible for any errors, omissions, or damages arising out of use of this information.

This work examines *inter alia* technical, legal and ethical dimensions of behavior regarding cybersecurity and Unmanned Aircraft Systems (UAS). It is not intended to turn counter terrorism, information technology, engineers or forensics investigator professionals or drone operator / pilots into lawyers. Many of the topics discussed will be concerned with the law and legal implications of certain behaviors. Every effort is made to provide accurate and complete information. However, at no time will legal advice be offered. This work is published with the understanding that the authors are supplying information but are not attempting to render professional services. Any reader requiring legal advice, should seek services of a lawyer authorized to practice in the appropriate jurisdiction. All scenarios discussed in this work are hypothetical in nature and not to be taken or construed to be actual occurrences.

The authors, publishers and associated institutions specifically represent that all reasonable steps have been taken to assure all information contained herein is from the public domain and to the greatest extent possible no information of a confidential or classified nature is set forth herein. Additionally, this misuse, re-engineering, retransmission or republication of any content, information or concept contained herein shall not be permitted unless express written permission is granted by the authors, publishers and associated institutions. Additionally, any use of the aforesaid information by any party or intentionally disseminated to any third party or parties for any illegal or improper purpose is expressly forbidden.

Foreword To 1st Edition

by R. Kurt Barnhart, Ph.D., KSUP Associate Dean of Research

It gives me great pleasure to commend this work to you the reader after having spent a great deal of time with the manuscript in recent weeks. Although still in draft form at the time of my review, I can say with certainty that the breadth and quality of information you will find herein is unparalleled in the unclassified sphere. This book will fully immerse and engage the reader in the cyber-security considerations of this rapidly emerging technology we know as unmanned aircraft systems (UAS). Many of these same vulnerabilities affect unmanned technology across the board and regardless of mode, however the focus of this work is exclusively on those vehicles which operate in the National Airspace System (NAS).

Aircraft without on-board human pilots have been around in various forms longer than piloted aircraft. In 1783 Joseph-Michael and Jacques-Ètienne Montgolfier performed the first heavier-than-air vehicle flight in Annonay France. The passengers were a sheep, a pig, and a chicken (at least the chicken had a fighting chance if things went awry). It has, however, only been within the last couple of decades that this technology has burst onto the modern stage driven by the distinct technological advantages associated with eliminating the risks and limitations of protecting humans on-board. Advances in hardware and software have driven UAS capabilities far beyond what many imagined just a few short years ago. Today we stand at the precipice of a period in history where, looking forward, most vehicles in the air will not be occupied. As a result, given that we in the U.S. are constantly on the receiving end of withering cyber-attacks, a detailed treatment of this subject matter is of national importance as we protect and secure our national interests.

When noted cyber-security pioneer and lead author professor Nichols and I began to engage in a dialogue on this topic several years ago, it was clear that there were large and looming gaps in unmanned systems that had already been exploited on the international stage from a cyber-perspective. Many of those gaps remain unaddressed today. Understandably, commercial technology developers remain keenly focused on gaining a competitive advantage and delivering products to market albeit often without thorough cyber risk assessments and mitigations. This book will give system designers, users, and their management teams an introduction to what it will take to begin to close many of the vulnerabilities associated with UAS in order to produce systems that will serve the market better by being much more reliable, capable, and secure than they would be otherwise. This book takes advantage of the extensive knowledge of multiple working experts in the realm of cyber-security and they have each done an excellent job at uncovering and detailing the core issues at hand as we continue the march toward full

NAS integration of UAS in the not-to-distant future. Let's take a brief look at what the reader will find herein.

In Section one, "The UAS Playing Field" the reader will gain an understanding of the history and scope of UAS as a technology and will come to have a greater understanding of the UAS market and of the policies which both enable, and inhibit the deployment of the technology into the NAS. In chapter three, the final in section one, some of the key vulnerabilities associated with UAS are introduced and discussed.

In section two, "UAS Information Security, Intelligence, and Risk Assessment", the reader will gain a more detailed exposure to the vulnerabilities of the information necessary for UAS to operate and thereby will appreciate the differences between explicit, implicit, and derived security requirements. Chapter four concludes with a paragraph which says that "Communications may need to have confidentiality, integrity, and availability protected". How that is integrated into UAS design is of high importance. Chapter five examines types of, and sources of, intelligence data and discusses common attack/defense scenarios for UAS. Finally, section two concludes with case studies that highlight the vulnerabilities of UAS in the cyber-domain.

Section three is all about collision avoidance systems which are indeed the "heart and soul" of a fully integrated and useful system of unmanned aircraft. Sense and avoid (SAA) systems are discussed in depth along with one significant antagonist of SAA systems which is "stealth design". Finally this section concludes with a detailed discussion of a related system which is the 'smart skies' collaborative commercial project of which SAA is a critical component.

Section four primarily relates to the defense applications of Intelligence, Surveillance, and Reconnaissance (ISR), weapon systems security, and electronic warfare considerations and other information-centric operations. This section should not be dismissed by those without a focus on military applications as often it is the military that simply encounters technological vulnerabilities first given the dynamic operational environment they are associated with.

Section five looks at the data vulnerabilities of the various system components and explores the relationships and associated vulnerabilities of intra-system communication pathways. Chapter 14 delves into the realm of electronic warfare from a detailed perspective including a discussion of the intelligence information cycle as well as "jamming" operational vulnerabilities. This section concludes with discussion of current international threats and considerations related to still-emerging political scenarios where UAS technology is front and center.

As I conclude this overview of the work you are about to delve into I would encourage you to read this work along with a ready-copy of today's most current headlines. In doing so you will discover that the topics covered in this book are not only of interest today, but of critical importance to the future of us all.

Dona nobis pacem,

R. Kurt Barnhart, Ph.D.
Associate Dean of Research
Kansas State University Polytechnic

Salina, KS

Foreword To 2nd Edition

by Alysia Starkey, Ph.D., KSUP CEO & Dean

I am delighted to write the forward for the second edition *of Unmanned Aircraft Systems in the Cyber Domain: Protecting the USA's Advanced Air Assets*. The first edition was published in 2018 and quickly established itself as a must-have text for academic programs and individuals looking to expand their knowledge in this emerging cyber discipline. Thanks to that text, conversations at conferences and other professional networking events this year were easy and spirited. As soon as *Kansas State University* was spotted on my name tag, individuals asked if I knew of the book and were eager to share their experiences.

The fact that the second edition follows so quickly after the first is indicative of the rapid pace in which the manned and unmanned airspaces continue to converge and the need to fully understand the direct impact on the civilian market. During the past year, I watched and listened as the authors electronically discussed and passionately debated the topics of national interest presented in this text. Their collective intellect coupled with the depth of their professional experience challenged my assumptions and continued to do so as I read the completed second edition.

As an educator, I appreciate that the second edition is structured in the same way as the first. Each chapter starts with the student learning objectives found therein, include value-added graphics, and concludes with scenarios and discussion questions for in-class use. This brings increased efficiency when creating a course syllabus or developing content-related assessments. The second edition further expands on the topics presented in the first edition. Prior knowledge of manned and unmanned aviation regulations, military/civilian/commercial unmanned applications, and basic cybersecurity concepts is beneficial for the reader to fully engage with the material.

It is my expectation that this text will provide an effective learning experience and become a referenced resource for students and professionals working to secure the national airspace and reduce known and unknown threats with this developing technology.

Alysia Starkey, Ph.D.
Interim CEO and Dean
College of Technology and Aviation
Kansas State University Polytechnic
Salina, KS

Preface To 1st Edition

HISTORICAL PERSPECTIVE

Unmanned Aircraft Systems (UAS) In the Cyber Domain: Protecting USA's Advanced Air Assets is the working product of five talented authors to meet the needs of students enrolled in Kansas State University Polytechnic's (KSUP) graduate Certificate in UAS – Cybersecurity. The book also serves as one of the technical resources for the KSUP Professional Masters in Technology (PMT) offering in their UAS – Cybersecurity discipline.

Interest in UAS-Cybersecurity Certificate / PMT specialty programs developed from two directions; one internal and external to the college. Internally it dates to 2014, when the KSUP Associate Dean for Research and Executive Director of the UAS Research Laboratory, Dr Kurt C. Barnhart, met with Professor Nichols to discuss the possibility of state-of-the-art Cybersecurity Masters and / or Certificate program. These would meet the need for outside online programs to enhance the University profit structure. Associate Dean Barnhart in 2014 approved the concept of a Graduate Certificate in UAS – Cybersecurity and gave permission to move forward with its development. The program was placed under the purview of the College of Technology and Aviation. Final program approval was given by the KSU Board of Trustees in January, 2017. The five courses in the Graduate Certificate UAS – Cybersecurity program were also approved for the Professional Masters of Technology (PMT) in 2017.

In 2014, Professor Nichols had discussions with students and professionals in multiple schools and states inquiring about the prospect of an Unmanned Aircraft Systems – Cybersecurity Masters curriculum or graduate certificate program at KSUP, especially the on-line component. Their perception was that there was a market of not only freshmen / transfers / graduate students who might be interested in such a program, but a larger market of working professionals in need of skill advancement, and of a forum for the discussion of developments in the industry. They also felt that the college could anticipate financial assistance from federal, state, aviation, corporate, law enforcement, and defense organizations to get such a program launched. There was considerable enthusiasm and a general feeling that a cybersecurity concentration to defend UAS assets and their Command, Control, Communications, Computers, Intelligence, Reconnaissance and Surveillance (C4IRS) systems from cyber-attacks would serve the interests of the college and its students, as well as those of the security / defense industries.

The outside interests from the intelligence and aviation communities became acute after the 2011 RQ-170 incident where Iran was credited with its capture. In addition, in 2014, Iran claimed the downing of an Israeli Hermes 450 Drone over Natanz. Reports like these caused major government concerns. Better risk assessment and teaching active cyber defenses is required to protect UAS assets. Hence, the graduate Certificate program in UAS – Cybersecurity was born.

The new MPT / Certificate discipline in UAS – Cybersecurity is NOT about drone training like that of Embry-Riddle Aeronautical University. **Its mission is CYBERSECURITY protection of UAS / UAV / Drones as Information Assets in the Air, all the networked computer systems related to the Intelligence / Counter Intelligence functions, and their payloads.**

MISSION

A key concern is the safety of integration of UAS systems into the National Air Space (NAS). **A critical component of this safety is the hardening of UAS/ sUAS /UAVs to cyber-attacks.**

The focus of this new program is on leadership, planning, and state-of-the-art practice for professionals in UAS / UAV aviation concerned with protecting this advanced technology against cyber-attacks or hostile/ intentional control of Command, Control, Communications, Computers, Intelligence, Reconnaissance and Surveillance (C4IRS) systems, or Loss of Signal (LOS) to critical navigational components. This program applies to all UAS / UAV personnel preparing to act or working as pilots, operators, communications, payload, navigation, ground support, satellite coordination with assets, or air-to-air delivery.

The Graduate Certificate Program in Unmanned Aircraft Systems – Cybersecurity requires five three-hour credit courses for certification. Each course is required to reflect current knowledge and practice in terms of cybersecurity, Information Security (INFOSEC), Communications Security (COMSEC), and Risk Assessment (RA) as applied to both safe integration of UASs into the National Airspace (NAS) and deployment for global Counter Terrorism operations (CT).

All courses in the proposed certificate focus on knowledge and skills to understand UAS / UAV issues related to UAS cyber security. If students desire to complete a Professional Masters in Technology (PMT), four courses from this certificate can be applied as electives towards the professional Master's Degree in College of Technology and Aviation.

The certificate program has one concentration – cybersecurity. CyberSecurity (in the context of cyber-conflicts) is defined in this document as, "the broad *tree* of investigation and practice devoted to cybercrimes, computer forensics, Information Assurance, Information Security (INFOSEC), Communications Security (COMSEC), and especially Cyber Counter Intelligence (CCI)" (Nichols, 2008). Cyber Counter Intelligence indicates the involvement of computer-based sensitive information, or information operations for three distinct sciences operating in the cyber realm: Cyber Counter Sabotage (CCS), Cyber Counter Terrorism (CCT), and Cyber Counter Espionage (CCE). (Nichols, 2008) In this book, Cybersecurity is limited to the prior three investigation areas. Computer *forensics* is the discipline that combines elements of law and computer science to collect and analyze data from computer systems, networks, wireless communications, and storage devices in a way that is admissible as evidence in a court of law (US-CERT, 2015)

The primary concerns of the graduate certificate program are protection of UASs / Small UAS (sUAS) / Unmanned Aircraft Vehicles (UAVs) from cyber-attacks, through negligent or hostile means, and teaching cyber security risk assessment principles to practitioners involved with UAS operations on land, sea, air, or satellite platforms. The impact of Loss of Signal (LOS) or intentional interference in UAS communications or navigation systems cannot be overstated. At the lowest end of the scale is the risk of a downed vehicle, mid-range risk is collision and failure to sense and avoid other vehicles or commercial / military traffic, and at the top of the risk scale is the hostile takeover of a payload to be used against US or US interests. It is not "good enough" to operate, fly or support UASs. Professionals must be concerned with protection of their charges.

Unmanned Aircraft Systems (UAS) *In the Cyber Domain: Protecting* USA's *Advanced Air* Assets is the authors attempt to provide some of the raw materials / tools for our students at a reasonable cost. (Free download like the MIT Open courseware project under a CCL open license arrangement.)

UAS – CYBERSECURITY CERTIFICATE PROGRAM COURSES

COT 680. Unmanned Aircraft Systems and Risk Assessment. (3) Fall. This course is an introductory course in Unmanned Aircraft Systems (UAS) history, elements, US aviation regulations, operations, use of geospatial data, automation and safety issues; detect and avoid systems, sensors and payloads, and human factors. Special attention to UAS Cyber Security Risks, Threats, Impact, Vulnerabilities, and Countermeasures will be identified. Various risk assessment equations will be used for qualitative risk analysis of threats so identified.

COT 682. Open Source Cyber Surveillance / Intelligence. (3) Fall. One of the key public concerns for safe integration of UAS into the NAS is privacy. This course questions the technical gaps, Intelligence Community (IC) assumptions, and important legal issues related to open source cyber surveillance / intelligence with emphasis on UAS activities/ deployment. Topics addressed include the responsible, legal, and ethical use of data and information gathered from the use of unmanned, semiautonomous systems, web data mining, social networks, and other modern technological systems.

COT 684. Advanced Topics in Cyber Data Fusion and Cyber Counter Intelligence. Prerequisites: three of four courses in the sequence. (3) Spring. This course is scenario-based applying cyber surveillance techniques and analysis of collected data to realistic, terrain-oriented problems. Topics include the digital soldier and sailor, 360-degree battlefield awareness and the use of unmanned, semiautonomous technologies. Risk assessment and cyber security countermeasures are the "glue" to successful implementation of data fusion techniques. Various risk assessment equations and other methods will be used for qualitative risk analysis of identified cyber threats. Cyber Counter Intelligence technology is applied to cases.

COT 686. Risk Management for UAS Operators, Pilots, and Ground Personnel. (3) Spring. UAS operators, pilots, and ground personnel must be committed to safety if the goal of UAS integration into NAS is to be accomplished. The best tool for assessment and determination of safest possible flight is risk management. This course introduces three risk assessment tools for UAS operators, pilots, and ground personnel to manage the workloads associated with each phase of flight.

COT 688. Sense and Avoid Technologies in UAS. (3) Summer / fall. This course is an advanced course in Sense and Avoid (SAA) technologies for UAS. SAA is extremely important concept and is the main obstacle for wider application of UAS in non-segregated airspace related to traffic safety in civilian and military/ defense domains.

TARGET AUDIENCE

Clearly, the students in the UAS -Cybersecurity Certificate and MPT programs, along with KSU's Aviation and Technology Department and UAS Research Laboratory, are the targets for this book. Cyber attacks and hostile control of UAS should not be underestimated.. It is as real as cyber attacks on computers, networks, personal identities, intellectual property loss, and delivery of cyber weapons on the battlefield. The larger audience are UAS operators, pilots, and ground personnel, owners and computer network analysts to manage the workloads associated with each phase of flight in any service: military, commercial, or recreational. Those concerned with UAS communications, navigation, payload, battery, sense and avoid, emergency components, satellite links, ground station links, materials construction and risk assessment / management associated with novel designs may well benefit from our textbook. All are factors in the vulnerable cyber domain.

STRUCTURE OF THE BOOK

Several themes covered in this text:

- C4ISR, Payload recovery, communications interference in the many different platforms,
- SAA and navigational functions and their interactions in the NAS (i.e. vulnerabilities)
- Protecting UASs from hostile intent in the Cyber Domain, and
- SCADA systems and how they may be exploited and protected in UAS vehicles.

Unmanned Aircraft Systems (UAS) In the Cyber Domain: Protecting USA's Advanced Air Assets is divided into five sections:

Section 1: The UAS Playing Field

Unmanned Aircraft Systems (UAS) – Defining UAS Cyber Playground

Chapter 1 A view of the UAS Market

Chapter 2 UAS Law – Legislation, Regulation and Adjudication

Chapter 3 Understanding Hostile Use and Cyber-Vulnerabilities of UAS: Components, Autonomy vs. Automation, Performance Trade-offs, SCADA and Cyber Attack Taxonomy

Section 1 above is concerned with the basic components and taxonomy of UAS that are vulnerable to cyber influence.

Section 2: UAS Information Security, Intelligence and Risk Assessment

Information Security (INFOSEC), Intelligence and Risk Assessments

Chapter 4 INFOSEC – Protecting UAS Information Channels & Components

Chapter 5 Intelligence and Red Teaming

Chapter 6 Case Studies in Risk for UAS

Section 2 above introduces the concepts and tools of Risk Assessment, Open Cyber Intelligence / Reconnaissance, network security, INFOSEC and vulnerability analysis. The use of Attack / Defense scenarios is introduced.

Section 3: UAS Heart & Soul – Sense and Avoid (SAA) Systems / Stealth

Sense and Avoid (SAA) – Heart of the UAS Package & Stealthy Design, its Soul

Chapter 7 SAA Sensors, Conflict Detection, and Resolution Principles

Chapter 8 Designing UAS systems for Stealth

Chapter 9 Smart Skies Project

Section 3 above focusses on the Sense and Avoid systems and common approaches to reduction of risk for failure of those systems. It also studies the brilliant Smart Skies project with speculations as to how the systems could be breached.

Section 4: UAS Weapons & ISR &IO

Payloads – UAS Delivery Systems

Chapter 10 UAS Intelligence / Reconnaissance / Surveillance Technologies (ISR)

Chapter 11 UAS Weapons

Chapter 12 UAS System Deployment and Information Dominance (ID)

Section 4 above concentrates on the unclassified UAS weapons systems, EW and IO systems, Information Dominance (ID) and surveillance technologies – all that can potentially be breached via cyber means.

Section 5: Computer Applications & Data Links – Exposing UAS Vulnerabilities via Electronic Warfare (EW) & Countering with Low Probability Intercept Signals (LPI)

UAS Vulnerabilities and Electronic Warfare (EW)

Chapter 13 Data – Links Functions, Attributes, & Latency

Chapter 14: Exposing UAS Vulnerabilities via Electronic Warfare (EW) & Countering with Low – Probability Intercept Signals (LPI)

Section 5 above is concerned with the attributes, functions, latency features of UAS communications links on ground, air, sea, and satellite.

Section 6: UAS / UAV Hostile Use & Countermeasures

Adversary UAS / Drone Hostile Use

Chapter 15: Africa – World's First *Busiest* Drone Operational Proving Ground – Where Counter-Terrorism and Modernization Meet

Chapter 16: Chinese Drones in Spratly Islands, and Threats to USA forces in Pacific

Section 6 above steps into the headlines of today. Part of the material comes from Professor Nichols' presentations to the public about hostile use of drones.

As our book goes to press, more potent examples of UAS Cyber intrusion (globally) may arise and will be included as time permits. In the meantime, the authors suggest that interested readers follow www.globalincidentmap.com or www.aviation.globalincidentmap.com both track the current global terror and non-terror incidents involving planes, and UAS.

Randall K Nichols, DTM
Professor of Practice
Director, Unmanned Aircraft Systems (UAS) – Cybersecurity Certificate Program
Managing Editor / Author
Kansas State University Polytechnic Campus &
Professor Emeritus – Cybersecurity, Utica College

Linkedin Profile:
http://linkedin.com/in/randall-nichols-dtm-2222a691

Illi nunquam cedunt.
"We Never Yield"

Bibliography

Nichols, R. K. (2008). *Cyber Counterintelligence & Sensitive Compartmented Information Facility (SCIF) Needs – Talking Points,*. Utica College, Chair Cybersecurity. Utica New York: Private Memo to R. Bruce McBride. Retrieved September 5, 2008

US-CERT. (2015, August 27). *Computer Forensics.* Retrieved from US-CERT: https://www.us-cert.gov/sites/default/files/publications/forensics.pdf

Preface to 2nd Edition

Summary

It has been less than a year since the first edition of *Unmanned Aircraft Systems in the Cyber Domain: Protecting USA's Advanced Air Assets* was published. Three different factors have spurred the authors into updating their textbook. First, unmanned aircraft technology has seen an economic explosion in production, sales, testing, specialized designs and friendly / hostile usages of deployed UAS / UAVs / Drones. There is a huge global growing market and entrepreneurs know it. Small UAS companies have been reproducing like rabbits. Only the FAA has been stumbling block trying to balance UAS safe integration into the National Airspace against hundreds of thousands new recreational and commercial operators testing their meddle in the skies. FAA's best efforts surround its decision to register UAS and provide a process for Part 107 Certification. Certification brings sanity and education into a chaotic public market in the US.

Second, hostile use of UAS is on the forefront of DoD defense and offensive planners. They are especially concerned with SWARM behavior. The author presented at several international C-UAS conferences which were attended by commercial, educational and military organizations for the purpose of hardening USA air assets against hostile drone activities. These were serious conversations and workshops – many of them behind closed doors and interacting with military brass.

Third, UAS technology was outpacing our first edition. Everyday our group read / discussed new UAS developments in navigation, weapons, surveillance, data transfer, fuel cells, stealth, weight distribution, tactics, GPS / GNSS elements, SCADA protections, privacy invasions, terrorist uses, specialized software and security protocols and more. As authors we felt compelled to address at least the edge of some of the new UAS developments. It was clear that we would be lucky if we could cover some of the more interesting and priority technology updates. The 2nd Edition adds six more chapters (see below) to harvest information on important advances in the UAS theater. We were privileged to bring on Captain John P Hood (US Army) as our military advisor and co-author.

Here is an outline of topics in the new chapters in our 2nd Edition:

Section 7: Technology Updates

Chapter 17: High – Altitude Platforms (HAPS) – A Promise not Reached

Student Learning Objectives
Introduction

Chapter 17 looks at the promise of UAS High Altitude Platforms (HAPS). It follows a similar investment path as that of UAM (Urban Air Mobility) systems for transportation. Lots of money, lots of new technology, lots of players, and failure to complete the mission.

Chapter 18 is an interesting look at Counter Unmanned Aircraft Systems (C-UAS), large scale UAS, and restrictions that the DoD and government has to suffer to extinguish UAS threats.

Chapter 19 presents the research formulations / Intellectual Property of Professor Nichols which were presented at two 2018 conferences. [1] It discusses the technology behind use of loud ultrasonic sound at specific frequencies to disrupt the MEMS components driving the rotors of a Hostile UAS, forcing the aircraft into a destructive path. It works best with SWARMS because the number and organization can be matched by the LRAD weapons. Chapter 19 also presents the novel idea that the same frequencies that can be used to down a UAS can also be used to identify friend or foe (IFF) by creating a searchable library of sound frequency signatures. Cur-

1. Prof Nichols was the Invited Keynote Speaker and Panel Moderator, (29-30 March 2019) 1st UAS CON for Law Enforcement and First Responders, speaking on Drone Wars: Threats, Vulnerabilities and Hostile Use of Unmanned Aircraft Systems (UAS) and Small UAS (sUAS), and Acoustic Defensive Countermeasures against SWARMS, Hazard Community & Technical College, Hazard, KY.He also was an Invited Speaker and Panelist (13-14 March 2019) 7th Annual DoD Summit, speaking on: Hardening USA Unmanned Systems Against Enemy Countermeasures, Alexandria, VA.

rently IFF units are too expensive and require too many SCADA and power communications to be included in SUAS / mid-level UAS. Prof. Nichols and his team are seeking grant / funding for testing at a national anechoic chamber.

Chapter 20 addresses the legal and regulatory conditions in the US that UAS operators / owners and defense planners face. Globally, restrictions are much lighter than in the US. It is a mess that FAA and others need to solve for the industry to grow in a challenging multi-issue environment.

Chapter 21 brilliantly addresses the Chinese Land / Sea New Silk Road Strategy and how UASs are being deployed for ISR operations and people control as well as interference with other nations assets. It presents a disturbing picture and one that should be taken to heart. The reader should also engage in self – learning by reading two seminal texts on the subject: 1) Brenner, J. (2011) *America the Vulnerable: Inside the Threat Matrix of Digital Espionage, Crime and Warfare.* New York: Penguin Books; and 2) Corr, A., Editor. (2018) *Great Powers, Grand Strategies: The New Game in the South China Sea.* Annapolis: The Naval Institute Press.

Chapter 22 presents a subject rarely discussed in public or regulatory offices – ethics. It looks at UAS and AI interfaces and how they present a real problem for society and act as a market barrier for an expanding UAS market. Several tough ethical cases are presented for evaluation.

We trust our 2nd edition will enrich our students and readers understanding of the purview of this wonderful technology we call UAS.

Best

Randall K Nichols, DTM
Professor of Practice
Director, Unmanned Aircraft Systems (UAS) -Cybersecurity Certificate Program
Managing Editor / Author
Kansas State University Polytechnic Campus &
Professor Emeritus – Cybersecurity, Utica College

LinkedIn Profile:
http://linkedin.com/in/randall-nichols-dtm-2222a691
Illi nunquam cedunt.
"We Never Yield"

Acknowledgments

Books such as this are the products of contributions by many people, not just the musings of the authors. *Unmanned Aircraft Systems in the Cyber Domain: Protecting USA's Advanced Air Assets*, 2nd Edition, has benefited from the review of numerous experts in the field, who gave generously of their time and expertise. In addition to named subject matter experts, this book was reviewed by sources in the two federal agencies who must remain anonymous. Their contributions were especially helpful in not releasing protected information, classified or deemed exportable categories. We will name only a few and clearly miss some special friends whose contributions were noteworthy. For this we apologize in advance and beg your forgiveness.

There are several people we would like to shout out a special thank you for your guidance, support and experience from Kansas State University / Kansas State University Polytechnic (KSU / KSUP): Dr. Richard Myers, President KSU; Dr. Kurt C. Barnhart, Associate Dean of Research and Executive Director of the UAS Research Laboratory KSUP; Dr. Alysia Starkey, Acting Dean & CEO of KSUP; Dr. Terri Gaeddert, Director of Academics, School of Integrated Studies (SIS) KSUP; Dr. Donald V. Bergen, prior Director of Graduate Studies KSUP; Fred Guzek, Professor and current Director of Graduate Studies KSUP; Dr. Kurt Caraway, Executive Director UAS, Dr. Michael Most. (Retired) UAS Department Chair, Dr. Mark J. Jackson, Professor, SIS KSUP; Dr. Saeed Khan, Professor, SIS KSUP; Professor Raju Dandu; Dr. Katherine Jones, KSUP Research and Library; Rachel Miles, Assistant Professor, Hale Library KSU; Lisa Shappee, Director, KSUP Library; Beth Drescher, Grant Specialist KSUP; Charlene Simser, Professor and Coordinator of Electronic Publishing at New Prairie Press, Chad Bailey, Instructor SIS KSUP, Professor Troy Harding; and especially Joel Anderson, KSU OVPR and Research Director.

Next comes our writing team: Dr Julie J. C. H. Ryan, CEO, Wyndrose Technical Group, is hands down the best subject matter expert (SME) in the Information security field. Dr. Hans C. Mumm is an expert in leadership and UAS weapons – a lethal combination. Dr. Wayne C. Lonstein, Esq., a previous Dragon (Nichols 'student) has gained recognition (licenses and certifications) in both law and cybersecurity. Professor Candice C. Carter, a Dragoness who is the creator of a cybersecurity program at Wilmington University and travels globally closing specialized cybersecurity breaches in major corporations. Capt. John Paul Hood, US Army, (our military advisor and previous Dragon) stepped up for a chapter in the 2nd edition. Professor Nichols is author / developer of six Masters and Certificate programs in Cybersecurity at Utica College and KSUP with five decades of experience.

Our textbook has been developed to replace two expensive textbooks in four of his graduate classes in the KSU graduate UAS Cybersecurity Certificate program and the KSU Professional Masters in Technology specialty. We would have failed our mission without our editor Aris

Theocharis. Many times, we growled under our breath for the changes required knowing always, Aris was right.

Finally, E. Montine Nichols deserves a commendation for her help on the final drafts and copy edit work for our book. Several KSUP UAS pilot – students helped with the "student view," and made valid suggestions for improvement, Randall Mai, Jeremy Shay, Vincent Salerno, Senior Airman in Kansas Air National Guard, John Boesen, (our handwriting expert); Diana K. Nichols, Josh Jacobs and Jordan McDonald. Special thanks go out to Devon S. Carter for website ideas and Kira C. Miller who professionally developed (more like "nailed") our cover art.

Randall K Nichols, DTM
Professor of Practice
Director, Unmanned Aircraft Systems (UAS) – Cybersecurity Graduate Certificate Program
Managing Editor / Author
Kansas State University Polytechnic Campus &
Professor Emeritus – Cybersecurity, Utica College

List Of Contributors

Professor Randall K. Nichols, DTM (Managing Editor* / Author)

Randall K. Nichols is Professor of Practice in Unmanned Aircraft Systems (UAS) – Cybersecurity at Kansas State University Polytechnic (KSUP) in Salina, Kansas. Nichols serves as Director, graduate UAS- Cybersecurity Certificate program at KSUP. Nichols is internationally respected, with 50 years of experience in leadership roles in cryptography, counterintelligence, INFOSEC, and sensitive computer applications. Throughout his career, Nichols has published seven best-selling textbooks. Nichols has provided counsel to the United States government and is certified as a federal subject matter expert (SME) in both cryptography and computer forensics. His most recent work involves creating master and certificate graduate – level programs for KSU and Utica College. To wit:

- Author/ Developer: MPT/ MS / Certificate in Unmanned Aerial Systems (UAS) -Cybersecurity
- Author/ Developer: BS Unmanned Aerial Systems (UAS) -Cybersecurity
- Retired Chair and Program Developer: MS – Cybersecurity –Intelligence and Forensics
- Retired Chair and Program Director: BS – Cybersecurity and Information Assurance
- Co-Author / Developer: MPS – Risk Assessment and Cybersecurity Policy
- Author / Developer: MS Cyber Surveillance and Warfare

Previously, Nichols was COO of INFOSEC Technologies, LLC, a consulting firm specializing in Counter-Terrorism, Counter-Espionage, and Information Security Countermeasures to support its 1700 commercial, educational and U.S. government clients.

Nichols served as CEO of COMSEC Solutions, a Cryptographic / Anti-virus / Biometrics Countermeasures Company, which was acquired by a public company in 2000. He served as Vice President of Cryptography and Director of Research of the acquiring firm.

Nichols served as Technology Director of Cryptography and Biometrics for the International

Computer Security Association (ICSA), President, and Vice President of the American Cryptogram Association (ACA).

Areas of Expertise / Research Interests

- Counterterrorism / Counter- Intelligence /Counterespionage / Computer Security
- Countermeasures Asymmetric Warfare and Attack / Defense Scenarios against National Critical Infrastructure
- Computer Forensics and Cryptography SME & Federal Expert Witness (Federal Criminal Cases: Treason / Espionage)
- Risk Assessment / Threat Analysis / Vulnerabilities Analysis / Countermeasures
- Cybersecurity / Surveillance Technologies: Aerial, Infrared, Visual, Ultraviolet, Radio, Radar & Sonar
- SCADA – Advanced Cyber-weapons Creation / Deployment / Deployment / Defense
- UAS- Integrating Unmanned Aircraft Systems into National Airspace System
- Designing Acoustic Countermeasures against hostile -actor UAS SWARMS & developing dual purpose IFF sound libraries.

Contact Prof. Randall K Nichols, DTM at 717-329-9836 or profrknichols@ksu.edu.

*Direct all inquiries about this book to Prof. Randall K. Nichols, DTM at profrknichols@ksu.edu

Dr. Hans C. Mumm (Co-Author)

Dr. Hans C. Mumm holds a Doctor of Management with a concentration in Homeland Security from Colorado Technical University (CTU) and an MS in Strategic Intelligence from American Military University (AMU). He gained notoriety during Operation Iraqi Freedom as the officer in charge of the "Iraqi Regime Playing Cards; CENTCOM'S Top 55 Most Wanted List" which was touted by the Defense Intelligence Agency (DIA) as one the most successful Information Operations (IO) in the history of Defense Intelligence Agency (DIA). Dr. Mumm is the former Division Chief for Cyber Security at the Office of The Director of National Intelligence (ODNI) programming and executing a budget of over $140M. Dr. Mumm has earned twenty-three personal military ribbons/medals including six military unit medals/citations, and two Directors Awards, from the DIA. In 2016 he was awarded the People of Distinction Humanitarian Award as well as being granted a US Patent and Trademark for How to Harmonize the Speed of Innovation and

Change with the Human Spirit's Need for Leadership. In 2005, Dr. Mumm was recognized as one of the "Ten Outstanding Young Americans," and in 2003 he was awarded the National Defense PAC "American Patriot Ingenuity Award" for his service during "Operation Iraqi Freedom."

He co-authored an international best-selling book titled "Lightning Growth" which is a follow up to his best-selling book in 2015 titled "Applying Complexity Leadership Theory to Drone Airspace Integration."

He is a published researcher in both the scientific and social science arenas and has won grants and contracts to further test and evaluate his original research. He has notable experience in research and systems engineering which includes contracts for UAV research and the creation of an advanced multiple fuel system which operated the world's first and only helicopter that can fly on five separate fuels without engine modifications. His research extends into emerging and disruptive technology for offensive and defensive missions supporting US and coalition operations. His UAV and robotics expertise has focused on determining the specific uses, exceptions, and allowances for robotics operations; including studying the unintended consequences, future use, and misuse of such technologies. Dr. Mumm's presentations and publications support his research into autonomous systems in the virtual and physical worlds. Additionally, he serves as an adjunct professor at California University of Pennsylvania (CALU) instructing Homeland Security courses in the Criminal Justice Department.

Contact Information: Dr. Hans C. Mumm, 703-303-1752, hans@hansmumm.com. www.HansMumm.com

Wayne D. Lonstein, Esq. CISSP (Co-Author)

Wayne Lonstein holds a Bachelor of Arts Degree in Political Science from Wilkes University, a Bachelor of Science Degree in Cyber Forensics and Information Security from Syracuse University – Utica Collage, A Master of Science Degree in Homeland Security with a concentration in Information Security from The Pennsylvania State University and a Juris Doctor Degree from Pace University School of Law. Additionally he holds a CISSP Certification from The Pennsylvania State University. He is a member of the state bars of New York, New Jersey, Massachusetts an Pennsylvania as well as being admitted to over 30 United States District Court Bars, The Court of Veterans Appeals, United States Tax Court and the bar of the United States Court of Appeals of the 2nd, 3rd and 5th Circuits.

In addition Mr. Lonstein has practiced law nationally since 1987 in the area of technology, intellectual property, sports and entertainment and has litigated over 2000 cases. He is also a member of the New York State Magistrates Association and has served as a Magistrate Judge in the Town of Wawarsing, New York since 1989.

He a member of Signal law PC, the Co- Founder and CEO VFT Solutions is a member of the Forbes Technology Council and has authored numerous articles including: "Why Industry and Government Leaders Need to Realize Vulnerabilities of the Cloud"

Published on June 16, 2017 on LinkedIn; 'Identifying The Lone Wolf Using Technology," on LinkedIn, Published on July 3, 2015; "Are Social Media Companies Using ToS And Safe Harbor To Profit From Infringement, Crime And Terror?," Forbes.com, April 28, 2017; "Weaponizing Social Media: New Technology Brings New Threat," Forbes.com, July 7, 2017; 'Pay No Attention To That Man Behind The Curtain': Technology vs. Transparency," Forbes.com, October 17, 2017; and "Drone Technology: The Good, The Bad And The Horrible," Forbes.com, January 10, 2018.

Julie J.C.H. Ryan, D.Sc. (Co-Author)

Julie J.C.H. Ryan, D.Sc., is the CEO of Wyndrose Technical Group, having retired from academia in 2017. Her last position in academia was Professor of Cybersecurity and Information Assurance from the U.S. National Defense University. Prior to that, she was tenured faculty at the George Washington University and a visiting scholar at the National Institute for Standards and Technology (NIST).

Dr. Ryan came to academia from a career in industry that began when she completed military service. Upon graduating from the U.S. Air Force Academy, Dr. Ryan served as a Signals Intelligence Officer in the Air Force, and then as a Military Intelligence Officer with the Defense Intelligence Agency. Upon leaving government service, she worked in a variety of positions, including systems engineer, consultant, and senior staff scientist with companies including Sterling Software, Booz Allen & Hamilton, Welkin Associates, and TRW/ESL supporting a variety of projects and clients.

She is the author /co-author of several books, including *Defending Your Digital Assets Against Hackers, Crackers, Spies, and Thieves* (McGraw Hill 2000), and a Fellow of the American Academy of Forensic Sciences (AAFS). At Wyndrose Technical Group, she focuses on futures forecasting and strategic planning with an eye on technology surprise and disruption.

Candice Carter (Co-Author)

Ms. Candice Carter is a cybersecurity expert with over 15 years of hands-on experience in the areas of counterterrorism, counterintelligence and criminal cyber investigations. She conducts Classified/Unclassified briefings in the areas of Terroristic Cyber Capabilities using Social Media and Counterterrorism for the Intelligence Community (IC). Ms. Carter conducts research and constructs Asymmetric Warfare and Attack / Defense Scenarios against National Critical Infrastructure. She is the Team Lead and for NASA Aeronautics Research Institute for *Transformative Vertical Flight* (TVF) *Commercial Intra-City On-Demand* VTOL group. Ms. Carter is an invited speaker for key organizations including BSides London and (ISC)2 Security Congress. She is an Assistant Professor/Chair MSc Cybersecurity program at the Wilmington University. Ms. Carter holds a MSc Cybersecurity Forensics and Intelligence from Utica College, Utica , NY and a PMT Cybersecurity UAS (expected 2019) from Kansas State University.

Aris Theocharis (Co-Editor)

Aris has 30+ years of IT experience and earned a BS in Cybersecurity from Utica College, Utica, NY while working full time. He has provided editing skills for Professor Nichols for 10 years now. His approach is all encompassing, as opposed to strict grammar rules. Reading ease, topic flow, clarity, and being succinct are the focus.

Kurt Barnhart, Ph.D. (Foreword To 1st Edition)

Dr. Barnhart is Professor and currently the Associate Dean of Research at Kansas State University Salina. In addition, he established and serves as the executive director of the Applied Aviation Research Center. He oversees the Unmanned Aerial Systems program office. Dr. Barnhart previously served as the Head of the Aviation Department at Kansas State University.

Dr. Barnhart is a member of the graduate faculty at K-State. He is eminently qualified with: 1) a commercial pilot certificate with instrument, multi-engine, seaplane and glider ratings; 2) a certified flight instructor with instrument and multi-engine ratings; 3) an airframe and power plant certificate with inspection authorization.

Dr Barnhart's educational pedigree is outstanding: an A.S. in Aviation Maintenance Technology from Vincennes University, a B.S. in aviation administration from Purdue University, an MBAA from Embry-Riddle Aeronautical University, and a Ph.D. in educational administration from Indiana State University.

Dr. Barnhart's Research agenda is focused in aviation psychology and Human Factors as well as the integration of Unmanned Aircraft Systems into the National Airspace System. His industry experience includes work as a R&D inspector with Rolls Royce Engine Company where he worked on the RQ-4 Unmanned Reconnaissance Aircraft development program, as well as serving as an aircraft systems instructor for American Trans-Air airlines. Formerly, Dr. Barnhart was an Associate Professor and Acting Department Chair of the Aerospace Technology at Indiana State University where he was responsible for teaching flight and upper division administrative classes. Courses taught include Aviation Risk Analysis, Citation II Ground School, King Air 200 Flight, Air Navigation, Air Transportation, Instrument Ground School and many others.

CPT John-Paul Hood USA (Co-Author)

CPT John-Paul Hood is a researcher focused on the development of future counter unmanned aircraft technologies, theories and best practices for both government and civilian applications. CPT Hood has commanded in the US Army Field Artillery with a background specializing in the coordination and delivery of conventional / smart munitions as well as achieving desired battlefield effects through the integration of lethal and non-lethal assets. CPT Hood holds a BS in Geospatial Information Systems from the United States Military Academy, West Point NY and a Professional Masters in Technology UAS (expected 2019) from Kansas State University.

Dr. Alysia Starkey (CEO & Dean Kansas State University Polytechnic; 2nd Ed. Foreword)

Dr. Starkey is a Professor and currently serves as the Interim CEO and Dean for the Kansas State University Polytechnic Campus. As Dean, she oversees the College of Technology and Aviation academic programs and campus research centers. Dr. Starkey holds an A.A. in Social Work from Colby Community College, a B.S. in Psychology from Fort Hays State University, a M.L.S. from University of North Texas, and a Ph.D. in Curriculum and Instruction from Kansas State University. Joining Kansas State Polytechnic in June 2002 as a technical services/automation coordinator and assistant professor, Starkey was promoted to library director and associate professor in 2007, and to assistant dean of continuous improvement and distance education in 2010. She was named associate dean of academics and promoted to full professor in 2014. She gained the additional duties of interim CEO and Dean in June 2018 and continues in that capacity today.

Abbreviations: Acronyms

The following terms are common to the UAS industry, general literature or conferences on UAS/UAV/Drone systems.

A /Aref	Amplitudes of source and reference points, see Eq-20-6, 7
AA	Anti-aircraft / Adaptive Antennas
AAA	Anti-aircraft artillery
AAIB	Air Accidents Investigation Board
AAM	Air-to-air missile
AAV	Autonomous air vehicle
A/C	Aircraft
ACAS	Airborne collision avoidance system / Assistant Chief of the Air Staff
ACL	Agent communication language / Autonomous control levels
ACS	Airborne control station (system)
ACTD	Advanced Concept Technology Demonstration
AD	Ansar Dine terrorist group
A/D	Attack / Defense Scenario Analysis
ADAC	Automated Dynamic Airspace Controller
ADC	Air data computer
ADF	Automatic direction finder/finding
ADS	Air Defense System (USA)
ADS-B	Automatic Dependent Surveillance – Broadcast systems
ADT	Air Data Terminal

AEW	Airborne early warning
AF	Adaptive Filtering
AFCS	Automatic flight control system
AFRICOM	US Africa Command
AGM	Air- to- surface missile
AGARD	Advisory Group for Aerospace Research and Development (NATO)
AGM-65	Maverick (USA) is an air-to-surface missile (AGM) designed for close air support. It is the most widely produced precision-guided missile in the Western world, and is effective against a wide range of tactical targets, including armor, air defenses, ships, ground transportation and fuel storage facilities
AHA	Autopilot Hardware Attack
AHRS	Attitude and heading reference system
AI	Artificial intelligence
AIAA	American Institute of Aeronautics and Aerospace
AIC	Aeronautical Information Circular
AIP	Aeronautical Information Publication
AIS	Automated Identification System for Collision Avoidance
AJ	Anti-Jam
AM	Amplitude Modulation / al-Mourabitoun terrorist group
ANSP	Air Navigation Service Provider
AO	Area of Operations
AoA	Angle of Attack
APEC	Asia Pacific Economic Cooperation
APG	Asia-Pacific Gateway
APKWS	Advanced precision kill weapon system
AQ	Al-Qaeda Terrorist Group – "the Base"
AOA	Aircraft Operating Authority
AQIM	Al-Qaeda in the Islamic Maghreb

Ar	Receive antenna effective area, m^2
AR	Aspect ratio
AR drone	AR stands for "Augmented Reality" in AR *drone*. AR *Drone* can perform tasks like object recognition and following, gesture following
ARM	Anti-Radiation Munitions
ARS	Airborne Remote Sensing
ARW	Anti-radiation weapons
AS	Airborne Sensing Systems
ASB	Advisory Service Bulletin
ASEA	Active electronically scanned arrays
ASEAN	Association of Southeastern Asian Nations
ASL	Airborne Systems Laboratory
ASMS	Automated Separation Management System
ASTM	American Society of Testing and Materials
ASTER	Agency for Science, Technology and Research
ASW	Anti-submarine warfare
AT	Aerial target
ATC	Air Traffic Control
ATM	Air Traffic Management
ATR	Automatic Target Recognition
ATS	Air Traffic Service
AUDS	Anti-UAV Defense System
AUV	Autonomous Underwater Vehicle
AUVSI	Association for Unmanned Vehicle Systems International
AV	Air Vehicle
AWSAS	All Weather Sense and Avoid System
B	IF equivalent bandwidth, Hz
BAMS	Broad Area maritime surveillance

Backhauling	Intermediate links between core network or internet backbone and small subnets at the edge of the network
Bandwidth	Defined as the Range within a band of wavelengths, frequencies or energy.Think of it as a range of radio frequencies occupied by a modulated carrier wave, assigned to a service over which a device can operate. Bandwidth is also a capacity for data transfer of electrical communications system.
BDA	Battle Damage assessment
BER	Bit error rate
BLOS	Beyond line-of-sight
BNF	Bind and Fly – with custom transmitter
BRI	Belt and Road Initiative (Chinese)
BR&T	Boeing Research and Technology
BSR	Bilinear Signal Representation
BSs	Base Stations
BVR	Beyond visual range
c	Speed of light ~ (3 x 108 m/s) [186,000 miles per sec] in vacuum named after Celeritas the Latin word for speed or velocity
c	Speed of sound (344 m/s) in air
C	Combined methods of CR [Conflict Resolution]
C2 / C2W	Command and control / Command and Control Warfare
C3I	Command, control, communications and Intelligence
C4	Command, control, communications and computers
C4ISTAR	Command, control, communications, computers, intelligence, surveillance, target Acquisition and reconnaissance
CA	Collision Avoidance / Clear Acquisition (GPS) / *Cyber Assault (aka CyA)*
CAA	Control Acquisition cyber attack
CAS	Close Air Support / Common situational awareness
CASA	Civil Aviation Safety Authority
C of A	Certificate of Airworthiness

CAP	Civil Air Publication / Combat Air Patrol
CAT	Collision Avoidance Threshold
CC / CyC	Cyber Crime
CCCI/II	*Classical Cryptography Course Volume I/II* (Nichols R. K., *Classical Cryptography Course Volume I / II*, 1996)
CCE	Cyber Counter Espionage
CCI	Command control interface / *Cyber Counterintelligence*
CCS	Cyber Counter Sabotage
CCT	Cyber Counter Terrorism
CD	Conflict Detection
CDL	Common data link
CDMA	Code division multiple access
CDR	Collision detection and resolution systems (automated SAA in UAS)
CEA	Cyber electromagnetic activities
CETC	Chinese Electronics Technology Group
CF	Computer Forensics
CFTA	Continental Free Trade Area
CFT	Certificate of flight trials
CI / CyI	Cyber Infiltration
CIA	Confidentiality, Integrity, Availability / Central Intelligence Agency
CIN	Common Information Network
CIR	Color Infrared – artificial standard where NIR bands shifted so that humans can see the infrared reflectance
C/N	Carrier to Noise ratio in HAPS, => C/ N_0
CM / CyM	Cyber Manipulation
CN3	Communications / navigation network node
CNO	Chief Naval Operations
CNPC	Control and non-payload links
COA	Certificate of Waiver or Authorization

COB	Chief of the Boat
COMINT	Communications intelligence
COMJAM	Communications Jamming
COMSEC	Communications Security
CONOP(S)	Concept(s) of Operations
CONUS	Continental United States
COS	Continued Operational Safety
COTS	Commercial off-the-shelf
CPA	Closest Point of Approach
CPA Spoof	CPA spoof involves faking a possible collision with a target ship
CPL	Commercial pilot's license
CPRC	Communist Party of the Republic of China
CR	Conflict Resolution / Close range / Cyber Raid (aka CyR)
CRH	Coaxial rotor helicopter
C_{RX}	Received Signal Power, watts
CS	Control station
CSDP	Common Security and Defense Policy missions (EU)
CSfC	Commercial Solutions for Classified Program
CSIRO	Commonwealth Scientific and Industrial Research Organization
CT	Counter Terrorism / Counter Terrorism Mission
CTOL	Conventional take-off and landing
C-UAS	Counter Unmanned Aircraft Systems (defenses / countermeasures)
CUAS	CSIRO Unmanned Aircraft Systems
CV	Collision Volume
CW / CyW	Cyber Warfare
D	Distance from transmitter in Range equation (Adamy D. -0., 2015)
DA	Danger area

Danger Close	Definition www.benning.army.mil/infantry/magazine/issues/2013/May-June/Myer.html Nov 14, 2013 – 1) *Danger close* is included in the "method-of-engagement" line of a call-for-fire request to indicate that friendly forces are close to the target. ... *Danger close* is a term that is exclusive from risk estimate distance (RED) although the RED for 0.1 percent PI is used to define danger close for aircraft delivery. Pi = Probability of incapacitation. 2) Definition of "*Danger close*" (US DoD) In close air support, artillery, mortar, and naval gunfire support fires, it is the term included in the method of engagement segment of a call for fire which indicates that friendly forces are within close proximity of the target.
DARO	Defense Airborne Reconnaissance Office
DARPA	Defense Advanced Research Projects Agency
DAS	Detection by Acoustical Signature
dB	decibels
DC	Direct current
DCPA	Distance between vessels approaching CPA
DDD	Dull, dangerous, and dirty
DDOS	Distributed Denial of Service cyber attack
DE	Directed Energy
DEFCON	DEFCON is the world's longest running and largest underground hacking conference
DE / EMP /EP	Directed energy / Electromagnetic pulse
DEW	Directed – energy weapons
DF	Direction finding
DFCS	Digital Flight Control System
DHS	Department of Homeland Security
DIME	Diplomatic, information, military and economy
Dj	Jammer location – to-target receiver location distance, in km, FM 34-40-7
DJ	Data Jamming / Drone Jammer

DJI	Popular and functional Chinese made drone series: Mavic, Phantom, Ryze, Matrix, Spark, Enterprise, Inspire, Tello {However, banned by USA Army} (Newman, 2017)
DL	Downlink in HAPS
DLA	Date last accessed (usually a web reference)
DLI	Data Link interface
DNA	Deoxyribonucleic acid – genetic key to human life
DoD	Department of Defense
DOF	Degrees of Freedom
DOS	Denial of Service cyber attack
DPM	Direct power management / Dynamic Power Management
DPRK	*Democratic* People's Republic of Korea
DSA	Detect, sense and avoid / Dynamic Sense-and-Act
DSSS	Direct sequence spread spectrum
Dt	Enemy transmitter location -to- target receiver location, in km, FM 34-40-7
DT	Directional transmission
DTDMA	Distributed Time Division Multiple Access network radio system
DTED	Digital terrain evaluation data
DTH	Direct-To-Home
DTRA	Defense Threat Reduction Agency
DUO	Designated UAS operator
EA	Electronic Attack
EARSC	European Association of Remote Sensing Companies
EAS	Equivalent airspeed
EAU	East Africa union comprising of Israel and six East African states, Kenya, Ethiopia, Tanzania, Uganda, Rwanda and South Sudan
(E_b / N_0)	Thermal noise power spectral density ratio
ECCM / EP	Electronic counter-countermeasures / Electronic Protection

ECM	Electronic countermeasures
ECR	Electronic combat reconnaissance
EDC	Estimated Date of Completion
EHS	Enhanced surveillance
EIRP	Effective Isotropic radiated power
Electrolaser	Electroshock weapon that is also a DEW. Uses lasers to form electrically conductive laser-induced plasma charge
ELINT	Electronic Intelligence
ELT	Emergency locator transmitter
ECM	Electromagnetic compatibility
EM	Electromagnetic
EMI	Electromagnetic interference
EMP	Electromagnetic pulse
EMR	Electromagnetic Radiation
EMS	Electromagnetic Spectrum
EMSVIS	Electromagnetic Spectrum Visible Light
EMW	Electromagnetic Waves
EO	Electro-optical (sensing) / Earth Observation
ERP_J	Effective radiated power of the jammer, in dBm
ERPS	Effective radiated power of the desired signal transmitter, in dBm
ESM / ES	Electronic support measures / Electronic warfare support / Earth station
EU	European Union
EUNAVFOR	European Union Naval Force's anti-piracy naval mission
EUTM	Somalia Military training mission in Somalia
EVTOL	Electric Vertical Take-off and Landing
EW	Electronic warfare
F	Field theory methods of CR
F	*Fundamental frequency* is defined as the lowest frequency of a periodic waveform

f	Frequency, cycles / second RRE)
F_0	Resonant frequency of string, Hz see Eq. 20-5
F	Frequency in MHz, FM 34-40-7
FAA	Federal Aviation Administration
FACE	Future Airborne Capability Environment
FAR	False Alarm rates
FBL	Fly-by-Light, a type of flight-control system where input command signals are sent to the actuators through the medium of optical-fiber
FBW	Fly-by-Wire: Predetermine flight mission path based on GPS coordinates
FCS	Flight control systems / Flight Control Station
FDF	Frequency Domain Filtering
FDM	Frequency division multiplexing
FHSS	Frequency hopping spread spectrum
FIR	Far Infrared (25-40) to (200-350) um
FIRES	Definition (US DoD – JP 3-0) the use of weapon systems to create a specific lethal or nonlethal effect on a target
FL	Flight level
FLIR	Forward-looking Infrared
FMS	Flexible manufacturing system
Follow-Me	UAS autopilot automatically follows operator
Fom	HAPS Figure of merit in upload /download link
FoV	Field of View
FFoV	Forward Field of View
FRAGO	Fragmentary Order – to send timely changes of existing orders to a subordinate
FPGA	Field programmable gate array
FS	Fixed service
FSS	Fixed satellite service
FW	Fixed wing

G	Geometric methods of CR
G5S	G5 Sahel (G5S) Joint Force, has membership of five states; Burkina Faso, Mali, Mauritania, Niger, and Chad
gAR	Receiving Antenna Gain as a Factor
GBU	Guided Bomb Unit
GCHQ	Government Communications Headquarters (Britain)
GCS	Ground Control Station
GDPR	European Union's (EU) General Data Protection Regulation
GDT	Ground data terminal
GEO	Geostationary Earth orbit satellite
GeoFence	A geofence is a virtual perimeter for a real-world geographic area
GLOW	Gross lift-off weight for a missile / rocket
GNSS	Global Navigation Satellite System
GPS	Global Positioning System / Geo Fencing
GPS/INS	Use of GPS satellite signals to correct or calibrate a solution from an inertial navigation system (INS). The method is applicable for any GNSS/INS system.
GPSSPOOF	Hack of GPS system affecting UAS commands
GPWS	Ground proximity warning system
G_R	The receiving antenna gain in the direction of the desired signal transmitter, dBi
G_{RJ}	Receiving antenna gain in the direction of the jammer, in dBi
GS	Ground segment of HAPS
GSE	Ground support equipment
GSHM	Ground Station Handover Method
GSM	Global System for Mobile Communications
GT	Game Theory methods of CR
G/T	Ratio of the receive antenna gain to system noise temperature
(G /Ts) dB	Represents the figure of merit of the HAPS receiver, in dB

GT	Gain of the transmit antenna, dB
GTA	Ground-to-Air Defense
Harmonic	Frequency, which is an integer multiple of the fundamental frequency
H	Elevation of the jammer location above sea level, feet, FM 34-40-7
HAE	High altitude endurance
HALE	High altitude – long endurance
HAPS	High Altitude Platforms (generally for wireless communications enhancements)
HAPS UAVs	UAVs dedicated to HAPS service (example to communicate via CNPC links)
HEAT	High-explosive anti-tank warhead
HITL	Human in-the-loop
HMI	Human machine interface
HPA	High power amplifier
Ht	Elevation of enemy transmitter location above sea level, in feet, FM 34-40-7
HUD	Heads-up display
HUMINT	Human intelligence (spy's)
HVT	High value target (generally, for assassination)
I	Sound intensity, W x m^{-2} [Source strength S / $4\pi r^2$] (Uni-wuppertal, 2019)
IA	*Information Assurance* / Intentional cyber warfare attack
I-actors	Intentional Cyber Actors
IAS	Indicated air speed
ICAO	International Civil Aviation Organization
I.C.B.C.	International Center for Boundary Cooperation (China)
ICGs	Information centers of gravity
ICS	Internet Connection Sharing
ID	Information Dominance / Inspection and Identification
IEDs	Improvised Explosive Devices

IEEE	Institute of Electrical and Electronics Engineers
IEWS	Intelligence, electronic warfare and sensors
IFF	Identification, friend or foe (see chapter 19)
IFR	Instrument flight rules
I&I	Interchangeability and Interoperability
IIT	Intentional Insider Threats
Imaging Sensors	ARS sensors that build images
IL	Intensity level of sound measured, dB, Eq. 20-2
IMINT	Imagery intelligence
IMM	Interacting-multiple-models tracker
INS	Inertial navigation system
IMU	Inertial Measurement Unit
INFOSEC	*Information Security*
IO	Information Operations
IOC	Intergovernmental Oceanographic Commission
IOR	India Ocean Region
IoT	Internet of things
IPL	Insitu Pacific Limited
IR	Infrared
IRST	Infrared search and tracking
IS	Information Superiority
ISIS	*Islamic State of Iraq and al Sham* (ISIS)
ISR	Intelligence, Reconnaissance and Surveillance UAS Platform
ISTAR	Intelligence, surveillance, target acquisition and reconnaissance
ITU	International Telecommunications Union – Standards Organization
ITU-R	International Telecommunications Union – Radio Sector
IW	Information Warfare

JAGM	Joint-Air-to-Ground Missile
JAUS	Joint architecture for UAS
JDAM	Joint direct attack munitions
JFO	Joint fires observer
JP	Joint Publication – followed by military identifier
JDAM	Joint Direct Attack Munition
JNIM	Jama'at Nusrat al-Islam wal-Muslimin
JOPES	Joint Operation and Planning System / Execution System
JP	Joint Publication
J/ S	= the ratio of the jammer power to the desired signal power at the input to the receiver being jammed in dB
JTAC	Joint Terminal Attack Controller
JTIDS	Joint Tactical Information Distribution System (JTIDS) is an L band DTDMA
K	Boltzmann's constant (Noise component, RRE) (1.38 x 10 -23 J/K), Kelvin
K	for jamming frequency modulated receivers (jamming tuner accuracy), FM 34-40-7
KAMIKAZE	Means "Divine Wind," Tactic best known for Japanese suicide A/C attacks on Allied Capital Vessels in WWII. UAS TEAMS or SWARMS could be directed in the same way.
KM	Katiba Macina Groups
L	λ / 2 in Eq. 20-5
LAANC	Low Altitude Authorization and Notification Capability

LASER	"A laser is a device that emits light through a process of optical amplification based on the stimulated emission of electromagnetic radiation. The term "laser" originated as an acronym for "light amplification by stimulated emission of radiation". A laser differs from other sources of light in that it emits light coherently, spatially and temporally. Spatial coherence allows a laser to be focused to a tight spot, enabling applications such as laser cutting and lithography. Spatial coherence also allows a laser beam to stay narrow over great distances – collimation, enabling applications such as laser pointers. Lasers can also have high temporal coherence, which allows them to emit light with a very narrow spectrum. i.e., they can emit a single color of light. Temporal coherence can be used to produce pulses of light as short as a femtosecond. Used: for military and LEO devices for marking targets and ,measuring range and speed." (Gould, R.G. 1959)
Laser JDAM	Laser Joint Direct Attack Munition – dumb bombs, all weather precision –guided munitions. Guided by an integrated inertial guidance system.
Laser rangefinder	Scope to assist targeting of munitions. Countermeasure: laser-absorbing paint
LGWs	Laser-guided weapons
Latency	Processing difference between time interval signal is transmitted and signal is received
LCDR	Lieutenant Commander
L/D	Lift to drag ratio
LDCM	Low Duty cycle methods
LEO	Low Earth Orbit Satellite
LGB	Laser-guided bomb, a guided bomb that uses semi-active laser guidance to strike a designated target with greater accuracy than an unguided one
LGTF	Liptako-Gourma task force (LGTF) established by Burkina Faso, Mali, and Niger to secure their shared border region
LIDAR	Light (Imaging) Detection and Ranging
LFS	Free-Space Loss as a Factor
LIPC	Laser-induced plasma channel
LJ	Propagation loss from jammer to receiver, in dBi

LMM	Lightweight Multi-role Missile (by Thales)
LOS	*Line-of-sight / Loss of Signal / Loss of Separation*
LOSAS	Low cost Scout UAV Acoustic System
LPA	Log periodic array
LPI	Low Probability of Intercept
LR	Long range
LRAD	Long Range Acoustic Device (Weapon) (Yunmonk Son, 2015)
LRCS	Low radar cross section
LRE	Launch and recovery element
LRF	Laser rangefinder
LS	Losses existing in the system (lumped together), dB (RRE)
LS	The propagation loss from the desired signal transmitter, in dBm
LSDB	Laser Small Diameter Bomb
LST	Laser spot trackers
LTA	Lighter than Air (airship) / Low noise amplifier
LTE /LTE+	Long Term Evolution – refers to mobile telecommunications coverage
LWIR	Long wave Infrared (sensor or camera)
M	Mass in Eq. 20-5
MA	Multi-agent methods of CR
MAD	Magnetic anomaly detection / Mutually Assured Destruction (International Nuclear Policies in 50s-70s)
MAE	Medium-altitude endurance
MAGTF	Marine air-ground task force
MALDRONE	Malware injected into critical SAA for UAS
MALE	Medium-altitude, long endurance UAS
MALE-T	Medium altitude long endurance – tactical UAS
MAME	Medium altitude, medium endurance
MASINT	Measurement and Signal Intelligence

MATS	Mobile Aircraft Tracking System
M-AUDS	Mobile Anti-UAV Defense System
MAV	Micro-air vehicle
Maverick	AGM -65 (USA) Missile
MCE	Mission control element
MCM	Mine countermeasures
MCU	Master Control Unit (SCADA)
MDR	Missed Detection Rates
MEB	Marine expeditionary brigade (14,500 marines and sailors)
MEMS	Micro-electromechanical systems (see chapter 19)
MEO	Medium Earth Orbit satellite
MFD	Multi Function display
MGTOW	Maximum gross takeoff weight
MHT	Multiple-hypotheses-testing
MIM	Man in the Middle cyber attack
MINUSMA	Multidimensional Integrated Stabilization Mission in Mali
MIR	Mid Infrared 5 to (25-40) um
MIT	Massachusetts Institute of Technology
MMI	Man-machine interface
MORS	Military Operations Research Society
MPI	Message-passing interface
MPO	Mission payload operator
MR	Medium range
MRE	Medium-range endurance
MRZR LMADIS	A Light Marine Air Defense Integrated System. System mounted on a Polaris MRZR diesel tactical combat vehicle. Comprised of two vehicles – one a command node and the other a sensor node. Once the threat is detected , the LMADIS uses jamming to disrupt the signals of the drone.
MS	Mobile service

MSL / AGL	MSL altitudes are measured from a standard datum, which is roughly equal to the average altitude of the ocean. So, an aircraft traveling 5,000 feet directly above a mountain that's 3,000 feet tall would have an altitude of 5,000 feet Above Ground Level (AGL) and 8,000 feet MSL.
MSR	Maritime Silk Road (China)
MTCR	Missile Technology Control Regime
MTI	Moving target indication
MTOM	Maximum take-off mass
Modulation	Signal Modulation is the process of varying one or more properties of a periodic waveform, called the carrier signal, with a modulating signal that typically contains information to be transmitted
MTOW	Maximum takeoff weight of an aircraft at which the pilot can attempt to take off, due to structural or other limits
MTS	Multi Spectral Targeting System
MTTR	Multitarget tracking radar/ Mean time to repair
MUAV	Mini-UAV or maritime UAV
MUJAO	Movement for Unity and Jihad in West Africa
MUM	Manned-unmanned teaming
MWIR	Midwave Infrared
MW	Microwave towers
N	Available Noise power, watts for HAPS
N	Terrain and ground conductivity factor, FM 34-40-7, where 5 = very rough terrain with poor ground conductivity; 4 = moderately rough terrain with fair to good ground conductivity; 3 = Farmland terrain with good ground conductivity; 2 = Level terrain with good ground conductivity. The elevation of the jammer location and the enemy transmitter location does not include the height of the antenna above the ground or the length of the antenna. It is the location deviation above sea level.
NAC	Network Access Control
NACA	National Advisory Committee on Aeronautics
NAS	National Airspace (USA)

NAV	Nano-air vehicle / NAV data message for GPS systems
NBC	Nuclear, biological and chemical warfare
NCO	Network-centric operations
NCW	Network Centric Warfare
NDRC	National Development and Reform Commission (China)
NEC	Network enabled capability
NGO	Non-Governmental Organization
NIEM	National Information Exchange Model
NIR	Near Infrared
NLOS	Non-line-of-sight
NMAC	A NMAC is defined as an incident associated with the operation of an aircraft in which a possibility of collision occurs as a result of proximity of less than 500 feet to another aircraft, or a report is received from a pilot or a flight crewmember stating that a collision hazard existed between two or more aircraft.
NMLA	National Movement for Liberation of Azawad (Tuareg Rebellion)
NO	Numerical Optimization methods of CR
NOLO	No onboard live operator (USN)
NOTAM	Notice to airmen
NPS	National Park Service
NSA	National Security Agency (US)
NTSB	National Transportation Safety Board
NTT	Non-Threat Traffic
NULLO	Not using live operator (USAF)
O	Other methods of CR
OEM	Original equipment manufacturer
OIO	Offensive Information Operations
OLOS	Out-of-the-line-of-sight
OODA	Decision Loop: Observe, Orient, Decide, Act
OPA	Optionally piloted aircraft

OPAV	Optionally piloted air vehicle
OPSEC	Operations Security
OSI	Open systems interconnection
OTH	Over-the-horizon
P	Isotropic source of an electromagnetic pulse of peak power, MW
PANCAS	Passive Acoustic Non-Cooperative Collision Alert System
PCAS	Persistent close air support
PEIRP	Transmitter effective isotropic radiated power, watts
PFMS	Predictive Flight Management System
PEMSIA	Partnership in Environmental Management of the Seas of East Asia
PGM	Precision guided missile
PHOTINT	Photographic intelligence (usually sky – ground)
PII	Personal Identifiable Information
PIM	Position of intended movements/Previously intended movements
PIT	Proximity Intruder Traffic
P_j	Minimum amount of jammer power output required, in watts, FM 34-40-7
PL	Power level, dB, Eq. 20-1
PLA	Chinese People's Liberation Army
PLAN	People's Liberation Army Navy (China)
PLC	Programmable Logic Controllers (SCADA)
PMIAA	Permissions Management: Identification, Authentication and Authorization
PNF	Plug and Fly with custom transmitter, receiver, battery and charger
PO	Psychological Operations
POS	Position and Orientation System
POV	Point of View
PPP	Precise Point Positioning

PPS	Precise positioning service (GPS)
PRC	People's Republic of China (China)
PSD	Power Spectral Density
PREACT	*Partnership for Regional East Africa Counterterrorism* (PREACT)
PRF	Pulse repetition frequency codes
PRM	Precision Runway Monitor
PSH	Plan-symmetric helicopter
PSR	Primary Surveillance Radar
P_t	Power output of the enemy drone, in watts, FM 34-40-7
PW / PSYWAR	Psychological Warfare
PWO	Principal Warfare officer
P(Y)	Precise Signal (GPS)
QOS	Quality of Service in HAPs
QUAS	QUT UAS (see below)
QUT	Queensland University of Technology
R	1 /Tb is the bit rate (b/s) in link equation
R^4	Energy density received at detected target range, R, nm
RA	Resolution Advisory
RAC	Range air controller
RADAR	Radio Detection and Ranging
RAST	Recovery, assist, and traverse
RB	Rule-based methods (Conflict Resolution)
RBW	Red-Breasted Woodpecker
RCE	Remote Code Execution
RCO	Remote-control operator
RCS	Radar cross-section
RCTA	Surf Radio Technical Commission for Aeronautics
RF	Radio Frequency

RGB	Red Green Blue for VIS camera
RGT	Remote ground terminal
Rician PDF	Rician probability density function
RIMPAC	Rim of the Pacific Exercise – Maritime
RL	Ramp launched
RMS	Reconnaissance management system /Root-mean-square
RN	Ryan-Nichols Qualitative Risk Assessment Equations 17-2, 17-3
RNRA	Ryan – Nichols Attack / Defense Scenario Risk Assessment for Cyber cases
ROA	Remotely operated aircraft
ROC	Republic of China (Taiwan)
RPA	Remotely piloted aircraft
RPH	Remotely piloted helicopter
RPV	Remotely piloted vehicle
RR	Radio regulations
RRE	Radar Range Equation
RSA	RSA (Rivest–Shamir–Adleman) -authors of early public –key cryptographic system
RSTA	Reconnaissance, surveillance and target acquisition
RTA	Dubai Roads and Transport Authority
RTF	Off-the-shelf, Ready-to-Fly
RTK	Real Time Kinematic
RTS	Remote tracking station/Request to send/Release to service
RTU	Remote Terminal Unit
RUAV	Relay UAV
RWR	Radar warning receiver
S	Intensity at surface of sphere
SAA	Sense and Avoid / *Sense and Act Systems*; replaces *See and Avoid function* of a human pilot

SAASM	Selective Availability Anti-Spoofing Module
SAE	Society of Automotive Engineers
SAM	Ace-to-Air Missile
SAMPLE	Survivable autonomous mobile platform, long-endurance
SAP	Systems Applications and Products also the name of a company
SAR	Synthetic aperture radar / Search and rescue- especially using helicopters
SAS	Safety Assurance System
SATCOM	Satellite communications
SCADA	Supervisory Control and Data Acquisition systems
SCHEMA	Security Incident Identification
SCIF	Sensitive Compartmented Information Facility
SCS	Shipboard control system (or station) / Stereo Camera System / South China Sea
SE	Synthetic environment
SECDEF	Secretary of Defense
Shadowing	Airframe shadowing – UAV- Ground signal degradation during maneuver
SEZ	Special economic zones
SHM	Simple harmonic motion – represented by sine wave
SHORAD	Short Range Air Defense systems
SIGINT	Signals Intelligence
Signature	UAS detection by *acoustic*, optical, thermal and radio / radar
SJM	Salafi-Jihad Movement
SKASaC	Seeking airborne surveillance and control
SKYNET	Fictional artificial intelligence system that becomes self-aware
SM	Separation Management
SMC	Single moving camera
SME	Subject matter expert

SMR	Single main rotor
S/N	S/N = is one pulse received signal to noise ratio, dB; Signal to Noise ratio at HAPS receiver
SOA	Static Obstacle – Avoidance system
SPL	Sound pressure level, dB = 20 Log p / p_o [measured pressures to reference pressure] See Eq. 20-3,4; 6-7
SPS	Standard position service (GPS)
Spoofing	A Cyber-weapon attack that generates false signals to replace valid ones
Spot Sensors	ARS sensors that measure single locations without image library
SQL	SQL Injection – common malevolent code injection technique
SR	Short range
SRL	Systems readiness level
SSA	Static Sense-and-Act
SSP	Smart Skies Project
SSR	Secondary Surveillance Radar
SST	Self-Separation Threshold
STANAG 4856	Standard interfaces of UAV Control System for NATO UAV
STK	Satellite tool kit
STOL	Short take-off and landing
sUAS	Small Unmanned Aircraft System
SUAVE	Small UAV engine
SWARM	High level, dangerous collaboration of UAS, UUV, or unmanned boats
SWAT	Special Weapons and Tactics (police / paramilitary)
SWAP	Size, weight and power
SWIR	Shortwave infrared, 1400-3000 nm, 1.4 -3.0 um wavelength range

SZ	Safety Zone is defined as the horizontal and vertical separation criteria which form a cylindrical airspace volume around the UAS. In figure 3-2 that volume is defined by 1000 ft radius and 200 ft height. It is assumed that initially the UAS is in the center with 100 ft above and below the A/C.
T	In Range equation & environment, strength of a received signal, function of square or fourth power of distance, d, from transmitter (Adamy D. -0., 2015)
T	Time, sec (RRE)
T	Tension in Eq.20-5
TA	Traffic Advisory
TAC	Target air controller
TACAN	Tactical air navigation
TAR	Antenna noise temperature, Kelvin
TAS	True airspeed
TBO	Time between overhauls
TC	Type certificate
TCAS	Traffic alert and collision avoidance system
TCPA	Time to reach Closest Point of Approach
T_e	Effective input noise temperature, Kelvin
TEAM (UAS)	High level, dangerous collaboration of UAS, UUV, or unmanned boats; differs from SWARM in that it has a UAS Team Leader, (TL) where SWARM does not. TL directs the UAS team and is the primary counter UAS target to disrupt.
TETRA	Terrestrial Trunked Radio for terrestrial terminals / services
Thermobaric	Metal augmented charge
TIR	Thermal infrared = 8000 – 15000 nm, 8 -15 um
TL	Team Leader
TO	Take-off
Tort	A tort is an act or omission that gives rise to injury or harm to another and amounts to a civil wrong for which courts impose liability.
TP	Trajectory Prediction

TRANSCOM	U.S. Transportation Command networks
TRL	Technology readiness level
TS	Measured noise temperature, Kelvin units above absolute zero / Top Secret classification
TSTCP	Trans-Sahara Counterterrorism Partnership. TSCTP partners include Algeria, Burkina Faso, Cameroon, Chad, Mali, Mauritania, Morocco, Niger, Nigeria, Senegal, and Tunisia.
TT & C	Telemetry, tracking and command
TUAV	Tactical UAV
UA	Unmanned Aircraft (non-cooperative and potential intruder)
U-Actors	Unintentional Cyber Actors
UAE	United Arab Emirates
UAM	Urban Air Mobility (vehicle)
UAPO	Unmanned Aircraft Program Office
UAS	Unmanned Aircraft System
UASC	Unmanned aircraft system commander
UASIPP	UAS Integration Pilot Program
UAS-p	UAS pilot
UAV	Unmanned aerial vehicle
UAV-p	UAV pilot
UBR	Uplink bit rate, Mb/s
UCAR	Unmanned combat armed rotorcraft
UCARS	UAV common automated recovery system
UCAV	Unmanned combat air vehicle
UCWA / UA	Unintentional cyber warfare attack
UGCS	Unmanned Ground Control Station
UGS	Unmanned ground-based station
UGV	Unmanned ground vehicle
UHF	Ultra High Frequency, 300 MHz – 3 GHz
UIT	Unintentional Insider Threats

UL	Upload link
UMTS	Universal Mobile Telecommunications System
UN	United Nations
UNESCO	United Nations Educational, Scientific and Cultural Organization
UNICEF	United Nations Children's Fund
USD	Unmanned surveillance drone
UTM	Unmanned Traffic Management
UTV	Unmanned target vehicle
UUV	Unmanned underwater vehicle
UUNs / DUNSs	Urgent / deliberate universal needs statements
V	Visible
VFR	Visual flight rules
VIKI	Virtual Interactive Kinetic Intelligence
VLA	Very light aircraft
VLJ	Very Light Jet
VLAR	Vertical launch and recovery
VLOS	Visual Line of Sight
VMC	Visual Meteorological Conditions
VNIR	Visible light and near infrared 400 – 1400 nm, 0.4 – 1.4 um wavelength range
Voloport	Landing site for Volcopter
VTOL	Vertical take-off and landing
VTUAV	Vertical take-off UAV
WEF	World Economic Forum
WEZ	Weapon Engagement Zone
WRC	World Radio Conference Standards Organization
XO	Executive Officer of Naval vessel
ZIGBEE or KILLERBEE	Sniffing / penetration tools specific to UAS

Greek Symbols

λ	Wavelength in Hz, c / f where c= speed of light 344 m/ s and f = frequency, Hz.
Σ	Radar Cross Sectional Area, m^2

Sources plus Bibliography below:

Austin, R, (2010) *Unmanned Aircraft Systems*: UAVS *Design, Development and Deployment*, West Sussex, UK: Wiley, [Condensed with additions from eleven-page "Units and Abbreviations Table." Pp. ix-xxix] Additional sources generated from / specific to Chapter development / discussion.

Cyber terminology from: Nichols, R. K. (Sept. 5, 2008) *Cyber Counterintelligence & Sensitive Compartmented Information Facility* (SCIF) *Needs – Talking Points &* (*Randall* K. *Nichols J.* J., 2018) *&* (*Nichols* R. K., *Hardening US Unmanned Systems Against Enemy Counter Measures*, 2019) *&* (*Randall* K. *Nichols D. , Chapter 20 Acoustic CM & IFF Libraries* V SWARMS *Rev* 1 05142019, 2018) *&* (*Randall* K. *Nichols and Lekkas*, 2002)*&* (NIST, *September* 2012)

Alford, L. D., Jr., USAF, Lt. Col. (2000) Cyber Warfare: Protecting Military Systems *Acquisition Review Quarterly*, spring 2000, V.7, No. 2, P, 105, http://www.Dtic.Mil/Dtic/Tr/Fulltext/U2/A487951.Pdf

Bibliography

49 U.S. Code §40103, 49 U.S. Code §40103 Sovereignty and use of airspace (U.S. Code July 5, 1994).

Abramson, E. (2016). *Ethical Dilemmas in the Age of* AI. Retrieved from Abramson, E. – knowmail.me/blog: https://www.knowmail.me/blog/ethical-dilemmas-age-ai/

Adamy, D. -0. (2015). EW 104 EW *against a New Generation of Threats*. Boston: Artech House.

Adamy, D. (2001). EW 101 A *First Course in Electronic Warfare*. Boston, MA: Artech House.

Adamy, D. (2001). EW 101 A *First Course in Electronic Warfare*. Boston: Artech House.

Adamy, D. (2004). EW 102 A *Second Course in Electronic Warfare*. Boston: Artech House.

Adamy, D. (2009). EW 103 *Tactical Battlefield Communications Electronic Warfare*. Boston: Artech House.

Adamy, D.-9. (1998, Jan). Lesson 4: the basic link for all EW functions. (electronic warfare)(EW Reference & Source Guide). *Journal of Electronic Defense, Jan 1998 Issue.*

Administrator. (2015, June 15). *Standing Wave and Fundamental Frequency.* Retrieved from Electronics Hub: https://www.electronicshub.org/?s=fundamental+frequency

Administrator. (2019, May 17). *Harmonic Frequencies.* Retrieved from electronicshub.org: https://www.electronicshub.org/harmonic-frequencies/

Alejandro Aragon-Zavala, J. L.-R.-P. (2008). *High-Altitude Platforms for Wireless Communications.* Chichester, West Sussex, UK: John Wiley & Sons.

Alford, L. (2000). Cyber Warfare: Protecting Military Systems. *Acquisition Review Quarterly.*

Anon. (2019). *Saudi Arabia grants citizenship to robot Sophia.* Retrieved from dw: Saudi Arabia grants cihttps://www.dw.com/en/saudi-arabia-grants-citizenship-to-robot-sophia/a-41150856

Asimov, I. (1950). *"Runaround". I, Robot (The Isaac Asimov Collection ed.).* New York City: Doubleday.

Atherton, K. D. (2019). Can the Pentagon sell Silicon Valley on AI as ethical war? . C4ISRNET.

Austin, R. (2010). *"Design for Stealth", Unmanned Aircraft Systems UAVS Design Development and Deployment.* New York: John Wiley and Sons.

Brown, E. F. (Dec 2008). Airborne Communication Networks for Small Unmanned Aircraft Systems. *Proc. IEEE, vol* 96, *no* 12, pp. 2008-17.

Burch, D. (2015). *RADAR for Mariners.* New York: McGraw-Hill.

Cameron, J. &. (Director). (1991). *Terminator 2: Judgement Day* [Motion Picture].

Chapman, A. (2019, May 31). *GPS Spoofing.* Retrieved from Tufts University – Tech Notes 2017: https://sites.tufts.edu/eeseniordesignhandbook/files/2017/05/Red_Chapman.pdf

Cornell University Legal Information Institute. (2019, June 5). *But-for test.* Retrieved from law.cornell.edu: https://www.law.cornell.edu/wex/but-for_test

Cornell University Legal Information Institute. (2019, June 5). *Intervening Cause.* Retrieved from law.cornell.edu: https://www.law.cornell.edu/wex/intervening_cause

Cornell University Legal Information Institute. (2019, June 5). *Personal Jurisdiction.* Retrieved from law.cornell.edu: https://www.law.cornell.edu/wex/personal_jurisdiction

D, G. a. (2010). *Broadband Communications via High Altitude Platforms.* New York City, NY: John Wiley & Sons.

Daniel-Cornel TĂNĂ, S. (2018). The Impact of the Development of Maritime Autonomous Systems on the Ethics of Naval Conflicts. *Annals: Series on Military Sciences*(2), 118-130.

Deloitte Center for Government Insights analysis. (2018, June 18). *The future of regulation. Principles for regulating emerging technologies.* Retrieved from Deloitte Insights: https://www2.deloitte.com/insights/us/en/industry/public-sector/future-of-regulation/regulating-emerging-technology.html

Dewey, D. (2017, July 14). *Drone crashes into LDS temple in Utah; raises questions of airspace rules.* Retrieved from eastidahonews.com: https://www.eastidahonews.com/2017/07/drone-crashes-lds-temple-utah-raises-questions-airspace-rules/

Diversity of citizenship; amount in controversy; costs, 28 U.S. Code §1332 (United States Congress June 25, 1948).

DJI. (2019, June 5). *DJI Enterprise.* Retrieved from Enterprise DJI.com: https://enterprise.dji.com/civil-protection

DLSR Pros. (2019, June 3). *Best Drones (UAVs) for Firefighting in 2019.* Retrieved from dlsrpros.com: https://www.dslrpros.com/dslrpros-blog/best-drones-firefighting-2019/

DoD. (2018). *Dictionary of Military Terms.* Retrieved from JCS.Mil: http://www.jcs.mil/doctrine/dod_dictionary/

DoD. (2018). *Joint Publication (JP) 3-01 Countering Air and Missle Threats.* Washington, DC: DoD.

DoD-02. (2018). *Information Operations (IO) in the United States.* Retrieved from JP 3-13 : http://www.dtic.mil/doctrine/new_pubs/jp3_13.pdf

DoD-03. (2015). *Unmanned Systems Roadmap 2013 to 2038.* Retrieved from DTIC: http://www.dtic.mil/dtic/tr/fulltext/u2/a592015.pdf

Drones, Q. S. (2017, July 11). *quadcopters-have-hit-the-sound-barrier/.* Retrieved from quadstardrones.com: https://quadstardrones.com/2017/07/11/quadcopters-have-hit-the-sound-barrier/

EARSC. (2015). A Taxonomy for the EO Services Market: enhancing perception and performance of the EO service industry. EARSC Issue 2.

Entokey, a. G. (2019, May 16). *entokey.com/acoustics-and-sound-measurement/.* Retrieved from entokey.com/acoustics-and-sound-measurement/: https://entokey.com/acoustics-and-sound-measurement/

ESA-ESTEC Contract 162372/02/NL/US. (September 2005). STRATOS: *Stratospheric Platforms a definition study for* ESA *Platform, Final Report*, 1-34. ESA-ESTEC .

European Union. (2019, May 2019). *About the regulation and data protection.* Retrieved from ec.europa.eu: https://ec.europa.eu/commission/priorities/justice-and-fundamental-rights/data-protection/2018-reform-eu-data-protection-rules_en

Federal Question, 28 U.S. Code §1331 (United States Congress June 25, 1948).

FEMA. (2013). *Lessons Learned from the Boston Marathon Bombings: Preparing for and Responding to the Attack.* Retrieved from www.fema.gov: http://www.fema.gov/media-library-data/20130726-1923-25045-1176/lessons_learned_from_t

Filippo Santoni de, S. &. (2018). Meaningful Human Control over Autonomous Systems: A Philosophical Account. *Frontiers in Robotics and* AI. doi:10.3389/frobt.2018.00015

Fleetwood, J. (2017). Public Health, Ethics, and Autonomous Vehicles. *American Journal of Public Health*, 107(4), 632-537.

Fortuna, C. (2017, 12 02). *Autonomous Driving Levels 0–5 + Implications.* Retrieved from cleantechnica.com: https://cleantechnica.com/2017/12/02/autonomous-driving-levels-0-5-implications/

Gelfand. (2004). "*Physical Concepts*", *Hearing an Introduction to Psychological and Physiological Acoustics, 4th ed.* New York City.

Gelfand, S. A. (2009). *Essentials of Audiology, 3rd Edition.* Stuttgart, DE: Thieme.

Giordano, N. (2009). *College Physics: Reasoning and Relationships.* New York City, NY: Cengage Learning. pp. 421–424.

Gould R. Gordon (1959). "The LASER, Light Amplification by Stimulated Emission of Radiation". In Franken, P.A.; Sands R.H. (eds.). *The Ann Arbor Conference on Optical Pumping, the University of Michigan, 15 June through 18 June* 1959. p. 128.

Guardbaum, S. (1994). The Nature of Preemption. *Cornell Law Review*, 767, 771.

Harris Aerial. (2019, June 5). *Carrier HX8 Sprayer Drone.* Retrieved from harrisaerial.com: https://www.harrisaerial.com/carrier-hx8-sprayer/

Heinman, C. (2019). *Hearing Loss Tests Patrient D v-105*. Carlisle, PA: Brown Optical Hearing Aid Service.

Henderson, T. (2017). The Doppler Effect – Lesson 3, Waves. *Physics tutorial. The Physics Classroom*. Retrieved from Henderson, Tom (2017). "The Doppler Effect – Lesson 3, Waves". Physics tutorial. The Physics Classroom. Retrieved September 4, 2017.: Henderson, Tom (2017). "The Doppler Effect – Lesson 3, Waves". Physics tutorial. The Physics Classroom. Retrieved September 4, 2017.

Hern, A. (2017, 1 12). *Give robots 'personhood' status, EU committee argues*. Retrieved from The Guardian: www.theguardian.com/technology/2017/jan/12/give-robots-personhood-status-eu-committee-argues

Hubbard, R. K. (1998). *Boater's Bowditch*. Camden, MA: International Marine.

Ibrahim, A. (2019). *Optimization Methods for User Admissions and Radio Resource Allocation for Multicasting over High Altitude Platforms* . Memorial Univeresity of Newfoundland , Canada: River Publications.

IEEE . (2017). The IEEE Global Initiative on Ethics of Autonomous and Intelligent Systems Announces New Standards Projects. *Telecom Standards*, 27(11), pp. 4-5. .

Jesus Gonzalo, D. L. (2018, March 15). On the Capabilities and Limitations of High Altitude Pseudo-Satellities. *Progress In Aerospace Sciences*, 37-56. doi:https://doi.org/10.1016/j.paerosci.2018.03.006

Johnson, O. &. (2012). *Ethics: Selections from Classic and Contemporary Writers*. Boston, MA: Cengage Learning.

Jones, T. (2017). *International Commercial Drone Regulation and Drone Delivery Services*. Santa Monica: The Rand Corporation.

Kanowitz, S. (2019, 05 15). *Toward the deployment of ethical* AI. Retrieved from Government Computer News. : Kanowitz, S. (2019). Toward the dephttps://gcn.com/articles/2019/05/15/ethical-ai-idc.aspx?s=gcntech_200519

Kirk, J. (2015, August 5). *sounds-can-knock-drones-sky*. Retrieved from www.computerworld.com.au/article/581231: https://www.computerworld.com.au/article/581231/sounds-can-knock-drones-sky/

Knight, W. (2018). Nine charts that really bring home just how fast AI is growing. *MIT Technology Review* .

Legal Information Institute – Cornell University. (2019, May 31). *Strict Liability* . Retrieved from Legal Information Institute: https://www.law.cornell.edu/wex/strict_liability

LRAD. (2019, May 189). *LRAD 450XL Datasheet.* Retrieved from LRADX: http://www.lradx.com/wp-context/uploads/2015/05/LRAD_datasheet_450XL.pdf

MacGregor, D. S. (2018). *Colorado Causes of Action: Elements, Defenses, Remedies, and Forms.* Denver: Bradford Publishing Co. .

Macnamara, T. M. (2010). *Introduction to Antenna Placement & Installation.* New York City, NY : John Wiley & Sons.

Mahon, J. (2012). *Classical Natural Law Theory* St. Thomas Aquinas (1227-1274) – *the "Angelic Doctor" Lecture.* Retrieved from Mahon, J. (2012). Classical Natural Law Theory St. Thomas Aquinas (1227-1274) – thePhilosophy of Law. : Mahon, J. (2012). Classical Natural Law Theory St. Thomas Aquinas (1227-http://home.wlu.edu/~mahonj/PhilLawLecture1NatLaw.htm

Marbury v. Madison, 5 U.S. 137 (United States Supreme Court February 23, 1803).

Matolak, R. S. (April 2015). Initial Results for Airframe Shadowing in L-band and C-band Air -Ground Channels. *Proc. Integrated Commun,, Navigation, and Surveillance Conf*, (pp. pp. 1-8).

McCullogh v. Maryland, 17 U.S. 316 (United States Supreme Court March 6, 1819).

Merriam-Webster. (2019, May 17). Merriam-Webster Online Dictionary.

Merriam-Webster, Inc. (2019). *Definition of Ethics.* online: Merriam-Webster, Inc. Retrieved from Definition of Ethics. (2019a). Online: Merriam-Webster, Incorporated.: Definition of Ethics. (2019a). Online: Merriam-Webster, Incorporated.

Middleton, C. (2018). *SAP launches ethical A.I. guidelines, expert advisory panel.* Retrieved from internetofbusiness.com: Middleton, C. (2018). SAP launches ethical A.I. guidelines, expert advisory panel. Retrieved from https://internetofbusiness.com/sap-publishes-ethical-guidelines-for-a-i-forms-expert-advisory-panel/

Misselhorn, C. (2018). Artificial Morality. Concepts, Issues and Challenges. *Society*, 55(2), 161-169.

Mohorcic, D. G. (2010). *Broadband Communications via High Altitude Platforms.* New York City, NY: John Wiley & Sons.

Monahan, K. (2004). *The Radar Book: Effective Navigation and Collision Avoidance.* Anacortes, WA: Fineedge Publications.

Muspratt, A. (2018, November 22). *New global drone standards proposed.* Retrieved from Defence iQ: https://www.defenceiq.com/defence-technology/news/new-global-drone-standards-proposed

Goddemeir, K. D. (June 2015). Role-based Connectivity Management with Realist Air to Ground Channels for Future Applications. *IEEE Vehic. Tech. Mag.* Vol 10, *no* 2, pp. 79-85.

National Conference of State Legislatures. (2018, September 10). *Current Unmanned Aircraft State Law Landscape.* Retrieved from NCSL.org: http://www.ncsl.org/research/transportation/current-unmanned-aircraft-state-law-landscape.aspx

NBC Today Show. (2018, May 9). *How peeping drones could be spying on you without you knowing it.* Retrieved from Today.com: https://www.today.com/video/how-peeping-drones-could-be-spying-on-you-without-you-knowing-it-1229001795967

Newman, L. H. (2017, August 7). THE ARMY GROUNDS ITS DJI DRONES OVER SECURITY CONCERNS. Retrieved from WIRED: https://www.wired.com/story/army-dji-drone-ban/

Nichols, R. K. (1996). *Classical Cryptography Course Volume I / II.* Laguna Hills, CA: Aegean Park Press.

Nichols, R. K. (1996). *Classical Cryptography Course, Volume I.* Laguna Hills, CA: Aegean Park Press.

Nichols, R. K. (2018). *Unmanned Aircraft Systems (UAS) In the Cyber Domain: Protecting USA's Advanced Air Assets. 1st Ed.* Manhattan, KS: New Prairie Press.

Nichols, R. K. (2019, March 14). Hardening US Unmanned Systems Against Enemy Counter Measures. *7th Annual Unmanned Systems Summit.* Alexandria, VA, USA: PPTX presentation , self.

Nichols, R. K., Mumm, H. C., Lonstein, W. D., Ryan, J. J., & and Carter, C. (2018). *Unmanned Aircraft Systems (UAS) in the Cyber Domain: Protecting USA's Advanced Air Assets.* Manhattan, KS: New Prairie Press (NPP) eBooks. 21.

NIST. (September 2012). *Guide for Conducting.* Washington, DC: GPO.

North Carolina Department of Transportation. (2019, May 30). *Law & Regulations.* Retrieved from NCDOT.GOV: https://www.ncdot.gov/divisions/aviation/uas/Pages/laws-regulations.aspx

Osseiran, A. (Dec 2014). Scenarios for 5G Mobile and Wireless communications: the vision of the METIS Project. *IEEE Communications Magazine*, Vol 52, *no* 5, pp. 26-35.

O'Sullivan, J. L. (1845). The Great Nation of Futurity. *United States Magazine and Democratic Review* Vol 6 Issue 23, pp. 426-430.

Pierson. (2019, May 16). *tuning-fork-waves-sound.* Retrieved from airfreshener.club – Pierson Education: https://airfreshener.club/quotes/tuning-fork-waves-sound.html

Porter, J. D. (2019, June 8). *jdporterlaw.com/intellectual-property-law/*. Retrieved from jdporterlaw.com: http://www.jdporterlaw.com/intellectual-property-law/

Possel, M. (2017). Waves, motion and frequency: the Doppler effect. *Einstein Online, Vol. 5. Max Planck Institute for Gravitational Physics, Potsdam, Germany.*

Pricewaterhousecoopers, LLP. (2018). *Skies without limits – Drones- taking the UK's economy to new heights.* London: Pricewaterhousecoopers, LLP.

PROTECTION OF CERTAIN FACILITIES AND ASSETS FROM UNMANNED AIRCRAFT, H. R. 302 (United States Congress January 3, 2018).

Proyas, A. (Director). (2004). *I, Robot. In. Hollywood,* CA. [Motion Picture].

Ramzy, A. &. (2008). *Tainted-Baby-Milk Scandal in China.* Retrieved from content.time.com/time/world/article/: http://content.time.com/time/world/article/0,8599,1841535,00.html

Randall K. Nichols and Lekkas, P. C. (2002). *Wireless Security: Threats, Models, Solutions.* New York City, NY: McGraw Hill.

Randall K. Nichols, D. (2018). Chapter 20 Acoustic CM & IFF Libraries V SWARMS Rev 1 05142019. In R. K. Nichols, H. C. Mumm, W. D. Lonstein, & J. S. Hood, *Unmanned Aircraft Systems* (UAS) *in the Cyber Domain: Protecting* USA's *Advanced Air Assets, 2nd ed.* Manhattan, KS: NPP Press.

Randall K. Nichols, D. (2019 for publication). Unmanned Aircraft Systems in the Cyber Domain: Protecting USA's Advanced Air Assets, 2nd ed. In H. M. Randall K. Nichols, *Chapter 18 Audiology, Acoustic Countermeasures against Swarms and Building IFF Libraries* (p. 2nd ed.). Manhattan, KS: For Publication, NPP.

Randall K. Nichols, J. J. (2018). *Unmanned Aircraft Systems* (UAS) *in the Cyber Domain: Protecting* USA's *Advanced Air Assets.* Manhattan, KS: New Prairie Press.

Rappaport, T. (2014). *Millimeter Wave Wireless Communications.* New York City, NY: Prentice Hall.

Ricker, D. (2017, July 1). *Navigating drone laws has become a growing and lucrative legal niche.* Retrieved from ABA Journal: http://www.abajournal.com/magazine/article/drone_law_attorneys

Said Emre Alper, Y. T. (December 2008). Compact Angular Rate Sensor System Using a Fully

Decoupled Silicon-on-Glass MEMS Gyroscope. JOURNAL OF MICROELECTROMECHANICAL SYSTEMS, VOL. 17, NO. 6.

Sanchez, M. (2019, June 4). *No Drones*. Retrieved from Unspalsh.com: https://unsplash.com/photos/oMqswmrie4Y

Schroeder, A. (2018, February 1). *Localizing Humanitarian Drones: Robotics & Disaster Response from the Maldives to Malawi*. Retrieved from medium.com: https://medium.com/radiant-earth-insights/localizing-humanitarian-drones-robotics-disaster-response-from-the-maldives-to-malawi-a1f362432cb1

Signia. (2019, May 16). *Signia Hearing Aids*. Retrieved from Signia Hearing Aids – Hear across America: www.signiausa.com

Singer v. City of Newton, 284 F. Supp. 3d 125 (U.S. District Court Massachusetts September 21, 2017).

Sood A.K. & Enbody, R. (2014, December 19). *https://www.georgetownjournalofinternau-s-military-defense-systems-the-anatomy-of-cyber-espionage-by-chinese-hackers*. Retrieved from georgetownjournalofinternationalaffairs.org/online-edition: https://www.georgetownjournalofinternationalaffairs.org/online-edition/u-s-military-defense-systems-the-anatomy-of-cyber-espionage-by-chinese-hackers

Sovereignty and use of airspace, 49 U.S. Code § 40103 (United States Congress July 5, 1994).

Staff. (2008). FINAL ACTS WRC-07. *World Radiocommunication Conference*. Geneva: ITU.

Staff. (2012). FINAL ACTS WRC-12. *World Radiocommunication Conference*. Geneva: ITU.

Staff. (2016, April 17). *Equal Loudness Contours*. Retrieved from Gutenberg Organization: http://central.gutenberg.org/article/WHEBN0001046687/Equal-loudness%20contour

Staff. (2019). FINAL ACTS WRC-15. *World Radiocommunication Conference*. Geneva.

Stone, Z. (2007, 11 7). *Stone, Z. (2017). Everything You Need To Know About Sophia, The World's First Robot Citizen. Retrieved from https://www.forbes.everything-you-need-to-know-about-sophia-the-worlds-first-robot-citizen*. Retrieved from Forbes: https://www.forbes.com/sites/zarastone/2017/11/07/everything-you-need-to-know-about-sophia-the-worlds-first-robot-citizen/#1667784246fa

Studios, D. D. (2017). Boaters Ref. USA.

sUAS News. (2018, March 2). *RAS Consulting & Investigations hire Jeff Parisse to offer sophisticated UAS security and surveillance services*. Retrieved from suasnews.com: https://www.suas-

news.com/2018/03/
ras-consulting-investigations-hire-jeff-parisse-offer-sophisticated-uas-security-surveillance-services/

Sun, W. M. (June 2015). Unmanned Aircraft Systems:Air-Ground Channel Characterization for future applications. *IEEE Vehic. Tech Mag.* Vol 10, No 2 , pp. 79-85.

T.C. Dozer, D. A. (2008). High Altitude Platforms for VHDR in-theater communications. *IET Seminar on Military Satellite Communications Systems.*

The Shepard News Team. (2018, September 12). *Liteye Receives Follow-on Contract for C-AUDS – DB – Digital Battlespace.* Retrieved from Aerospace, Defense and Security News and Analysis – Shephard Media, The Shepard Press, Ltd: www.shephardmedia.com/news/digidigital-battlespace/liteye-receives-follow-contract-c-auds

Toomay, J. (1982). RADAR *for the Non – Specialist. London; Lifetime Learning Publications.* London: Lifetime Learning Publications.

TRS, S. (2018, July 10). *Tontechnic-Rechner-Sengpielaudio.* Retrieved from Tontechnic-Rechner-Sengpielaudio Calculator: www.sengspielaudio.com/calculator-wavelength.htm

UAV Coach. (2019, May 30). *Drone Laws in South Carolina* (2019). Retrieved from UAVcoach.com: https://uavcoach.com/drone-laws-south-carolina/

United States Constitution Article VI, Sec.2 (United States of America September 17, 1787).

Uni-wuppertal. (2019, May 15). *Inverse Square Law, General.* Retrieved from hydrogen.physik.uni-wuppertal.de/hyperphysics/: http://hydrogen.physik.uni-wuppertal.de/hyperphysics/hyperphysics/hbase/forces/isq.html

Urban, T. (2018). *Teach Your Robots Well: Will Self-Taught Robots Be the End of Us?* Retrieved from www.worldsciencefestival.com: Urban, T. (2018). Teach Your Robots Well: Will Self-Taught Robots Be the End of Us? Retrieved from https://www.worldsciencefestival.com/programs/teach-robots-well-will-self-taught-robots-end-us/

Usenix.org. (2019, 6 9). *MEMS, Drones, & Sound Sourcing.* Retrieved from Usenix.org: www.usenix.org

WebFinance, Inc. (2019). *Definition of Ethics. (2019b).* online: Online: WebFinance, Inc.

Weise, E. (2017, August 23). *could-hackers-behind-u-s-navy-collisions.* Retrieved from USATODAY: https://www.ruidosonews.com/story/tech/news/2017/08/23/could-hackers-behind-u-s-navy-collisions/594107001/

Wong, C. (2017). *Top Canadian researcher says AI robots deserve human rights*. Retrieved from Wong, C. (2017). Top Canadian researcher says AI robots deserve human rights. Retrieveit-business.ca: Wong, C. (2017). Top Canadian researcher says AI robots deserve human rhttps://www.itbusiness.ca/news/top-canadian-researcher-says-ai-robots-deserve-human-rights/95730

Wordpress. (2012, 08 29). *The True Sign of Intelligence*. Retrieved from deepthinkings.wordpress.com: http://deepthinkings.wordpress.com/2012/08/29/the-true-sign-of-intelligence/

Wright, T. (2017, August 11). *You've Been Warned: Keep Your Drones Away From Military Bases*. Retrieved from Air & Space, Smithsonian: https://www.airspacemag.com/daily-planet/keep-your-drones-away-military-base-180964451/

Wyvern, T. (2018). *National Critical Intelligence Estimate: Counter Unmanned Aircraft Systems (C-UAS) in the US*. Salina, KS: KSUP.

Xiaoyang Liu, C. L. (2016). High Altitude Platform Station Network and Channel Modeling Performance Analysis. *Mathematics and Computer Science. Vol. 1, No 1*, pp. 10-16. doi:Xiaoyang Liu, Chao Liu, Wanping Liu, Xiaoping Zeng. High Altitude Platform Station Network and Channel Modeling Performance Analys10.11648/j.mcs.20160101.13

Zeng, R. Z. (May 2016.). Wireless communications with unmanned aerial vehicles: opportunities and challenges. *IEEE Communications Magazine.vol.* 54, *no.*5, pp. 36-42.

Yong Zeng, R. Z. (2016). Wireless Communications with Unmanned Aerial Vehicles: Opportunities and Challenges. *IEEE Communications Magazine*, 36-42.

Yunmonk Son, H. S. (2015, August 12-14). Rocking Drones with Intentional Sound Noise on Gyroscopic Sensors. *Proc. 24th Usenix Security Symposium*. Washington, DC: USENIX. Retrieved from https://www.usenix.org/conference/usenixsecurity15/technical-sessions/presentation/son

Zetter, K. (2015). So, The NSA Has An Actual SKYNET Program . *WIRED Magazine(Online)*. . Retrieved from Zetter, K. (2015). So, The NSA Has An Actual SKYNET Program WIRED Magazine(Online).

DETAILED TABLE OF CONTENTS

TABLE OF FIGURES

Chapter 13: Data Links Functions, Attributes and Latency

Chapter 14: Exposing UAS Vulnerabilities via Electronic Warfare (EW) *and Countering with Low Probability Intercept Signals* (LPI)

TABLE OF TABLES

TABLE OF EQUATIONS

of pressure squared because ($I \approx p2$).
Eq. 19-4 Convenient form of Eq. 19-3 recognizes that $\log x^2 = 2 \log x$. So, $SPL = 20 \log p / p_o$.
Eq. 19-5 Formula for the string's resonant frequency Fo
Eq. 19-6 Sound Pressure Level (SPL) & Sound Amplitude- derives the attack distance, d
Eq. 19-7 SPL as a f ($SPL_{ref} - 20 \log (d / d_{ref})$)

SECTION I

THE UAS PLAYING FIELD

Chapter 1: A View of the UAS Market

Student Learning Objectives – The student will gain an understanding of the current UAS marketplace and barriers to UAS growth. Students will be introduced to the UAS cybersecurity issues in the UAM sector.

Marketplace History

Consider the US marketplace for UAS. The US UAS marketplace is comprised of three UAS sectors: commercial, consumer, and military. The UAS marketplace has evolved from the military to commercial rapidly over the last four years. Technological advancements for unmanned space vehicles has contributed to the rapid growth and opened new markets for UAS sales. Projections are for the UAS market to reach approximately $12 billion in 2021.[1]

Is the UAS marketplace growing faster than we can secure? UAS technology growth in the consumer sector has a 40% compound annual growth rate.[2] Commercial UAS registration was mandated after April 2016. FAA reported that they received about 1,000 registrations a week. In 2017, UAS registrations exceeded 150% from 2016.[3] See Figure 1-1. In 2015, the first year of –the UAS market growth, security researchers at DEF CON[4] successful knocked an A.R. drone[5] out of the sky.

Figure 1-1 Consumer sUAS registration as of December 31, 2017

1. Meola, 2017. (Tech Musings, 2009)
2. IBID
3. (United States Department of Transportation, 2015 -2018)
4. DEF CON is the world's longest running and largest underground hacking conference.
5. AR drone. AR stands for "Augmented Reality" in AR drone. AR Drone can perform tasks like object recognition and following, gesture following.

Source Figure 1-1: United States Department of Transportation (2015 -2018, August 4). *Federal Aviation Administration.* URL: Retrieved June 2018, from Federal Aviation Administration: https://www.faa.gov/news/updates/?newsId=83395 (*United States Department of Transportation,* 2015 -2018)[6],[7]

UAS Marketplace Drivers

Key drivers in the development of the US UAS marketplace are: FAA regulations, public acceptance, national / state infrastructure, and advancements in technology. The FAA regulatory environment, state law and international laws concerning UAS will be presented in Chapter 2. The key goal for all UAS regulations is the safe integration of UAS into the national airspace.

Public Acceptance

Before the retail marketplace can grow there needs to be a public acceptance of the UAS services. In 2016, the Inspector General of the United States Postal Service Office administered an online survey regarding UAS among 18-75-year-old residents in all fifty states and the District of Columbia.[8] The findings paint a picture of uncertainty of nationwide acceptance. Overall, the three generational groups (Baby Boomers,[9] Generation X,[10] and Millennials[11] think it is too soon for UAS delivery. Millennials are the largest U.S. population that would accept the concept of drone delivery. Forty-eight percent of Millennials believe it would be "safe" to integrate UAS systems into the National Airspace (NAS). This compares to 47 % of Baby Boomers that believe delivery of a package by UAS would be "unsafe."[12] All three generational groups chose Amazon as their trusted brand for drone delivery.[13] "Baby Boomers" traditionally are more conservative than other generations. For this reason, the US UAS marketplace growth may be delayed.

6. IoT Internet of things
7. Green circles represent hub-densities of UAS registrations.
8. (United States Postal Service Office of Inspector General, 2016)
9. Wikipedia (2018) Baby Boomers are generally defined as those born between 1945 and 1964. That would make the generation huge (71 MM) and encompass people who were 20 years apart in age.
10. IBID, Generation X, or Gen X, is the demographic cohort following the baby boomers and preceding the Millennials. There are no precise dates for when Generation X starts or ends. Demographers and researchers typically use birth years ranging from the early-to-mid 1960s to the early 1980s.
11. IBID, anyone born between 1981 and 1996 (ages 22 to 37 in 2018) will be considered a Millennials
12. Safe and unsafe are relative terms. No definition was given in the survey. This presents a problem for the researchers because of the vagueness of the terms.
13. Op cit. (United States Postal Service Office of Inspector General, 2016)

Figure 1-2 U.S. Postal Service Survey Results

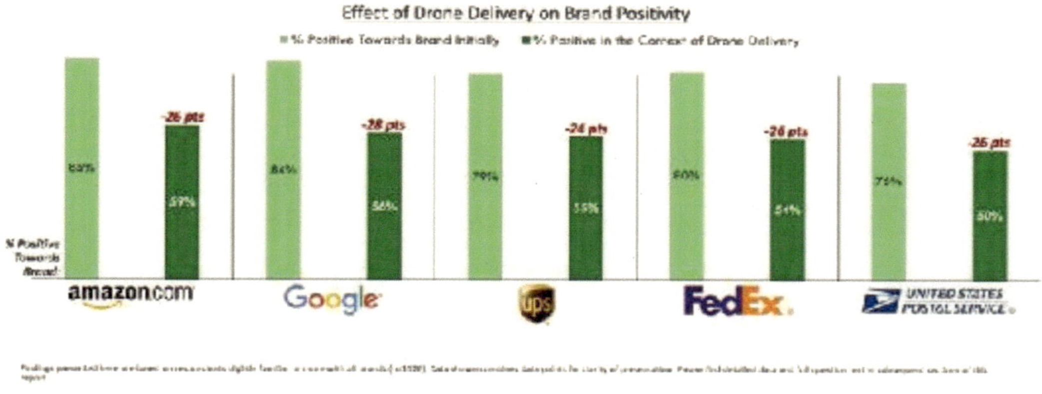

Source: United States Postal Service Office of Inspector General (2016, October 11). Retrieved June 2018, from Public Perception of Drone Delivery in the United States: https://www.uspsoig.gov/sites/default/files/document-library-files/2016/RARC_WP-17-001.pdf

Infrastructure Influence

Along with government regulations, infrastructure in the U.S. has been a roadblock when it comes to growth of the UAS market in the U.S. The U.S. aviation infrastructure ranks low among other countries around the globe.[14] According to the *Air Help* global survey,

With poor infrastructure, the UAS industry has an uphill battle in building a diverse marketplace in the U.S. To mitigate falling behind in the global UAS commercial market private corporations have taken the U.S. infrastructure issue into their own hands. "Amazon, Google, Boeing and General Electric have joined forces with NASA to build out a privately funded and operated air-traffic control network separate from current federal system." (Pasztor, 2018) One reason for the private industry taking the lead, is it is recognized that FAA is not able to process requests from commercial drone operators fast enough and/or reject a large amount of applications. The current air-traffic control system is not compatible with the Low Altitude Authorization and Notification Capability (LAANC) that is needed to safely operate commercial UAS traffic in a highly populated area. U.S. will need to allocate more money and resources in UAS infrastructure to keep up with the globe.

Restricting Drones Flight

Some states have taken a negative view of UAS activity and have legislated restrictions on their use. See Figure 1-3.

14. (Markovich, 2015)

There are areas of restricted air space that UAS / drones are not permitted to enter or operate. It could be restricted for protection of the U.S. government or stealing plays of a major league football team. While a majority of UAS manufacturers have built in GPS *Geofencing.*[15] this is not a "silver bullet" solution. Different methods of preventing drones in restricted areas are being tested. Airspace Systems has created a drone with sensors and artificial intelligence, to prevent unwanted drones.[16]

Figure 1-3 State of Florida Drone Signage

Source: State of Florida Drone Signage. Dixon, D. (August 1, 2017) The Florida Times-Union, Jacksonville. Geofencing Stops Drones in Their Tracks. http://www.govtech.com/public-safety/Geofencing-Stops-Drones-in-Their-Tracks.html

The defensive drone, named Avlayan, is capable of high speed travel and uses a Kevlar net to capture the threatening drone away from the area. See Figure 1-4. Another drone blocking innovation is the SkyDroner camera, it detects and disables drones that enter an area.

Commercial

Commercial UAS use in the U.S. has primarily been limited due to FAA regulations. As a result, U.S. companies have gone abroad to Europe, Africa, and Asia to test their drone functionality and begin commerce. Recently the FAA began to change regulations in favor of the UAS com-

15. "A geo-fence is a virtual perimeter for a real-world geographic area. A geo-fence could be dynamically generated—as in a radius around a point location, or a geo-fence can be a predefined set of boundaries (such as school zones or neighborhood boundaries). The use of a geo-fence is called geo-fencing, and one example of usage involves a location-aware device of a location-based service (LBS) user entering or exiting a geo-fence. This activity could trigger an alert to the device's user as well as messaging to the geo-fence operator. This info, which could contain the location of the device, could be sent to a mobile telephone or an email account." (Wiki-G, 2018)
16. (Mannes, 2016)

mercial market, however the change remains behind other countries. The U.S. commercial market consists of the following sectors: movie / film, agriculture, humanitarian aid, and retail industries. Retailers, like Amazon and FedEx, are exploring UAS delivery of goods to and services to consumers. UAS used for humanitarian efforts are moving at an accelerated pace through the FAA approvals. The U.S. remains behind in the global UAS race in the commercial market. The focus in the U.S. for commercial UAS remains on safety and pollution. The topic of security of the UAS is limited to military, law enforcement and intelligence organizations. *The subject of security remains unaddressed at a formal level in the development of commercial UAS globally.* Critical security factors of UAS could be used to develop the marketplace, allowing the U.S. to be an example for other countries.

Figure 1-4 Airspace Systems Interceptor autonomous aerial drone

Source: Drone-catchers Emerge on a New Aerial Frontier. (March 21, 2017). In Silicone Valley & Technology https://www.voanews.com/a/drone-catchers-emerge-new-aerial-frontier/3776609.html

Retail

In December 2016, Amazon successfully completed one of the first ever commercial deliveries in Cambridge, UK.[17] See example Figure 1-5. The U.S. based company chose to arrange the first delivery in the United Kingdom because of their regulations are ahead of the U.S. Amazon's first U.S. delivery occurred in March 2017, which is behind other countries. In 2016, China's largest ecommerce retailer began deliveries to customers. In 2016, Dominos started delivering pizzas to New Zealanders via UAS marketplace.

17. (Global Market Insights, Inc., 2018)

Figure 1-5 Amazon Prime Air

Source: Bishop, T. Geek wire (December 14, 2016). Video: Amazon makes first Prime Air drone package delivery, offers glimpse of new design https://www.geekwire.com/2016/video-amazon-makes-first-prime-air-drone-package-delivery-offers-glimpse-new-aircraft-design/

Overall, Africa is leading the UAS commercial market, they established a drone corridor and working on building out the first drone port.[18]

Zipline, a U.S. based UAS company, delivered blood to remote area of outside of Rwanda in 2016. Rwanda was chosen by the sky ambulance company because of the lack of established infrastructure. By the end of 2018, Zipline is projected to be the world's largest drone delivery service. Zipline's commercial drones, backed by UPS, can travel 500 flights a day up to 80 mph. It is expected Zipline will conduct 2,000 medical deliveries a day by the Tanzania government. Zipline's currently planning to launch business in the U.S. by the end of 2018. The issue of security does not seem to be a Zipline priority. (Figure 1-6) This might be a grave concern in a country / continent which has so many terrorist groups in the field.

18. The UAS African connection is much more complex than just retail entities. See Chapter 15: Africa – World's First Busiest Drone Operational Proving Ground – Where Counter-Terrorism and Modernization Meet

Figure 1-6 Zipline drone testing package drop

Source: Shankland, S. CNET (April 2, 2018). Zipline's second-gen drones speed its medical delivery business. https://www.cnet.com/news/zipline-new-delivery-drones-fly-medical-supplies-faster-farther/

In May 2018, the FAA announce their UAS Integration Pilot Program (UASIPP), which is a program comprised of ten companies that would be paired with local governments to conduct commercial drone testing. Amazon, the world's largest online retailer, and DJI were not chosen to take part in the pilot.[19] *China's DJI is the world's top drone manufacture, owning 70% of the global marketplace.*[20] Among the ten chosen companies to participate in the FAA test program are Apple, Intel, Uber, and Zipline.

UAS For Hire – Urban Air Mobility and E-VTOL

Often compared to *The Jetsons*, Urban Air Mobility (UAM) remains a far-reaching idea for the public. UAM can be a manned or unmanned flying vehicles that transport one to six people, without delay, unlike automobile traffic. However, the reality of UAM is closer than the public would chose to believe. Uber announced Uber Elevate in Fall of 2016. Consult Figure 1-8. Elevate is a project that will make UAM a "common household name". Initially the project started with the idea of unmanned air vehicles landing on rooftops. Since the introduction of Elevate, it has evolved into so much more.

19. (Shepardson & Dastin, 2018)
20. (Borak, 2018)

1-7 Uber Elevate

Source: Stewart, J. (2018, May 8), *Wired*. Retrieved July 2018, from Wired Transportation: https://www.wired.com/story/uber-unveils-flying-taxi/

Uber was one of the companies approved for commercial testing by the FAA as part of the UASIPP. This decision was a significant for the U.S. in the UAS race. Upon completion, Uber's project Elevate will implement intra-city taxis that will be unmanned aerial vehicles. The technology is often referred to as Electric Vertical Take-Off and Landing (E-VTOL). This on-demand urban air taxi is fully electric and takes off and lands vertically. Uber is planning by 2023, to have a network of E-VTOLs distributed by vertiports. (Figure 1-7) The vertiports will have multiple takeoff and landing pads. The pads will be equipped with charging infrastructure. By 2030, Dubai is planning to be using E-VTOL to mobilize 25% of the population.[21]

21. (Stewart, 2018)

Figure 1-8 Uber Flying Skies

Source: Stewart, J. (2018, May 8), *Wired*. Retrieved July 2018, from Wired Transportation: https://www.wired.com/story/uber-unveils-flying-taxi/

The consumer will be able to use their mobile device to request an Uber Elevate aircraft. After arriving at the vertiport, Uber's version of an UAV airport, the consumer will be weighted as they check in and assigned an aircraft.

The consumer would then proceed to the assigned pad for departure.[22] (See Figure 1-8) The author believes, that if the U.S. government continues to support Uber UAS creativity, the USA would be a front -runner in the inter-city taxi industry. See Figure 1-9 for a look at the future Uber UAM station.

22. (Stewart, 2018)

Figure 1-9 Uber UAM Station Check-in

Source: Stewart, J. (2018, May 8), *Wired*. Retrieved July 2018, from Wired Transportation: https://www.wired.com/story/uber-unveils-flying-taxi/

It is estimated there are fifty companies creating E-VTOL aircrafts globally. The designs vary from amount of motors, sensors, and propellers. As of 2017, the EHang 184 E-VTOL conducted numerous tests for Dubai's Roads and Transport Authority (RTA). Passengers have been flown in China and Dubai. The aircraft is being developed in China and in Redwood City, California. The EHang 184 is a single passenger aircraft that has safety features such as sensors that prevent it taking off in a storm. See Figure 1-10.

Figure 1-10 EHang 184 E-VTOL

Source: EHang 184. Electric VTOL News by Vertical Flight Society (September 8,2018). Retrieved 09082018 from EVTOL News at: http://evtol.news/aircraft/ehang/

In comparison, the Volocopter holds two passengers and boasts eighteen electric drives. It is estimated the Volocopter can move 1,000 passengers an hour by landing on rooftops, or as the company calls them "Volo-ports".[23] The aircraft uses a fly-by-light (FBL) communications network that includes Gyroscopes, acceleration sensors, magnetic field measurement sensors, and manometers.

UAS for hire is expanding to explore transporting more people for further distances. The Lilium is a five passenger VTOL, with the ability to move five times faster than a car. The unmanned taxi is estimated to transport from Manhattan to New York's JFK airport in five minutes verse the fifty-five minutes by car.

E-VTOL Cybersecurity

Cybersecurity protection of E-VTOL data is not respected as an issue – even though millions of customers have suffered breaches of personal data by well-respected companies. NASA is leading a series of workgroups to research areas of pollution, and safety. Only recently has NASA

23. (III, 2018)

integrated cybersecurity with the addition of one researcher. The hype to bring UAM to the marketplace has blinded the industry of reviewing the cybersecurity implications. The industry should learn from others that have come before them. Let's look at the Tesla company.

In 2016, Tesla launched the first self-driving car. The automaker has been in the electric car business since 2008. They only became popular in 2015, selling over 50,000 vehicles.[24] It took less than a year from record sales for the car to be compromised. *Hackers were able to take remote control of a Tesla Model S from twelve miles. They were able to operate the brakes, door locks, dashboard computer screen and other electronically controlled components.*[25] [26]

Additionally, another group from University of South Carolina made the car's autopilot believe it was seeing objects that were not actually there, causing the car to brake prematurely (Solon, 2016). Tesla does run a bug bounty program, however their plan for ensuring security of the car was flawed. In 2017, the Tesla Model X was again compromised, turning on the brakes remotely and getting the doors and trunk to open and close while blinking the lights in time to music streamed from the car's radio.[27] See Figure 1-11. Both compromises were completed by Keen Security Lab of Shanghai, China. As a result, Tesla released a statement, "They actively encourage this type of research so that it can prevent potential issues from occurring. The risk to customers from such exploits is very low and Tesla has not seen a single customer ever affected by it."[28] [29]

24. (Shahan, 2016)
25. (Solon, 2016)
26. This was an impressive hack and should give intelligence personnel a pause.
27. (Weise, 2017)
28. (Weise, 2017)
29. This was clearly a legal CYA for incompetence in their security cyber-protections of their own product.

Figure 1-11 Model X 2017

Source: Tesla (September 8, 2018) Tested: Driving the Tesla Model X w/ Autopilot! Youtube.com via https://www.f-sport.lt/play/5670542/tested-driving-the-tesla-model-x-w-autopilot.html

Agriculture

Agriculture is the fastest growing industry in the UAS commercial market. Drones equipped with sensors are used for monitoring crops and providing soil analysis. Outside of the U.S. UAS are used to apply fertilizer, pesticides, and irrigate crops. Farm animals can be tracked, and health monitored by data collected by drones. A leading agriculture drone provider, Agribotix, has created FarmLens. The cloud-based solution provides extensive data to the farmer regarding their crops. This product is part of a partnership with AgrAbility, enabling farmers with disabilities to continue in their profession. See Figure 1-12 for temperature study data for an environmental permitting process performed by drone and analyzed by Nero software.

Architecture and Construction

The construction sector is projected to be the largest commercial buyer of UAS, totaling $ 11.2

billion globally by 2021.[30] Architects use drones for site scanning, feasibility studies, and presenting the client with the first virtual images of what the site will look like at completion. The designers can use thermal imaging and multispectral cameras to provide critical data about the land prior to building. Builders can use drones during the construction process to track progress, monitor assets, and capture issues prior to inspection. Inspectors are able to use drones to monitor for compliance to building regulations. After the construction is complete, the collection of drone data can be used in selling the property. Currently, real estate agents are using drones to enhance their business by providing aerial views of property for their consumers.

Figure 1-12 Nero temperature data collection

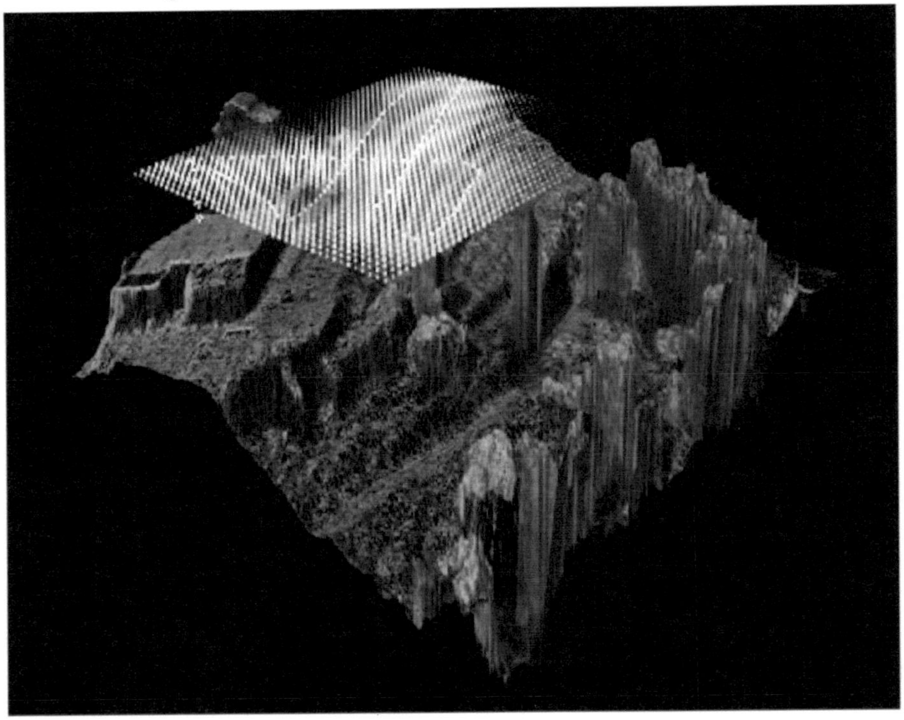

Aerial average temperature data gathered by drone and environmental sensing platform. Image © NERO / Noumena

Source: NERO / Noumena (12-14th of February 2016). Temperature Data Captured Through Environmental Sensing Platform (Smart Citizen Kit) And Drone, Mapping Location: Ctra. BV-1415 (Horta-Cerdanyola), km 7,08290 Cerdanyola del Vallès, Barcelona – Spain https://noumena.io/11545-2/

Chinese drones

30. (Wood, 2018)

The global consumer market for drones continues to grow. Since the development of the DJI drone in by Frank Wang 2006, as part of his graduate thesis, the company holds 70% of the consumer marketplace. (Gan, 2018) See Figure 1-13. In 2017, DJI had $2.7 billion in sales comprised of 30% of its revenues from China, the US, and Europe, respectively, and 10% from South America.[31] Their manufacturing capabilities, infrastructure, and logistics is the key to China remaining a market leader in the consumer space. The U.S. has outsourced much of their manufacturing, therefore weakening their ability to produce and get to market in a timely manner. Additional, China has a large test market, supporting infrastructure, and consumer support. The U.S. continues to find it difficult to keep up in all areas.

Figure 1-13 DJI Founder Frank Wang

DJI founder Frank Wang

Source: DJI. In *Fortune*, The Unicorn List. Retrieved September 09, 2018 http://fortune.com/unicorns/dji-16/

The U.S. consumer uses the drone primarily for photography and film. There are a number of drone hobbyists as well. Overall the impact to consumer grow has been stifled in the U.S. for same reasons as commercial market. The increasing regulations on UAS, across all sizes, has changed the consumer market in the U.S. The FAA has implemented the same risk assessments across all drones, limiting the marketplace. A consumer drone can be treated the same as an emergency services drone.

The inability to change the process to recognize distinct UAS markets in the U.S. will continue

31. (Gan, 2018)

to have dramatic impact. The marketplace for UAS is as wide as the imagination, however the U.S. barriers of regulation, infrastructure, and technology advancements will place them behind less developed countries.

Discussion Questions

1. What are the impacts to the U.S. if the country is unable to keep up with the growth of UAS?
2. How can security be included in the development of a product without having an impact in the race to market?
3. What do you think about the security legacy issues for UAS components? How could It be improved by building in security up-front AND controlling the manufacturing supply chain? See: Dr Julie J.C.H. Ryan's PDF class handout on "*An Exploration of Information Security Aspects* Julie J.C.H. Ryan" [Also available by contacting Professor Nichols at KSU.]

Bibliography

Bishop, T. (2016, December 14). *Amazon makes first Prime Air drone package delivery*. Retrieved from Geek Wire: https://www.geekwire.com/2016/video-amazon-makes-first-prime-air-drone-package-delivery-offers-glimpse-new-aircraft-design/

Borak, M. (2018, January 3). *Tech Node*. Retrieved June 2018, from World's Top Drone Seller made $2.7 billion: https://technode.com/2018/01/03/worlds-top-drone-seller-dji-made-2-7-billion-2017/

Dimock, M. (2019, January 17). *Defining generations: Where Millennials end and Generation Z begins*. Retrieved from Pew Research Center: https://www.pewresearch.org/fact-tank/2019/01/17/where-millennials-end-and-generation-z-begins/

Dixon, D. (2017, Aug 1). *State of Florida Drone Signage*. Retrieved from Government Technology: http://www.govtech.com/public-safety/Geofencing-Stops-Drones-in-Their-Tracks.html

Gan, J. (2018, July 02). *Harvard Business Review*. Retrieved July 2018, from Harvard business Review Innovation: https://hbr.org/2018/07/the-factors-behind-chinas-growing-strength-in-innovation

Geo-Fencing Definition. (2019, July 18). Retrieved from Whatis.TechTarget.com: https://whatis.techtarget.com/definition/geofencing

Global Market Insights, Inc. (2018, February 26). *Global Market Insights*. Retrieved June 2018, from Commercial Drone / UAV Market worth over $17bn by 2024: https://www.gminsights.com/pressrelease/unmanned-aerial-vehicles-UAV-commercial-drone-market

III, W. B. (2018, June 1). *Aviation Today*. Retrieved July 2018, from Avionics Magazine June/July: http://interactive.aviationtoday.com/avionicsmagazine/june-july-2018/avionics-for-2020s-evtol-and-supersonic-aircraft/

Mannes, J. (2016, November 18). *Tech Crunch*. Retrieved June 2018, from sensors and machine intelligence, to autonomously intercept threatening drones at high speeds and carry them away from large crowds: https://techcrunch.com/2016/11/18/airspace-systems-interceptor-can-catch-high-speed-drones-all-by-itself/

Markovich, S. (2015, May 28). *Council on Foreign Relations*. Retrieved July 2018, from U.S. Aviation Infrastructure: https://www.cfr.org/backgrounder/us-aviation-infrastructure

Meola, A. (2017, July 13). *Business Insider*. Retrieved June 2018, from Drone market shows positive outlook with strong industry growth and trends: http://www.businessinsider.com/drone-industry-analysis-market-trends-growth-forecasts-2017-7

Pasztor, A. (2018, March 9). *Wall Street Journal*. Retrieved July 2018, from Business Wall Street Journal: https://www.wsj.com/articles/amazon-google-others-are-developing-private-air-traffic-control-for-drones-1520622925

Rizzo, C. (2018, June 6). *Travel and Leisure*. Retrieved July 2018, from Airlines and Airports: https://www.travelandleisure.com/airlines-airports/most-least-punctual-airports-airhelp-study

Shahan, Z. (2016, January 5). *Clean Technica*. Retrieved July 2018, from Clean Technica Transport: https://cleantechnica.com/2016/01/05/tesla-surpasses-50000-sales-in-2015/

Shankland, S. (2018, April 2). *Zipline's second-gen drones speed its medical delivery business*. Retrieved from CNET : https://www.cnet.com/news/zipline-new-delivery-drones-fly-medical-supplies-faster-farther/

Shepardson, D., & Dastin, J. (2018, May 9). *Reuters*. Retrieved July 2018, from Technology News: https://www.reuters.com/article/us-usa-drones-companies/u-s-drone-program-taps-apple-passes-over-amazon-chinas-dji-idUSKBN1IA2WC

Solon, O. (2016, September 20). *The Guardian*. Retrieved July 2018, from The Guardian US World Envionment: https://www.theguardian.com/technology/2016/sep/20/tesla-model-s-chinese-hack-remote-control-brakes

Stewart, J. (2018, May 8). *Wired*. Retrieved July 2018, from Wired Transportation: https://www.wired.com/story/uber-unveils-flying-taxi/

Tech Musings. (2009, December 18). *Pingdom*. Retrieved June 2018, from Worst Internet disasters of the decade: https://royal.pingdom.com/2009/12/18/worst-internet-disasters-of-the-decade/

Tesla . (2018, September 8). *Tested: Driving the Tesla Model X w/ Autopilot!* . Retrieved from F Sport: https://www.f-sport.lt/play/5670542/tested-driving-the-tesla-model-x-w-autopilot.html

United States Department of Transportation. (2015 -2018, August 4). *Federal Aviation Administration*. Retrieved June 2018, from Federal Aviation Administration: https://www.faa.gov/news/updates/?newsId=83395

United States Postal Service Office of Inspector General. (2016, October 11). *United States Postal Service Office of Inspector General*. Retrieved June 2018, from Public Perception of Drone Delivery in the United States: https://www.uspsoig.gov/sites/default/files/document-library-files/2016/RARC_WP-17-001.pdf

Weise, E. (2017, July 28). USA *Today*. Retrieved July 2018, from USA Today Tech: https://www.usatoday.com/story/tech/2017/07/28/chinese-group-hacks-tesla-second-year-row/518430001/

Wood, H. (2018, April 4). *Archinect*. Retrieved July 2018, from Archinect: Features – Drones for Architects: https://archinect.com/features/article/150058176/drones-for-architects-new-capabilities-for-the-construction-sector-how-to-get-started-and-how-to-navigate-the-law

Chapter 2: UAS Law - Legislation, Regulation, and Adjudication

Student Learning Objectives – A set of considerations relating to the history, methods and objectives of regulating Unmanned Aerial Systems (UAS). Since the skies are already crowded, drones adds complexity to the environment. Skies crowded with manned and unmanned aircraft require regulations to ensure safe operation in a controlled environment. Much like the rules of the road, airspace travel needs some degree of uniform operations of aircraft, responsibility, and consequences for violating them. The history of regulation for automobiles, trains, and other modes of travel provide over 150 years of historical data from which to learn. Students will be introduced to the general principals of legislative, regulatory, judicial processes, and their impact on UAS operation.

Law & Technology – The Tortoise and the Hare

In the Hare & the Tortoise, the Greek Fabulist Aesop wrote about the interplay between speed and measured deliberate progress. As the fable goes the Hare teased the Tortoise for being slow and not making progress to which the Tortoise challenged the Hare to a race to prove that in fact his progress would be much faster than the hare believed. The Hare accepted the challenge. They agreed that the Fox would set the distance and act as judge. As the race started the Hare sped out of sight, while the Tortoise slowly but surely ambled down the road. Given his large lead the Hare decided to take a roadside nap to mock the Tortoise. When the Hare awoke the Tortoise had slowly but surely passed the sleeping Hare and was nearing the finish line. The Hare attempted to use his great speed to catch the Tortoise, but it was too late. The Tortoise won the race. Aesop's moral was, "the race is not always to the swift" (Schlichting, 1993).

Figure 2-1 Tortoise and Hare

Source: Schlichting, M., S: (1993). *Aesop's Fable the Tortoise and the Hare, as retold by Mark Schlichting*. Novato, CA: Broderbund Company.

The relationship between technology and the law is complex and does not always result in outcomes satisfactory to parties on either side of the issue. Students should keep this inherent compromise in mind as we examine the interplay between the law and UAS technology. We will focus upon the relationship between US legal system and technology. This is not to say that other nation's legal systems are less sophisticated than the American system. Rather, it is in recognition of the vast amount of information to be examined and the relatively short judicial history of 250 years m XXX takes it easier study.

Transportation in the United States – Lessons from the Past Help Guide the Future

In the early 1900's the automobile went from an extravagant acquisition of the elite to a mass produced stable in a relatively brief period. In 1908 Henry Ford began selling the Model-T. The initial cost of the vehicle was $825 and approximately 10,000 vehicles were sold. In 1913 Henry Ford introduced the motorized assembly line method of mass manufacturing reducing the Model-T chassis manufacturing time from 12.5 hours to 1.5 hours and the cost of the vehicle to $575 (Ford Motor Company, 2018). Within sixteen years Ford had sold over ten million Model-T automobiles (The Saylor Foundation, 2008). At the same time other companies such as General Motors were also producing millions of automobiles annually (Figure 2-2).

With millions of new automobiles came new challenges. Roads needed to be financed and built. Auto operation needed to be regulated to ensure public safety and operation predictability. Coded laws were required to ensure that liability could be addressed through the legal process.

Unregulated motor vehicle operation in the late 19th and early 20th century could, and in fact did, result in traffic anarchy. In 1917 alone, the city of Detroit, Michigan, with approximately 65,000 autos on the road, had 7,171 accidents and 168 fatalities (Loomis, 2015).

Figure 2-2 Ford Motor Company Production Plant

Source: Ford Motor Company, (2018). Company Timeline – Ford.Com. Detroit, Michigan: Ford Motor Company

Mass production of the automobile created substantial problems for the government. How could a safe and effective system of roadways and rules be developed without impeding the evolution of automobiles?

This was the challenge of the day and in the first two decades of the 1900's large cities struggled to design infrastructure and regulate motor vehicle operation to stem the tide of mass carnage. Traffic fatalities increased by 3,000 percent between 1901 and 1923.

In fact, many cities were so devastated by the substantial number of children and adults killed, they erected monuments to the fallen (Norton, 2008). Figure 2-3

Figure 2-3 (National Safety News 1922)

Source: Ford Motor Company (2018). Company Timeline – Ford.Com. Detroit, Michigan: Ford Motor Company.

Unlike the introduction of the railroads in in the 1800's, automobiles presented a greater challenge. Trains and trolleys were limited to travelling on rails, which made the design of infrastructure and regulation of operation less challenging[1] By 1920 it was clear that affordable automobiles were the conveyance of the future. People were migrating away from the mass transit staples of trains, busses and trolleys, and opting instead for personalized on-demand, motorized transportation.

The concept of regulating innovative technology requires a deep understanding as well as recognition of how overzealous regulation can stifle innovation. There is a delicate balance that must be struck depending upon circumstances. Imposing too much regulation can chill innovation, yet too little can lead to harm. For example, a regulator may define "success" as avoidance

1. This is not to say that the introduction of rail transport in was safe by any means. Pedestrians, other trains, animal drawn carriages and animals themselves were killed in large numbers due to being run over by a train.

of a catastrophe. This is known as the "precautionary principle" of regulation, while an entrepreneur would likely view it as the least favorable outcome of regulation and one that makes it very difficult to innovate (Fenwick, 2017).

Regulation of the Automobile

As early as 1906 it was clear that the evolution from horse drawn carts to automobiles required rules to protect the public and create an ordered traffic system. In 1906 Xenophon Huddy, a New York lawyer wrote the first known law legal treatise on regulation of the automobile. Huddy wrote of the need to embrace the modern technology, but also warned that rules of manufacture, operation, vehicle safety and ownership would be needed to protect the public. (Huddy, 1906)

The first codified embodiment of automobile and driving laws were adopted by New York City in 1906. They were called the "Rules for Driving," drafted by aristocrat William Phelps Eno and started to regulate operation on automobiles by focusing on speed, turns, and yielding to pedestrians and emergency vehicles. Eno viewed these rules to, "substitute order for where chaos now reigns almost supreme" and believed the most logical way to enforce these rules was through the already existing police agencies (Norton, 2008).

Slowly nations, states, and local governments began to adopt their own laws and regulations, as well as design and safety principles for the creation and expansion of the highway network. By the 1920s most cities and states had enacted regulations sufficient to ameliorate the carnage of 1900-1915.

Figure 2-4 Flights Everywhere

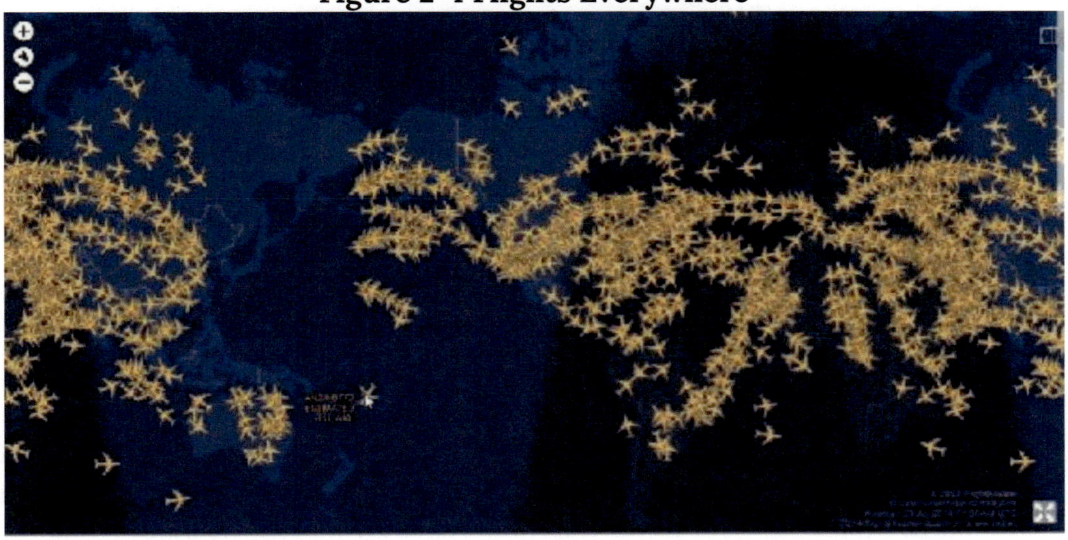

Source: Kekatos, M. DailyMail.com (September 09,2018) Flightaware. The great getaway begins: 6.4 million travelers set off for the holidays as incredible flight tracker map shows how busy the skies are above the US. https://www.dailymail.co.uk/news/article-5207639/Flight-tracker-map-shows-6-4-million-traveling-plane.html

The Next Transportation Challenge – Aviation

Commercial aviation followed closely on the heels of automobiles in the first two decades of the Twentieth Century (Figure 2-4)

The National Advisory Committee on Aeronautics (NACA) was created by Congress and the bill signed by President Woodrow Wilson. The NACA created rules to establish order and safety in aviation (Marshall, 2012)

Figure 2-5 Jet Setting

GRAND CANYON | TWA Flight 2 and United Airlines Flight 718

Source: Noland, D & Peterson, B. (August 4, 2017) Popular Mechanics. 12 Plane Crashes That Changed Aviation: Out of these tragedies arose major technological advances in flight safety that keep air travel routine today. https://www.popularmechanics.com/flight/g73/12-airplane-crashes-that-changed-aviation/

The advent of jet propulsion made trans-continental and trans-oceanic passenger travel a reality. The age of "jet setting," as it was known in the 1960's, led to a massive increase in air traffic, (Figure 2-5) which in turn heightened the need for regulations and oversight. Due in part to a tragic midair collision in the Grand Canyon, 1958 President Dwight Eisenhower signed legislation establishing what is still known today as the Federal Aviation Administration (FAA).

Concurrently, other nations were also enacting aviation regulations to maintain order and safety in the skies. Rules for safe air transit of passengers and cargo had to be balanced with increasing crowding, from simultaneous increases of military and private aviation. Great Britain enacted the Aerial Navigation Act of 1911, with oversight vested in the Board of Trade. However, with the outbreak of World War I in 1913, that responsibility was transferred to the Secretary of State for War. After the Armistice ending WWI the Department of Civil Aviation was created (Chaplain, 2011).

Since aviation provided the ability to travel great distances rapidly, with few physical restrictions or borders, the next logical step was global aviation regulation and air traffic control. The treaty of Versailles in 1919 addressed the issue by creating the Convention for the Regulation of Aerial Navigation, establishing global standards and rules for aviation over and between nations. The convention specifically provided, "every aircraft of a contracting State has the right to cross the air space of another State without landing. In this case it shall follow the route fixed by the State over which the flight takes place. However. For reasons of general security, it will be obliged to land if ordered to do so by means of the signals provided" (Treaty of Versailles, 1919).

The Convention was the first attempt to balance competing interests related to global navigation, the need for free airspace transit over a nation, to assist commercial aviation develop, with increased risk of attack by a non-military aircraft.

Even though World War II erupted less than two decades later, the need for civil aviation uniformity, predictability, and safety was still paramount in the minds of many. In to achieve this goal, near the end of WWI a convention was held in Chicago called the International Civil Aviation Conference, which resulted in the creation of the International Civil Aviation Organization (ICAO), with fifty-two nations being initial members. The ICAO promulgated rules that apply to global civil aviation in international airspace defined as, "more than 12 miles from the sovereign territory of a country as well as some domestic airspace by virtue of incorporation into a contracting state own regulatory scheme" (Franzese, 2009).

Though each member nation established its own aviation regulatory body and laws, the provisions of the ICAO were meant to create an acceptable level of uniformity and airspace access, while permitting the growth of commercial air travel. But, aircraft and crews from distant regions had to be able to effectively operate and communicate with air traffic control across the globe. Additionally, global aircraft registration, pilot licensing, and safety still needed to be developed.

The first and most pressing issue relating to civil aviation globally was to develop a uniform language requirement so that aircrews could communicate effectively with air traffic and ground controllers no matter their native language. Without language standards, neither pilots nor air traffic controllers who did not speak the same language would be able to communicate. The lack of a uniformity in aviation communication, coupled with the exponential growth in civilian air travel, would result in greater risk of collisions, crashes, and other tragedies. Recognizing this issue, the ICAO in1951 adopted an international requirement that air traffic controllers and air crew must communicate in English to reduce the risk of incidents caused by an inability to understand each other (MacKenzie, 2010).

With thousands of aircraft airborne at any given time, the chance of collisions is a constant, especially in congested skies over major metropolitan areas. Couple that with time of day, weather, terrain, and the risk of mid-air, or even ground collisions, increases exponentially. Unfortunately, simply requiring air traffic controllers and air crew to speak English proved insufficient, as it became clear that language proficiency would also be required. In 1977 it is believed that the lack of language proficiency was a primary cause of the largest loss of life in aviation history. Two 747 aircraft crashed on the runway in Tenerife, Canary Islands, Spain, resulting in 583 fatalities. (Figure 2-6)

Figure 2-6 Two 747 aircraft crashed on the runway

Source: Smith, P. (March 27, 2017) Daily Telegraph. The true story behind the deadliest air disaster of all time. https://www.telegraph.co.uk/travel/comment/tenerife-airport-disaster/

Aviation Design and Manufacture Standards

Aviation safety must extend beyond the flight crew and ground control operations. The design and manufacturing of aircraft also need regulatory oversight. As part of its mission the FAA was empowered by Congress to establish standards for the design and manufacture of commercial aircraft. For a commercial aircraft to enter service in the US, regardless of its origin, the design and manufacture must receive prior approval from the FAA. Currently there is a five-step process to receive approval from the FAA, starting with the original design of the aircraft, leading all the way through to the issuance of an airworthiness certificate, to allow it to commence passenger service (Federal Aviation Administration, 2009).

Development, design, and manufacture standards are a complex process which requires a great deal of collaboration between the FAA, and stakeholders, other nations, the ICAO, and advisory organizations. Three of the most prominent advisory organizations include the Radio Technical Commission for Aeronautics (RCTA) the Society of Automotive Engineers (SAE) and the American Society of Testing and Materials (ASTM) (Marshall, 2012).

Aircraft design and manufacturing standards are developed and regulated to guide the aircraft manufacturers in building aircraft within or operating foreign manufactured aircraft within the airspace of a country. Many nations have, by agreement, adopted the same or very similar commercial aviation design and manufacturing standards to allow uniform standards compliance. For example, an aircraft manufactured by Boeing in its North Charleston, South Carolina plant, is inspected for compliance by the FAA. Due to international uniformity in standards enforced by the FAA, they can certify compliance to the aviation authorities of other countries (Federal Aviation Administration, 2016). (Figure 2-7)

Not only do design and manufacturing standards require compliance, they are also monitored by the government authority throughout the manufacturing and flight testing process. Most, if not all, western aircraft manufacturing facilities are manned by FAA officials or other authorities, to ensure compliance with applicable standards.

Some of the areas in which design and manufacturing standards are established and enforced are through collaboration between Original Equipment Manufacturers (OEM's) and the various stakeholders previously discussed.

1. Aircraft design, structures, avionics.
2. Electrical and mechanical systems.
3. Power plants.
4. Equipment.
5. Engineering flight testing (Government of Canada, 2017).

Figure 2-7 Boeing Company

Source: Polek, G. (May 14, 2010) Boeing Halts Delivery of 787 Dreamliner Parts. https://www.ainonline.com/aviation-news/air-transport/2010-05-14/boeing-halts-delivery-787-dreamliner-parts

Regulation is not only vital to safety in the design and manufacture of commercial aircraft, it may be most important when it comes to oversight of commercial air carriers. Day to day operation, crew training, aircraft maintenance and overall operation safety are key components to safety and reliable operation. The FAA has adopted what is known as the Safety Assurance System (SAS) to better manage the certification and operation surveillance or air carriers.

The objectives of the SAS is defined as:

- Verify an applicant can operate safety (sic) and comply with regulations and standards before issuing a certificate and approving or accepting programs.
- Conduct periodic reviews to verify that a certificate holder continues to meet regulatory requirements when the environment changes, and
- Validate the performance of a certificate holder's approved and accepted programs for Continued Operational Safety (COS) (Federal Aviation Administration, 2018).

While all the regulatory considerations discussed are vital, it should not be assumed that the regulators themselves do not require oversight. Towards this end the US Department of Transportation, under whose aegis the FAA operates, has established an Office of Inspector General, created by the Inspector General Act of 1978.

The Inspector General of each agency of the executive branch of the US government is charged with the duties of protecting the interests of the public and taxpayers. They are supposed to be non-partisan ombudsmen who detect waste, fraud, and abuse in their respective agency. The Inspector General is the watchdog of the watchdog (Brian, 2010).

The importance of the function of the Inspector General cannot be overstated. For example, in 2017 the FAA Inspector General issued a report of its investigation into Edward Carl Hernandez, an FAA Designated Airworthiness Representative, charged with aircraft parts fraud. Hernandez pled guilty to falsifying his certification of airworthiness of a hydraulic adapter fitting that he had never accessed or inspected (Federal Aviation Administration, 2018). Other more widespread abuses have been uncovered through whistleblowers who report irregularities and illegalities in the aviation industry. One such instance, which was particularly troubling, was the claim that Boeing was installing defectively manufactured airframe parts on 737 aircraft where it was also alleged that FAA inspectors conducted a less than thorough review and investigation of the claims. This exact reason is why an independent overseer is needed to prevent too cozy a relationship between the regulators and the regulated (Graves, 2006).

The Inspector General plays an integral role in maintaining transparency and trust in the FAA, through oversight and regulation of UAS. In 2012 the FAA Inspector General commenced an audit of the FAA's oversight of domestic UAS operation, citing its belief that over 10,000 active systems would be in operation by 2017 (Guzzetti, 2012). Keep this important function in mind in the discussion of UAS regulation and the stresses their popularity will place upon regulators.

Regulated Activity Coexisting with Unregulated Activity

There was no formal regulation of airspace in 1903 when the Wright Brothers first flew their aircraft in Kitty Hawk, North Carolina. Just as was the case with the introduction mass production of automobile, the number of aircraft and air traffic increased rapidly. As discussed earlier the beginnings of air regulation were related to the advent of the air mail operations. It was clear early on regulation would greatly enhance safety. Between 1922 and 1925 the fatality rate for the air mail service was one per 789,000 miles flown while the rate for fledgling, non-regulated commercial fliers was one per 13,500 miles flown (Messier, 2016).[2]

The Aeronautics Branch of the Department of Commerce began to establish pilot licensing and training as well aircraft design and manufacturing safety certification, which led to safer aviation. Regulation and oversight still needed to balance safety with commercial viability. From the Wright Brothers to Charles Lindbergh to Amelia Earhart the rich tradition of private aviation remains. Consequently, aviation regulation had to be able to accommodate less regulated private aviation operating within increasingly crowded skies, simultaneously with increased commercial and military aviation.

2. Commercial flight data available for 1924 only.

A tragic example of the danger occurred in September 1978, when Pacific Southwest Airlines Flight 182, a Boeing 727 collided with a private Cessna 172 over San Diego, California. The accident resulted in the death of 144 people including nine people on the ground. Figure 2-8. The investigation pointed to many failures, not the least of which were air traffic control procedures relating to minimum safe separation between aircraft and the complexity of the interface between private commercial aircraft (National Transportation Safety Board, 1979).

Regulating Unmanned Aerial Systems – Smaller Aircraft Larger Problems

In January 1982, the FAA introduced National Airspace System (NAS) Plan to help enhance aviation safety domestically and to comply with aircraft classification systems created by the ICAO.

Figure 2-8 Tragedy in Pacific South West, 1978

Source: National Transportation Safety Board (1979). Aircraft Accident Report – Pacific Southwest Airlines Flight, 182. Washington, DC: National Transportation Safety Board.

Broadly speaking there are five general classes of controlled airspace in the NAS as defined by the FAA they are (See Figure 2-9):

1. **"Class A Airspace** – Generally that airspace from 18,000 feet MSL up to and including FL 600, including the airspace within twelve nautical miles of any coast of the forty-eight contiguous States and Alaska, and designated international airspace beyond twelve nautical miles of the coast of the forty-eight contiguous States and Alaska, within areas of domestic radio navigational signal or ATC radar coverage, and within which domestic procedures are applied.

Unless otherwise authorized, all persons must operate their aircraft under Instrument Flight Rules (IFR)." (FFA-PH, 2018)

2. **"Class B Airspace** – Generally, that airspace from the surface to 10,000 feet MSL surrounding the nation's busiest airports in terms of IFR operations or passenger enplanements. The configuration of each Class B airspace area is customized and consists of a surface area and two or more layers (some Class B airspace areas resemble upside-down wedding cakes) and is designed to contain all published instrument procedures once an aircraft enters the airspace.

An ATC clearance is required for all aircraft to operate in Class B Airspace, and all aircraft that so cleared receive separation services within the airspace. The cloud clearance requirement for VFR operations is "clear of clouds". Arriving or transiting aircraft must obtain an ATC clearance prior to entering Class B airspace on the appropriate frequency and relation to geographical fixes shown on local Class B aeronautical charts. Departing aircraft require a clearance to depart Class B airspace and should advise clearance delivery of their intended altitude and route of flight.

Unless otherwise authorized by ATC, aircraft must be equipped with an operable two-way radio capable of communicating with ATC on appropriate frequencies for that Class B airspace. Also, unless otherwise authorized by ATC, the aircraft must be equipped with an operable radar beacon transponder with automatic altitude reporting equipment.

There are currently twelve airports with Class B airspace where the pilot must hold at least a private pilot certificate to take off and land. At other Class B airports, a student pilot or recreational pilot who seeks certification may take off and land if certain requirements are met. The student or recreational pilot must receive ground and flight instruction from an authorized instructor and receive an endorsement from that instructor stating the student or recreational pilot is proficient to conduct solo operations at the specific Class B airport & airspace.

Mode C Veil A Mode C transponder with altitude reporting is required within thirty nautical miles of a Class B airport from the surface to 10,000 feet MSL. An aircraft that was not originally certificated with engine driven electrical system or which has not subsequently been certified with a system installed may conduct operations within a Mode C veil provide the aircraft remains outside Class A, B, or C airspace, and below the altitude of the ceiling of a Class B or Class C airspace area designated for an airport or 10,000 feet MSL, whichever is lower." (FFA-PH, 2018)

3. **"Class C Airspace** – Class C Airspace is generally defined as from the surface to 4,000

feet above the airport elevation (charted in MSL) surrounding those airports that have an operational control tower, are serviced by a radar approach control, and have a certain number of IFR operations or passenger enplanements. Although the configuration of each Class C airspace area is individually tailored, the airspace usually consists of a 5 NM radius core surface area that extends from the surface up to 4,000 feet above the airport elevation, and a ten NM radius shelf area that extends no lower than 1,200 feet up to 4,000 feet above airport elevation.

No specific pilot certification is required to operate in Class C airspace. A two-way radio and unless otherwise authorized by ATC an operable radar beacon transponder with automatic altitude reporting equipment is required.

Two-way radio communication must be established with the ATC facility prior to entry and thereafter maintain those communications while in Class C airspace. Pilots of arriving aircraft should contact the Class C airspace ATC facility on the publicized frequency and give their position, altitude, radar beacon code destination, and request Class C service.

Radio contact should be initiated far enough in advance from the Class C airspace boundary to preclude entering Class C airspace prior to establishing two-way radio communications. If the controller responds to a radio call with, "aircraft call sign, standby" radio communications have been established and the pilot can enter the Class C airspace.

If conditions prevent immediate provision of Class C services, the controller will inform the pilot to remain outside the Class C airspace until conditions permit the services to be provided.

It is important to understand that if the controller responds to the initial radio call without using the aircraft call, radio communications have not been established and the pilot may not enter the Class C airspace." (FFA-PH, 2018)

4. "**Class D Airspace** – is generally that airspace from the surface to 2,500 above the airport elevation (charted in MSL) surrounding those airports that have an operational control tower. The configuration of each Class D airspace area is tailored and when instrument procedures are published, the airspace will normally be designated to contain the procedures.

No specific pilot certification is required. Unless otherwise authorized by ATC, an operable two-way radio is required.

Two-way radio communication must be established and maintained with the ATC facility providing ATC services prior to entry while in Class D airspace. Pilots of arriving aircraft should contact the control tower on the publicized frequency, giving their position, altitude, destination, and any request(s). Radio contact should be initiated and established far enough in advance to preclude entering the Class D airspace boundary without it.

If the controller responds to a radio call with, "aircraft call sign, standby," radio communications have been established and the pilot can enter the Class D airspace. If conditions prevent immediate entry into Class D airspace, the controller will inform the pilot to remain outside the Class D airspace until conditions permit entry." (FFA-PH, 2018)

5. **"Class E Airspace** – Generally, if the airspace is not Class A, B, C, or D, and is controlled airspace it is Class E airspace. There are no specific pilot certification or equipment requirements to operate in Class E airspace. Special VFR operations are permitted, but clearance must be obtained from the controlling facility." (FFA-PH, 2018)

6. **"Class G Airspace –** is the portion of the airspace that has not been designated as Class A, B, C, D, or E. It is therefore designated uncontrolled airspace. Class G airspace extends from the surface to the base of the overlying Class E airspace." (Federal Aviation Administration, 2016). (FFA-PH, 2018)

Routine access by UAS to the NAS is highly favored by the many stakeholders who design, manufacture and operate. Estimates of 200,000 hours of domestic UAS operation in 2012 were projected to increase well beyond one million hours by 2020 UAS (Marshall, 2012).

UAS will need to be capable of interacting with ATC systems and commands to operate at all levels of airspace simultaneously with manned aircraft. (Marshall, 2012)

Figure 2-9 FAA Airspace Classification

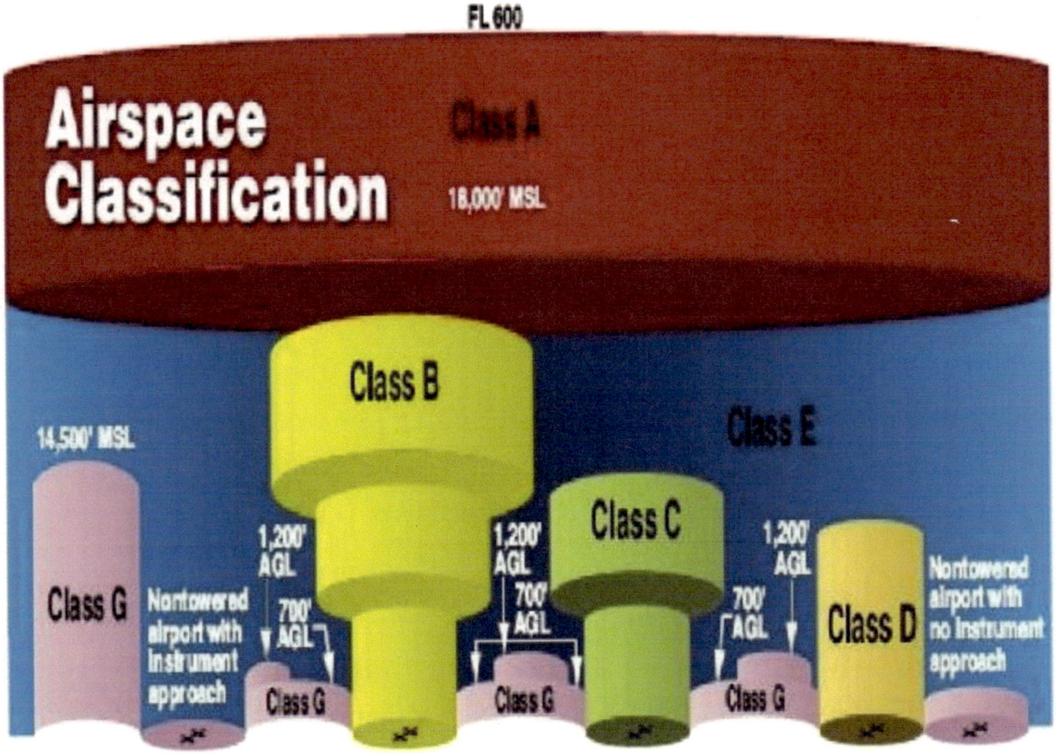

Source: Federal Aviation Administration. (2016). Classes of Airspace. Washington DC: United States Department of Transportation. Retrieved from Types of Controlled Airspace: https://www.faasafety.gov/gslac/ALC/course_content.aspx?cID=42&sID=505&preview=true

For example, long endurance high altitude UAS are currently capable of operating in all classes of airspace up, and including Class A. Having UAV's travelling at speed approaching the speed of sound in the same airspace as commercial and military jetliners presents numerous safety concerns. Imagine a Global Hawk UAV operating in Class A airspace losing ability to communicate with its ground station and pilot. The risk of collision with commercial aircraft presents a challenge to regulators which, if not successfully addressed, could lead to catastrophic consequences. Alternatively, a civilian operator of a small UAV in Class G Airspace without mandated ATC communication presents far less, but still import risk, to the safety of the NAS.

Regulating UAS Operation in the NAS

UAS operations present a different challenge, since there may or may not be a ground-based pilot operating the vehicle at any point in the flight. Even with ground-based control there is no pilot onboard to report, communicate, and execute commands. The easiest and safest way to regulate UAS operation in NAS would be to ban the practice. Although effective, it is not a practical solution, since it would stifle technology and allow other UAS "friendly" nations an economic and military advantage.

Since the Department of Defense (DoD) was the primary operator of UAS in the NAS in the late 20th century its collaboration with the FAA in the development of approval criteria for Certificates of Authorization (COA) was logical. As the FAA was established to oversee the operation of manned aircraft in controlled airspace the advent of UAS presented new challenges. First and foremost, the ATC system relies upon communication between air crew and ground controllers. As technology such as radar, anti-collision, and ground proximity warning systems were introduced, the safety and reliability of aviation increased. Much of the innovation was designed at increasing pilot and controller information and communication.

The result was the issuance of FAA and DoD joint order 7610.4, special military operations in 2001. Once issued pursuant to FAA Order 7210.3, the COA or waiver would permit an operator of a UAV in the NAS outside restricted or warning areas (Marshall, 2012).

For the DoD to receive a COA or waiver for UAS operation in NAS it had to supply the following data and information:

1. Detailed description of the intended flight operation, including the classification of airspace to be used,
2. UAS physical characteristics (configuration, length, wingspan, gross weight, means of propulsion, fuel capacity, color, lighting, etc.),
3. Flight performance characteristics (top speed, cruise speed, maximum altitude, rate of climb, range/endurance, means of recovery etc.),
4. Method of pilotage and proposed method to avoid other traffic,
5. Coordination procedures,
6. Communication procedures,
7. Route and altitude procedures,
8. Lost link/ mission abort procedures, and
9. A statement by DoD proponent that the unmanned aircraft is airworthy.

The issuance of a COA or waiver was an effective way of ensuring ATC and other aviation operators had prior notice of the nature and extent of a UAS operations in the NAS a system had to be created to effectively apply and disseminate COA's or waivers once approved.

Under the provisions of joint order 7610.4 the FAA, in conjunction with the military, were able to study the safety, reliability and efficacy of operating UAS in the National Airspace System, so that by 2005 the FAA issued Interim Operational Approval Guidance for UAS entitled AFS 400 (Federal Aviation Administration, 2005).

Civilian UAS Operations – Striking Legislative Balance

While FAA recognized the development of civilian UAS in the NAS, it still required civilian operators to follow the existing airworthiness certification process. In 2006 it took the additional step of creating the Unmanned Aircraft Program Office (UAPO) in 2006.

In 2007 the UAPO created an online COA application for public operators; police, fire, meteorological, etc. The old process was manual. The system is automated, thereby allowing swifter approval. Networked transmission of COA issuance to required stakeholders to further study UAS operations.

Figure 2-10 Sample COA

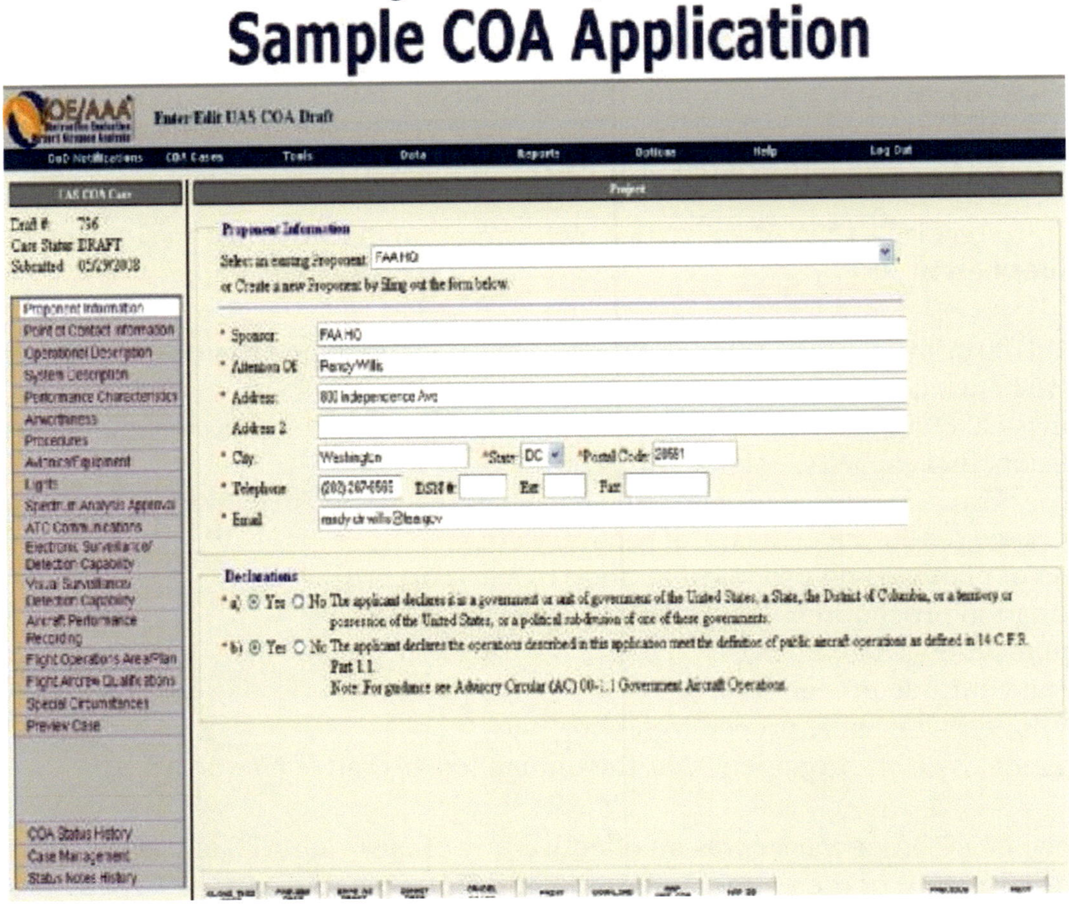

Source: Federal Aviation Administration. (2016, July 6). Aircraft Certification - Bilateral Agreements. Retrieved from Federal Aviation Administration. https://www.faa.gov/aircraft/air_cert/international/bilateral_agreements/

In 2007 the UAPO created an online COA application for public operators; police, fire, meteorological, etc. The old process was manual. The system is automated, thereby allowing swifter approval. Networked transmission of COA issuance to required stakeholders to further study UAS operations.

With increased reports of near-misses between UAV's and commercial aviation as the size, speed and number of UAV's increases in the NAS, so does the likelihood of a catastrophic accident. Since such an occurrence could be devastating; caution must be emphasized when regulating their expanded presence in all classes of airspace. Figure 9 details the damage potential even when a small UAV collides with a commercial jetliner. In this situation, a LAM Mozambique on January 5th, 2017 flight TM 1715, a Boeing 737 with eighty-six passengers and crew onboard was approaching Tete airport in northwestern Mozambique. Crew members and passenger reported a "bang" that was believed to be a bird strike. Upon examination after the aircraft had landed, the authorities inspected the damage and determined that the jet had collided with a drone, as there was no organic residue present at or near the damaged. Figure 2-11.

Although disaster was averted this incident calls needed attention to the risk of cohabitated airspace. This risk is heightened with a small UAV with a small radar signature. From the perspective of a regulatory body the natural inclination is to heavily regulate private UAV operation, possibly to the extent of retarding development. Perhaps regulating body should consider a balanced and proportional legislative response to risk.

UAS and Constitutional Rights

Some believe that when seeking legislative balance, given the risk for catastrophic consequences, should tip in favor of more comprehensive regulation. Some believe that civilian operation of UAS at altitudes under 200 feet is best left to the state or local government where the flight is taking place.

Figure 2-11 Drone Crash into Commercial Airline

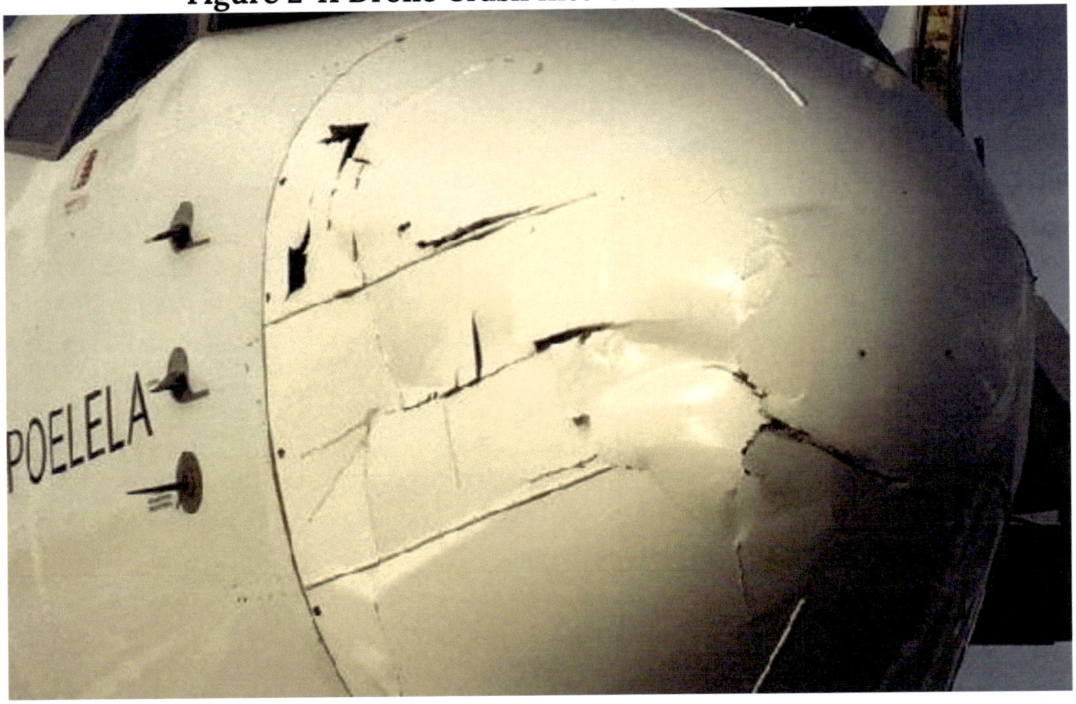

Source: Cuskelly, C. (January 6, 2017) UK Express. Drone CRASHES into Boeing 737 passenger jet coming into land. http://www.express.co.uk/travel/articles/751165/drone-boeing-737-planecrash-Mozambique. Also See: https://youtu.be/2jzx8BpDuHE

Representative Jason Hill introduced the Drone Innovation Act (H.R. 2930) in 2017, which provides for new a new designation of "local airspace," defined as under 200 feet in altitude.

Some of the highlights of Drone Innovation Act include:

1. Standardize reasonable time, manner, and place limitations and restrictions across the nation.
2. Aid states in adopting Unmanned Traffic Management (UTM) and making limitations publicly available to all users.
3. Create an environment that is friendly to innovation and fosters rapid integration of UAS.
4. Prevents the FAA from authorizing the operation of an unmanned aircraft in local airspace above a property without permission of the owner.
5. Preserves State and Federal statutes and common law rights
6. Restricts the Secretary from impeding State and local government authority to define local property rights as they apply to UAS
7. Ensures UAS can reach navigable airspace.
8. Provides for the use of UAS on your property or right of way (Drone Innovation Act , 2017)

HR 2930 addresses several constitutional principles ordinarily not associated with UAS manufacturing, operation, regulation, and responsibility. First and foremost are provisions of the U.S. Constitution and Bill of Rights. How does a document ratified in 1789 apply to UAS legislation and regulation? Here is an example:

Scenario 1: Farmer 1 in Iowa wants to use a UAV to inspect his crops for infestation, hydration, and monitor growth. Neighboring farmers use bi-planes to spray insect repellant on their crops. Suppose that FAA regulations prohibit a private citizen from operating an UAS within 500 feet of another property where commercial or private fixed wing aviation occurs more than once every thirty days? See: Figure 2-12 and Figure 2-13

The Fifth Amendment of US Constitution provides in the portion applicable to this scenario that "No person shall...nor be deprived of life, liberty, or property, without due process of law; nor shall private property be taken for public use, without just compensation." (United States of America, 1791).

The Fourteenth Amendment specifically applies the protections of the Fifth Amendment through what is known as the "equal protection clause" which extends these protections to citizens of individual states.

Figure 2-12 Scenario 1 Part 1

Source: Anderson, C. (December 24, 2015) MIT Technology Review. Agricultural Drones: Relatively cheap drones with advanced sensors and imaging capabilities are giving farmers new ways to increase yields and reduce crop damage. https://www.technologyreview.com/s/526491/agricultural-drones/

Some of the issues raised:

- By enacting this regulation has the FAA deprived farmer 1 of the right to maintain their private property as they see fit?
- Does this prohibition deprive farmer 1 of the liberty in that the UAS flight at low levels will be conducted solely upon farmer 1's property?
- Does Farmer 1 have any ability to appeal this regulation or have it modified so that they can use UAS on their farm thereby depriving them of due process of law?

- Do the protections afforded farmer 2 in prohibiting farmer 1 from UAS operation deprive her of equal protection" under the law? Since it puts farmer 1 at a competitive disadvantage to farmer 2, simply because they each chose to employ different technology?

Figure 2-13 Scenario 1 Part 2

Source: Hammond, K. (October 1997). Crop duster.

- Can farmer 1 bring an action against the FAA? In which court? The Federal Court for the District of Iowa would have jurisdiction (officially conferred power to render decisions and judgment involving the subject of the case) because the claim presents what is known as "federal question" jurisdiction. It could also be brought in an Iowa State Court, since the dispute arises in Iowa where the State Court has concurrent jurisdiction with the Federal Court.
- Should farmer 2 be named as a defendant, since they will be adversely affected should farmer 1 win?
- If the FAA regulation is a violation of the constitutional rights of farmer 1, what is the remedy? Can the court award farmer 1 monetary compensatory damages (an order declaring that the FAA law is either unconstitutional as written or as applied to Farmer 1)? If so, how much?

Now let us address some of the legal implications which can result from the operations of a UAS.

Scenario 2: Homeowner 1 lives at 1814 Todd Road in Manhattan, Kansas. Homeowner 2 lives next door at 1812 Todd Road and has a "no trespassing" sign posted. Homeowner 1 is a customer of Amazon and has signed up for its Prime-air service. (Figure 2-14 Scenario 2)

Figure 2-14 Scenario 2

Source: Mumm, H. (May 11, 2018) Author created collage using Google Maps and Amazon partial figure, Scenario 2, partial. https://24tv.ua/amazon_tag1532/

Homeowner 1 orders peanuts from Amazon, with thirty-minute prime-air home delivery. While calculating the UAV route the operator studies ground and aerial images and discovers trees that might inhibit delivery. The pilot creates a course between the trees on Todd Road and the home of Homeowner 1. The UAV travels over the front porch of homeowner 2 (Figure 13).

The Amazon UAV arrives to make the delivery to 1814 Todd Road. Due to an unforeseen software glitch in the GPS when the number 4 appears in an address after the number 1 it treats it as the number 2. The UAV delivers its payload to 1812 Todd Road, the residence of homeowner 2.

Homeowner 2 has a relative looking to relocate, due to a serious and life-threatening allergy to peanuts. This person forgot to pack their Epi-pen, which could prevent them from going into anaphylactic shock from a severe allergic reaction. As fate would have it no one else is home when the UAV delivers its payload. Having never witnessed this modern technology the relative wanders outside and picks up the package. Unfortunately, Amazon package has left peanut residue all over the box.

The relative of homeowner 2 immediately has a serious allergic reaction. Without his Epi-pen, they succumb a blockage of their airway.

Some of the Legal Issues Raised:

- Did the Amazon UAV trespass on the property by landing, albeit unknowingly, violate Kansas Statutes Annotated (KSA) 21-3721, criminal trespass, or commit the tort of trespass?[34]
- Was homeowner 1 contributorily negligent in signing up for Amazon prime-air delivery when a reasonable person might foresee the delivery landing at his neighbor's home, which had a posted no-trespassing sign?
- Did the events which occurred because of the Amazon UAV pilot deciding to travel via the air space of homeowner 2 amount to a violation of the COA or other certification?
- Did the flight through over homeowner 2's home constitute the crime of criminal trespass as defined in KSA 21-3721?
- Could Amazon argue that it was unaware of the requirements of KSA-3721 when it sells signs on its marketplace specifically advising of the statute? (Figure 14)
- Was Amazon criminally responsible and/or civilly liable for the death of homeowner 2's relative, since it was reasonably foreseeable that the UAV might mistakenly deliver peanuts, which are widely known to cause serious allergic reactions in many, to the wrong address?
- Do the "Terms of Service" agreed to homeowner 1 and Amazon have any effect against legal or criminal claims brought by homeowner 2, the relative, or the survivors?

3. A tort is an act or omission that gives rise to injury or harm to another and amounts to a civil wrong for which courts impose liability.
4. It is illegal to enter or remain on any land, non-navigable body of water, structure, vehicle, aircraft or watercraft, or nuclear or healthcare facility without the owner or other authorized person's consent, according to Kansas Statute 21-3721. It's considered trespassing if there are signs or postings, or if there are fences, locks or other means of enclosed, shut or secured passages of entry. Criminal trespassing in Kansas is a Class B nonperson misdemeanor, which carries a penalty of no less than 48 consecutive hours of imprisonment and up to six months in jail.

Figure 2-15 No Trespassing

Source: Private property sign by Elljay. Licensed under CC0. https://pixabay.com/en/private-property-sign-gate-private-1109273/

- Was Amazon criminally responsible and/or civilly liable for the death of homeowner 2's relative, since it was reasonably foreseeable that the UAV might mistakenly deliver peanuts, which are widely known to cause serious allergic reactions in many, to the wrong address?
- Do the "Terms of Service" agreed to homeowner 1 and Amazon have any effect against legal or criminal claims brought by homeowner 2, the relative, or the survivors?
- Should the FAA investigate through the National Transportation Safety Board (NTSB)? If violations of laws or customs by Amazon and parties institute proceedings to revoke some or all their COA's or permits?

Common Law Fills the Technology Gap

Consider the many other legal ramifications and consequences in both scenarios. Is it possible to legislate and foresee every possible contingency or event that may cause harm from the operation of a UAS? Certainly not, so how can a broad range of possible accidents or intentional acts be causing harm to be addressed? Common Law. "Common law is the system of deciding cases that originated in England and which was later adopted in the U.S. Common law is based on precedent (legal principles developed in earlier case law) instead of statutory laws. It is the traditional law of an area or region created by judges when deciding individual disputes or cases. Common law changes over time" (U.S. Legal, Inc., 1997)

The concept of common law is especially useful when dealing with new and rapidly changing technology. Broadly applicable common law theories such as the theory of negligence is a perfect example of how the common law can bridge the gap. Just as in Aesop's fable of the tortoise and the hare, the concept of common law finds its adaptability to changing times and technology in the slow and steady way it has developed.

While the courtrooms of England of the 1600 & 1700's were not the place where UAV litigation occurred, cases were being heard regarding safe design, manufacture, and operation of modern technology of the times. For example, common law established the same principles of negligence discussed in scenario 2 above. In the English Court of Common Pleas in the mid-1700's "common carriers" such as horse drawn coaches and ships were could be held liable under the theory of negligence for injuries caused by their activity. If the claimant could prove the carrier was negligent in their transport they would be held financially responsible for damages sustained (Kaczorowski, 1990). History contains many common themes that should be examined before new legislation is enacted. When it comes to the creation of laws to address modern technology, the adage "everything old is new again" takes on new meaning.

UAS Manufacturing and Design Standards

As discussed in relation automobiles and manned aircraft codes are essential tools in ensuring the UAS produced and operating in the NAS have certain essential technologies and capabilities which render them airworthy and capable of safely co-exist in public airspace. Just as Huddy consulted many stakeholders ranging from automobile engineers, designers and manufacturers to police, accident victims, and medical professionals for his book Rules of Driving, legislators, regulators and jurists should consult informed stakeholders in the UAS regulatory process. Groups ranging from educators to pilots to the ASTM should be the source of determining what areas relating to UAS require regulation.

Regarding design and manufacturing standards, enough time has passed that a large population of individuals and groups can consult with the FAA in their promulgation. For example, an engineer may suggest that what a standard for UAS anti-collision lights adequate to avoid collisions while a regulator may suggest mandating more, such as onboard radar. While onboard radar for all classes of UAS may make sense as a utopian concept, the cost, weight and size of the technology may make engineering an affordable solution out of reach for all but commercial operators. Yes, it would certainly effective but could never justify such a farfetched notion. Such is overzealous caution in the regulation process. Perhaps Larry Downes of the Georgetown Center for Business and Public Policy put it best when he observed:

"That solution – to stay the course, to continue leaving tech largely to its own correctives – is cold comfort to those who believe tomorrow's problems, coming up fast in the rear-view mirror, are both unprecedented and catastrophic. Yet, so far there's no evidence supporting shrill predictions of a technology-driven apocalypse. Or that existing safeguards – both market and legal – won't save us from our worst selves. Nor have tech's growing list of critics proposed anything more specific than simply calling for "regulation" to save us. Perhaps that's because effective remedies are incredibly hard to design" (Downes, 2014).

Conclusions

As was the case with the automobile, airplane, personal computer and so much other innovative technology, there will always be much fear caused by speculation and fear of the unknown. Just ten or twenty years ago most people would never dream of, much less trust the idea of doing their banking online, having their doctor keep their medical records electronically or using a mobile phone to unlock the front door to their home. Concepts which were once considered futuristic, science fiction seem to be becoming reality with greater speed than ever. Still it is prudent to proceed with caution in areas of unknown technology because there may be hidden, or unknown risks associated with the adoption or use of the technology on a large scale.

Figure 2-16 Three Mile Island

Source: United States Nuclear Regulatory Commission. (2018, June 21). *Backgrounder on the Three Mile Island Accident*. Retrieved from USNRC: https://www.nrc.gov/reading-rm/doc-collections/factsheets/3mile-isle.html#summary

A perfect example occurred not too many decades ago in Pennsylvania. The Three Mile Island nuclear power plant experienced what was called a "partial meltdown" on March 29, 1979 which caused the release of radiation into the surrounding area. (Figure 2-16) While ultimately the amount of radiation released was considered minimal by the Nuclear Regulatory Commission, a long hard look at nuclear energy was undertaken and new, far more stringent design, manufacturing and operational standards were adopted (United States Nuclear Regulatory Commission, 2018). The result of the regulations were in some ways beneficial and in some way not. As an immediate matter the use and popularity of nuclear power in the United States became significantly diminished. This was also fueled by the more devastating and significant meltdown which occurred in 1986 in the then Soviet Union city of Chernobyl. Eventually nuclear power plants were re-engineered and designed with the lessons learned from these accidents in mind to create a safer way of harnessing nuclear energy to produce electricity.

The same lessons learned through aviation accidents and the tragic loss of life have resulted common sense regulations, based upon knowledge acquired from these painful events which have led to the safest period of commercial passenger aviation in history.

The challenge for the students of today who may be the legislators, regulators and adjudicators of UAS law in the future is to strike a balance between enough regulations to ensure public safety without overregulation. Too little oversight and safety may fall victim to profit and market competition, too much and safety can also be negatively impacted by inhibiting the development of innovative technologies which may enhance the safety of UAS in the future. This paradigm has been a truism throughout the industrial revolution right through today and it is a safe assumption that the future will yield comparable results.

Discussion Questions

1. At what altitude over private property, if any should the owner have right to privacy that allows them to request the UAV depart or be held liable for trespass? Would your answer be different if the UAV were equipped with highly sensitive audio and visual recording technology or wireless data interception technology?

2. Should Federal, State or Local governments mandate that all non-military UAS operators be required to carry liability insurance for damage or injury caused (including invasion of privacy) prior to allowing any UAS to be purchased or operated by any civilian or private entity? If so should the liability insurance requirements vary depending upon the class of UAS and its potential of causing damage or injury?

3. Law enforcement is called by a homeowner who says that a drone crash landed in his backyard, there has been no contact from anyone claiming to own the UAV nor did the homeowner see any anyone in proximity to their property appearing to operate or control it. Further the homeowner says there is a box below the drone which appears to be its payload leaking some sort of bubbling liquid. How should the homeowner be advised? Should law enforcement treat the UAV as an unidentified suspicious package presenting a Hazmat or other safety threat?

4. The local army navy store has a sale on used surplus military equipment. Upon visiting the store, you recognize one of the items as a case of Perdix miniature military drones capable of performing swarm attacks on designated targets. Based upon your knowledge of the technology you understand the lethality of the threat they pose. What if any legal obligation do you have to inform the store owner of the fact the items pose a danger? Suppose the store owner dismisses your concern and says he is going to sell them anyway. Do you have an obligation to report him to the police or other governmental authority? Should you decide not to do so, and the drones are eventually used to attack a school and students are injured or killed. Are you criminally liable? Should there be inventory control and disposal laws which prohibit the sale of surplus UAS technology, even if modified for civilian use, to the public?

Bibliography

Brian, D. (2010, May 4). Inspecting For Trust: The Role Of Inspectors General. (B. Naylor, Interviewer) Retrieved from https://www.npr.org/templates/story/story.php?storyId=126511407

Chaplain, J. (2011). Safety Regulation The First 100 Years. *Journal of Aeronautic History.*

Downes, L. (2014). How More Regulation for U.S. Tech Could Backfire. *Harvard Business Review.*

Drone Innovation Act , H.R. 2930 (United States House of Representatives – 115th Congress June 16, 2017).

Federal Aviation Administration. (2005, September 16). Memorandum – AFS 400 UAS Policy 5-01. *Unmanned Aerial Systems Operations in the U.S. National Airspace System – Interim Operational Approval Guidance.* Washington, DC: Federal Aviation Administration.

Federal Aviation Administration. (2009, August 9). *Original Design Approval Process.* Retrieved from Federal Aviation Administration: https://www.faa.gov/aircraft/air_cert/design_approvals/orig_des_approv_proc/

Federal Aviation Administration. (2016, July 6). *Aircraft Certification – Bilateral Agreements.* Retrieved from Federal Aviation Administration: https://www.faa.gov/aircraft/air_cert/international/bilateral_agreements/

Federal Aviation Administration. (2016). *Classes Of Airspace.* Washington DC: United States Department of Transportation. Retrieved from Types of Controlled Airspace: https://www.faasafety.gov/gslac/ALC/course_content.aspx?cID=42&sID=505&preview=true

Federal Aviation Administration. (2018, March 14). *Air Carrier Oversight.* Retrieved from https://www.faa.gov/about/initiatives/atos/oversight/

Federal Aviation Administration. (2018, January 25). *Former South Florida FAA Designates Airworthiness Representitive Charged With Aircraft Parts Fraud.* Retrieved from Investigations: https://www.oig.dot.gov/library-item/36272

Fenwick, M. D. (2017). Regulation Tomorrow: What Happens When Technology Is Faster than the Law. *American University Business Law Review,* 55.

FFA-PH. (2018, August 26). *Faa.gov/regulations_policies/handbooks_manuals/aviation/phak/media/pilot_handbook.pdf.* Retrieved from FAA: https://www.faa.gov/regulations_policies/handbooks_manuals/aviation/phak/media/pilot_handbook.pdf

Ford Motor Company. (2018). *Company Timeline – Ford.Com.* Detroit, Michigan: Ford Motor Comany.

Franzese, L. C. (2009). Sovereignty In Cyberspace: Can It Exist? *Air Force Law Review.*

Government of Canada. (2017, September 13). *Transport Canada.* Retrieved from National aircraft Certification: https://www.tc.gc.ca/eng/civilaviation/certification/menu.htm

Graves, F. G. (2006, April 17). Boeing Parts and Rules Bent, Whistle-Blowers Say. *the Washington Post.*

Guzzetti, J. B. (2012, October 22). *Audit Initiated of FAA's Oversight of Unmanned Aircraft Systems.* Retrieved from New Audit Announcements: https://www.oig.dot.gov/library-item/29314

Hammond, K. (1997, October 18). *Crop Duster, ID k7803-2.* Retrieved from USDA: http://www.ars.usda.gov/is/graphics/photos/oct97/k7803-2.htm

Huddy, X. P. (1906). *The Law of Automobiles.* Albany, NY: Matthew Bender and Company.

Kaczorowski, R. J. (1990). The Common Law Background of Nineteenth Century Tort Law. *Ohio State Law Journal*, 1129-1130.

Loomis, B. (2015, April 26). 1900-1930: The years of driving dangerously. *The Detroit News.*

MacKenzie, D. (2010). ICAO A *History of the International Civil Aviation Organization.* Toronto: University of Toronto Press.

Marshall, D. M. (2012). *Introduction To Unmanned Aircraft Systems* . Boca Raton FL: Taylor & Francis.

Messier, D. (2016, March 3). A *Closer Look at Early Aviation Safety & Regulation.* Retrieved from Parabolic Arc: http://www.parabolicarc.com/2016/03/03/early-aviation-safety/

National Transportation Safety Board. (1979). *Aircraft Accident Report – Pacific Southwest Airlines Flight* 182. Washington, DC: National Transportation Safety Board.

Norton, P. D. (2008). *Fighting Traffic: The Dawn of the Motor Age in the American City.* Cambridge, MA: MIT Press.

Schlichting, M. (1993). *Aesop's Fable The Tortoise and the Hare, as retold by Mark Schlicting.* Novato, CA: Broderbund Company.

Tennekes, H. (1997). *The Simple Science of Flight – From Insects to Jumbo Jets.* Cambridge MA: MIT Press.

The Saylor Foundation. (2008). *Scientific Management Theory and the Ford Motor Company.* Washington, DC: Saylor Academy.

Treaty of Versailles. (1919). Convention Relating to the Regulation of Aerial Navigation . *Treaty of Versailles.* Paris.

U.S. Legal, Inc. (1997). *Common Law And Legal Definition.* Retrieved from US Legal: https://definitions.uslegal.com/c/common-law/

United States Nuclear Regulatory Commission. (2018, June 21). *Backgrounder on the Three Mile Island Accident.* Retrieved from USNRC: https://www.nrc.gov/reading-rm/doc-collections/fact-sheets/3mile-isle.html#summary

United States of America. (1791, December 15). Constitution of the United States. *Fifth Amendment* . Washington, DC: United States Government.

Chapter 3: Understanding Hostile Use and Cyber-Vulnerabilities of UAS: Components, Autonomy v Automation, Sensors, SAA, SCADA and Cyber Attack Taxonomy

Student Learning Objectives

An introduction to the *Problem* of countering hostile use of UAS against U.S. national defense interests will be introduced. The student will be able to identify critical components of an Unmanned Aircraft System (UAS), identify potential cyber vulnerabilities and understand the taxonomy of UAS operations that may be compromised against USA interests. FAA rules and US laws lag the UAS technology growth. In this chapter, UAS critical components, sensors, and levels of self-reliance are identified. These components are viewed in terms of Sense and Avoid (SAA) and SCADA environments. These components are then incorporated -into the Cyber – Attack Taxonomy. (Nichols R.-0. , 2016)

What Is The Counter -UAS Problem?

The risk of successful terrorist attacks on USA Air Defense Systems (ADS) via UASs is **greater** because of improving commercial capabilities and accessibility. Advanced small drones, capable of carrying sophisticated imaging equipment and significant payloads, are readily available to the public. A range of terrorist, insurgent, criminal, corporate, and activist threat groups have demonstrated their ability to use civilian drones and gather intelligence. How does the country defend against a growing UAS threat? This is also known as the counter – UAS *Problem*. General James D Mattis, SECDEF summed up the *Problem* succinctly:

> "Unmanned Aircraft are being developed with more technologically systems and capabilities. They can duplicate some of the capabilities of manned aircraft for both surveillance/ reconnaissance and attack missions. They can be small enough and / or slow enough to elude detection by standard early warning sensor systems and could pose a formidable threat to friendly forces." (Chairman, 2012)

Operational Protection from Hostile UAS Attacks – A Helicopter View

"According to LCDR Boutros of the Navy War College, developing technologies do not paint a pleasant picture of counter – UAS problem (Boutros, Operational Protection 2015). UAS has seen a widespread proliferation among both state and non-state actors. This is a cause for concern to US Operational Commanders." (Boutros, 2015) General James D Mattis, SECDEF concluded:

> "The proliferation of low cost, tactical unmanned aerial systems demand we think about this potential threat now… we must understand the threat these systems present to our joint force and develop the tactics, techniques and procedures to counter the problem." (Chairman, 2012) (Myer, 2013)

It can be argued from the quantity and diversity of production that China is the current leader in this technology. China is thoroughly exercising its UAS muscles in the Spratly Islands.

Over 90 countries and non-state actors have UAS technology. Many of these actors foster terrorism. "Most of the UAS systems, except for China, Russia, USA, Turkey, Saudi Arabia, and Iran inventories are low-technology, Intelligence, Surveillance, and Reconnaissance (ISR) platforms." (Boutros, 2015) Experts believe that by 2025 China will produce over 50% of UAS systems. (Yan, 2017) China's commercial drone market to top 9B USD by 2020. The market value would be tripled to 180 billion yuan by 2025, according to the guidelines made by the Ministry of Industry and Information Technology. The estimate was much higher than a forecast by an iResearch report last year, which said the overall market of UAVs, commonly known as drones, could reach 75 billion yuan by 2025 in China. (Yan, 2017)

Iran has supplied long range, low technology Ababil UAS weapons systems to Syria and Sudan, and to extremist groups like Hezbollah, Hamas, and ISIS. Hezbollah's inventory is estimated at over 200 UAS, which concerns the Israeli military commanders. (Zwijnwenburg, 2014)

Joint Publication (JP) 3-01 identifies friendly assets that an adversary may attack during a campaign using UAS. A Theater Commander must plan for counter – UAS actions against air defense sites, logistics centers, and national critical infrastructure. (Boutros, 2015) "Due to their small size and unique flying signatures, many UAS are difficult to detect, identify, track, and engage with current joint air defense systems. The increasing proliferation of global UAS has exposed a critical vulnerability in the protection function of operational commanders, requiring joint efforts to include intelligence, Electronic Warfare (EW), cyber warfare, (CW) and FIRES." (Boutros, 2015)

But UAS are not invincible. Neutralizing threats or mitigating risk includes active and passive defense methods with kinetic and non-kinetic FIRES.[1] (US DoD – JP 3-0, 2012)

Countering UAS Air Threats

Advanced UAS can carry large payloads great distances. US Predator and Global Hawk UAS, "Chinese Pterodactyl and Soring Dragon counterparts, and Iranian Ababil can carry at least 500 Kg payloads greater than 300 km." (Boutros, 2015) "They can be armed or unarmed, with ISR payloads, communications relays, Over-The-Horizon (OTH) target acquisition, and precision strike capabilities." (Boutros, 2015)

"Shorter range, tactical, small/micro UAS may not have the distance or payload capacity of more advanced systems, but they can impact a campaign (or US Homeland Defense) in equally serious ways. Because of their size, their heat signatures are almost nonexistent. They easily evade detection. They offer more freedom of action. They can be launched from within US air defense zones and fly to their targets in less time than it takes for a coordinated response." (Boutros, 2015) **[Nightmare alert: Imagine a swarm of UAS carrying small potent binary bomb payloads attacking a US Carrier at port less than one mile away from the UAS launch point.]** The enemy can effectively balance space, time, and force (arguably frequency too). (Beaudoin, 2011) "Small UAS (sUAS) can perform short-range ISR, be outfitted with explosive charges or chemical and biological agents for aerial dispersion, or simply fly over troops or civilians to demoralize." (Boutros, 2015) **[Nightmare alert: Given the effectiveness of enemy use of IEDs in Iraq and Afghanistan, a mobile, airborne version would take the Problem to an entirely new level!]** (Nichols R.-0. , 2016)

Vulnerabilities Perspective

"sUAS are vulnerable to kinetic and non-kinetic outside influence in four different areas; their link to a ground station, the ground station itself, the aircrafts various sensors, and cyber weapons." The military recognizes the first three factors, the authors will concentrate on the fourth.

"In 2009 Iraqi insurgents successfully hacked into US Reaper drones, crashing them." (Boutros, 2015) (Horowitz, 2014). "In September of 2011, ground control stations at Creech AFB were infected by a virus, temporarily grounding the entire UAS fleet." (Boutros, 2015) (Hartman, 2013) UAS onboard sensors can be manipulated in many ways. "High intensity light directed at an optical sensor can blind it. GPS receivers can be cyber-spoofed, which consists of transmitting a stronger, but false, GPS signal to a receiver, resulting in inaccurate navigation. Influencing

1. FIRES definition (US DoD – JP 3-0) the use of weapon systems to create a specific lethal or nonlethal effect on a target.

the local magnetic field can have adverse effects on both onboard hard drives and sensors that require magnetic orientation to operate correctly." (Boutros, 2015) (Hartman, 2013) The object is to better understand UAS subsystems, to facilitate exploiting their weaknesses.

The author's contention is that: *The hostile technology of remote-controlled warfare is difficult to control or abort; the best defense (counter – UAS) is to address the root drivers of these threats.* ***The threat- roots are SAA and SCADA***. Chapter 3 UAS landscape includes automation, collaboration, conventional vulnerabilities and countermeasures, commercial UAS primer, SAA Attack / Defense (A/D) A/D Issues and SCADA vulnerabilities.

Conventional Vulnerabilities of Air Defense Systems (ADS), Attacks By sUAS and Countermeasures

A simplified, non-classified view of the US Air Defense System (ADS) against a hostile UAS attack occurs in two stages:

1. Early Detection and Identification of "Danger Close" (Myer, 2013)[2]
2. Applied appropriate countermeasures with secondary goal of restricted collateral damage.

The traditional ADS family of tools for Detection include:

1. Active Radar Surveillance – generate waves, use rebound echoes on UAS to locate, estimate distance, approach speed, size, penetration vector and short-term trajectory, and
2. Passive Monitoring – covers electromagnetic spectrum via visible, thermal infrared, radio waves on common communications channels.

When considering hostile UAS defense planners need to consider several issues. The US ADS is optimized for missiles and aircraft deployed at high altitude and speeds. ADS data fusion (detection, identification, weapon lock-on, execute countermeasures) works better with larger targets, not very small ones like UAS / sUAS. US ADS is effectively reactive for longer ranges. Close reactive engagements are sub-optimal. US ADS are not optimal for sUAS /UAS. (Nichols R.-0. , 2016)

2. Danger Close Definition www.benning.army.mil/infantry/magazine/issues/2013/May-June/Myer.html Nov 14, 2013 – 1) Danger Close is included in the "method-of-engagement" line of a call-for-fire request to indicate that friendly forces are close to the target. ... Danger close is a term that is exclusive from risk estimate distance (RED) although the RED for 0.1 percent PI is used to define danger close for aircraft delivery. Pi = Probability of incapacitation. 2) Definition of "danger close" (US DoD) In close air support, artillery, mortar, and naval gunfire support fires, it is the term included in the method of engagement segment of a call for fire which indicates that friendly forces are within close proximity of the target.

There are clear vulnerabilities of the US ADS to UAS:

- "sUAS can be launched into action close to target(s), less than 1 mile.
- sUAS exhibit a small Radar signature. The detection phase is hindered.
- Reactive dictates quick response near target. This is not always possible.
- sUAS / UAS are designed for slow, low flight. Low flying sUAS avoids Radar identification.
- sUAS / UAS electric motors are both quiet and have limited thermal signature. This makes for difficult detection for noise.
- sUAS /UAS operate in urban areas. Urban sphere presents additional problems and potential collateral damage." (Nichols R.-0. , 2016)

Conventional Countermeasures Against sUAS /UAS:

There are two families of conventional countermeasures used to disrupt /destroy hostile UAS/ sUAS systems (Regulatory ~ locked in firmware GPS No-Fly Zones, Registration, FAA rules excluded).

Active Measures – Designed to incapacitate, destroy the sUAS/UAS threat in a direct way (*Ground-to- Air Defense* (GTA), missiles or, acoustical gun, or simple cyber rifle)

However, there are some defensive issues to be considered:

- GTA efficiency against sUAS, reactive targets is reduced, even less efficient in urban zones where public at risk.
- Simultaneous attacks on multiple fronts (See Team or Swarm formats, Tables 3-1 and 3-2) very difficult to apply and defense measures are mitigated.

> UAS countermeasures research are improving. The goal is to increase ability of GTA to react and improve capabilities to a defined to a saturation limit. Team formation allows decoys and shields. Swarm formation is easier to detect. Arrival of a cloud of robot drones is hard to mask, but tough to neutralize. Commercial company Liteye has developed an Anti-UAV Defense System (AUDS) which are able to detect, track, and disrupt sUAS operation by pulsed, brief focused broadcast of direction frequency jamming. Liteye has also developed a mobile version call M-AUDS. (Liteye, 2018) China has developed a "5-sec" laser weapon to shoot down sUAS at low altitude (500 m) with a 10KW high energy laser beam. Its range is 1.2 mi and handles sUAS speeds up to 112 mph. (Nichols R.-0. , 2016)

Passive – Designed to protect indirectly; physical protections around target, decoys, shields, organized roadblocks, nets, jamming of sensors of the aggressor, GPS total or partial cyber-*Spoof* of signals. Passive countermeasures have some positive outcomes. Decoys can be effective if the ADS knows what the sensors employed for sUAS Kamikaze attack and how they are used in the SAA subsystem. Communication jamming is effective against level 1 & 2 drones (Table 3-1) which require pilot interaction. It can disrupt inter-drone communications required for either team or swarm formations. Sensor Jamming – especially GPS signals – giving false GPS information, camera/gimbal dislocation, and heading sensor demagnetization is effective regardless of automation.

The 2011 Iranian incident taught US ADS planner's lessons about passive spoofing waypoints and Loss of Signal (LOS) via GPS. LOS is an emergency condition. sUAS/UAS have programmed responses. One of those responses may be," return to waypoint". Two types of spoofs were executed. A complete spoof uses the friendly SAA to estimate course, ground-speed, time to target to force a LOS and final waypoint change. A partial spoof reports false positions, during LOS and changes waypoints for perceived emergency conditions. Both spoofs are difficult to detect & effective (Editor, 2012)

Aggressor Counter-Countermeasures Specific to UAS Deployment – Swarm

The authors contend that a UAS swarm attack is practically unstoppable *unless* the defender (US ADS) exhibits strong collaboration and ability to match/identify the SWARM locations in a timely matter. This requires combined active & passive measures. This portends the ADS computer networks must process, detect, identify, and target information (and make critical decisions) significantly faster and more effectively than their enemies. Cost is an additional vulnerability factor. Swarms can be assembled, delivered, and targeted in a relatively inexpensive weapons package. A swarm can use local counter jamming on target nets. (Nichols R.-O. , 2016) The subjects of SWARM research, tactics and defenses, will be examined in later chapters.

Autonomy vs. Automation

Table 3-1 shows the normal five levels of automation that characterize UAS systems with examples of commercial vehicles. NASA presents a more detailed level of automation breakdown based on the OODA (Observe, Orient, Decide and Act) decision loops. (Barnhart, 2012) However, Table 3-1 should suffice to understand the cyber-purview. Level 1 lave and Level 2 Automated (minimal) are commonly found on UAS sold at Amazon, Walmart, and similar outlets. The pilot makes all the decisions and has complete control of flying orders. Level 3 steps up the navigation capabilities using an a priori mission plan.

Levels 4 and 5 add higher level decision making capabilities; collision avoidance without human intervention, complex mission planning in all weather conditions, expert systems intelligence without human intervention i.e. Artificial Intelligence (AI) and advanced Sense and Avoid systems (SAA). Level 5 is not commercially available; many designers are well on their way to a fully operational Level 5 UAS.

Table 3-1:

UAS Automation Scale

Level 1: Slave – assisting piloting, reaction to disturbance

Level 2: Automated – maintains its flying orders and receives higher level orders

For Levels 1 and 2 are common, require pilot intervention and continuous communication link; reasonable prices < $1500 US, small, weight < 10lbs: Drone Parrot, Quad Flyer GAUI

Level 3: Automated Navigation (a priori mission plan)

For Level 3 micro-UAS premium (< $20,000 US): Dragonfly, Microdrone Gmbh, Fly-n-Sense, Mikrokopter

Level 4: Response from contextual data *Collision Avoidance* (CA) (w/o human intervention)

For Level 4 minimum knowledge of surrounding environment, reacts to events, perform CA, uses active SAA, requires mission plan

Level 5: Decision-Maker (expert system) from contextual data: navigation in unknown environment, complex missions, coordination and collaboration of signals

For Level 5 AI, decision making with heavy networked computer support, perceptive sensors for space and time, complex mission in unknown environments, capable of intelligent adjustments including mission rescheduling, key word- adaptive control

Levels 4 and 5 are confined to laboratories. (Nichols R.-0. , 2016)

Table 3-2 UAS Collaboration shows four types of possible UAS collaboration. At the lower end of a threat scale is the isolated UAS or a small group of UAS. The advantages lie in a specific mission, which may be piloted or autonomous. They carry light payloads and are affordable. They are easy to assemble in the field. An example is the Raven used by US Special Forces. The disadvantage (countermeasure applied) is to identify the pilot or leader vehicle and destroy/disable it. A UAS attack team is particularly effective against divided attack targets, Level 3 allows automatic navigation, synchronized actions, and limited updated mission information. With increased team members, synchronization is not guaranteed. Disabling part of the UAS Team does not guarantee that mission failure. The real vulnerability of the UAS team is the Chief. All synchronization and updates go through the Chief. Disable/destroy the Chief and the Team is rendered useless. Determining who the Chief is critical.

Far more dangerous is the Swarm configuration especially in the higher levels of autonomous engagement. Swarms have several advantages. They are efficient based on numbers, they demonstrate emergent large group behaviors and reactions. Even not controllable or automated, they show a decentralized intelligence – think shoal of fish with evolving local rules. UAS

Swarms are a highly resistant form, not changing based on survivability of members. There is no hierarchy like a team. Destroy part of the swarm and the rest will continue their mission without abatement.

The two known countermeasures are: 1) Disrupt / **Change the Strategic Global View of Swarm (its only real vulnerability) and 2) force defender collaboration**. (Nichols R.-O. , 2016)

China appears to be the leader in innovative UAS swarm intelligence, through the efforts of the Chinese Electronics Technology Group Corporation (CETC). (Kania, 2017)

Table 3-2 UAS Collaboration

Type 1: Isolated Individual UAS

Advantages: piloted or autonomous w/ specific mission to perform. Small, easy to assemble, affordable, light payloads.

Countermeasures: Stop, Disable or Destroy Pilot, Threat removed.

Type 2: Group of Individual UASs (Isolated with own mission but not coordinated)

Advantages: sphere of action may be different for each mission, increased numbers, and increases success of attacks by defenses saturation

Countermeasures: Stop, Disable, Discover and Deter or Destroy Pilot(s), Threat(s) may be removed.

Type 3: Team of UASs (All members assigned specialized tasks and coordinated by Chief)

Advantages: Particularly effective against divided attack targets, Level 3 allows automatic navigation, synchronized actions, but no update to mission plans based on field activities.

Disadvantages: Level 4 (w/o humans) yields surrounding reactions but may lose synchronization between team members. Level 5 permits continuous updates, communications, commando style.

Countermeasures: Stop, Disable or Destroy Team members. Determine behavior logic and intervene. Survival of team members is critical to defense actions. Threat mitigated.

Type 4: UAS Swarm (Uniform mass of undifferentiated individual's w/o Chief at level 4 or 5)

Advantages: Efficient based on numbers, emergent large group behaviors and reactions, not controllable or automated, decentralized intelligence – think shoal of fish w/ evolving local rules; highly resistant form, not changing based on survivability of members, no hierarchy

Countermeasures: Disrupt / **Change the Strategic Global View of Swarm (its only real vulnerability). Defender collaboration. (Kania, 2017)**

Figure 3-1 Drone Crash into 737-700 passenger jet while landing at Mozambique

Source: Cuskelly, C. (January 6, 2017), UK Express. http://www.express.co.uk/travel/articles/751165/drone-boeing-737-planecrash-Mozambique. Also See: https://youtu.be/2jzx8BpDuHE

Commercial Small Unmanned Aircraft Systems (sUAS) Overview

There is a natural tendency to think that small unmanned aircraft systems present no threat, especially to US defenses. They are simply recreational or commercial toys. But they present a threat to National Airspace (NAS) – especially near airports. Figure 3-1 shows the results of a sUAS crashing into a jetliner in 2016.

USA FAA Part 107 special rule forbids use of sUAS within a five-mile radius of an airport. (FAA, 2018)

Table 3-3 shows some of the available options and each year more capabilities are being added. Imaging, camera capabilities, weather-proofing, and payloads all can be used to gather intelligence, provide reconnaissance or deliver a lethal payload. They are radar resistant and deploy with a very small heat signature, so they can be in close target quickly, before defenders can activate countermeasures.

Table 3-3 Commercial sUAS Parameters

- "**Flying Characteristics** Available as **RTF** (off-the-shelf Ready to Fly); **BNF** (Bind and Fly -with custom transmitter); **PNF** (Plug and Fly with custom transmitter, receiver, battery, and charger). RTF and BNF – no prior flight experience required.
- **Models** most rotary multicopter – quad (4), hexa (6) octo (8) variants. Fixed wing used for deployments in agriculture, public safety, emergency response and ISR (Intelligence, Surveillance, and Reconnaissance) many fully customizable to achieve specific capabilities, flight time, payload capacity, programmable flight, maximum speed and weather hardening.
- **Average SUAS flight time** 18 minutes, average range approximately one mile, cost $600 US, dry conditions" (Angelov, 2012)

"Specifications affecting hostile UAS operations

- **Payload capacity** function (weight and size more than gimbal, camera, battery) LIDAR or infrared or experimental sensors require larger capacity and subject to easier detection.
- **Range** function (signal transmission, LOS, image relay distance, battery and power constraints).
- **Weather Proofing** function (limited operating conditions, mostly dry. Upgradable to near military grade to operate in extreme conditions) Retrofit to harden for weather is a trade-off for weight, cost, flight time and payload capacity unless no of rotors increases.
- **Imaging** function (available medium –high resolution cameras of > 12 megapixels, with still and video) Infrared and LIDAR installable.
- **Automated and Programmable Pilot / Follow Me** settings function (predetermined flight mission path based on GPS coordinates (Fly-by-wire). Some with Follow Me autopilot settings enable the SUAS to automatically follow the operator." (Angelov, 2012)

Airborne Sensing Systems

There are two technologies available for airborne sensing of other aircraft; cooperative and non-cooperative. Cooperative technologies receive radio signals from other aircraft's onboard equipment. Two requirements for cooperative behavior. First ATC Transponder, which responds to ground-based secondary radar interrogations for air traffic control (ATC) usage. Traffic Alert Collision Avoidance System (TCAS) uses the same technology in FAA classes of airspace. Second is the Automatic Dependent Surveillance – Broadcast systems (ADS-B). ADS-B technology uses the Global Positioning System (GPS) or alternative navigational source to make broadcasts of its own aircraft position, velocity, and data required to avoid collisions. (Angelov, 2012) Table 3-4 shows typical sensor coordinate systems. The first three cooperate with each other, the latter five are non-cooperative technologies. (Angelov, 2012)

Table 3-4 Typical Sensor Coordinate Systems

Sensor Technology	Coordinate System
Active interrogation of Mode A/C transponder	Relative range, altitude
TCAS	Relative range, altitude
ADS-B	Latitude, longitude, altitude, velocity
Electro-Optical	Bearing (azimuth and elevation)
Laser /LIDAR	Relative range
Onboard radar	Relative range, Bearing (azimuth & and elevation)
Ground-based radar	Range and bearing from ground-reference
Acoustic	Bearing

Sensor Parameters

Sensor technologies use standard parameters to provide a basis for comparison and ISR performance. Table 3-5 Standard Sensor Parameters shows the base set:

Table 3-5 Standard Sensor Parameters

Sensor	Function
"Field of View	Describes angular sector within sensor making measurements. Outside this field of view, sensor is blind.
Range	Distance measured by sensor, within which some good probability of detection of targets
Update Rate	Interval at which sensor provides measurements
Accuracy	Uncertainty of position measurement – usually single dimension
Integrity	probability that measurement falls beyond some normal operation limit
Data Elements	Cooperative sensors – specific data to enhance ISR platform, ex: trajectory, identity, intent" (Angelov, 2012)

SAA Critical Control Systems include circuitry to affect UAS movement, landing, control of direction, detection, and correction of the aircraft. Many of these functions are incorporated into a UAS Autopilot, if capable.

Autopilot

Table 3-6 shows the common components found in UAS autopilots. These provide the means for UAS to affect movement, control, communications, detection, emergency operations, battery, waypoint delivery, and payloads.

Table 3-6 Common components found in UAS autopilots

- "Main Program/Processor: processing sensor data & implementation of control of UAV
- Magnetometer: measuring direction
- GPS: determine global position
- Airspeed/Altimeter: measure air speed & altitude
- UAV Wireless Communication: communicating with ground station
- Power System: provides power to UAV
- Inertial Measurement Unit: measures movement of UAV
- Boot Loader Reset Switch: loads programs into main program board
- Actuators: receives commands from main processing board & moves control surfaces
- Manual Flight Control: overrides autopilot & gives control of UAV control surfaces to ground station" (Clothier R. R., 2011) (Boutros, 2015)

(Clothier R. F., 2010) describes an ambitious SAA scientific research project involving multiple vendors, research teams, multiple aircraft, control stations, global communications and payloads known as the Smart Skies project. (Clothier R. F., 2010) Table 3-7 shows just a few of the SAA technologies that Smart Skies employed.

Table 3-7 SAA systems include (Smart Skies Project)

- "Low Cost Scout UAV Acoustic System (LOSAS)
- Passive Acoustic Non-Cooperative Collision-Alert System (PANCAS)
- Beyond Line of Sight Combat Identification (BLOS)
- Optical Collision Avoidance
- Common Control System
- Conventional Surveillance: Radar & Beacon Transmitters
- Acoustic Sensing
- Radio Emission Sensing
- Electro-Optical Sensing (EO)

(Clothier R. F., 2010)

SAA Subsystems

The purpose of a Sense and Avoid systems (SAA) function is replace a human pilot, to detect and resolve certain hazards of flight. "These hazards consist traffic or objects presenting a risk of collision. Air traffic includes anything that flies. Other hazards include terrain and obstacles (buildings, towers, power lines.) Aircraft preservation is not the primary reason for SAA. SAA must operate in emergency and diversionary events as well as throughout normal operations." (Angelov, 2012)

The main components of SAA configurations are the aircraft and systems onboard, the off-board control station, and the communication link(s) between these. The key distinctions are:

1. "Whether the SAA surveillance system consists of sensors located onboard the aircraft, off-board, or both, and
2. Whether the SAA decisions are made at the off-board control station or onboard the aircraft by its automation."

In both cases, there is volume of air above the ground stations in which the UAS flies. (See Figure 3-2) "In one configuration, all decisions are made on the ground. In another configuration, sensors and decisions are located aboard the aircraft, the ground station is monitor, and depending on the level of autonomy" (Table 3-1 above) the UAS detects targets, declares threats and selects maneuvers. In yet another configuration example, "sensors are located onboard or ground, but threat and maneuver decisions (human) are made via the ground station and communicated via communication links to the UAS." (Angelov, 2012)

SAA Services and Sub-Functions

The SAA function supplies two services. These are:

1. A Self-separation service to act BEFORE a collisions avoidance maneuver is needed, and could support earlier, gentler maneuvers, and
2. "The collision avoidance service that attempts to protect a small collision zone and usually is achieved by means of a late, aggressive maneuver" (Angelov, 2012)(see Figure 3-2 Self -Separation and Collision Volume). (Angelov, 2012)

To achieve these services, the following list of SAA design sub-functions is required:

- "Detect – various hazards and objects such as other aircraft, weather, terrain
- Track – detected motion of object, determination of position and trajectory
- Evaluate – tracked objects, against criteria that would need a SAA maneuver
- Prioritize – tracked objects based on evaluation and tests performed

- Declare – that paths of object and own aircraft reach a point of decision for movement
- Determine – specific maneuver based on geometry of encounter
- Command – own aircraft to perform maneuver
- Execute – commanded maneuver" (Angelov, 2012)

Figure 3-2 Self -Separation and Collision Volume

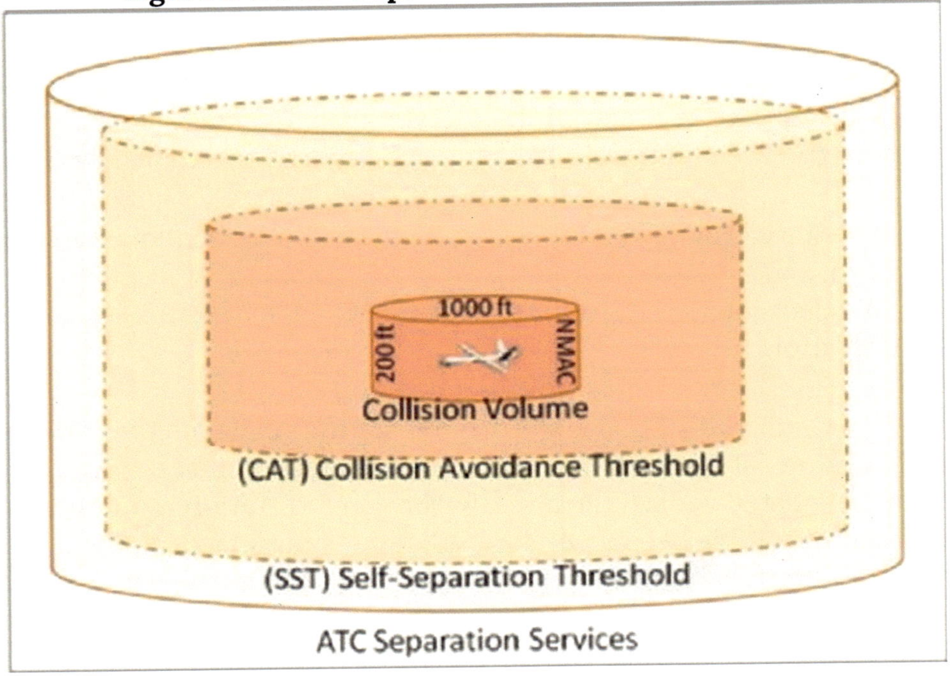

Source: Angelov, P. (2012). *Sense and avoid in UAS research and applications.* Hoboken: NJ.

Figure 3-3 shows the process of decisions required when two or more aircraft or UAS are on collision courses. (Yu, 2015)

Figure 3-3 Decision Process to Avoid Collision of Two Aircraft

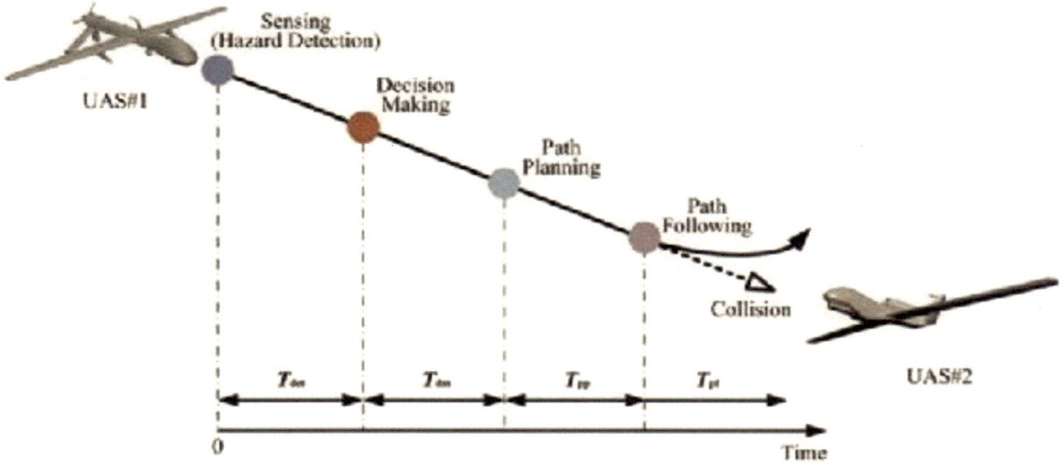

Source: Yu, X. (2015). Sense and avoid technologies with applications to unmanned aircraft systems: Review and prospects. *Progress in Aerospace Sciences*, 74, 152-166.

Low Hanging Fruit

Many textbooks detail UAS SAA systems ad nauseum. Angelov's *Sense and Avoid in* UAS is by far the best. The reader is directed to it as the aviation source. (Angelov, 2012) What makes the present book unique is its concentration on hostile UAS activities and speculative cyber -vulnerabilities and attacks on the US ADS. The security fault "low hanging fruit" in UAS systems is SCADA.

SCADA

There are hundreds of millions of SCADA systems. They are used to control every practical machine you can imagine. SCADA stands for **Supervisory Control and Data Acquisition.** SCADA started in the 1940's to control manufacturing processes such as flow rates, temperatures, valves, pressure, density, chemical, mechanical processes of all kinds. See Figure 3-4 for Legacy SCADA system for Chemical Plant. (Nichols R., Nov 28-30, 2006)

SCADA systems have improved significantly over the decades in all areas except one – **SECURITY**. SCADA systems are a security sieve. Figures 3-4 and 3-5 show examples of SCADA Architectures. (Nichols R., Nov 28-30, 2006) An interesting example are the automated/computerized systems in modern cars.

Figure 3-4 for Legacy SCADA system for Chemical Plant.

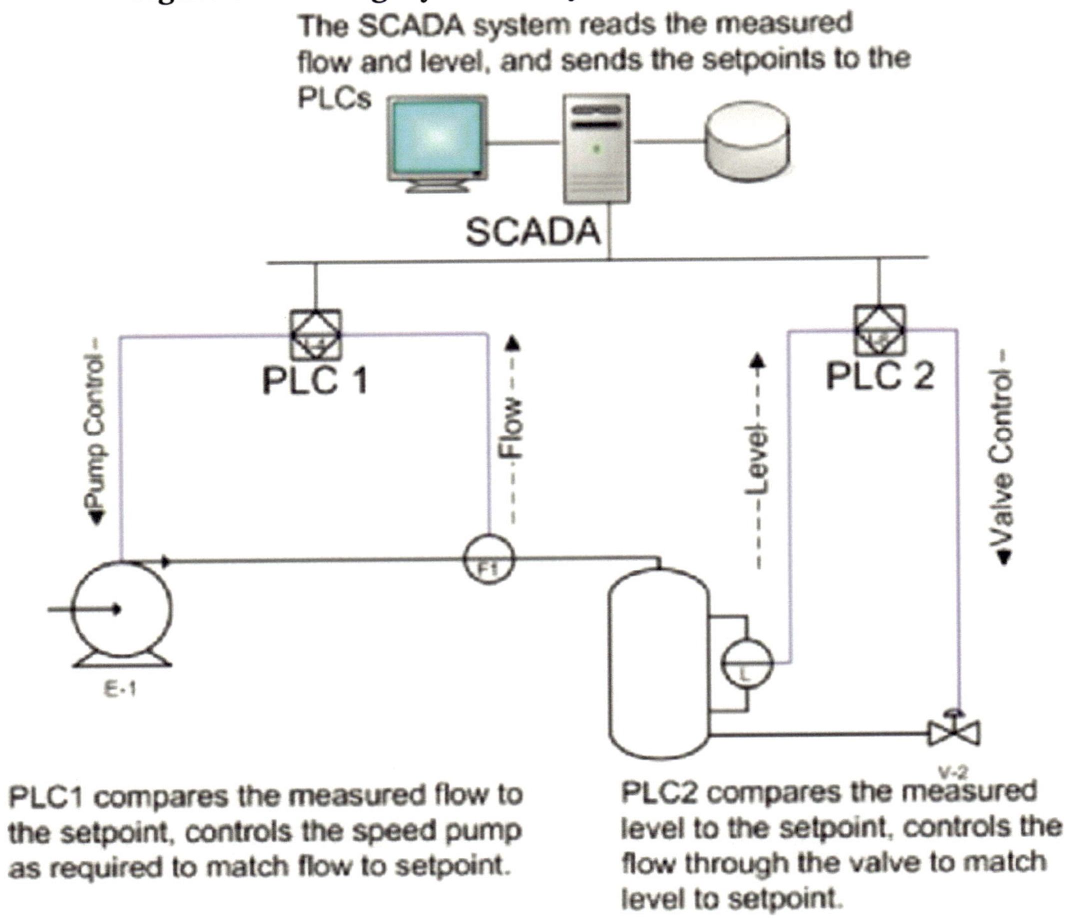

Source: Nichols, R.-0. (2016, March 29). NCIE UAS SAA Final Rev 4. 2016 INFOWARCON *conference presentation April* 4-7, Nichols, R.K. et. al. (3-29-2016) Presentation to INFOWARCON April 4-7 on NCIE UAS SAA Final Rev 4, presented to 2016 INFOWARCON conference, Memphis TN. Available as PPTx presentation download from author or in CANVAS. Memphis, TN, USA: INFOWARCON16

Everything is controlled by SCADA; tires, engine, seat belts, safety bags, oil pressure, even door locks. However, cyber hackers can exploit SCADA to disable a car remotely, with the driver still in it! Greenburg, Wired (2015). Hackers Remotely Kill a Jeep on the Highway—With Me in It. (Greenburg, 2015)

"UAS ARE JUST FLYING SCADA MACHINES!" (Nichols R.-0. , 2016) Table 3-8 SCADA shows the principle functions that apply to all SCADA systems, especially UAS.

Table 3-8 SCADA

- Supervisory Control and Data Acquisition (SCADA) systems facilitate management with remote access to real-time data

    - Channel to issue automated or operator-driven supervisory commands to remote station control devices
    - A human–machine interface (HMI) is responsible for data presentation to human operator
    - Composed by a console that makes it possible to monitor & control process
    - Remote terminal units (RTUs) are microprocessor-controlled electronic devices that interface sensors to SCADA by transmitting telemetry data
- Is a process control system for computerized real-time monitoring and control?
- Typically consists of:
    - Master Control Unit (MCU)
    - Remote Terminal Unit (s) (RTU)
    - Communication Links
- Supervisory system is responsible for:
    - Data acquisition
    - *Control activities on process*
- Programmable logic controllers (PLCs) are final actuators used as field devices
- Communication infrastructure connecting supervisory system to RTUs
- Various process & analytical instrumentation
- RTU's Alarm Systems
    - Doors
    - Battery Backup
    - Low Power/Loss of Power Alarm
    - Power Protection
    - Passwords for Keypads, PC ports
    - Log Alarm (or Event) When Local User Plugs PC in or Signs On
    - Log Event when Local User Changes Values

Figure 3-5 for Corporate SCADA system

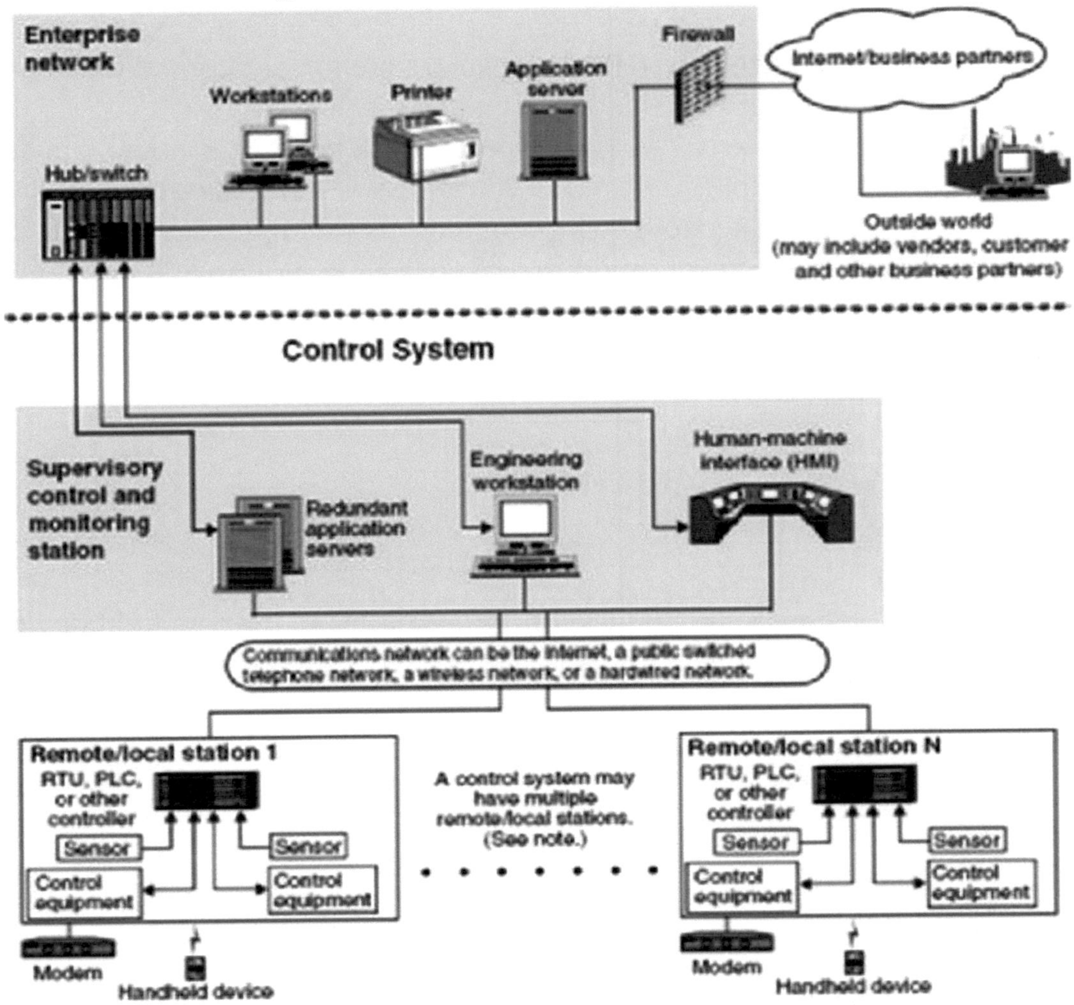

Source: Nichols, R.-0. (2016, March 29). NCIE UAS SAA Final Rev 4. 2016 INFOWARCON *conference presentation April* 4-7, Nichols, R.K. et. al. (3-29-2016) Presentation to INFOWARCON April 4-7 on NCIE UAS SAA Final Rev 4, presented to 2016 INFOWARCON conference, Memphis TN. Available as PPTx presentation download from author or in CANVAS. Memphis, TN, USA: INFOWARCON16

Figure 3-6 UAS SCADA System Internals

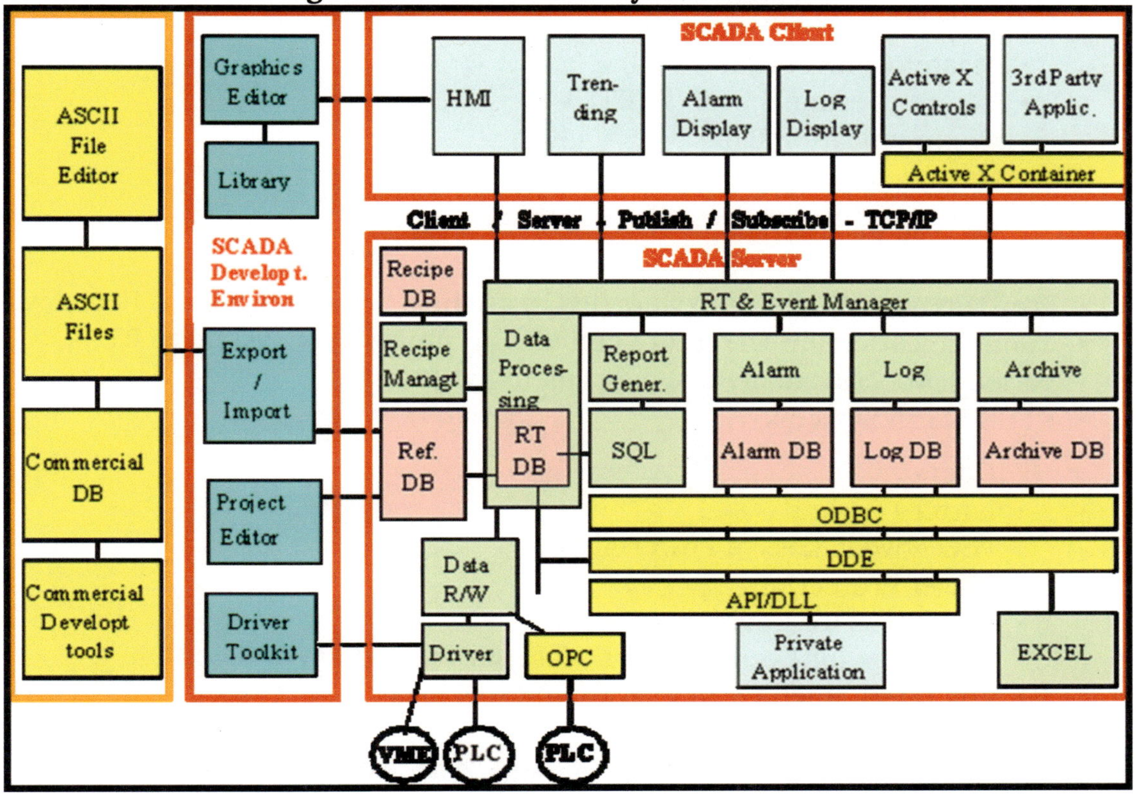

Source: Nichols, R.-0. (2016, March 29). NCIE UAS SAA Final Rev 4. 2016 INFOWARCON *conference presentation April* 4-7, Nichols, R.K. et. al. (3-29-2016) Presentation to INFOWARCON April 4-7 on NCIE UAS SAA Final Rev 4, presented to 2016 INFOWARCON conference, Memphis TN. Available as PPTx presentation download from author or in CANVAS. Memphis, TN, USA: INFOWARCON16.

Table 3-9 SCADA shows the principle functions that apply to all SCADA systems, especially UAS.

Table 3-9 SCADA Functions

- Supervisory Control and Data Acquisition (SCADA) systems facilitate management with remote access to real-time data
    - Channel to issue automated or operator-driven supervisory commands to remote station control devices
    - A human–machine interface (HMI) is responsible for data presentation to human operator
    - Composed by a console that makes it possible to monitor & control process
    - Remote terminal units (RTUs) are microprocessor-controlled electronic devices that interface sensors to SCADA by transmitting telemetry data

- Is a process control system for computerized real-time monitoring and control
- Typically consists of:
  - Master Control Unit (MCU)
  - Remote Terminal Unit (s) (RTU)
  - Communication Links
- Supervisory system is responsible for:
  - Data acquisition
  - *Control activities on process*
- Programmable logic controllers (PLCs) are final actuators used as field devices
- Communication infrastructure connecting supervisory system to RTUs
- Various process & analytical instrumentation
- RTU's Alarm Systems
  - Doors
  - Battery Backup
  - Low Power/Loss of Power Alarm
  - Power Protection
  - Passwords for Keypads, PC ports
  - Log Alarm (or Event) When Local User Plugs PC in or Signs On
  - Log Event when Local User Changes Values

SCADA systems have plenty of cyber related vulnerabilities. Most are connected to computers. Those vulnerabilities multiply when connected to the Internet. SCADA systems differ from the IT structures. See Figure 3-7. (Shapiro, 2006) Table 3-10 Sample SCADA Design Vulnerabilities apply to all systems including UAS. (Nichols R. , Nov 28-30, 2006)There are so many design flaws and vulnerabilities in SCADA systems that the US government has a special SCADA testing lab in Utah and has published copious recommendations to improve security. (NTSB, 2009)

Figure 3-7 IT Systems Vs Control Systems (Kilman, 2003)

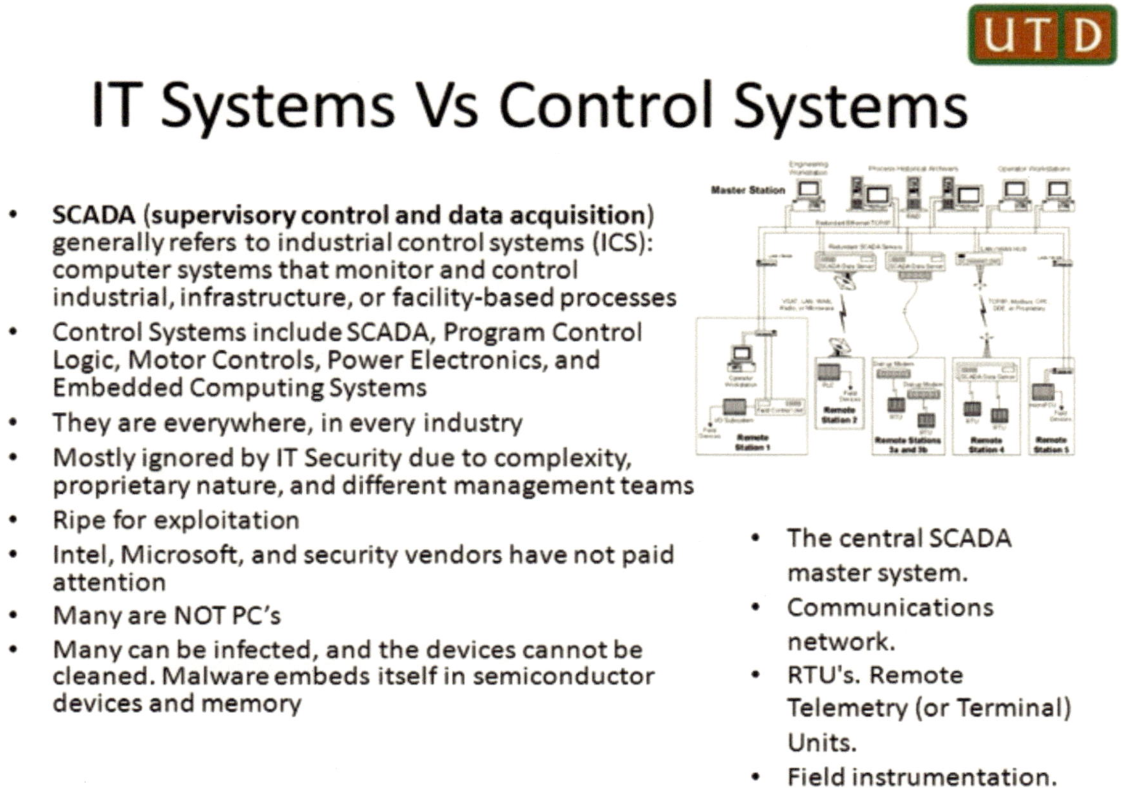

Source: Kilman, D. & Stamp, J. (2003), CT: *Framework for SCADA Security Policy.* Albuquerque, NM: Sandia National Laboratories. Retrieved from Energy.gov: https://www.energy.gov/sites/prod/files/Framework for SCADA Security Policy.pdf

Table 3-10 Examples of SCADA Design Vulnerabilities

- "Ease of operation outweighs security
- Commonly set up on operating systems with known vulnerabilities
- Poor authentication systems in place
- Remote access allowed for maintenance &/or IT support
- Interconnectivity to vulnerable corporate networks
- Weak access control lists on firewalls
- Proper Network Access Control (NAC) is most crucial to prevent unauthorized connection within network
- First target of compromise for an attacker
- No use of standard IT defense software
- Wireless technology common
- System connect to unsecured remote processors

- SCADA software not designed with robust security features
- Public information often available on specific systems
- Poor physical security on remote access points
- No use of standard IT defense software
- Wireless technology common
- System connect to unsecured remote processors
- SCADA software not designed with robust security features
- Public information often available on specific systems
- Poor physical security on remote access points" (Kilman, 2003)

Attack Vectors

A brief overview of UAS Attack Vectors (by no means the exhaustive list) is demonstrated in Table 3-11. (Nichols R.-0. , 2016)

Table 3-11 Common Attack Vectors

"Common Vectors

- Backdoors & holes in network perimeter
- Protocol vulnerabilities
- Attacks on field devices through cyber means
- Database attacks
- Communications hijacking & Man-in-the-middle attacks
- Cinderella attack on time provision & synchronization
- Bogus input data to controller introduced by compromised sensors &/or exploited network link between controller & sensors
- Manipulated & misleading output data to actuators/reactors from controller due to tempered actors/reactors or compromised network link between controller & actuators
- Controller historian changes – feed forward control
- Distributed Denial of Service – missing deadlines of needed task actions
- Backdoors and holes in network perimeter
- Vulnerabilities in common protocols
- Attacks on field devices through cyber means
- Database attacks
- Communications hijacking and Man-in-the-middle attacks
- Cinderella attack on time provision and synchronization
- To a control engineer, possible attacks can be grouped into following categories:
- Bogus input data to controller introduced by compromised sensors and/or exploited network link between controller and sensors

- Manipulated and misleading output data to actuators/reactors from controller due to tempered actors/reactors or compromised network link between controller and actuators
- Controller historian
- Denial of Service – missing deadlines of needed task actions

Attacks on Software:

- No Privilege Separation in Embedded Operating System
- Buffer Overflow
- Structured Query Language Injection

Possible UAS Attack Hardware / Software

- SkyJack©[3]
- Aircrack-ng©[4]
- Node-ar-drone©
- Raspberry Pi©
- Parrot AR. Drone -2©
- Alfa© AWUS036H wireless adapter
- Edimax© EW-7811Un wireless adapter
- Snoopy©[5]

3. Skyjack Drone hack. Drone that flies around seeking Seeks wireless signal of any other drone in area. Forcefully disconnects wireless connection of true owner of target drone. Authenticates with target drone pretending to be its owner. Feeds commands to it and all other zombie drones SkyJack primarily a Perl application which runs off a Linux. Detect drones by seeking out wireless connections from MAC addresses.
4. Aircrack-ng© To put wireless device into monitor mode to find drones and drone owners. De-authenticate true owner of drone. Once de-authenticated, connect as drone waiting for owner to reconnect.
5. Snoopy is Software that can hack into Wi-Fi and steal data – attached to drones. Comprised of various existing technologies. Uses Distributed tracking and profiling framework. Runs client-side code on any device that has support for wireless monitor mode. Collects proberequest and uploads to a central server. Exploits handsets looking for wireless signal. Most leave their device Wi-Fi setting on Spoof network available to Wi-Fi searchers to use. Once connected to rogue network, data is stolen. Differs from other rogue access points in way data is routed. Traffic is routed via an OpenVPN connection to a central server. Able to observe traffic from all drones in field at one point. Traffic manipulation only done on server. Allows basic data exploration and mapping.

Attacks on Communication Stack

- Network Layer
- Transport Layer
- Application Layer

Auxiliary tools:

- Password Theft
- Wireshark
- Man-In-the-Middle Attacks
- Trojan Horse Virus
- Distributed Denial of Service Attacks" (Nichols R.-0. , 2016)

Cyber – Attack Taxonomy

UAS SCADA systems susceptible to a broad range of cyber and network specific attacks on the SAA modules in the aircraft and communication structures from the ground or satellite links. These represent system threats and vulnerabilities of the UAS structure, increasing the risk of hostile use or takeover. (Nichols R. , Nov 28-30, 2006)A UAS Cyber Attack Taxonomy is an organized view of potential cyber threats to UAS assets. *The Taxonomy is a list of agents that increase risk of a successful attack on US UAS ADS assets.* Remember the authors' contention, the risk of success of terrorist attacks on USA Air Defense Systems (ADS) via UASs is higher because of improving commercial capabilities and accessibility.

A qualitative view of information risk (*also a measure of cyber-attack lethality*) in a system such as SAA or computer network is expressed as:

Risk = (Threats x Vulnerabilities x Impact / Countermeasures) Eq. 3-1

And at time state 0, this equation can be reduced to

Risk = function (Threats / Countermeasures) Eq. 3-2

(Nichols R.-0. , 2016)

At time state =0, where Vulnerabilities & Impact are constants and drop out of the equation.

Threats are real, and if applied in the absence of appropriate countermeasures, will increase the likelihood of a successful cyber-attack. Vulnerabilities are weaknesses in the system that a threat may or may not exploit. Vulnerabilities essentially in the system, ab initio. Threats can be mitigated or improved based on the attack circumstances. Impact is an after-the-fact accounting of the cyber-attack. No matter what the magnitude, it is a constant. Countermeasures are a host of technologies that can be applied to mitigate threats and reduce Risk. Increased Threats means increased Risk. Increased Countermeasures means decreased Risk. In practice, these equations require a qualitative legend to make comparable cases. Conversely, decreased threats means decreased Risk and decreased countermeasures means increased Risk. (Nichols R.-0. , 2016)Some authors use Vulnerabilities to assess Risk. (Garcia, 2006) Therefor our cyber-attack taxonomy must work for either Risk approach. There are many approaches to evaluating Risk. The authors choose the simplest approach to understand the attack vectors.

Espionage

Let us define some of its elements. First, we have **espionage** by foreign hackers. One of the main culprits are the Chinese who both encourage stealing state secrets have been quite successful. (Brenner, 2011) Hackers have successfully attacked U.S. Transportation Command (TRANSCOM) networks, "Stole designs for advanced U.S. weapons systems, including:

- F-35 Joint Strike Fighter,
- F/A-18 Fighter Jet,
- Patriot Missile System,
- RQ-4 Global Hawk Drones,
- P-8 Poseidon Reconnaissance Aircraft,
- UH-60 Black Hawk Helicopter,
- Littoral Combat Ship,
- Army's Terminal High Altitude Area Defense Missile Defense System
- Navy's Aegis Ballistic Missile Defense Program,
- compromised White House computer systems for espionage against military offices, targeting: Interoffice communication;
- Nuclear codes,
- took temporary control of National Oceanic and Atmospheric Administration's (NOAA) weather satellites, and
- Civil Reserve Air Fleet systems by compromising computers defense contractors." (Sood A.K. & Enbody, 2014)

Adding to our taxonomy, there are three Vulnerability classes exploited in UAS cyber-attacks: "**Software-based, Insider Threats, and Hardware-based.** Software and hardware vulnerabilities are result of poor coding practices and dearth of security understanding in developers. Hard-coded passwords exist because programmers prefer an effortless way for recovery purposes such as debugging. Vulnerabilities have drastic impact on security of software." (Sood A.K. & Enbody, 2014) "Developers or programmers with malicious intent may implant computer hardware with code as part of a targeted attack. Insider threat vulnerabilities are difficult to assess due to human element involved. Multi-layer defenses are required to combat vulnerabilities in software/hardware world." (Sood A.K. & Enbody, 2014)

Software – Based Vulnerabilities

"Military UAS defense systems deploy widely used software in their network devices: Operating systems, open source software, routers, radio frequency devices, Internet Connection Sharing (ICS) and SAA SCADA." (Sood A.K. & Enbody, 2014) UAS ground system network software may have the standard vulnerabilities; "hardcoded passwords, backdoors in firmware, insecure protocols, Remote Command Execution (RCE), default passwords for Human-Machine Interfaces (HMIs), Insecure authentication and authorization, malicious hardware, critical infrastructure systems have hardcoded passwords embedded in firmware which may allow attackers to gain complete access to system." (Sood A.K. & Enbody, 2014) It doesn't end there. Other software-based vulnerabilities: "Backdoors exist for support or remote access purposes, Hardcoded passwords easily obtained by: Reverse engineering firmware, analyzing functional components," (Sood A.K. & Enbody, 2014) Remote Code Execution (RCE) which is an attacker's ability to execute attacker's commands on target machine or target process remotely. Another RCE vulnerability is a software bug that gives attacker way to execute arbitrary code or ability to trigger arbitrary code execution from one machine on another. (Nichols R.-0. , 2016)

Unfortunately," Remote Code Execution (RCE) can be triggered by exploiting security flaws in:

Operating system components, browsers," ICS, SCADA, routers, Microsoft Office, Adobe Reader, and Java. Remote Code Execution (RCE) is a powerful threat to UAS and supporting computer systems. "Attackers exploit security issues; buffer overflows (stack, heap, integer), use-after free errors, race conditions, memory corruption, privilege escalations and dangling pointers."

Remote Code Execution (RCE) vulnerabilities keeps growing and RCE vulnerabilities allow "attackers to execute arbitrary code on compromised systems, drive-by downloads, spear phishing attacks." (Sood A.K. & Enbody, 2014)

ICS/SCADA is particularly vulnerable to remote code execution vulnerabilities. Another form is SQL injections, "which exploits weaknesses in web applications to allow attackers' queries to be executed directly in backend database" and allow attackers to extract sensitive information such as credentials, emails, critical documents, intelligence. "Data stolen using SQL injection can provide critical information for advanced UAS targeted attacks." (Sood A.K. & Enbody, 2014)

The final group in the software- based vulnerabilities set is "insecure authentication and file uploading flaws. These allow remote attackers to access critical systems by exploiting weak authentication design and uploading malicious code or firmware. This security issue persists due to inability of systems to implement granular control through proper authentication and authorization checks. File uploading attacks exploit a system's inability to determine type of files being uploaded on server." (Sood A.K. & Enbody, 2014)

Insider Threat Vulnerabilities

"Insider threats involve a malicious employee or contractor that steals sensitive organizational assets or otherwise diminishes integrity of organization." (Sood A.K. & Enbody, 2014) **Insider threats may be intentional or unintentional**. The outcome is the same, increased risk.

Intentional Insider Threats (IIT): is "when contractors or employees turn malicious," their motives: revenge, personal grudge, and /or greed. Regarding **Unintentional Insider Threats** (UIT), "the US military defense outsources jobs to contractors, which presents a softer target than military facilities. The result is compromised contractors' assets allows attackers to steal intelligence from military defense networks. The contractor acts as a proxy to provide entry for attacker, even though contractor has no intention to conduct spying / stealing assets." (Sood A.K. & Enbody, 2014)

"Edward Snowden worked as a contractor for Central Intelligence Agency and NSA. Snowden leaked copious amounts of information, including information of Government Communications Headquarters (GCHQ) and National Security Agency (NSA) project, 'Anarchist; Hacking of Israeli drone feeds; 'The Drone Papers' – The Intercept and Leaks regarding drone usage in Afghanistan, Yemen, and Somalia." (Sood A.K. & Enbody, 2014)

Hardware-based Vulnerabilities

The US sometimes picks the wrong vendors to supply its UAS critical hardware. Hardware imported from China includes backdoor access to hardware after deployment. "Exported Chinese manufacturing units compromised military-grade FPGA computer chips, circuits, and counterfeit devices, such as scanners." "Zombie Zero malware has been implanted in software of scanner hardware manufactured in China as part of attack targeting shipping and logistics industries, especially printers. When scanners are connected to networks they provide platforms for compromising networks. Counterfeit devices and circuits developed in China for U.S. military and defense contractors to be used in warships, missiles, airplanes and

UAS." (Sood A.K. & Enbody, 2014) (Threat to all nations that receive hardware preinstalled with malware.) (Nichols R.-0. , 2016)

"Hardware based vulnerabilities observed in actual attacks on military defense systems (Army) and applications include the following; backdoors and hardcoded passwords, compromised

GPS Satellite Communication (SATCOM) systems," SCADA systems vulnerable to buffer overflows, and compromised GPS SATCOM systems. The Navy had its share of hardware-based threats; Remote Code Execution – "XMLDOM Zero-day vulnerability was exploited to attack U.S. Veterans of Foreign Wars' website, SQL injections, Royal Navy website hacked, U.S. Army website hacked, insecure protocols, spoofing and hijacking and attacks to spoof GPS communication to control U.S. drones." (Sood A.K. & Enbody, 2014)

Wireless attacks are the most generic form of hacking. "Strategies to compromise a system's ability to be controlled by rightful owner include:

- Password Theft
- Wireshark
- Man-In-the-Middle Attacks
- Trojan Horse Virus
- Gain Scheduling
    - Fuzzing
    - Digital Update Rate,
    - Distributed Denial of Service,
    - Buffer Overflow." (Rani, 2015)

Password cracking/theft. "Even complex passwords can be cracked using software tools such as dictionary attacks and brute force attacks". Cracking programs are easy to obtain. "Statistical methods such as *Aircrack-ng*" which is a "Wired Equivalent Privacy and Wi-Fi Protected Access Pre-Shared Key cracking program is freeware." (Rani, 2015)

Wireshark. A robust tool to analyze and capture packets for wireless networks. It provides valuable and sensitive information of transmitted packets. Wireshark allows effortless access to "client system and gain control over it." It displays a list of available interfaces. A victim's username and password can be obtained on a Graphical User Interface (GUI).

Another cyber-goodie is the *Man-in-the-Middle* (MIM) *attack*: "Attacker gains control of sensitive data by furtively modifying communication link between two parties. End users are usually unaware of manipulation performed by attacker." (Rani, 2015) "Attacker forms fake connections between server and client. "Forms of MIM attacks are:

- URL manipulation
- Rogue Domain Name Server
- Address Resolution Protocol poisoning
- Duplication of Media Access Control
- False Emails" (Rani, 2015)

Trojan Horse Virus. It is malicious program or software that causes detrimental effects. It can monitor traffic over a network, damage hard drives, leverage of a security glitch, multiply/replicate like rabbits, give attacker an access to system remotely, "Trojan can continuously delete files and ultimately demolish Operating System and cannot be easily identified by anti-virus programs." (Rani, 2015)

Gain Scheduling. A special form of cyber- attack. There are four classes: fuzzing, digital update rate, Distributed Denial of Service (DDoS), buffer overflow. "Gain scheduling is used to control non-linear UAS systems and hybrid UAS systems." (Rani, 2015) "UAS will need different gains for control depending on states of UAS: Mass, Altitude, Speed, Flaps down." "In a hybrid system, a system is assumed to have multiple modes of operation corresponding to take off, landing, and cruising." "UAS will have different gains for controlling the vehicle. Control gains are often pre-computed and trusted. Gain parameters are coded into on-board autopilots. Without strict monitoring of software, an override of gains goes undetected. Changes to gains or gain scheduling logic can cause decreased performance in autopilot or instability in UAS." (Rani, 2015) Attacker can manually override safety systems for an hour, shut down system, and take control. Not much is required; IP address of SCADA Server, path to server and then a Trojan or back door can be installed. (Nichols R.-0. , 2016)

Gain Scheduling attack methods vary depending on UAS SAA systems. "There are generally five methods of attacks:

- Sensor spoofing to cause mode confusion,
- Overriding gains through hacking,
- Infinite switching between gains, will cause loss of control,
- Causing Denial of Service (DOS) between controller gain block, and
- UAS controller block by overloading the on-board processor." (Kim, 2012)

Fuzzing. "In UAS autopilot system, random inputs with expected distributions are not uncommon, and Gaussian noise inputs are routinely accounted for however, unexpected, invalid, or completely random inputs can cause unknown behaviors. Attacker can access any data flow between components and corrupt them with bogus values." "Attackers use malicious fuzzing to discover SAA vulnerabilities." Intentional white-box and black-box fuzzing tests can determine system robustness of Guidance, Navigation and Control (GNC) algorithms. Avionic and UAS programmers develop GNC for unmanned rotatory and fixed-wing aircraft. GNC algorithms are key to counter-terrorist activities and targeting High Value targets (HVT). "Consequences for a fuzzing attack: Aircraft instability, Process lock-up and/or invalid outputs to next SAA/GNC process. Common fuzzing attack methods include buffer overflow attacks, sending malicious packets with invalid payload data to UAS, and adding malicious hardware between components." (Kim, 2012)

Digital Update Rate. UAS autopilots are digital computers. Inputs/outputs to/from the autopilot are discretized. (Zaharia, 2012) "Any continuous inputs to autopilots are converted to digital inputs through discrete sampling. If the autopilot was designed with a continuous controller, it is also converted to a discretized form. For a discretized system, as sample time increases system becomes unstable/uncontrollable. For data collection, longer sampling periods will increase probability of data aliasing." (Kim, 2012) Hackers, knowing the above, have created a few methods cyber-attacks; "changing sampling time of analog-to-digital converters through buffer overflow, hardware manipulation and Denial of Service (DOS) or Distributed [networked] Denial of Service (DDOS) attacks that prevent processor from running controller or navigator at desired update rate." (Kim, 2012)

Distributed Denial of Service (DDoS). "A large-scale intrusion method performed by a host source which causes detrimental effects to legitimate users by withholding services. This either causes system to shut down completely or rapidly drain system resources such as computing power and bandwidth." (Rani, 2015) "Once attacker gains access new tools can be installed to enable control of host. Infected systems continue to look for other vulnerable systems and attack them too. Distributed Denial of Service (DDoS) attacks results in a master-slave process where affected victim is controlled by the attacker." The attacker comes into possession of controls of many systems. "Resources of system are exhausted rapidly, and flooding of packets occurs on victim end. Attacker then removes all traces that could lead to locating the source of attack by using spoofed Internet Protocol address preventing victim to permeate illegal traffic targeted towards them." (Rani, 2015)

The question is how is a DDoS attack initiated and how related to UAS? Think team or swarm configuration controlled at a ground-station network.

"Steps in initiating a complex DDoS attack:

Compilation of vulnerable agents

- Network first scanned for vulnerable agents for attacker to compile list of agents to attack
- Possible to attack systems by setting up automated software to scan network and take over vulnerable agents

Defuse

- Flaws in security and agent vulnerability are misused by attacker
- Software codes used to automatically attack and disband owner from controlling system
- Actions are taken by attacker to safeguard code planted from being Distributed Denial of Service (DDoS)

Connection

- Protocols such as TCP or UDP used to connect with numerous agents and plan attacks accordingly via scheduling
- Attacks can be performed on either single or multiple agents

Intrusion

- Features of victim such as port numbers are conformed
- Attacker launches attack and alters properties
- In favor of attacker since alteration of packets would lead to complications in identification of source of attack" (Rani, 2015)

"Distributed Denial of Service (DDoS) Scanning attack techniques:

- Random
- Local subnet
- Hit-list
- Permutation
- Topological

Distributed Denial of Service (DDoS) System Shutdown for at least five minutes:

    - UAS Operations can no longer monitor or control process conditions
    - SCADA locks up, must be rebooted
    - When comes back on-line, locks up again
        - Items Needed for DDoS:
        - Ability to flood server with TCP/IP calls
        - IP Address of SCADA Server
        - Path to server" (Rani, 2015)

Buffer Overflow SCADA Attack. Exploiting a buffer overflow vulnerability is to overwrite some control information to modify flow of control in a program, exploiting and compromising control information to give control to the attacker's code. It involves overwriting return address stored on stack to transfer control to code placed in buffer or past end of buffer. On hardware implementations, stack grows downwards towards lower memory addresses.

When a function is called, the address of instruction following function call is pushed onto stack, so that the function knows where to return control when it is finished. While the called function is being executed, it allocates local variables on stack, variables and buffers, allocated at lower memory addresses than return address. The process permits writes to base address of buffer plus offset to return address, overwrites value of return address.

In a buffer overflow SCADA attack, if the return address is overwritten with some arbitrary value, control will be transferred either to invalid location in memory or location that does not contain valid executable code. This results in segmentation fault. If the overflow SCADA attack

return address is overwritten in a clever manner, it can transfer control to code inserted by the attacker in a buffer overflow. Attacker can place executable code into the buffer, then overwrite return address with the address of buffer. Control will then be transferred to executable attack code contained in buffer. (Nichols R.-0. , 2016)

Control Acquisition Attack (CAA). The objective of attacker in a CAA is to assume direct control of unmanned vehicle. A proven example is to use GPS spoofing to shift flight path of UAS to suit purposes of attacker with limited control. If the attacker can gain complete control of unmanned vehicle, the possibility of a man-in-the-middle attack is promulgated. The attacker could send falsified data to original controller to make it appear that vehicle is behaving normally, when it is being controlled by attacker. CAA is undetectable attack that is especially dangerous. (Nichols R.-0. , 2016)

General Attack Possibilities

Autopilot Hardware Attack. The attacker can link directly to UAS autopilot to give control over UAS and/or tactical data. "They can corrupt data stored on-board autopilot or install extra components that corrupt data flow. Hardware attacks affect UAS survivability, control and allow tactical data to be collected. Hardware attacks can be carried out during maintenance, storage, manufacturing, and delivery phases." (Kim, 2012)

Wireless Attack. "Attacks are carried out through wireless communication channels. They can occur if an attacker uses wireless communication channels to alter data on-board UAS autopilot. Worst case scenario is if an attacker can break encryption of communication channel. Attacker can gain full control of UAS if communication protocol is known and then can use a buffer overflow or like corrupt data onboard or initiate some event." (Kim, 2012) Another type of attack is sensor spoofing. The attacker send false data through on-board sensors of UAS autopilot. The danger of wireless attacks they can be carried out from afar while UAS is being operated. (Nichols R.-0., 2016)

Control System Security. "Attacks that prevent hardware/CPU from normally

Include buffer overflow exploits through some input device, forced system resets to load malicious code, and hardware changes or additions to system." (Kim, 2012)

Application Logic Security. These attacks use malicious manipulation of sensors or environment, providing false data to control system. The control system behaves as programmed without fault, but some or all inputs to system are corrupted. Attacks include sensory data manipulation

Vehicle/system component state data manipulation, navigational data manipulation, command and control data manipulation, to gain complete control over connected drone and gained access to corrupt pre-programmed flight path. Since the communication channel between ground station and flight controller of drone is based on an unsecured protocol, a hacker can easily attack and gain real-time access to drone. (Nichols R.-0. , 2016)

Conclusions

Unmanned Aircraft Systems represent some of US most advanced air assets. They are critical to the US ADS. Because they were designed on top of open communications, SCADA controlled, and interact with the Internet for ground, air, sea, and satellite support, plus use COT software/hardware, they are vulnerable to a host of cyber-attacks to control their destiny. Based on the sheer diversity and number of potential cyber weapons that can be used at every stage of the UAS mission, and the significant growth of UAS for defense purposes, the risk of hostile use against US ADS is steadily increasing. Couple this with a huge growth and sophistication in the UAS commercial technologies supporting and global sectors requiring UAS, the threat of hostile use cannot be minimized.

Discussion Topics

1) You are a terrorist. Contemplate the Risk and Success of a hostile take-over or destruction of a friendly UAS by the following four methods:

> 1A – Spoofing GPS satellite communication
>
> 1B – Corrupting authentication of ground station communication signals
>
> 1C – Hijacking a UAS by hacking internal ASIC chip communication
>
> 1D – ADS-B Spoofing / Jamming

What cyber tools from the Chapter 3 Cyber- Attack Taxonomy would you use to accomplish the 1A- 1D goals?

2) Discuss the cyber counter measures that could be used to terminate/disrupt/destroy a SWARM attack of fifty or more UAS on an ADS facility.

3) U.S Navy Vessel Collisions in the Pacific:

Facts:

In 2017, there were chain of incidents/collisions involving four U.S. Warships and one U.S Submarine. On 20 August, the guided-missile destroyer USS John S McCain collided with the 600-foot oil and chemical tanker Alnic MC at 0624 JST. Ten sailors died. On 17 June, the destroyer USS Fitzgerald collided with the ACX 30,000-ton container ship at 1330 JST, leaving seven dead. Records show that the ACX turned sharply right at the time of collision. The route of the destroyer is not shown on maps because commercial tracking data does not include military ships. Damage to the starboard side of the USS Fitzgerald indicates it would have been on a bearing of approximately 180 deg. (South). *The captain of the Philippine-flagged container ship accused the Navy destroyer of failing to heed warning signs before the crash.* On 9 May, the guided-missile cruiser USS Lake Champlain collided with a South Korean fishing boat off the Korean Peninsula. There were no injuries. On 31 January, the guided-missile cruiser USS Antietam ran aground dumping more than 1000 gallons of oil into Tokyo Bay. On 18 August, the ballistic- missile submarine, USS Louisiana collided with the Navy Offshore Support Vessel in the Strait of Juan de Fuca. No injuries.

US Navy Response

In all five incidents, the US Navy blamed their field leadership for not responding in an appropriate manner. Court marshals and relief from duty are the punishments of the day. The Navy blames funding, lack of readiness, and lack of training. Investigations and maintenance "holds" have been initiated. Reading between the lines, this response would imply that the Skipper/XO/COB and at least five watch sailors on each Naval vessel (roughly –forty to fifty personnel including bridge staff) were judged incompetent. Many US Navy careers were finished. Further, this would also imply that all five vessels radar, emergency positioning alert systems, AIS, sonar, and long-range collision avoidance equipment must have been functioning perfectly, without a catastrophic failure or interference of any kind.

OR...

The Case for a Cyber Weapon

To this researcher, there appears to be valid evidence to support the theory that at least two of the US Navy Warships above AND the commercial vessels they struck were the on the wrong end of a cyber-weapon. They were receiving wrong GPS generated positional information. *The Cyber Weapon may have been deployed by UAS off a small, nearby vessel by an adversary*. The author believes, that the subject cyber weapon is an advanced modular entity that can spoof the GPS signals received by all vessels in its range. Vessels given misleading data will make incorrect decisions in terms of navigation and emergency responses – potential leading to collisions and deaths.

Contemplate and comment on the viability of the researcher's cyber-weapon theory.

Bibliography

Adamy, D. -0. (2015). EW 104 EW *against a New Generation of Threats*. Boston: Artech House.

Adamy, D. (2001). EW 101 A *First Course in Electronic Warfare*. Boston, MA: Artech House.

Adamy, D. (2009). EW 103 *Tactical Battlefield Communications Electronic Warfare*. Boston, MA: Artech House.

Adamy, D. L. (2004). EW 102 A *Second Course in Electronic Warfare*. Boston: Artech House.

Adamy, D.-9. (1998, Jan). Lesson 4: the basic link for all EW functions. (electronic warfare)(EW Reference & Source Guide). *Journal of Electronic Defense, Jan 1998 Issue*.

Alford, L. (2000). Cyber Warfare: Protecting Military Systems. *Acquisition Review Quarterly*.

Angelov, P. (2012). *Sense and avoid in UAS research and applications*. Hoboken: NJ.

Army, U. (1992, November 23). US Army Field Manual FM 34-40-7. *Communications Jamming Handbook*.

Austin, R. (2010). *"Design for Stealth", Unmanned Aircraft Systems UAVS Design Development and Deployment*. New York: John Wiley and Sons.

Barker, W. (2003, August). SP 800-59 *Guidelines for Identifying an Information System as a National Security System*. Retrieved from NIST: https://csrc.nist.gov/publications/detail/sp/800-59/final

Barnhart, R. K. (2012). *Introduction to Unmanned Aircraft Systems*. New York: CRC Press.

Beaudoin, L. e. (2011). Potential Threats of UAS Swarms and the Countermeasures Need. ECIW.

Boutros, D. (2015, May 15). US *Navy War College*. Retrieved from Operational Protection from Unmanned Aerial Systems: http://www.dtic.mil/dtic/tr/fulltext/u2/a621067.pdf

Brenner, J. (2011). *America the Vulnerable: Inside the New Threat Matrix of Digital Espionage, Crime, and Warfare*. New York: Pilgrim Press.

Burch, D. (2015). *RADAR for Mariners*. New York: McGraw-Hill.

C4ISystems. (2013). *basics-of-information-operations*. Retrieved from Blogspot: http://c4isys.blogspot.com/2013/11/basics-of-information-operations-24.html

Chairman, U. (2012, March 23). Countering Air and Missile Threats, final coordination, JP 3-01. CJCS.

Clothier, R. (2017, April 02). *The Smart Skies Project: Enabling Technologies for UAS Operations in Non-segregated Airspace*. Retrieved from QUT ePrints: http://eprints.qut.edu.au/40465/3/40465.pdf

Clothier, R. F. (2010). *The Smart Skies Project: Enabling technologies for future airspace*. . Clothier,

R.A., Frousheger, D., Wilson, M., (2010). The Smart Skies Project: Enabling technologies for future airspace. Australian Research Center for Aerospace Automation, Commonwealth Scientific and Industrial Research Organization, Boeing Research an. Australian Research Center for Aerospace Automation, Commonwealth Scientific and Industrial Research Organization.

Clothier, R. R. (2011). The Smart Skies project. *IEEE Aerospace and Electronic Systems Magazine*.

DoD. (2018). *Dictionary of Military Terms*. Retrieved from JCS.Mil: http://www.jcs.mil/doctrine/dod_dictionary/

DoD-01. (2018). *JP 1-02*. Retrieved from Department of Defense Dictionary of Military and Associated Terms: www.dtic.mil/doctrine/new_pubs/jp1_02.pdf

DoD-02. (2018). *Information Operations (IO) in the United States*. Retrieved from JP 3-13 : http://www.dtic.mil/doctrine/new_pubs/jp3_13.pdf

DoD-03. (2015). *Unmanned Systems Roadmap 2013 to 2038*. Retrieved from DTIC: http://www.dtic.mil/dtic/tr/fulltext/u2/a592015.pdf

Editor. (2012, April 22). *RT Question More*. Retrieved from Iran starts cloning of American spy drone: https://www.rt.com/news/iran-spy-drone-copy-667/

FAA. (2018, February 1). *Part 107 Rule for sUAS*. Retrieved from Fly under the Special Rule for Model Aircraft: https://www.faa.gov/uas/getting_started/model_aircraft/

Filbert, F. &. (2014, (July – August). *Joint Counter Low, Slow, Small Unmanned Aircraft Systems Test. Fires PB644-14, no* 4. Washington: DoD.

Fitts, R. (1980). *The Strategy of Electromagnetic Conflict*. Los Altos, CA: Peninsula Publishing.

Garcia, M. (2006). *Vulnerability Assessment of Physical Protection Systems*. Albuquerque: Sandia National Laboratories,BH.

Greenburg, H. (2015). *Hackers Remotely Kill a Jeep on the Highway—With Me in It*. Retrieved from Wired : https://www.wired.com/2015/07/hackers-remotely-kill-jeep-highway/

Hartman, K. a. (2013). The Vulnerability of UAVs to Cyber Attacks – An Approach to the Risk Assessment. 2013 5th *International Conference on Cyber Conflict* . Tallin: NATO CCD COE Publications.

Horowitz, M. C. (2014). *Droning On: Explaining the Proliferation of Unmanned Aerial Vehicles*. University of Pennsylvania and Texas A&M Universities. University of Pennsylvania and Texas A&M Universities.

Hubbard, R. K. (1998). *Boater's Bowditch.* Camden, MA: International Marine.

Kania, E. (2017, July 6). Swarms at War: Chinese Advances in Swarm Intelligence. China Brief Volume: 17 Issue 9. *China Brief Volume*: 17 *Issue* 9.

Kilman, D. &. (2003). *Framework for SCADA Security Policy.* Albuquerque, NM: Sandia National Laboratories. Retrieved from Energy.gov: https://www.energy.gov/sites/prod/files/Framework%20for%20SCADA%20Security%20Policy.pdf

Kim, A. G. (2012, June). *Cyber Attack Vulnerabilities Analysis for Unmanned Aerial Vehicles.* Retrieved from Infotech@Aerospace.com: https://www.researchgate.net/publication/268571174_Cyber_Attack_Vulnerabilities_Analysis_for_Unmanned_Aerial_Vehicles

Liteye. (2018, August 25). AUDS. Retrieved from Liteye Corporation: http://liteye.com/products/counter-uas/auds/

Marshall, D. M. (2016). *Introduction to Unmanned Aircraft Systems, 2nd Edition.* New York: CRC Press.

Merrick, K. (2016). Future Internet. 10.3390/*fi*8030034 *Review*, 8(3), p. 34.

Moir, I. a. (2006). *Military Avionics Systems.* New York: Wiley Aerospace Series.

Monahan, K. (2004). *The Radar Book: Effective Navigation and Collision Avoidance.* Anacortes, WA: Fineedge Publications.

MORS. (2018). *Military Operations Research Society* . Retrieved from http://www.mors.org/meetings/oa_definition.htm

Myer, G. (2013, May-June). *Danger Close Definition.* Retrieved from US Army Magazine: www.benning.army.mil/infantry/magazine/issues/2013/May-June/Myer.html

NASA. (2018). *Unmanned Aircraft Systems* (UAS) *Integration in the National Airspace System* (NAS) *Project.* Retrieved from NASA: https://www.nasa.gov/feature/autonomous-systems

Nichols, R. K. (2008, September 05). Counterintelligence & Sensitive Compartmented Information Facility . (SCIF) *Needs – Talking Points.*

Nichols, R. (Nov 28-30, 2006). Cyber Terrorism, Critical Infrastructure, & SCADA Presentation. In R. Nichols (Ed.), *Defense Threat Reduction Agency Conference.* Shirlington VA: Utica College, Utica NY.

Nichols, R.-0. (2016, March 29). NCIE UAS SAA Final Rev 4. 2016 INFOWARCON *conference presentation April* 4-7, Nichols, R.K. et. al. (3-29-2016) Presentation to INFOWARCON April 4-7 on

NCIE UAS SAA Final Rev 4, presented to 2016 INFOWARCON conference, Memphis TN. Available as PPTx presentation download from author or in CANVAS. Memphis, TN, USA: INFOWARCON16.

NTSB. (2009, September 16). *National SCADA testbed Documents and Media.* Retrieved from National SCADA Testbed Fact Sheet: https://www.energy.gov/sites/prod/files/oeprod/DocumentsandMedia/NSTB_Fact_Sheet_FINAL_09-16-09.pdf

Pettit, R. (1982). *ECM and ECCM Techniques for Digital Communication Systems.* Belmont, CA: Lifetime Learning Publications .

Rani, C. M. (2015). Security of unmanned aerial vehicle systems against cyber-physical attacks. *The Journal of Defense Modeling and Simulation: Applications, Methodology, Technology.*

Shapiro, J. (2006, February 14). *Slideplayer.com.* Retrieved from Cybersecurity: http://slideplayer.com/slide/4545982/

Singer, P. W. (2010, February 25). Will Foreign Drones One Day attack the US? . *Newsweek.*

Sood A.K. & Enbody, R. (2014, December 19). *https://www.georgetownjournalofinternau-s-military-defense-systems-the-anatomy-of-cyber-espionage-by-chinese-hackers.* Retrieved from georgetownjournalofinternationalaffairs.org/online-edition: https://www.georgetownjournalofinternationalaffairs.org/online-edition/u-s-military-defense-systems-the-anatomy-of-cyber-espionage-by-chinese-hackers

Toomay, J. (1982). *RADAR for the Non - Specialist. London; Lifetime Learning Publications.* London: Lifetime Learning Publications.

TRS, S. (2018, July 10). *Tontechnic-Rechner-Sengpielaudio.* Retrieved from Tontechnic-Rechner-Sengpielaudio Calculator: www.sengspielaudio.com/calculator-wavelength.htm

Wiley, R. G. (1993). *Electronic Intelligence: The Analysis of Radar Signals, 2nd ed.* Norwood, MA: Artech House.

Wilson, M. (2012). The Use of Low-Cost Mobile Radar Systems for Small UAS Sense and Avoid. *Sense and Avoid in UAS Research and Applications.*

Yan. (2017, December 23). *China's commercial drone market to top 9 bln USD by 2020.* Retrieved from Xinhuanet: http://www.xinhuanet.com/english/2017-12/23/c_136847826.htm

Yu, X. &. (2015). Sense and avoid technologies with applications to unmanned aircraft systems: Review and prospects. *Progress in Aerospace Sciences*, 74, 152-166.

Zaharia, M. D. (2012). *Discretized Streams: An Efficient and Fault-Tolerant Model for.* Retrieved from UNIX Org: https://www.usenix.org/system/files/conference/hotcloud12/hotcloud12-final28.pdf

Zwijnwenburg, W. (2014, October 8). *ZwijnwenbDrone-tocracy? Mapping the Proliferation of Unmanned Systems.* Retrieved from Sustainable Security.org.

Readings

Abbot, C, Clarke, M, Hathorn, S, Hickie, S. (2016) *Hostile Drones: The Hostile -Use of Drones by Non-State Actors against British Targets.*

Aeronautical Decision Making (2015) (PDF) Professor Handout.

Ang, C. and Costello, M. (n.d.). When Firmware Modifications Attack: A Case Study of Embedded Exploitation. Columbia University, NY

Anonymous (n.d.). 21 Steps to Improve Cyber Security of SCADA Networks, Department of Energy.

Anonymous. (2012). Advanced Threats in the Enterprise: Finding an Evil in the Haystack with RSA ECAT (Rep.). RSA.

Anonymous, (2011). Keeping Track of Unmanned Aircraft by Overcoming "Lost Links", MITRE.

Anonymous (2004). National Communications System Technical Information Bulletin 04-1. Supervisory Control and Data Acquisition (SCADA) Systems. Virginia, Chantilly.

Anonymous (n.d.). Unmanned Aircraft Systems: Perceptions and Potential, Arlington, VA, Aerospace Industries Association.

Anonymous, (2009). Unmanned Air Systems: The Future is Now, New York, Access Intelligence LLC.

Barnhart, R.K., Hottman, S.B, Marshall D.M., and Shappee, E. (2012) *Introduction to Unmanned Aircraft Systems*. New York: CRC Press.

Bartelds, T., Capra, A., Hamaza, S., Stramigioli, S., Fumagalli, M., (2015). Compliant Aerial Manipulators: Towards a New Generation of Aerial Robotic Workers, IEEE.

Beaudoin, L. et.al (2011) *Potential Threats of UAS Swarms and the Countermeasures Need.* European Conference on Information Warfare and Security (ECIW), Tallin, Estonia.

Birnbaum, Zackary (2012). Behavior based analytics for securing cyber-physical systems. Binghampton University.

Brenner, J. (2011) *America the Vulnerable: Inside the New Threat Matrix of Digital Espionage, Crime, and Warfare*. NY: Pilgrim Press.

Bristeau, P., Callou, F., Vissiere, D., and Petit, N. (2011). The Navigation and Control Technology inside the AR. Drone Micro UAV. Proceedings of the 18th IFAC World Congress, 1477-1484. Retrieved February 11, 2016.

Boutros, D.A, LCDR. (2015) *Operational Protection from Unmanned Aerial Systems: Drones, UAS Proliferation, ISR Platforms, UAS Capabilities and Vulnerabilities, Current Protective Measures, Protection Shortfalls*. UAS Navy War College, Newport, RI: Progressive Management Publications.

Brodsky, D. W. (2012). An Analysis of the Vulnerabilities of Unmanned Aircraft Systems to Cyber Attack. Retrieved February 13, 2016.

Bunker, R. J. (2015). Progressive Management Publications (United States, Department of Defense, U.S. Army).

Christensen, R. (1997). Effect of technology integration education on the attitudes of teachers and their students. Doctoral dissertation, Univ. of North Texas. Based on Russell, A. L. (1995) Stages in learning modern technology. Computers in Education, 25(4), 173-178.

Class Lecture [Personal interview]. Randall Nichols (2016, February 23).

Class Lecture [Personal interview]. Randall Mai (2016, February 23).

Claveau, D., (2015). Autonomous UAS Controlled by Onboard Smartphone, International Conference on Unmanned Aircraft Systems.

Compton, M.D., (2009). Improving the Quality of Service and Security of Military Networks with a Network Tasking Order Process, Wright-Patterson Air Force Base, OH, Air Force Institute of Technology.

Delves, P. and Angelov, P. (2012). Sense and Avoid in UAS: Research and Applications (2nd Edition), John Wiley and Sons.

Dekker, M. (2013). Transforming Information Security: Future-Proofing Processes (pp. 1-13, Rep.). RSA.

Derynck, R. (2004). SCADA system security threats, vulnerabilities and solutions. IEE Seminar on Developments in Control in the Water Industry.

Elkaim, G. H., Pradipta Lie, F. A., and Gebre-Egziabher, D. (2012). Principles of Guidance, Navigation and Control of UAVs [Scholarly project]. Retrieved March 11, 2016, from https://users.soe.ucsc.edu/~elkaim/Documents/UAV_GNC_chapter.pdf

Federal Aviation Administration, (2011). Unmanned Aircraft Operations in the National Airspace System, U.S. Department of Transportation, Unmanned Aircraft Systems Group.

Finke, C., Butts, J., Mills, R., and Grimalia, M. (2013). Evaluation of a Cryptographic Security Scheme for Air Traffic Control's Next Generation Upgrade. Retrieved March 9, 2016.

Firmware [Personal interview]. Danny Mersch (2016, January 26).

Fonash, P., Schneck, P., (n.d.). Cybersecurity: From Months to Milliseconds, US Department of Homeland Security, Outlook.

Garcia, M.L. (2006) *Vulnerability Assessment of Physical Protection Systems*. Sandia National Laboratories. Albuquerque, NM: BH.

Gardinier, M. (2013). The Critical Incident Response Maturity Journey (Rep.). RSA.

Giancarmine, F. and Domenico, A. (2015) Radar Electo-Optical Data Fusion for non-cooperative UAS Sense and Avoid Retrieved February 6, 2016

Gowda, C. R. (2015). System Security, Threat Detection and Prevention Measures of Autonomous Systems. Retrieved February 19, 2016.

Hartman, K. and Steup, C (2013). *The Vulnerability of UAVs to Cyber Attacks – An Approach to the Risk Assessment.* 2013 5th International Conference on Cyber Conflict NATO CCD COE Publications, Tallinn. Retrieved February 5, 2016.

Heitner, K. (2014). Cyber Threats within Civil Aviation, Utica College. Retrieved February 17, 2016

House, Commonwealth of Virginia, (2013). Unmanned Aircraft Systems Protocols for use by Law Enforcement Agencies, Virginia Department of Criminal Justice Services.

Idaho National Laboratory, (2006). Control Systems Cyber Security: Defense in Depth Strategies, Department of Homeland Security.

Horowitz, M.C. & Fuhrmann, M. (2014) *Droning On: Explaining the Proliferation of Unmanned Aerial Vehicles*. University of Pennsylvania and Texas A&M Universities.

IKANOW. (2015). Managed Service Providers: The Talent and Tech to Beat Back Cyber Attacks (Rep.).

IKANOW. (2015). Threat Intelligence Ranking (Rep.).

Incze, M.L., Sideleau, S.R., Gagner, C., Pippin, C.A., (n.d.). Communication and Collaboration Among Heterogeneous Unmanned Systems Using SAE JAUS Standard Formats and Protocols, Newport, RI, Naval Undersea Warfare Center Division.

Jackson, A., (n.d.). Operational Reconnaissance and Intelligence Gathering is Critical to Battle Success, Audio Visual Innovations.

Jackson, B. A., Lostumbo, M. J., Frelingeer, D. R., and Button, R. W. (2008). Evaluating novel threats to the homeland: Unmanned aerial vehicles and cruise missiles. Santa Monica, CA: RAND National Defense Research Institute.

Jacobs, T., Coffey, J., (2015). Arctic Shield 2015 Unmanned Aircraft Systems Test Plan and Operational Assessment, NOAA UAS Program Office.

Kania, E. (6 July 2017) *Swarms at War: Chinese Advances in Swarm Intelligence*. China Brief Volume: 17 Issue 9. The Jamestown Foundation.

Kaspersky Lab (2015). Equation group: Questions and Answers. Retrieved February 24, 2016

Kim, A., Wampler, B., Goppert, J., Hwang, I., and Aldridge, H. (2012). Cyber Attack Vulnerabilities Analysis for Unmanned Aerial Vehicles. Infotech@Aerospace 2012. Retrieved February 13, 2016.

Kilman, D. & Stamp, J. (2003) Framework for SCADA Security Policy. Albuquerque, NM: Sandia National Laboratories. https://www.energy.gov/sites/prod/files/Framework for SCADA Security Policy.pdf

Kindervater, K.H., (2015). Lethal Surveillance: Drones and the Geo-History of Modern War, University of Minnesota.

Knapp, E. D., and Langill, J. T. (2015). About Industrial Networks. Industrial Network Security, 9-40.

Lacher, A., Zeitlin, A., Maroney, D., Markin, K., Ludwig, D., Boyd, J., (2010). Airspace Integration Alternatives for Unmanned Aircraft, the MITRE Corporation.

Ledbetter III, T., Munoz, C., (2010). Outgoing USAF Intel Chief Sees Need for More Autonomy in UAS Platforms, Arlington, Inside Washington Publishers.

Lemay, A., (2013). Defending the SCADA Network Controlling the Electrical Grid from Advanced Persistent Threats, University of Montreal.

Li, L., Heymsfield, G., Schaubert, D.H., McLinden, M.L., Creticos, J., Perrine, M., Coon, M., Cervantes, J.I., Vega, M., Guimond, S., Tian, L., Emory, A., (2016). The NASA High-Altitude Imaging Wind and Rain Airborne Profiler, IEEE.

Loukas, G., Gan, D., Vuong, T., (2013). A Review of Cyber Threats and Defense Approaches in Emergency Management, London, UK, University of Greenwich.

Loukas, G. (2015). Cyber-physical attacks: A growing invisible threat (First Ed.). Waltham, MA: Butterworth-Heinemann.

Lugabihl, J., Owen, D., Rand, T. A., and Tran, P. M. (2013). Taking Charge of Security in a Hyper-connected World (Rep.). RSA.

Lyu, Y., Pan, Q., Zhao, C., Zhu, H., Tang, T., Zhang, Y., (2015). A Vision Based Sense and Avoid System for Small Unmanned Helicopter, Key Laboratory of Information Fusion Technology, Ministry of Education, Northwestern Polytechnical University.

Mackie, J., Spencer, J., and Warnick, K. F. (2014). Compact FMCW radar for a UAS Sense and Avoid system. 2014 IEEE Antennas and Propagation Society International Symposium (APSURSI).

Maddox, S., Stuckenberg, D., (2015). Drones in the U.S. National Airspace System: A Safety and Security Assessment, Harvard Law School.

Matolak, D.W., Sun, R., (2015). Air-Ground Channel Characterization for Unmanned Aircraft Systems: The Near-Urban Environment, IEEE.

McCallie, Donald. (2011) Exploring Potential ADS-B Vulnerabilities in the FAA's Nextgen Air Transportation System. Air Force Institute of Technology. http://www.hsdl.org/?abstractand-did=697737

McCallie, D., Butts, J., and Mills, R. (2011). Security analysis of the ADS-B implementation in the next generation air transportation system [Scholarly project]. Retrieved March 9, 2016.

Meixall, B., Forner, B., (2013). Out of Control: Demonstrating SCADA Exploitation, Cimation.

Mukherjee, A., (2015). First Ever Anti-Drone Weapon that Shoots Down UAVs with Radio Waves, Death Ray.

Mumm, H. (2015). Applying Complexity Leadership Theory to Drone Airspace Integration. Melbourne, Florida: Motivational Press

Mumm, H. (2015, December 4). Managing the Integration and Harmonization of the National Airspace for Unmanned and Manned Systems. Lecture presented at DuPont Summit in the Historic Whittemore House, Washington D.C.

Nestler, V. (2015, October 15). The Age of Drones and Cybersecurity. Lecture presented at Centers of Academic Excellence TECH TALK in Capitol Technology University, San Bernardino, CA.

Nichols, R.K, and (Nov 28-30, 2006) Cyber Terrorism, Critical Infrastructure, and SCADA Presentation: Utica College, Utica NY. Defense Threat Reduction Agency Conference, Shirlington VA

Nichols, R.K. (2004, January 1). "Trust Me, Its Encrypted". Lecture presented at Rutgers University, New Brunswick, NJ.

Nichols, R.K., and Lekkas, P.C., (2002) Wireless Security: Models, Threats and Solutions, New York: McGraw Hill.

Nichols, R.K., Ryan, D.J., and Ryan, J.C.H., (2001) Defending Your Digital Assets against Hackers, Crackers, Spies and Thieves, McGraw-Hill.

Nichols, R.K. (2009, April 16). Terrorism – The Mutating Threat CYBERSECURITY – A Future in Crisis? Lecture presented at NYSETA Plenary in SUNYIT, Utica, NY.

Norris, A. (2013). Legal Issues Relating to Unmanned Maritime Systems, Monograph. Retrieved November 23, 2015.

O'Connor, T. (2011). Military Intelligence, Mega Links in Criminal Justice.

Oded, G. (2014). Understanding the threat to SCADA networks.

Oltsik, J. (2013). The Big Data Security Analytics Era is Here (Rep.). RSA.

Oltsik, J., McKnight, J., Gahm, J. (2010). Enterprise Strategy Group, "Assessing Cyber Supply Chain Security Vulnerabilities within the U.S. Critical Infrastructure."

Ott, J.T., (2014). Well Clear: General Aviation and Commercial Pilots' Perception of Unmanned Aerial Vehicles in the National Airspace System, San Jose State University.

Owen, M. and Duffy, S. (2014). Unmanned Aircraft Sense and Avoid Radar: Surrogate Flight Testing Performance Evaluation.

Paganini, P., (2013). Hacking Drones – Overview of the Main Threats, Infosec Institute.

Paganini, P. (2015). SCADA and Security of Critical Infrastructures. In InfoSec Institute. Retrieved February 11, 2016, from http://resources.infosecinstitute.com/scada-security-of-critical-infrastructures/

Pauner, C., Kamara, I., Viguri, J., (2015). Drones. Current Challenges and Standardization Solutions in the Field of Privacy and Data Protection, University Jaume I, Spain, Vrije Universiteit Brussel, Belgium.

Parker, D.B, (1991) Proceedings of the 14th National Computer Security Conference.

Petersen, J. (2001). Understanding surveillance technologies: Spy devices, their origins and applications. Boca Raton, FL: CRC Press.

Pitchford, M., (2013). What's needed to ensure safety and security in UAV software, Military Embedded Systems?

Radmanesh, M., and Kumar, M. (2016). Flight formation of UAVs in presence of moving obstacles using fast-dynamic mixed integer linear programming. Aerospace Science and Technology, 50, 149-160. Retrieved February 11, 2016.

Raffetto, M., (2004). Unmanned Aerial Vehicle Contributions to Intelligence, Surveillance, and Reconnaissance Missions for Expeditionary Operations, Monterey, CA, Naval Postgraduate School.

Ramasamy, S., Sabatini, R., Gardi, A., (2014). Avionics Sensor Fusion for Small Size Unmanned Aircraft Sense-and-Avoid, Melbourne, Australia, RMIT University – SAMME.

R.G., D.M., (n.d.). SCADA Security and Terrorism: We're not crying wolf, Internet Security Systems.

Sabatini, R. (2015). Airborne Lasers and Integrated Weapon Systems: Design, Development, Test and Evaluation, RMIT University.

Sabatini, R., Gardi, A., Richardson, M., (2014). A Laser Obstacle Warning and Avoidance System for Manned and Unmanned Aircraft, Melbourne, Australia, RMIT University – SAMME, Shrivenham, Swindon, UK, Cranfield University – DAUK.

Shaneck, M. (2003). An Overview of Buffer Overflow Vulnerabilities and Internet Worms [Scholarly project]. In University of Minnesota. Retrieved February 20, 2016, from http://www-users.cs.umn.edu/~shaneck/MarkShaneck_BufferOverflows.pdf

Shull, A.M., (2013). Analysis of Cyberattacks on Unmanned Aerial Systems, West Lafayette, IN, Purdue University.

Stockwell, T. M. (2012). Defending Against Data Breach: Developing the Right Encryption Strategy (Rep.). Linoma Software.

The National Intelligence Strategy, (2009).

Thiobane, F. (2015). Cyber Security and Drones, Utica College.

Tsang, R. (n.d.). Cyberthreats, Vulnerabilities and Attacks on SCADA Networks [Scholarly project]. Retrieved February 19, 2016.

Wapner, J., (2016). Medical Transport Drones Could Transform Health Care in Overcrowded Cities, India, Newsweek, Global Ed.

Wilhoit, K., (2013). The SCADA that Didn't Cry Wolf, Trend Micro.

Yagdereli, E., Gemci, C., and Aktas, A. Z. (2015). A study on cyber-security of autonomous and unmanned vehicles. The Journal of Defense Modeling and Simulation: Applications, Methodology, Technology, 12(4), 369-381

Yu, X., and Zhang, Y. (2015). Sense and avoid technologies with applications to unmanned aircraft systems: Review and prospects. Progress in Aerospace Sciences, 74, 152-166. Retrieved February 13, 2016.

Zeitlin, A., Lacher, A., Kuchar, J., and Drumm, A. (2006). Collision Avoidance for Unmanned Aircraft: Proving the Safety Case [Scholarly project]. Retrieved February 11, 2016.

Zhu, B. and Anthony, Joseph. (n.d.). A Taxonomy of Cyber Attacks on SCADA System. University of California, Berkley.

Secondary References

45 years ago: First message sent over the Internet. (2014, October 29). Retrieved December 3, 2015, from http://www.cbsnews.com/news/first-message-sent-over-the-internet-45-years-ago/

1950s and 1960s. (n.d.). Retrieved December 5, 2015, from https://sites.google.com/site/uavuni/1950s-1960s

Accenture Technology. (2015). Security for the Internet of Things: A Call to Action. Retrieved February 14, 2016, from http://www.slideshare.net/AccentureTechnology/security-for-the-internet-of-things-a-call-to-action

Acoustic Sensors for Unmanned Air Vehicles. (2016). Retrieved February 11, 2016, from http://www.sara.com/ISR/acoustic_sensing/LOSAS.html

ACPI, firmware and your security. (2014) Retrieved January 31, 2016, from http://www.markshuttleworth.com/archives/1332

Adamcaudill/Psychson. (n.d.). Retrieved January 31, 2016, from https://github.com/adamcaudill/Psychson

ADS, Inc. (2013). Tactical Communications: C4ISR Expanding the Grid. Retrieved from https://adsinc.com/c4isr-expanding-the-grid/

Advanced Drone Detection Technology. (2015). Retrieved December 3, 2015, from http://www.dronedetector.com/

Aerialtronics. (2016). Drones for a world of applications – Unmanned Aircraft Systems – Aerialtronics. Retrieved March 12, 2016, from http://www.aerialtronics.com/

Aircrack-ng. (2016). Documentation. Retrieved March 12, 2016, from http://www.aircrack-ng.org/

Aircraft Hull Insurance. (2012). Retrieved November 29, 2015, from http://financial-dictionary.thefreedictionary.com/

Air Informatics® LLC. (2015). Retrieved October 25, 2015 from http://www.airinformatics.com/e-Enabled_Definition.html/

Amazon's Drone Highway Concept. (2015, September 23). Retrieved December 3, 2015, from https://www.workinghomeguide.com/23979/amazons-drone-highway-concept

Anderson, C. (2016). DIY Drones. Retrieved March 12, 2016, from http://diydrones.com/

Anonymous. (n.d.). BlueScan – Scanner de dispositions Bluetooth. Retrieved March 12, 2016, from http://bluescanner.sourceforge.net/

Anonymous. (2012). Global Hawk Aircraft. Retrieved March 11, 2016, from http://www.nasa.gov/multimedia/imagegallery/image_feature_2362.html

Anonymous. (2015). IAI Unveils the Drone Guard: Drone Detection, Identification and Disruption Systems. Retrieved March 11, 2016, from http://www.israeldefense.co.il/en/content/iai-unveils-drone-guard-drone-detection-identification-and-disruption-systems

Anonymous. (2011). Keeping Track of Unmanned Aircraft by Overcoming "Lost Links" Retrieved March 11, 2016, from http://www.mitre.org/publications/project-stories/keeping-track-of-unmanned-aircraft-by-overcoming-lost-links

Anonymous. (2015). Sky for All: Air Mobility for 2035 and Beyond. Retrieved March 11, 2016, from https://herox.com/SkyForAll

Anonymous. (2015). SQL Injection Tutorial – w3resource. Retrieved March 11, 2016, from http://www.w3resource.com/sql/sql-injection/sql-injection.php

Arthur, C. (Ed.). (2009, December 17). SkyGrabber: The $26 software used by insurgents to hack into US drones. Retrieved December 5, 2015, from http://www.theguardian.com/technology/2009/dec/17/skygrabber-software-drones-hacked

B, W. (2009, June 8). Air Traffic Control System Vulnerable to Cyber Terrorism. Retrieved December 2, 2015, from https://www.globaldatavault.com/blog/air-traffic-control-system-vulnerable-to-cyber-terrorism/

Baugh, C. (2011). GPS Hacking Guide: Learn How to Hack Your GPS Device. Retrieved March 12, 2016, from http://www.brighthub.com/electronics/gps/articles/125230.aspx#imgn_0

BCN Drone Center. (2016). BCN Drone Center | UAV Test Site – UAV Remote Sensing Applications 6. Retrieved March 12, 2016, from http://www.barcelonadronecenter.com/index.php/uav-remote-sensing-applications-6

Beaudoin, L, Grademer, A, Avanthey, L, Vittori, V, Germain, V. (January 2011) Potential Threats of UAS Swarms and the Countermeasure's Need. Conference Paper retrieved 21 March 2016. https://www.researchgate.net/publication/261633504

Beaufort scale. (2011). Science/SOSE. Retrieved March 09, 2016, from http://cpsroom10.edublogs.org/2011/03/03/sun/#more-73

Bonggay, C. (2015, March 4). Commercial Drone Rules Around the World PrecisionHawk. Retrieved November 17, 2015, from http://media.precisionhawk.com/topic/commercial-drones-faa/

Breakthrough Technologies for National Security. (March 2015) Retrieved January 30, 2016, from http://www.darpa.mil/attachments/DARPA2015.pdf

C4ISR Definition. (n.d.). Retrieved March 09, 2016, from http://www.acronymfinder.com/

Calvo, K. (2015, October 29). So, You Want to Keep Track of All Your Drone Flights? Retrieved November 3, 2015, from http://voices.nationalgeographic.com/2015/10/29/so-you-want-to-keep-track-of-all-your-drone-flights/

Carpenter, R., and Anderson, A. (2006). The death of Schrödinger's cat and of consciousness-based quantum wave-function collapse. *Annales De La Fondation Louis De Brogli*, 31(1), 45-52. Retrieved December 2, 2015, from http://web.archive.org/web/20061130173850/http:/www.ensmp.fr/aflb/AFLB-311/aflb311m387.pdf

CARAC Activity Details. (2015). Retrieved November 9, 2015, from http://wwwapps.tc.gc.ca/Saf-Sec-Sur/2/NPA-APM/actr.aspx?id=17andaType=1andlang=eng

Castillo, A. (2015, November 10). A DMV for Drones? Inside the FAA's Clumsy Push to Regulate Flying Computers. Retrieved November 22, 2015, from https://reason.com/archives/2015/11/10/faa-versus-drones

Chesson, J. (n.d.). Cyber Crime. Retrieved December 3, 2015, from http://www.powershow.com/view/20b54-OGU4Z/Cyber_Crime_powerpoint_ppt_presentation

Chicago Protects Critical Infrastructure and Services with Security Connected. (2014). Retrieved December 2, 2015, from http://www.mcafee.com/us/case-studies/cs-city-of-chicago.aspx

Chinese drone manufacturer DJI building outpost and hiring in Palo Alto – Silicon Valley Business Journal. (2015, October 30). Retrieved October 31, 2015, from http://www.bizjournals.com/sanjose/blog/techflash/2015/10/chinese-drone-manufacturer-dji-is-building-a.html

Cole, C. (2012). Mapping drone proliferation: Big business vs. the MTCR. Retrieved March 11, 2016, from http://dronewars.net/2012/09/18/mapping-drone-proliferation-big-business-vs-the-mtcr/

COTS Journal. (2016). Retrieved March 11, 2016, from http://www.cotsjournalonline.com/

Coventor. (2016). MEMS Accelerometer Design and Simulation. Retrieved March 12, 2016, from http://www.coventor.com/mems-solutions/accelerometers/

Cravey, P. A. (2016). TRADOC Capability Manager for Unmanned Aircraft Systems. Retrieved March 12, 2016, from http://www.rucker.army.mil/usaace/uas/

Crowe, S. (2015). Boeing's Mid-Air Charging Lets Drones Fly Forever – Robotics Trends. Retrieved March 12, 2016, from http://www.roboticstrends.com/article/boeings_mid_air_charging_lets_drones_fly_forever/

Cyber Threats. (2015). Retrieved December 2, 2015, from http://thefc2.org/news/cyberthreatvectors.aspx

DARPA uses open systems to boost airpower. (2015). Retrieved January 30, 2016, from http://www.af.mil/News/ArticleDisplay/tabid/223/Article/582052/darpa-uses-open-systems-to-boost-airpower.aspx

Definition of Black Swan Theory. (2015). Retrieved December 2, 2015, from http://www.davemanuel.com/investor-dictionary/black-swan-theory/

DJI: The World Leader in Camera Drones/Quadcopters for Aerial Photography. (2015). Retrieved December 1, 2015, from http://www.dji.com/

Drones: Detect, identify, intercept and hijack. (n.d.) Retrieved January 31, 2016, from https://www.nccgroup.trust/uk/about-us/newsroom-and-events/blogs/2015/december/drones-detect-identify-intercept-and-hijack/

Drones vs. Radio-Controlled Aircraft: A Look at the Differences between the Two. (2015). Retrieved November 3, 2015, from https://rcflightline.com/drones-vs-radio-controlled-aircraft-a-look-at-the-differences-between-the-two/

DuPaul, N. (2014). Wireless Sniffer: Tools, Software to Detect Packet or Network Sniffers. Retrieved March 09, 2016, from http://www.veracode.com/security/wireless-sniffer

Dussalult, J. (2014, March 14). 7 commercial uses for drones. Retrieved December 1, 2015, from http://www.boston.com/business/2014/03/14/commercial-uses-for-drones/dscS47PsQdPneIB2UQeY0M/singlepage.html

EBumper4. (2016). Retrieved February 11, 2016, from http://www.panoptesuav.com/ebumper

Edwards, D. (2015). Flying-Swimmer (Flimmer) UAV/UUV. Retrieved December 1, 2015, from http://www.nrl.navy.mil/lasr/content/flying-swimmer-flimmer-uavuuv

Electro-Optical sensor payloads for small UAV's (2013) Retrieved February 3, 2016, from C:\Users\Carrie\Desktop\Payload research items\Electro-optical sensor payloads for small UAVs – Military and Aerospace Electronics.htm

EPSON. (2016). Measurement instrument | Application | Epson device. Retrieved March 12, 2016, from http://www5.epsondevice.com/en/applications/measurement_instruments/

Equation Group: The Crown Creator of Cyber-Espionage. (2015) Retrieved January 31, 2016, from http://www.kaspersky.com/about/news/virus/2015/Equation-Group-The-Crown-Creator-of-Cyber-Espionage

Essential Equipment: Reach of FLIR Surges Forward. (2014) Retrieved January 31, 2016, from http://www.aviationtoday.com/rw/public-service/police/Essential-Equipment-Reach-of-FLIR-Surges-Forward_82549.html#.VsNI5PkrLIU

FAA Part 107 Rule for sUAS: https://www.faa.gov/uas/getting_started/model_aircraft/

FLIR. (2016). Thermal Imaging Cameras for sUAS. Retrieved March 12, 2016, from http://www.flir.com/suas/content/?id=70733

Federal Aviation Administration. (2015, November 1). Retrieved November 30, 2015, from https://www.faa.gov/

Facts. (2015). Retrieved November 1, 2015, from http://knowbeforeyoufly.org/ – Facts/

Federal Register. (2015, February 3). Retrieved October 25, 2015 from http://www.gpo.gov/fdsys/pkg/FR-2015-02-03/html/2015-01918.html

Ferranti, M. (2015, October 19). US to require registration process for drones. Retrieved October 31, 2015, from http://www.cio.com/article/2994818/us-to-require-registration-process-for-drones.html

Foxx, A. (2015, October 19). Clarification of the Applicability of Aircraft Registration Requirements for Unmanned Aircraft Systems (UAS) and Request for Information Regarding Electronic Registration for UAS. Retrieved October 31, 2015, from https://www.federalregister.gov/articles/2015/10/22/2015-26874/clarification-of-the-applicability-of-aircraft-registration-requirements-for-unmanned-aircraft

Gabbert, B. (2015). Fire drones. Retrieved March 11, 2016, from http://fireaviation.com/tag/uav/

GAO Report: Unmanned Aircraft Systems. (2012). Retrieved March 09, 2016, from http://radionavlab.ae.utexas.edu/spotlight/279-gao-report-unmanned-aircraft-systems-september-2012

Geofencing: What is it and how does it Work? (2014). Retrieved December 3, 2015, from http://socialbrothers.net/2014/03/18/geofencing-what-is-it-and-how-does-it-work/

Gettinger, D. (2015). Domestic Drone Threats. Retrieved March 11, 2016, from http://dronecenter.bard.edu/what-you-need-to-know-about-domestic-drone-threats/

Glas, J. (2010). Frequency Hopping. Retrieved March 12, 2016, from http://www.wirelesscommunication.nl/reference/chaptr05/spreadsp/fh.htm

Glenn. (2012). SensePost | Snoopy: A distributed tracking and profiling framework. Retrieved March 11, 2016, from https://www.sensepost.com/blog/2012/snoopy-a-distributed-tracking-and-profiling-framework/

Goodrich, R. (2013). Accelerometer vs. Gyroscope: What's the Difference? Retrieved March 12, 2016, from http://www.livescience.com/40103-accelerometer-vs-gyroscope.html

Gorman, S., Dreazen, Y., and Cole, A. (2009, December 17). Insurgents Hack U.S. Drones. Retrieved December 2, 2015, from http://www.wsj.com/articles/SB126102247889095011

GPS Spoofing: Hijacking in the Digital Age. (2013) Retrieved January 30, 2016, from http://www.nolanwpeterson.com/gps-spoofing-hijacking-in-the-digital-age/\

Greenburg, A, Wired (2015). Hackers Remotely Kill a Jeep on the Highway—With Me in It. https://www.wired.com/2015/07/hackers-remotely-kill-jeep-highway/

Greenwood, J. (2015). The Phantom Menace – Weaponizing a consumer drone. Retrieved February 14, 2016, from https://www.4armed.com/blog/phantom-menace-weaponising-drones/

Guo, E. (2011). VT-SHA3 A. Retrieved March 09, 2016, from http://rijndael.ece.vt.edu/sha3

Hackers. (2015, March 10). Retrieved December 3, 2015,

from http://csrc.nist.gov/publications/nistir/threats/subsection3_4_2.html

Haines, B., (n.d.). Hackers and Airplanes, No Good Can Come from This. http://korben.info/wp-content/uploads/defcon/SpeakerPresentations/Renderman/DEFCON-20-RenderMan-Hackers-plus-Airplanes.pdf

Hard disk hacking -Parts on the PCB. (n.d.) Retrieved January 30, 2016, from https://spritesmods.com/?art=hddhackandpage=2

Help Net Security. (2016). Deep and Dark Web: Complexity and escalating cybercriminal activity – Help Net Security. Retrieved March 12, 2016, from https://www.helpnetsecurity.com/2016/02/22/deep-and-dark-web-complexity-and-escalating-cybercriminal-activity/

Hernandez, A. (2015, October 2). UAV. Retrieved November 17, 2015, from http://www.slideshare.net/AnthonyHernandezMPAB/uav-53447022

Holmlund, K. (2010). When the Eyjafjallajökull volcano in Iceland erupted in 2010, satellite imagery proved to be instrumental in helping track the movement of the large ash plume. Retrieved March 09, 2016, from http://www.eumetsat.int/website/home/Images/ImageLibrary/DAT_2187509.html

Howard, C. (2015). *Team of military pilots enter consumer drone market, launch* UAV *control software*. Retrieved November 16, 2015, from http://www.intelligent-aerospace.com/articles/2015/05/team-of-military-pilots-enter-consumer-drone-market-launch-uav-control-software.html

Howard, C. (2013). UAV Command, Control and Communications. Retrieved March 09, 2016, from http://www.militaryaerospace.com/articles/print/volume-24/issue-7/special-report/uav-command-control-communications.html

Huerta, M. (2013). *Integration of Civil Unmanned Aircraft Systems (UAS) in the National Airspace System (NAS) Roadmap*. Retrieved November 17, 2015, from http://www.faa.gov/uas/legislative_programs/uas_roadmap/media/UAS_Roadmap_2013.pdf

IEEE. (n.d.) Airborne Communication Networks for Small Unmanned Aircraft Systems. (n.d.). Retrieved March 09, 2016, from http://ieeexplore.ieee.org/ieee_pilot/articles/96jproc12/jproc-EFrew-2006127/article.html

I-HLS. (2015). Nanoradar system for UAVs collision avoidance set for 2016 – i-HLS Israel Homeland Security. Retrieved March 12, 2016, from http://i-hls.com/2015/10/nanoradar-system-for-uavs-collision-avoidance-set-for-2016

Information Security Office Shared Services. (2015). Retrieved December 2, 2015, from http://www.cityofchicago.org/city/en/depts/doit/provdrs/security_and_datamanagement/svcs/information-security-office-shared-services.html

Intelligence Debates. (2016). Retrieved March 11, 2016,

from http://intelligencesquaredus.org/

Insurance Coverage for Unmanned Aerial Vehicles – UAV. (2015). Retrieved November 9, 2015, from http://www.transportrisk.com/uavrcfilm.html

Iran starts cloning of American spy drone. (2012, April 22). Retrieved December 3, 2015, from https://www.rt.com/news/iran-spy-drone-copy-667/

James, L. (2014, February 8). How do Beacon and Geo-Fencing Actually Work? Retrieved December 3, 2015, from http://www.mobiledonky.com/blog/how-do-beacon-and-geo-fencing-actually-work

Jansen, B. (2014, June 5). Drones will be revolutionary, but hurdles remain. Retrieved December 1, 2015, from http://www.usatoday.com/story/money/business/2014/06/05/drones-national-research-council-faa-clarke-lauber/10002007/

Jarzombek, J. (2012). Supply chain Risk Management. Retrieved from http://csrc.nist.gov/scrm/documents/workshop_oct2012/jarzombek_ict_supply_chain_workshop_oct-15-2012.pdf

Jenkins, D., and Vasigh, B. (2013, March 1). The economic impact of unmanned aircraft systems integration in the United States. Retrieved November 9, 2015, from https://higherlogicdownload.s3.amazonaws.com/AUVSI/958c920a-7f9b-4ad2-9807-f9a4e95d1ef1/UploadedImages/New_Economic Report 2013 Full.pdf

Kamkar, S. (2013). SkyJack. Retrieved February 24, 2016, from http://samy.pl/skyjack/

Kargo, D. (2012). Dad's Toys. Retrieved December 5, 2015, from https://www.pinterest.com/pin/203365739393483039/

Karp, A. (2015, October 1). FAA warns of 'a million drones under people's Christmas trees'. Retrieved October 31, 2015, from http://www.suasnews.com/2015/09/38847/faa-warns-of-a-million-drones-under-peoples-christmas-trees/

Keller, J. (2014, July 24). U.S. spending on unmanned aerial vehicles (UAVs) to reach $15 billion by 2020, market researcher says. Retrieved December 1, 2015, from http://www.militaryaerospace.com/articles/2014/07/igi-uav-*forecast.html

King, S. T., Tucek, J., Cozzie, A., Grier, C., Jiang, W., and Zhou, Y. (2008). Designing and implementing malicious hardware [Scholarly project]. In Usenix. Retrieved February 24, 2016, from https://www.usenix.org/legacy/event/leet08/tech/full_papers/king/king_html/

Koebler, J. (2015, October 19). 8 Questions Raised by the FAA's Decision to Register Every Drone in the US. Retrieved November 22, 2015, from http://motherboard.vice.com/read/8-questions-raised-by-the-faas-decision-to-register-every-drone-in-the-us

Kumar, M. (2015). Hacking Team and Boeing Built Cyber Weaponized Drones to Spy on Targets. Retrieved February 14, 2016, from http://thehackernews.com/2015/07/boeing-drone-hacking.html

Kumar, M. (2015). How Drones Can Find and Hack Internet-of-Things Devices from the Sky. Retrieved February 14, 2016, from http://thehackernews.com/2015/08/hacking-internet-of-things-drone.html

LeMieux, J. (2012). Alternative UAV Navigation Systems. Retrieved March 12, 2016, from http://m.electronicdesign.com/embedded/alternative-uav-navigation-systems

Lerner, M. (2014, August 5). The Chilling Effect of Domestic Spying (T. DeWeese, Ed.). Retrieved December 2, 2015, from http://americanpolicy.org/2014/08/05/the-chilling-effect-of-domestic-spying/

Lewis, J., and Caplan, L. (2015, July 28). *Drones to satellites: should commercial aerial data collection regulations differ by altitude?*

Liles, S. (2008). From Information operations to cyber warfare and a new terrain. Retrieved March 09, 2016, from http://selil.com/archives/336

Lockheed Martin awarded contract to enhance US Navy's C4ISR collection and dissemination capabilities. (2014). Retrieved from http://uss-america.blogspot.com/2014/06/lockheed-martin-nyselmt-will-work-to.html

Louviere, G. (2006). Newton's 3 Laws of Motion. Retrieved March 12, 2016, from http://teachertech.rice.edu/Participants/louviere/Newton/

Mahoney, D. (2015, November 10). Aviation insurer offers ground rules for drones. Retrieved November 17, 2015, from http://www.businessinsurance.com/article/20151110/NEWS06/151119989/aviation-insurer-offers-ground-rules-for-drones-faa-federal-aviation?tags=|71|76|80|83|329|302

Malicious Code Information. (n.d.) Retrieved January 30, 2016, from http://www3.safenet-inc.com/csrt/malicious-code-more.aspx

Merriam-Webster Dictionary. http://www.merriam-webster.com/dictionary/cybersecurity/

Miasnikov, E. (2011). The Threat of the Use of Small UAVs by Terrorists: Technical Aspects. Retrieved March 12, 2016, from http://www.armscontrol.ru/UAV/em040711.htm

MIMO Radar. (2015). Retrieved December 3, 2015, from http://www.androcs.com/mimo_radar.html

Muncaster, P. (2016). CIOs: Hackers Hiding in Encrypted Traffic is Major Threat. Retrieved March 12, 2016, from http://www.infosecurity-magazine.com/news/cios-hackers-hiding-encrypted/

Mulrine, A. (2015, July 28). Robots in war: Ethical concern, or a help for social ills? Retrieved November 23, 2015, from http://www.csmonitor.com/USA/Military/2015/0728/Robots-in-war-Ethical-concern-or-a-help-for-social-ills

Murphy, M. (2015, November 5). The future of drones is apps. Retrieved November 9, 2015, from http://qz.com/540559/the-future-of-drones-is-apps/

National Institute of Advanced Industrial Science and Technology. (2012). Side-channel Attack Standard Evaluation Board (SASEBO) -SASEBO-R. Retrieved March 09, 2016, from http://satoh.cs.uec.ac.jp/SASEBO/en/board/sasebo-r.html

New 50″ Mini Telemaster RC Plane Kit Remote Control R/C Airplane 50in Balsa. (2015). Retrieved December 5, 2015, from http://www.ebay.com/itm/New-50-Mini-Telemaster-RC-Plane-Kit-Remote-Control-R-C-Airplane-50in-Balsa-/181948366992

NSA planted Stuxnet-Type Malware Deep within Hard Drive Firmware. (2016) Retrieved January 31, 2016, from http://thehackernews.com/2015/02/hard-drive-firmware-hacking.html

O'Donnell, A. (2016). 4 Secrets Hackers Don't Want You to Know. Retrieved March 12, 2016, from http://netsecurity.about.com/od/secureyourwifinetwork/a/4-Secrets-Wireless-Hackers-Do-Not-Want-You-To-Know.htm

Opinion: Jamming Is Needed Against Agile Radar Threat. (2014) Retrieved January 30, 2016, from http://aviationweek.com/defense/opinion-jamming-needed-against-agile-radar-threat

Paganini, P. (2013). Hacking Drones … Overview of the Main Threats – InfoSec Resources. Retrieved February 14, 2016, from http://resources.infosecinstitute.com/hacking-drones-overview-of-the-main-threats/

Paganini, P. (2014). Hack-Proof Drones Possible with HACMS Technology – InfoSec Resources. Retrieved February 14, 2016, from http://resources.infosecinstitute.com/hack-proof-drones-possible-hacms-technology/

Paganini, P. (2013). Hardware attacks, backdoors and electronic component qualification – InfoSec Resources. Retrieved February 24, 2016, from http://resources.infosecinstitute.com/hardware-attacks-backdoors-and-electronic-component-qualification/

Palermo, E. (2014, July 29). Drones Could Grow to $11 Billion Industry by 2024. Retrieved December 1, 2015, from http://www.livescience.com/47071-drone-industry-spending-report.html

Patel, D. (2014). How does Accelerometer and Gyro sensor work in digital gadgets? Retrieved March 11, 2016, from how does Accelerometer and Gyro sensor work in digital gadgets?

Peck, M. (2016). Army will test 3-D printed UAVs. Retrieved March 12, 2016, from http://www.c4isrnet.com/story/military-tech/uas/2016/02/24/army-3d-printed-uav-drone/80858028/

Peterson, S. (2011, December 9). Downed US drone: How Iran caught the 'beast' Retrieved December 3, 2015, from http://www.csmonitor.com/World/Middle-East/2011/1209/Downed-US-drone-How-Iran-caught-the-beast

Petrovsky, O. (2015). Attack on the drones: Security vulnerabilities of unmanned aerial vehicles. Retrieved February 23, 2016, from https://www.virusbulletin.com/conference/vb2015/abstracts/attack-drones-security-vulnerabilities-unmanned-aerial-vehicles

Perlman, A. (2015, November 22). DJI Introduces New Geofencing System for Its Drones. Retrieved December 3, 2015, from http://uavcoach.com/dji-introduces-new-geofencing-system/

Pomerleau, M. (2015, October 28). How to do air traffic control for drones. Retrieved November 3, 2015, from https://gcn.com/articles/2015/10/28/latas-drone-control.aspx

Posel, S. (2013). Anti-Drone Tech Corp To Begin Selling to Public While Working with DARPA – Top US and World News | Susanne Posel. Retrieved March 11, 2016, from https://occupycorporatism.com/anti-drone-tech-corp-to-begin-selling-to-public-while-working-with-darpa/

Powers, R. (2014). What Is a 15W – Unmanned Aerial Vehicle Operator? Retrieved March 11, 2016, from http://usmilitary.about.com/od/enlistedjobs/a/96u.htm

Purchase, D. (2015). Clearing the air: Study examines human-wildlife conflicts in crowded airspace. Retrieved March 11, 2016, from http://www.swansea.ac.uk/media-centre/news-archive/2015/clearingtheairstudyexamineshuman-wildlifeconflictsincrowdedairspace.php

Radic, D. (n.d.). 7.9. C4ISR, Basic Principles of C4 System. Retrieved March 09, 2016, from http://www.informatics.buzdo.com/p939-c4isr.htm#n1

Radio. (2002) Retrieved January 30, 2016, from http://www.encyclopedia.com/topic/radio.aspx

Ranjan, A. (2015). Ethacklesias. Retrieved February 25, 2016, from http://ethacklesias.blogspot.com/

Rakshasa: The hardware backdoor that China could embed in every computer. (2012) Retrieved January 30, 2016, from http://www.extremetech.com/computing/133773-rakshasa-the-hardware-backdoor-that-china-could-embed-in-every-computer

Reagan, J. (2014, December 15). 11 States Enacted New Drone Laws in 2014. Retrieved November 17, 2015, from http://dronelife.com/2014/12/15/11-states-enacted-new-drone-laws-2014/

Real Clear Technology. (2016). How to Hack a Smartphone with a Fake Fingerprint. Retrieved March 12, 2016, from http://www.realcleartechnology.com/video/2016/02/24/how_to_hack_a_smartphone_with_a_fake_fingerprint.html

Ribeiro, J. (2014, August 20). US senator to introduce proposal for mandatory drone geofencing. Retrieved November 17, 2015, from http://www.cio.com/article/2973586/us-senator-to-introduce-proposal-for-mandatory-drone-geofencing.html

Road to the Future Part 2. (n.d.) Retrieved January 30, 2016, from http://www.onfinalblog.com/2015/08/darpas-vision-of-joint-future.html

Rogawski, M., Gaj, K., and Homsirikamol, E. (2013). A high-speed unified hardware architecture for 128 and 256-bit security levels of AES and the SHA-3 candidate Grøstl ☆. Retrieved March 09, 2016, from http://www.sciencedirect.com/science/article/pii/S0141933113000847?np=y

RT. (2012). Drone hack explained: Professor details UAV hijacking. Retrieved March 12, 2016, from https://www.rt.com/usa/texas-professor-drone-hacking-249/

Satellite Missions. (n.d.). Retrieved March 09, 2016, from http://www.nesdis.noaa.gov/about_satellites.html

Schwarz, T. (2004). 1. How Input Can Be Bad. Retrieved March 12, 2016, from http://www.cse.scu.edu/~tschwarz/coen152_05/Lectures/BufferOverflow.html

Sifton, J. (2012, February 7). A Brief History of Drones. Retrieved December 1, 2015, from http://www.thenation.com/article/brief-history-drones/

Sense and Avoid for Unmanned Aerial Vehicles. (n.d.). Retrieved December 1, 2015, from http://www.frc.ri.cmu.edu/projects/senseavoid/technology.html

September 11th Flights. (2008, May 21). Retrieved December 2, 2015, from http://911research.wtc7.net/planes/sept11.html

Shapiro, J. (2015). Cybersecurity. Retrieved February 14, 2016, from http://slideplayer.com/slide/4545982/

Shukla, M., Chen, Z., and Lu, C. (2015). Distributed Drone Flight Path Builder System. Retrieved November 1, 2015, from http://europa.nvc.cs.vt.edu/~ctlu/Publication/2015/GIS-TAM-2015-Proceedings.pdf

SkyGrabber is offline satellite internet downloader. (2008). Retrieved December 2, 2015, from http://www.skygrabber.com/en/skygrabber.php

Skaves, P. (2015, April 1). Retrieved October 25, 2015 from http://www.cabaa.com/documents/FAA_Aircraft_System_Information_Security_Protection_Overview_4-1-2015_Jim_Skaves.pdf

Sky Safe. (2016). Sky Safe. Retrieved March 12, 2016, from http://www.skysafe.io/

Snow, C. (2014, February 6). The Yellow Brick Road of FAA Drone Regulations. Retrieved November 17, 2015, from http://droneanalyst.com/2014/02/06/the-yellow-brick-road-of-faa-regulations/

Sood, A. K., and Enbody, R. (2014). U.S. Military Defense Systems: The Anatomy of Cyber Espionage by Chinese Hackers. Retrieved February 13, 2016, from http://journal.georgetown.edu/u-s-military-defense-systems-the-anatomy-of-cyber-espionage-by-chinese-hackers/

Tesla, C. (2014, June 17). The past and the future of drones. Retrieved November 17, 2015, from http://www.tumotech.com/2014/06/17/the-past-and-the-future-of-drones/

The Drones Report: Market forecasts, regulatory barriers, top vendors, and leading commercial applications. (2015, May 27). Retrieved December 3, 2015, from http://www.businessinsider.com/uav-or-commercial-drone-market-forecast-2015-2

Timeline for Domestic Drone Integration. (2012, June 13). Retrieved November 28, 2015, from https://www.eff.org/document/timeline-domestic-drone-integration

The Drones Danger. (2015, August 19). Retrieved November 22, 2015, from http://www.askthepilot.com/the-drone-danger/

Thurber, M., (2012). ADS-B Is Insecure and Easily Spoofed, Say Hackers. Retrieved from http://www.ainonline.com/aviation-news/aviation-international-news/2012-09-03/ads-b-insecure-and-easily-spoofed-say-hackers

Tomiuc, E. (2012, January 31). Drones – Who Makes Them and Who Has Them? Retrieved December 1, 2015, from http://www.rferl.org/content/drones_who_makes_them_and_who_has_them/24469168.html

Tomkins, R. (2014, May 13). Frost and Sullivan forecasts five-year rise in spending for UAVs. Retrieved December 1, 2015, from http://www.upi.com/Business_News/Security-Industry/2014/05/13/Frost-Sullivan-forecasts-five-year-rise-in-spending-for-UAVs/5631400008680/

UAS / UAV Research and Advisors – Drone Analyst. (2015). Retrieved November 9, 2015, from http://droneanalyst.com/

UAV Communications Equipment. (n.d.). Retrieved from http://www.hse-uav.com/communications.htm

UAV Navigation. (2015). Sensor Fusion and Estimation. Retrieved March 12, 2016, from http://www.uavnavigation.com/support/kb/general/general-system-info/sensor-fusion-and-estimation

UAV Protect. (2014, December 23). Retrieved December 3, 2015, from http://www.uav-protect.com/

UAV Tracking Systems | UAV Tracking and Recovery by Marshall. (2015). Retrieved December 3, 2015, from http://www.unmannedsystemstechnology.com/company/marshall-radio-telemetry/

Ulanoff, L. (2016). Drone lands on moving car like it's no big deal. Retrieved March 09, 2016, from http://mashable.com/2016/01/21/drone-lands-on-car/#ixU6piag5aqX

Unmanned System Common Control System (CCS). (2016). Retrieved February 11, 2016, from http://www.navair.navy.mil/index.cfm?fuseaction=home.displayPlatform

United States Air Force RPA Vector. (2014) Retrieved January 30, 2016, from http://www.defenseinnovationmarketplace.mil/resources/USAF-RPA_VectorVisionEnablingConcepts2013-2038_ForPublicRelease.pdf

USA Army Magazine definition of Danger Close: www.benning.army.mil/infantry/magazine/issues/2013/May-June/Myer.html.

USAF Operating Next-Generation Remotely Piloted Aircraft for Irregular Warfare | Public Intelligence. (2011). Retrieved March 09, 2016, from https://publicintelligence.net/usaf-drones-in-irregular-warfare/.

Using Botnets to provide security for safety critical embedded systems – a case study focused on UAVs.

(Unknown) Retrieved January 30, 2016, from http://iopscience.iop.org.er.lib.k-state.edu/article/10.1088/1742-6596/633/1/012053/pdf.

Drone Crash into Boeing 737 in Mozambique. (6 Jan 2017) (UK Express, http://www.express.co.uk/travel/articles/751165/drone-boeing-737-plane-crash-mozambique. Also See: https://youtu.be/2jzx8BpDuHE

UST. (2015). UAV Navigation | Navigation and Guidance Systems for UAVs, ROVs, Ground Robots. Retrieved March 12, 2016, from http://www.unmannedsystemstechnology.com/category/supplier-directory/navigation-systems/

Vanden Brook, T. (2015). U.S. loses drone in Iraq. Retrieved March 11, 2016, from http://www.usatoday.com/story/news/world/2015/05/29/drone-isil-iraq/28177803/

Variable Vectoring Technique for Propeller-Powered UAVs – UAS VISION. (2016). Retrieved February 11, 2016, from http://www.uasvision.com/2016/01/15/variable-vectoring-technique-for-propeller-powered-uavs/

Vintage RC. (2012). Retrieved December 5, 2015, from http://www.rcuniverse.com/market/item.cfm?itemid=813625

Wallace, R. (2015, June 16). SciTech Tuesday-First V-1 rockets launched June 1944. Retrieved December 5, 2015, from http://www.nww2m.com/2015/06/scitech-tuesday-first-v-1-rockets-launched-june-1944/

Welch, G. (2016). The Kalman Filter. Retrieved March 12, 2016, from http://www.cs.unc.edu/~welch/kalman/

Welcome to Homeland Surveillance and Electronics LLC Unmanned Aerial Vehicle (UAV) Website! (2015, September 17). Retrieved November 2, 2015, from http://www.hse-uav.com/

Williams, M. (2015, July 28). Amazon proposes drone superhighways in sky. Retrieved December 3, 2015, from http://www.pcworld.com/article/2953952/government/amazon-proposes-drone-superhighways-in-sky.html

What is cyberterrorism? (2010, May 1). Retrieved December 3, 2015, from http://searchsecurity.techtarget.com/definition/cyberterrorism

What is Remote Code Execution? How to Hack Websites – Hacking-Sec. (2014). Retrieved February 24, 2016, from http://www.hackingsec.in/2014/02/what-is-remote-code-execution-how-to.html

What is SCADA (supervisory control and data acquisition)? (2005, September 1). Retrieved December 3, 2015, from http://whatis.techtarget.com/definition/SCADA-supervisory-control-and-data-acquisition

Whitlock, C. (2014, January 22). Crashes mount as military flies more drones in U.S. Retrieved December 1, 2015, from http://www.washingtonpost.com/sf/investigative/2014/06/22/crashes-mount-as-military-flies-more-drones-in-u-s/

Wolff, C. (n.d.). Radar Basics. Retrieved March 12, 2016, from http://www.radartutorial.eu/20.airborne/ab08.en.html

Staff (2017) Xinhua news. See: http://www.xinhuanet.com/english/2017-12/23/c_136847826.htm

Zaharia, M., Das, T., Li, H., Shenker, S. and Stoica. (nd) DLA: 07252108 from https://www.usenix.org/system/files/conference/hotcloud12/hotcloud12-final28.pdf

SECTION II

UAS INFORMATION SECURITY, INTELLIGENCE AND RISK ASSESSMENT

Chapter 4 INFOSEC – Protecting UAS Information Channels & Components

This chapter provides an overview of the basic concepts of information security to provide a common set of terms and concepts for discussion and analysis. This is only an overview: students interested in learning more should be aware that this discussion is cursory and there is a wealth of knowledge out there to be discovered.

Student learning objectives. After reading this chapter, students should be able to do the following:

- Identify, describe, and explain the three basic security policy questions
- Define the three commonly used security attributes
- Explain how security requirements can be systematically derived
- Identify and describe the three security engineering phases
- Explain the detection timeline
- Differentiate between the classes of problems that need to be detected
- Describe the types of activities that should be triggered by a detection event
- Extrapolate types of security challenges for UAS
- Identify how UAS information security challenges can be analyzed

Basic Concepts in Information Security

Information security is as old as information itself. Information has value; where the oasis is located, how many warriors are in the opposing force, how to safely prepare medicines, etc. Information security is not just about keeping secrets, although that is an important aspect. In this section, we will describe the basic concepts of information security and define important terms. Examples will illustrate concepts. Keep in mind that this is an abstraction of a very real problem space. What you will be exploring is a model of information security. Like any model, it simplifies and approximates reality to aid conceptualization. Just like an organizational chart does not capture the complexity of office politics or a data flow diagram does not capture the quality of the data, a security model has limitations. Its usefulness lies in that it gives you a way of understanding and analyzing something enormously complex.

Policy Questions

Before any analysis can begin, you must know your starting point and your goal. We refer to that as starting at first principles. **We can start with the most obvious question; what infor-**

mation needs to be protected? To answer that question, you require substantial information. Depending on the size of the system or organization, distinct types of information can exist and require *varying levels of protection*. It seems like such a simple question, but it is deceptive in its simplicity. For example, consider yourself a target for a security analysis. What information requires protection? Your personally identifiable information (PII) is an obvious choice; after all, that is a target for identity thieves. What else requires protection? Usernames, passwords, bank account information, and medical records are all sensitive information. There may also be some relationships you would prefer to keep secret.

One side note: the information and systems considered in this question are not only the ones owned and controlled by the enterprise, but also *custodial files*. The responsibilities associated with having custody of data, equipment, or personnel require special consideration. When accepting custodial responsibility, you are tacitly or explicitly accepting the responsibility to exercise due care and control of those assets. These considerations must be considered in your analysis of the policy questions. In other words, it is not just the assets that establish your baseline operational capability that you need to consider, but also the set of assets that flow in and out of your enterprise. Two tools that can help you focus on where these challenges might lie are *data flow diagrams* and *functional decomposition diagrams*.

How Much Protection is needed?

Each asset has different security needs as well, so **another question needs to be asked; how much protection is needed?** Some information types require a great deal of protection, possibly layers of security. Others might require less. Another way of thinking about this question is what is the minimal amount of security required? Think of common types of information in your life. How much protection does your personal data (like your social security number) require as compared to information about your residence? For the majority (those who are not in the witness protection program, for example), personal data warrants more protection than their address and phone number. Understanding these distinctions helps determine how resources should be allocated in developing and ensuring protective mechanisms.

How long must the information be protected?

In addition to the question of 'how much?' **there is also the question regarding how long such protection must be maintained.** Some protections must remain in place for substantial time, while others can be allowed to expire relatively quickly. For example, suppose one proposes marriage to their partner, how long would that information need to be protected? It depends on the variables, but the author believes we can safely say that once married, the information is no longer sensitive and public access is acceptable. In contrast, consider an espionage agent that has infiltrated the highest level of an adversary government. How long should this information be protected? An argument could be made that the information should be protected perpetually.

To summarize, there are three policy questions that need to be addressed for competent security analysis. Both the positive and negative versions of these questions need to be considered.

These three questions are:

– What information requires protection? What information does not?

– How much protection does asset require? Conversely, what is the minimum amount of protection required?

– How long must security be kept in place? Conversely, how soon can it lapse?

There are risk implications to the answers, which will be explored in Chapter 6. Every decision is a risk decision, but decisions are informed by the availability of resources and operational circumstances. It is impossible to be risk-free, so the decisions on how and when risks are accepted need to be made thoughtfully.

Security Attributes

The previous section discussed protection of information. Various aspects of protection are called "security attributes." There are three commonly used security attributes in information security. They are **confidentiality, integrity, and availability,** abbreviated as CIA in many publications. (Some researchers have proposed emphasis on other security attributes; keep an open mind about what security attributes are important in your unique operational environment.)

Confidentiality refers to the need to keep information, operations, and transactions secret. This applies to more than simply data. There **are four distinct kinds of secrets** that need to be kept; ***actual secrets, relationship secrets, sources and methods secrets, and operational secrets***. We differentiate between these types because protecting them requires different approaches. A relationship secret is not protected the same way a sources and methods secret is protected. However, and this is a very important point, these types are not mutually exclusive. A secret may require several of these characteristics simultaneously.

Consider some examples of the types of secrets to understand this concept. An example of *an actual secret* might be something you did when you were seven years old. Perhaps you ate your sister's cookie or broke a lamp and blamed the dog. Keeping that information secret requires that access to the actual information be restricted; only those who are granted the highest level of trust are granted access to that information. The gate keeper considers the access request and either grants it or not, based on the adjudication of trust and need to know. The more people who know increases the probability that the secret will be breached, so the decision to grant access is made on a need to know basis.

Relationship secrets are secrets in which the relationship are what need to be kept secret. Some relationships are regulated; there are legal requirements to keep them secret. An example of this would be the relationship between two corporations considering a merger: before an agreement is made, the government requires that the relationship between the organizations for the purposes of negotiating the merger be kept secret. Other times, the relationship secret is a matter of protecting its illicit nature, such as an extramarital affair or membership in a crime syndicate. And still others, the relationship between types of information is what requires protection. For example, if a company has different pay scales by gender, the relationship between employee gender and pay would be regarded as a secret. A very famous example of this type of secret was the role of "Deep Throat" during the Watergate affair. It was obvious that the data being leaked was coming from an insider and the data itself was being made public. From the perspective of the participants, the most important secret was the relationship between the reporters and the insider. They took extreme measures to prevent the relationship from being discovered.

***Sources and methods secrets* refer to the need to protect sources of information and methods by which operational goals are achieved.** Protecting sources and methods secrets can be tricky. For example, consider an advanced secret imaging platform that provides extremely high-resolution images of targets. These images are useful, and secret, but an additional consideration is protection of the source. Both the existence of the source and any clues as to the existence of the source must be protected. The latter is the difficult part: steps must be taken to remove or obfuscate any hint that the source exists. For our example of the imaging platform, one way to protect the source would be to share only degraded versions of the images. Similarly, methods for achieving some goal might need protection. For example, the exact process of mixing ingredients in controlled humidity and temperature environments could result in a superior product. In this example, the ingredients for the product are known. It is the method that is the secret. Protecting this secret requires layering of operational protective measures.

Operational secrets are those secrets that are situationally based. The operational situation may vary significantly in time or geography. For example, it may be well known that a bombing raid is being planned, but it is the actual time of the attack that is the secret. Similarly, it may common knowledge that drones are in a general area, but the actual location at any point in time may be secret. These types of secrets are considered separately from the others because of their transient nature; as soon as the operational circumstances change, the secrecy requirements become moot.

Consideration of these distinct types of secrets is extremely helpful when attempting to answer the three policy questions. It is easier to address the 'what', 'how much', and 'how long' aspects when possessing a sophisticated understanding of what type of confidentiality challenge needs to be addressed.

Integrity refers to the need to ensure the unchanging and unchangeable nature of data and transactions. There are three types of integrity challenges to pay attention to: *data*, *transaction*, *and communications integrity*. During this discussion, we will also discuss the concept of *nonrepudiation* as an integrity concept.

Data integrity refers to the allowable and unallowable variances or changes in the actual data. This may seem like a strange concept: why would there be allowable variances? Shouldn't all data be kept pristine and whole? Like everything else, there are tradeoffs to consider. Consider image compression algorithms. Some are lossy, some are lossless. A lossy image compression algorithm loses data by design, to balance between image quality and file size. The higher resolution the image, the more bandwidth required to transmit it and the more storage is required to store it. It may be acceptable to lower data integrity for the image file if the result is operationally adequate. The key here is to understand what level of data integrity is required, what level of variance is allowable, and how to detect malicious variations to data integrity.

Transaction integrity refers to the unchanging and provable nature of a transaction. A well-formed transaction has several qualities; the initiator of a transaction provably initiates the transaction, the transaction is provably received by the intended and authorized recipient, and the transaction is provably unchanged during the exchange. For example, suppose one deposits a check into an account. The provability of the origination of the transaction is supported by one's signature on the check, identifying it for deposit to a specified account. The provability of the recipient of the transaction is supported by the bank receipt, further supported by the updated ledger entry showing the exact amount of the deposit. Each one of these steps could be subverted; there are many scams that do exactly that. For high transactional integrity significant protections are implemented to ensure a high degree of transaction correctness, consistency, and completeness. An example of a transaction that needs very high integrity protection would be the authentication protocol for sensitive area access.

A special case in integrity is referred to as *nonrepudiation*. The root word 'repudiate' means to deny. When someone repudiates something, they deny it. If either party can deny their participation in the transaction or its content, it would not be a well-formed transaction. Consider, what if one went to the bank, deposited a check, got a receipt, and then a few days later the bank claimed that it never happened? The bank repudiates the transaction. It is serious enough with one's own money but becomes even more serious when the transaction is a lawful order to execute a military mission. Neither party would want there to be an ability to repudiate the transaction. The person executing the order would want to prove that they are operating under a legitimate order from a superior, while the superior would want to prove that they ordered a lawful action in accordance with policy. As a result, engineering nonrepudiation into a system is an important consideration.

Communications integrity refers to the unchanging nature of data while it is being transmitted from one entity to another. In this aspect, our concern is that the message itself is not corrupted. Keep in mind that this aspect of integrity is not just for digital communications, but for all communications. The challenges to integrity can result from many different problems, including the system itself. A fun way to explore this problem is to play the game of telephone. In this game, many people sit in a circle. The first person whispers a message to the next person, as clearly as possible. The second person whispers the message to the third, the third to the fourth, and so on. Even if every single person tries as hard as they can to be clear and enunciate, the message at the end of the chain will bear little resemblance to the message that originated. The point being that to ensure the integrity of the communications, you need to pay attention to not only the message but also to the channel.

Availability refers to the need to be able to access and use data and systems when operationally required. Obvious? It is the most straight forward of the security attributes, but that does not mean that it is simple or trivial. What does it mean to have access when needed? For some organizations, it is simple. Maximum availability during operational hours and maintenance as required. But what if there is no down time? Then it becomes complicated. Consider a very large enterprise spanning multiple time zones; how does one calculate availability requirements across the entire enterprise? Is it 100% availability required? That is almost impossible without a very large investment. Even highly reliable phone systems only measure availability in "five nines", 99.999% available. When considering availability, one needs to account for the availability of many various aspects, including the data itself, the tools (including both computational and other, such as pens and paper), and the infrastructure components (including, again, both computationally based, such as networks, and physical, such as secure areas and safes).

In summary, there are three commonly used security attributes:

- Confidentiality;
- Integrity, with the special case of nonrepudiation, and
- Availability.

Security Phases

There are three security phases that we think about; protection, detection, and reaction/correction. There are two functions for these phases. The first use is to identify the viable solutions for the security requirements for an enterprise. The second use is to manage security operations for an enterprise.

Protection is the security engineering phase in which protective measures are implemented to meet security requirements. Protective mechanisms can consist of product technologies, processes, and methods. They come from the specialty areas of physical, personnel, administrative, electronic, communications, and computational security. No one protective measure can suffice to meet the needs of an enterprise, so it is important that the protective measures be integrated to meet the widest range of requirements within company resource limitations. Examples of protective mechanisms include locks, curtains, computer access controls, and cryptography.

Detection is the term in which the processes and methodologies associated with discovering potentially damaging situations. Detection works on a timeline and every activity on that timeline triggers the occurrence of one or more additional activities. The amount of elapsed time between activities is critical, but shortening that time requires additional resources. Figure 4-1 shows an abstraction of the detection timeline. Note that detection is closer to the middle of the timeline than expected. This is a crucial point: there may be a significant amount of time between the start of an event and the detection of phenomenology that indicates an event has occurred (or may still be underway).

Figure 4-1 The Detection Timeline

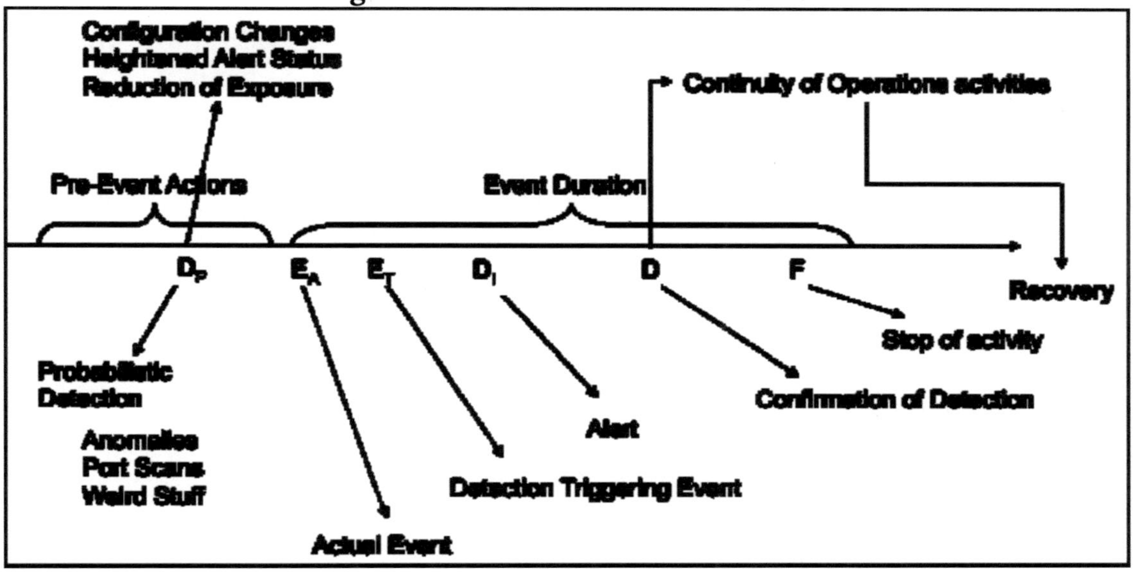

Source: Ryan, J.J.C.H., (summer, 2018). Detection Timeline.

Detection is the default status for fully operational systems. Once the system is designed, developed, and implemented, all the security controls are in place, hopefully operating correctly and consistently, and the security team is in place, on the alert for a negative incident to happen. This is detection. Once a negative incident happens, then the status transitions to reaction and correction, and eventually back to protection, and once more to detection. How fast the team can transition between phases is a function of training, experience, capabilities, and the severity of the problem(s).

A design consideration for detecting problems is the detection mechanism must be in place before the event occurs. In some cases, it is possible to detect problems without have specific detective mechanisms (like reading corporate secrets on the front page of the newspaper), but in most cases that simply is not true. If you are not collecting real time data or you do not have intrusion detection systems in place, you will not be aware that a problem is occurring in near real time nor will you be able to reconstruct the forensics evidence for evaluation. It does no good to install an alarm on a door after the door has been forced open and the Crown Jewels stolen.

There are *five categories of problems* that require detection. The first is the occurrence of problems against the Protected Class. The second category is the occurrence of problems against the Unprotected Class. The third category is the occurrence of problems on the Unknown Class. The fourth category is the misuse or abuse of privilege by insiders: The Insider Class. Finally, the fifth category of problem is the use of activities, methods, or technologies designed to reduce the probability of detection: The Counter-Detection Class. Each of these classes requires different approaches to detection, which is why consider each individually.

When you are designing a security architecture, scarce resources will require that you make compromises. Some problems can be protected against with a reasonable expenditure of resources, some cannot. Some problems may require extremely expensive protective mechanisms; other problems may have such a low likelihood of occurrence such problems are not addressed at all. *This choice creates the distinction between Protected Class detection challenges and Unprotected Class detection challenges.*

Protected Class

The Protected Class detection challenges fall into two subclasses. The first is *failure of the protective mechanism*: detecting the occurrence of problems even though the protective mechanisms are in place. This is significant; if one invests time and resources to address potential problems, those problems are clearly a priority (otherwise, why bother protecting against them?). If the problems occurs anyway, it means that the protective mechanisms have failed in some manner. Problem detection both alerts one to the fact that security has been compromised and the need to review the protections in place. The second subclass is *degradation of the protective mechanism*. Simply stated, protective mechanisms must be continually assessed to ensure they are correctly implemented, working correctly, and remain uncompromised. Locks rust, software updates can subvert exploit counter-measures, and people can get lazy. Ensuring protection functionality is critical.

The Unprotected Class

The Unprotected Class detection challenge is that the first line of defense against the problems that you chose not to protect against is, in fact, detecting the occurrence of the problem. The possibility of a data breach was then but intentionally ignored. Placing detective mechanisms in place to trigger a reaction to the problem is the only thing you can do to minimize potential damage.

Unknown Class

The final three classes are more complicated. For the Unknown Class, you do not know what you are seeking. So simply monitoring your environment for operational aberrations becomes your first detection method. It seems simple, but one must understand they are looking for unknown or newly conceived scenarios.

For example, consider a spike in help desk phone calls: one management reaction might be to shrug and think, "must be a glitch in the system," while a different reaction might be to alert and start analyzing what might be going on. Mindset makes all the difference in detecting the Unknown Class of problems.

Insider Class

This is true for the Insider Class as well: insiders are, by definition, trusted to operate within the enterprise. There are varying levels of trust, but each insider has at least some level of trust. Detecting abuse or misuse of that trust requires the same sort of awareness, proactive analysis, and operational surveillance that the Unknown Class requires, but with a difference; the detection activities should be engineered to be able to establish agency of purpose.

Counter – Detection Class

Finally, for the Counter-Detection Class, the focus of the detective efforts is to determine when protections and detective mechanisms are being subverted. The subversion can occur in many ways. In a way, this class of detection effort combines all the aspects of the other four problems. To tackle this problem, the use of red teams periodically testing the detective mechanisms is useful, as is constant analysis of detective mechanism performance.

The **Reaction/Correction** phase consists of the set of activities required to continue operating while executing the activities needed to fully recover operational capabilities. There can be significant overlap between detection activities and reaction/correction activities, particularly for complicated problem sets, so best practice is to have each activity have dedicated people, resources, and management structures. Reaction/correction activities include (at the very least) the following efforts: investigations, forensics analyses, business continuity, crisis communications, and business recovery.

To summarize, there are three engineering phases for security architectures:

- Protection
- Detection
- Reaction/Correction

Risk

All decisions are risk-based decisions, even if one is not actively considering risk when deciding. When walking down the street at 2 A.M., one makes a decision-based risk assessment associated with the activity. When a decision to purchase a certain type of computer system, one considers several types of risk, including the risk that the company may go out of business and leave the company with no support. Risk minimization is essential.

There are two elements to risk decisions: the assessment of risk in the current time state (Now Risk) and an assessment of potential changes to the risk elements over time (Future Risk). As an entity requiring evaluation, one can think of risk as a combination of the probability that a negative incident will happen and the impact that such a terrible thing would have. For example, the risk of your secrets being exposed is analyzed by determining the probability that implemented protective mechanisms could be defeated plus the impact of data exposure. Note that this use of the term risk is slightly different than as it is used in common parlance. Typically, the word risk is used informally as a substitute for probability without including the impact assessment. However, when assessing risk associated with a potential decision, including the impact assessment gives you the additional data needed to make the subsequent management decisions on where to expend resources to reduce risk.

Now Risk is an evaluation of the elements of risk using the information that exists currently. In the information security community, it is hard to impossible to get actual data on the probability of terrible things happening. So, it is common to perform the analysis by estimating the chances of terrible things happening by investigating the elements of Threats, Vulnerabilities, and Counter-Measures. Threats exploit vulnerabilities to do terrible things. Vulnerabilities are exploitable by threats. Counter-measures reduce the ability of threats to exploit vulnerabilities successfully.

Threats come in two varieties: natural and man-made. Natural threats include things like fires, hurricanes, floods, earthquakes, and the odd meteorite. Yes, security planning does need to address these elements. For example, if you are operating in a flood zone, you should make the risk decision to put your server farm on a floor that is above the flood plain. If you live in an earthquake-prone area, like Japan, you might want to invest in earthquake mitigation technologies. Man-made threats are problems that originate in the mind of humans. They can be realized though the actions of people or through the agents of people, including animals, robots, and software.

When considering the elements that constitute viable threats, there are some aspects we can tease out. A credible threat requires both capability and intent. If the potential threat has capability but not intent, we call that a trusted insider. If the potential threat has intent but no capability, there is very little the threat can accomplish. Intent can be either organic to the individual or programmed into the threat agent (through software, training, or design). Further, capability requires several elements. To be truly capable, a threat requires three traits:

- the knowledge, skills, and ability needed to act;
- the resources to plan, develop, and execute actions;

- and access to the target.

Understanding those elements informs as to how to design counter-measures; what can be done to reduce or eliminate any or all the components? For example, what can be done to reduce the probability of a trusted insider developing the intent to act maliciously? How to reduce the probability that a threat can acquire the resources needed to act? When considering these elements, it quickly becomes obvious that there are easy and difficult things that can be implemented. A very summary of what can be done to counter risk activities are listed here:

Intent:

- Motivate threat to not form intent
- Intimidate threat through implication of personal harm or danger
- Scare threat through implication of harm to reputation, livelihood, etc.
- Discourage action through psychological motivations

Resources: Not much.

Knowledge, Skills, and Ability:

- Limit knowledge of security details –keep the details and engineering data secret; keep plans and procedures secret
- Use technologies, tools, and equipment that require highly specialized skills or training
- Combine security elements in ways that limit the probability that a threat could easily develop the needed KSA
- Pay for the development of unique systems that are not commercial off the shelf

Access:

- Deny access through barriers
- Deny access by use of technology such as locks
- Deny access through checkpoints
- Control access through I&A procedures
- Minimize access available to all people
- Limit speed of access
- Limit speed of egress

Vulnerabilities, similarly, can be considered to come in two types: accidental and by design. Accidental vulnerabilities are the result of lack of understanding, sloppiness, or unintended consequences. The danger here is you do not know that they exist until the analysis is completed. Efforts must include a search for such problems. One method used by large corporations is to search for vulnerabilities by having "bug bounties". This may work well for some organizations but may be absolutely the wrong thing for other organizations. Vulnerabilities that exist by design, on the other hand, should be the result of risk-based decision making. These include decisions to connect to vulnerable networks to conduct business or using a less than optimal engineering solution because a better one is not affordable. These types of vulnerabilities must be monitored, and detective measures should be considered to alert on any attempt of a threat to exploit these vulnerabilities.

Taking the threat, vulnerabilities, and countermeasures into account gives one insight into possible attack probability. Considering the impact of what the effects of an attack might be provides a way to measure the relative importance of each potential. Decisions can then be made about what to do, or not do, regarding each aspect of risk in the environment.

Now Risk decisions are based on knowledge of what is known regarding the current and past. **Future Risk** analysis looks at the potential for change in the situation. Each element of a risk decision has a change potential that varies over time. To do a competent analysis of what may change that can cause a decision's effect to be moderated, it is necessary to do a futures analysis. For example, a decision to plagiarize a dissertation made in 1987 might have been a fully logical decision, while the impact of detection would have been very high, the probability of the plagiarism being detected was relatively low. Fast forward to the Internet age, where plagiarism checking is both automated and crowd-sourced, and the situation has changed tremendously. The decision that seemed logical in 1987 has been turned on its head because of the unforeseen technological evolution.

In considering the impact future decisions made now, one should take into consideration how difficult it would be to change a decision should it become necessary or desirable to do so. Some decisions simply cannot be changed, these are point decisions. The example of the plagiarized dissertation is a point decision. Once it is published, there is no taking it back. Other decisions can be changed, but at a cost. Some costs are manageable. If one hires a spy, they can be fired and then a cleanup. Some damage will be incurred, but the decision can be changed. Some decisions more difficult to change due to costs. For example, making the decision to standardize on a software suite for accounting and fiscal management has long term implications that can be pervasive. A more dramatic example of a long-term risk decision that can have substantial future risk variability might be the decision to fund a new aircraft carrier: a dramatic change in technology could make aircraft carriers obsolete.

To summarize, risk decisions are made now for future events and so all potential futures should be considered while deciding. Over the life of a decision, future events should be monitored to see what is emerging and what potential impacts of change might be.

Systems Engineering an Information Security Solution

Using the ten pieces to the puzzle presented above, one now has the tools to think through an architectural approach to developing and managing an information systems security solution. To review, the ten pieces of the puzzle are:

- The three policy questions
- The three security attributes
- The three security phases
- The risk decision elements

Building on that, one can derive security requirements for the enterprise and then systematically identify a set of processes, technologies, and engineering approaches to address them. All of these are done within a risk management envelope: all decisions are risk decisions.

Identifying Security Requirements for an Enterprise

Understanding what the security requirements are for an enterprise is the first step in developing and managing security solutions for that enterprise. One cannot possibly meet all requirements; that would take more resources and have more operational impact than can be tolerated. However, knowing which requirements are not being met gives one the ability to monitor the unmet needs. Putting together a solution is an exercise in engineering a system, or a system of systems. Further, every solution is time limited, so the system must be periodically re-evaluated, and the set of solutions revised, updated, and improved. This is not a point solution effort; it is a life-cycle effort.

There are **three basic types of requirements** in systems engineering. These are referred to as *explicit, implicit, and derived requirements*. These requirements are identified for the information security of a system by systematically considering the three policy questions against the three security attributes. By going through this exercise, one can concretely describe what is required to protect the information in an enterprise. Table 4-1, a 3 x 3 matrix, specifies what is needed in terms of policy questions and security attributes. In this table, the policy questions are abbreviated to make it easier to read. But remember that it is important to consider the full meaning of the questions for each cell in the matrix. Cells are renumbered to make it easier to refer to in the discussion about requirements. Refer to this table to explore examples of security requirements for a system.

Table 4-1			
	What?	**How Much?**	**How Long?**
Confidentiality	C-1	C-2	C-3
Integrity	I-1	I-2	I-3
Availability	A-1	A-2	A-3

When each cell is considered, requirements are identified. For example, in cell C-1, the full policy question consideration is:

> *What data, systems, and operational elements need to have aspects of confidentiality protected?*
>
> *What elements do not need confidentiality protection?*

Then, as one proceeds through C-2 and C-3, one can elaborate on how much (or how little) confidentiality protections are needed and how long they need to be kept in place (or when they can be allowed to lapse). While this is a systematic and straightforward process. There are many complexities in systems, addressed in previous examples, so a strong suggestion is that this exercise be conducted by a diverse team with multi-faceted knowledge of the system under analysis.

Explicit Requirements

The first requirements that will be identified will be the *explicit* requirements. These are the ones that are defined as things that are needed in a system, independent of any implementation or technology solution. For example, an explicit requirement that may be identified in cell C-1 could be that <u>the existence of a sensor must be kept secret</u>. That is an obvious type of requirement to specify, as it reflects a specific secrecy need.

Implicit Requirements

For all explicit requirements, *implicit* requirements also exist. These are the requirements that are implied by the need that is identified in the explicit requirement. What is implied by a need to keep the existence of a sensor secret? There are many implied requirements to consider. First and foremost, <u>the sensor must be hidden from observation</u>. Observation may be accomplished though vision, imaging, signals interception, and other means. It is important to then complete the analysis of the types of observation that the sensor might be subjected to. Another implied requirement is that <u>all individuals who observe or work with the sensor or the sensor products must be approved to do so</u>. Yet another implied requirement is that the <u>sensor products themselves be protected</u>.

In the matter of moments, three implied requirements have been created that come from the single explicit requirement; more can develop as required. This is the power of working through these analyses; by extrapolating what it would take to meet an explicit requirement, we identify the systemic elements that also need to be considered. Nothing lives in an isolated space; surrounding system components must be considered.

Derived Requirements

Beyond explicit and implicit requirements, there are *derived* requirements. Derived requirements are the things that are necessary conditions for the environment. These, unfortunately, are sometimes simply assumed in a system. That can lead to hilarious outcomes, such as delivering a fully-compliant system to a customer that is totally unusable because it lacks a user interface.

Derived requirements are important to consider; it is usually best to start by reviewing the implicit requirements. One of the implicit requirements for the secret sensor system was that <u>all individuals who observe or work with the sensor or the sensor products must be authorized</u>. There are many requirements that can be derived from this single implicit requirement. One is that <u>a system exists to vet individuals for trustworthiness</u>. Another is that <u>only vetted individuals are authorized to work or with the sensor</u>. Another is that <u>there are controls that check to make sure that only vetted and authorized individuals get access to the sensor and to the area in which the sensor is housed</u>.

There is power in this approach to identifying explicit requirements and then identifying the underlying implicit and derived requirements. Going through the exercise of thinking through all these things can not only help one understand the management needs of the environment, it can also point to possibilities for efficiencies in operational processes. At the very least, it provides one with a comprehensive understanding of what the security needs are for a system, a system of systems, or an enterprise.

Security Solutions Consideration

Once a set of requirements is created, one can start to parse out how to develop and implement solutions for each requirement. Note well: no enterprise can afford all potential solutions, so this will start out as a list of options, which would then need to be winnowed down based on enterprise priorities, operational realities, and resource availability.

Table 4-2 presents a very simplified view of starting the process. Simplified because information exists in many divergent phases. A trivial way of thinking about the information phases include processing, storage, and transmission. That is a normal abstraction, but it overlooks some very specialized states that may warrant separate consideration according to enterprise needs. A more layered approach to information states is that of input, output, processing, local communications, external communications, temporary storage, permanent storage, and display. In any event, when considering how to select security solutions, it is important to consider which states the solution works for and which ones it fails.

Table 4-2 Information Security Parameters and Process

	Protection	Detection	Reaction/ Correction
Confidentiality	C-P	C-D	C-RC
Integrity	I-P	I-D	I-RC
Availability	A-P	A-D	A-RC

Look at an example of how this table can be used to explore security solutions. For cell C-P, assume that requirements have been identified for protecting the confidentiality of information in the display state. Some technologies that can contribute to meeting this requirement include putting curtains on the windows and restricting access to the area. But these solutions would be useless if the operational environment were an economy seat on a commercial flight. In this case, other solutions would be considered. The crucial point? The solutions need to match the operational environment as well as meet the statement of requirement.

UAS Security Challenges

Now with a basic understanding of how to think about the information security challenges in an operational enterprise, consider what that mean for UASs. Again, any specific security solution will need to be unique to the enterprise; solutions appropriate for commercial delivery drones may not be sufficient for national security operations of a sensitive nature. Some elements that may require security consideration; communications, data processing systems, sensors, location, and control systems.

Communications may need to have confidentiality, integrity, and availability protected. Data processing systems may have high integrity needs as well as some availability needs. Sensors may need high confidentiality and integrity requirements. Location may be sensitive based on mission and so may have contextual security needs. Control systems may need to have strong integrity and availability needs.

Discussion Questions

Test your understanding of the material presented in this chapter by thinking through the following questions. Discuss them with other people. There are some subtleties that are fascinating to explore.

1. Explore the differences and similarities of confidentiality and privacy. How do the con-

cepts overlap? How are they distinctly different? What does that mean in terms of managing privacy?
2. How is the concept of integrity different from truthfulness?
3. How is the concept of ownership different from availability?
4. How does custodial responsibility for information translate into security requirements? What are some of the explicit, implicit, and derived requirements?
5. What security requirements are needed during the acquisition of sensitive equipment, material, or information? What are some of the explicit, implicit, and derived requirements?
6. What security requirements are needed for the management of systems, particularly information processing systems? Think about installing patches or software updates: what security controls are needed during these processes?

Bibliography

Ryan, J. (2018, August 26). *Dr. Julie J.C.H Ryan Research Page* . Retrieved from GWU SEAS: https://www2.seas.gwu.edu/~jjchryan/research.html

References

"Nichols, Randall K, Daniel J. Ryan, and Julie J.C.H. Ryan, *Defending Your Digital Assets Against Hackers, Crackers, Spies, and Thieves*, New York: McGraw Hill, ISBN: 978-0072122855, 2000" (Ryan, 2018)

"Ryan, Julie J.C.H., *Teaching Information Security to Engineering Managers*, Proceedings of the 33rd ASEE/IEEE Frontiers in Education Conference, Boulder, Colorado, November 2003" (Ryan, 2018)

U.S. Department of Defense (1970). "Security Controls for Computer Systems (U): A Report of the Defense Science Board Task Force on Computer Security". Rand Corporation: Santa Monica, California. Available online at https://csrc.nist.gov/csrc/media/publications/conference-paper/1998/10/08/proceedings-of-the-21st-nissc-1998/documents/early-cs-papers/ware70.pdf

"Pettigrew, J. Andrew, and Julie Ryan, Making Successful Security Decisions: A Qualitative Evaluation, IEEE Security Privacy Magazine, Vol 10 (1), 2012" (Ryan, 2018)

"Amin, Rohan, Julie Ryan, and Johan Van Dorp, Detecting Targeted Malicious Email Using Persistent Threat and Recipient Oriented Features, IEEE Security Privacy Magazine, 2012, vol 10n3 (64-71)" (Ryan, 2018)"

"Ryan, Julie J.C.H., Thomas A. Mazzuchi, Daniel J. Ryan, Juliana Lopez de la Cruz, Roger Cooke, Quantifying Information Security Risks Using Expert Judgment Elicitation, Computers and Operations Research, April 2012, v39 n4 (774-784)" (Ryan, 2018)"

"Ryan, Julie J.C.H. and Daniel J. Ryan, Performance Metrics for Information Security Risk Management, IEEE Security and Privacy, vol. 6 no. 5, Sep/Oct 2008, pp. 38-44" (Ryan, 2018)

"Ryan, Julie J.C.H., Use of Information Sharing Between Government and Industry as a Weapon, Journal of Information Warfare, Journal of Information Warfare" (Ryan, 2018)

"Ryan, Julie J.C.H., Cyber Security: The Mess We're In and Why It's Going to Get Worse, The Cyber Security Policy and Research Institute, Report GW-CSPRI-2100-4, April 11, 2011, 2010-2011 Seminar Papers" (Ryan, 2018)

"Ryan, Julie J.C.H. (ed), Leading Issues in Information Warfare and Security Research, Reading, UK: Academic Publishing International, Ltd, ISBN : 978-1-908272-08-9, 2011" (Ryan, 2018)

"Ryan, Julie J.C.H, Information Warfare: A Conceptual Framework, Seminar on Intelligence, Command, and Control (1996), ISBN : 1-879716-39-9,Proceedings of the 1996 Seminar on Intelligence, Command, and Control. Boston, MA: Harvard University Press, 1997" (Ryan, 2018)"

"Ryan, Daniel J. and Julie J.C.H. Ryan, Protecting the NII, Information Warfare: Protecting Your Personal Security in the Electronic Age, ISBN : 978-1560251323, (New York: Thunder's Mouth Press, 1994), 626" (Ryan, 2018)

Websites of Interest

JJCH Ryan Publications and other stuff, https://www2.seas.gwu.edu/~jjchryan/research.html

JJCH Ryan Curricular Materials, https://www2.seas.gwu.edu/~jjchryan/curricularmaterials.html

Proceedings: The 1st United States – Japan Critical Information Infrastructure Protection Workshop, https://www2.gwu.edu/~usjpciip/

Chapter 5 Intelligence & Red Teaming

This Chapter will introduce the basic concepts in intelligence and build upon these concepts to explore their role in attack/defend scenarios.

Student learning objectives. Upon completion of this chapter, students should be able to:

- Identify, describe, and explain the intelligence cycle.
- Identify common problems in intelligence data collection.
- Distinguish between reputation, reliability, and quality of intelligence sources.
- Describe the types of sources for intelligence data.
- Identify distinct types of open sources.
- Describe how open sources can be used to develop intelligence assessments.
- Describe the concept of attack/defend scenarios in intelligence requirements terms.
- Develop attack/defend scenarios for UAS.

Basic Concepts in Intelligence

The word intelligence can have several meanings. In one instance, it is an estimate of how capably a person can reason and think. In another, it refers to the data that has been collected for purposes of divining purpose, capability, or meaning. In yet another, it refers to the process by which information is collected and processed. In this chapter, intelligence refers to processes by which intelligence data are collected, processed, and analyzed.

The fundamental reason for having an intelligence process is to collect and evaluate information that informs personnel about the operating environment. Governments have intelligence functions to make better informed decisions. Businesses have intelligence functions to be more effective and competitive. This is nothing new; intelligence functions are as old as mankind itself. As expected, they have grown more sophisticated and capable, with the advances in both product and process technologies. By integrating such technologies into intelligence processes, the timeliness and the scale of the data collected is greatly improved. Further, the integration of information processing technologies into everyday life has vastly increased the opportunities for data collection.

The Intelligence Cycle

The functions related to intelligence occur in phases known as the intelligence cycle. In simple terms the cycle consists of four phases: **requirements**, **collection**, **analysis**, and **reporting**. This section will describe the phases general terms. Note that some organizations have specific terms for the functions that are addressed here.[1]

Figure 5-1 The Intelligence Cycle

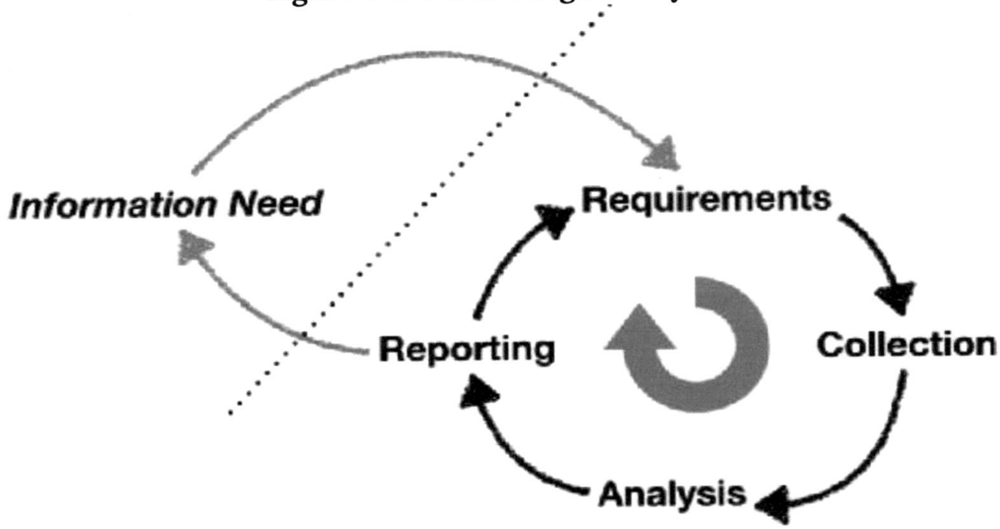

Source: Ryan, J.J.C.H. (summer, 2018). Intelligence Cycle Decisions Collage.

The **requirements** phase defines/identifies data of interest to the organization. For example, a business executive may identify the pricing strategy of competitors as desirable data. The executive would create a requirement statement for the business intelligence function, which would have the responsibility to try to meet the requirement. Similarly, a government leader may want to understand the military readiness of an adversary. That leader would, through established processes, require a statement of requirement to be generated and entered the intelligence community processes.

The aggregated set of requirements statements are the tasking for the intelligence process: and determines what data needs to be collected. When a requirement is generated, it is categorized in three ways; *importance*, *transience*, and *complexity*. These three categories help prioritize resource allocation to meet requirements. Obviously, the *importance* of the required data is something that should impact collection priority. Critical data needs should be met before a low priority data requirement. The *transience* of data is also a principal factor. Transience refers to how long the data will be available. The data may be fleeting in nature, like a radioactive emission, or may be available for collection only during a limited window. Finally, the *complexity* of the data needed affects how much planning must be done to assure successful data collection.

1. The Intelligence cycle is important and is covered in various degrees in more than one chapter of this book. Chapter 14 also covers the intelligence process in terms of vulnerabilities of UAS and EW.

Table 5-1 illustrates how these three characteristics can inform the prioritization of requirements. Please note: these are hypothetical examples for illustrative purposes only.

Table 5-1 Prioritizing Requirements Example

Requirement	Importance	Transience	Complexity	Priority
Competitor's pricing strategy for a contract	High	Medium	Low	1
Names of scientists on research effort	Medium	Low	Low	2
Amount of reserve cash on hand for investment	Low	Low	Medium	3

Source: Ryan, J.J.C.H. (2018)

Once requirements are prioritized, they are transferred to the **collection** function. The job of the collection function is just that, to collect data. Several types of data can be collected in diverse ways. To specify how the data will be collected, appropriate mechanisms need to be identified and allocated to the collection effort. I.E., *sources* of data and *methods* for data collection must be considered.

For example, collecting the data associated with a competitor's pricing strategy for a contract can be tricky. One strategy would be to infiltrate the organization and steal the strategy. This approach has obvious legal implications that make it less attractive option. Another method would be to go dumpster diving, sift through the garbage thrown out by the competitor, to extract clues and hints. This approach would only work if the garbage was both accessible and not shredded. A third alternative would be to send people to the local establishments to eavesdrop on conversations. Each of these approaches requires different collection strategies and resources.

The data collected may or may not, fully or partially, fulfill the stated requirement. Collected data that fully meets the requirement is complete and unambiguous. Data that only partially meets requirement may have some ambiguity or be missing some aspects. Data that does not meet any requirements is found to be useless; lacking in specificity and/or trustworthiness. The collected data is adjudicated during the **analysis** phase of the cycle. There are three steps in the analysis phase are *structural analysis*, *content analysis*, and *fusion*. Structural analysis consists mostly of describing the data. The descriptions can contain identifiers, such as index numbers and metadata, as well as actual descriptive data, such as statistical information. Structural analysis is important for both administration (record keeping) and post hoc studies conducted on the aggregated data. *Content analysis* is what most people think of as analysis, examining the actual data to derive meaning and insight. *Fusion* occurs when multiple types or multiple source data are analyzed together. The benefit of fusion is that increased nuance can be parsed out of the data.

Once the analysis is completed, the results of the analysis is reported. The **reporting** phase consists of several several types of reporting efforts. First, and most obviously, there is *results* reporting, the delivery of the analyzed data to the required personnel. Results reports can include more than the data itself. The results can also include ancillary information discovered that relates to the original need, an assessment of the implications of the data, and caveats associated with the estimated reliability of the information. A second type of reporting is *feedback* reporting. Feedback reporting provides information to the other phases of the cycle. Types of reports that might be included are work product flow rates, percentage of needs met, identification of derived data needs based on the analytical process, and quality assessments. Finally, *strategic intelligence* reports can be generated based on aggregated analysis of many different results reports over a significant portion of an area of interest. Figure 5-1 shows the intelligence cycle in graphic form.

Common Problems in Intelligence

The motto for an intelligence analyst should be to believe nothing fully until it verified through multiple and distinctly dissimilar sources. There are too many people trying to obfuscate and too many challenges in interpreting the actual meaning from many different scraps of data. There are also many challenges associated with collecting and analyzing intelligence that go beyond simple access to the information. In general, there are three primary problems that must be considered. These are *quality* of data, *bias* in data interpretation, and *circular reporting*.

Data quality is an obvious issue for intelligence processes. This general problem set includes counter-intelligence activities, such as deception. When estimating the quality of the collected data, both the source and any methods that the data has been subject to must be weighed. Sources of data should be rated for truthfulness and informativeness. This discussion includes all types of sources, from instruments to people.

Truthfulness is an estimate of how close the data is to actuality. This can be an interesting measure, in that the data may be a complete falsehood but an accurate replica of actual data. For example, a business may be setting a pricing strategy based on wildly inaccurate material cost estimates. Thus, the data collected that reveal the inaccurate cost estimates is a truthful set of data, even though it reveals false information. A better example would be sending out info on false target knowing signal interception is a certainty.

A source may be rated as having a reputation for providing truthful data, sometimes truthful data, or mostly untruthful data. While sources that have a reputation for providing truthful data are obviously prized, a source that is known for repeatedly providing untruthful data can also be useful for understanding the nature of the competitive landscape. Consider the repeated provision of untruthful data can hardly be considered an accident or coincidence. Ergo, the type and content of the untruthful data can provide insight into the adversary. All data is useful, just for different purposes.

Informativeness is the measure of how much added value the source provides to the intelligence process. High informative data sources provide a significant added value, while low informative sources provide little. Low informative sources are not necessarily bad; sometimes gathering a lot of low informative data sets can be extremely revealing in the aggregate. For example, a highly informative data source could provide you with the actual pricing strategy document for a company. Alternatively, low informative data sources could provide you with all the component elements that feed into a pricing strategy; salaries, qualifications, etc.

Combining truthfulness and informativeness provide a way of rating data sources. For scientific instruments, this is generally couched in terms like type 1 and type 2 error rates, as well as calibration. For publication sources, this might be discussed in terms of media bias, peer review, and impact factor. For human sources, terms such as reliability and reputation might be used.

Bias is a challenge for both human and algorithmic interpretation of data. Why algorithmic? Because humans write the algorithms. There are conscious and unconscious forms of bias, all of which must be addressed. A personal preference for uniformity of data may lead an analyst to treat various kinds of data differently, inadvertently skewing the results. Another type of bias may arise from a desire to see data that confirms suspicions. When such data is seen, it may be treated with more gravity than other data. Sufficiency introduces another type of bias, when analysts are satisfied that the data is complete, they may stop looking, even though they have only a small percentage of available data. Yet another form of bias can arise from the analyst overlooking subtle clues in favor of obvious data. Over-reliance of single source information also leads to interpretative bias. The bottom line is that data interpretation is a subtle art that must be approached carefully.

Circular reporting can sneak into intelligence processes in ways that make it difficult to recognize and manage. A circular reporting problem occurs when data that is collected and analyzed by one component of an intelligence function serves as input to another component, masquerading as new data. To the uninitiated, this may seem like a bizarre problem, but it happens too frequently in both very large intelligence functions and in news reporting. Examine a hypothetical news report. News service A reports that seven civilians have been killed in a suicide bomb attack. The wire services pick this up this report and publicize it. News service B then republishes the material but alters some of the wording. The wire services pick up this second report and repeat the process. News service A sees that report and takes it as confirmation of its original report, even though it is nothing more than its original report repackaged. This problem of circular reporting compounds the problems of bias in the analysis of intelligence data significantly and must be addressed.

Sources of Intelligence Data

Intelligence data has many diverse sources. Some of the more common sources are described in this section.

The earliest form of intelligence source is also the most easily understood. Human intelligence, or HUMINT, is intelligence data that is collected by human beings. HUMINT includes collectable information: stolen papers, intercepted letters, overheard conversations, observations, and physical artifacts. HUMINT need not be collected by specially trained agents, although that is a key role for them. HUMINT can come from travelers returning from a vacation, from chance encounters at conferences, or from interactions at parties. Because of the power of the human brain as a general processing system, HUMINT can be an amazingly important source of information.

Some technical data requires special technologies for collection. Humans are great at observing information in the visual light spectrum and hearing things within normal hearing ranges but are not much good at collecting infrared or encoded digital signals. For these types of data, special sources are required. Imagery Intelligence, or IMINT, uses imaging capabilities to collect data. Signals Intelligence, or SIGINT, is a combination of different intelligence sources. SIGINT includes Communications Intelligence (COMINT) and Electronics Intelligence (ELINT). COMINT is the collection and analysis of communications through various channels. This includes the content of the communications (the internals) and the external routing information; the return address, the intended recipient, the content and time of the transmission, etc. Sometimes referred to as metadata; the data that describes the facts of the communication. ELINT is like COMINT, in that it is the collection of signals electronically transmitted.

As information technology expands and becomes incorporated into everyday life, specialized forms of intelligence data collection and analysis have been developed to focus on the material from special sources. Some of these are obvious, such as Financial and Geographic Intelligence. But other sources are sufficiently specialized to warrant designation. These include Measurement and Signature Intelligences (MASINT) and Cyber Intelligence.

Countering these extremely specialized technical sources is Open Source Intelligence (OSINT). OSINT takes advantage of the Internet and other open sources of information, to build robust analytical portraits of targets of interest. For example, consider an OSINT analysis of a competitive business. The types of open source information may include employment ads, floor plans, financial filings, and leadership profiles. Consider a business that is running employment ads for quantum computer specialists, whose floor plans include laboratories, whose financial filings include speculative notices regarding technological risk associated with research, and whose leadership includes people who have long histories in the communications business. Are they more likely to be developing a quantum communications capability or a space vehicle?

The fact of the matter is that the residue of an enterprise's operations are an incredibly rich source of information. Open source data can be extracted from social media, conventional media sources, scientific publications, press releases, public speeches, financial filings, legal filings, official documents, and administrative documents. With the availability of powerful datasets at local libraries, it can be quite cost effective to leverage open source intelligence.

Combining data from different sources can assist in intelligence analysis. One use is detecting deception. Another of use is building a more complete and sophisticated knowledge of capabilities. Combining static imagery, video, and signals data can provide a structured understanding of the control and capabilities of a smart munition.

Understanding Attack/Defend as a Tool

It is one thing to collect information and analyze the capabilities and intentions of an adversary. Using that intelligence to create attack/defend scenarios takes it to the next level. There is tremendous training experience gained from engaging in attack/defend exercises. Building attack/defend scenarios forces one to determine the extent of one's knowledge base, potentially identifying latest information requirements. Practicing attack/defend scenarios can reveal strengths and weaknesses on both sides, which can lead to strategy development, tactics refinements, and organizational changes.

Such scenarios can be conducted with varying levels of abstraction. At one end of the abstraction scale are simulations. At the other level are real world exercises, that use real equipment in actual conflict conditions, albeit with constraints. In between are many different combinations of games and exercises.

Simulations include table top games, computer programs that mimic real world interactions, or mathematical algorithms that generate outcomes based on input values. For attack/defend exercises, table top exercises are quite common and even have their own acronym, TTX. A benefit of table top exercises is that complex scenarios that require human decisions can be worked through and discussed as the game progresses. A limitation of table top exercises is that there is typically little fidelity to real world conditions.

Real world exercises are conducted using actual equipment in conflict conditions with limitations and constraints for safety and security. The benefit of real world exercises is that participants significantly advance their knowledge and experience. The downside of real world exercises is that they tend to be enormously expensive, expend resources, take significant time to plan, and require extensive safety and security protocols.

Attack/defend scenarios can be viewed as a series of activities conducted by teams. The red team acts as the adversary, attacking opponent's capabilities. The blue team acts as the defenders, detecting and countering red team activities. Because these are exercises, it is important to have safeguards and strictly delineated boundary conditions. For example, if the attack/defend scenario includes cyber activities, a typical safeguard would be to ensure that there is no Internet connection. This prevents adversaries from spying on the exercise, but can also prevents accidents, such as the inadvertent penetration of a non-participatory network. While it may appear as if it could never happen in a well-managed system, the fact of the matter is humans make mistakes. Better safe than sorry.

Red Teaming

Red teams act as the adversary. What does that mean? One implication is that the red teams need to know technologically how to execute attacks on the target. If the red team mimicking a specific adversary, the members of the red team must also be knowledgeable about the strategies and tactics of the actual adversary. This can include methods of attack, cultural assumptions about how activities occur, and organizational constraints on command and control.

To technologically execute attacks on the target requires a set of activities. First, the target must be identified. In attack/defend scenarios, this might or might not be specified. Assuming it has not been specified, the first thing to do would be to conduct reconnaissance activities to identify potential targets. There are interesting implications here. Conducting reconnaissance is much more than simply a visual assessment. Depending on the type of target and its location, diverse types of intelligence sources might be needed to collect the information to identify potential targets. This challenge is particularly tricky when the potential targets are moving, such as airborne platforms.

Once the potential targets have been identified, then the capabilities and weaknesses need to be cataloged. This is done by probing and testing. For example, if the target of a red team exercise is a computer system on an unmanned aerial system, then one-way weaknesses can be potentially identified through analysis of attempts to connect to the system. Error messages can be useful, in that they can reveal information about why an error occurred.

After the capabilities and weaknesses of the targets are cataloged, an attack plan is created. This plan allocates resources in a prioritized pattern to execute a strategy to achieve a specified result. Resources include both human operators and tools, such as weapons. If the red team is mimicking an actual adversary, this plan must mirror the normal processes of the adversary. If not, then the red team is free to develop their own plan. Once the plan is in place, the red team executes the plan, possibly making real time adjustments in response to changes in conditions, discovery of new information, or to ward off defenses.

The most important result of the red team exercise is knowledge. It should be collected at every step of the exercise. All activities should be noted. This can accomplish through filming, recording, instrument readings, or taking notes. Generally, it is best to have one or more observers collecting the data. After the exercise has concluded, it is very important for the actual team members to immediately conduct a hot wash-up; a real time review of what was successful, what as not, and lessons learned. Their individual impressions and observations can be extremely valuable to developing the knowledge and capabilities of the enterprise.

Blue Teaming

The blue team is the defending team. Their challenge is as complex, perhaps more so, than the red team. In order to defend a target, the blue team needs to have full knowledge of the target, including technical capabilities and weaknesses, vulnerabilities, and operational patterns. There should be detection mechanisms in place to alert the blue team to adversarial actions. Additionally, the blue team needs to be on the lookout for stealthy or unexpected adversarial activities. Finally, the blue team needs to be able to stop adversarial actions and remediate any harm done.

Having full knowledge of each potential target is a daunting task. Obviously, not any one person can full knowledge of a complex system. This is where team composition becomes important. The team must possess a variety of capabilities, spanning technical to operational knowledge sets. Running many exercises can assist in identifying areas that need additional expertise as well as broaden experience. One possible problem in some systems is that the knowledge of the vulnerabilities and remedies may be extremely sensitive. It may be necessary for some blue teams to have compartmentalized knowledge areas; a common set, a sensitive set, and a closely held set. This becomes a challenge for both constituting the team and managing the interactions of the team members.

The detection mechanisms the blue team relies on should be ubiquitous. It does no good to have special detection mechanisms simply for exercises. The blue team needs to be operating in as close to a real environment as possible. The conduct and result of the exercise should be a substantial input to planned improvements in the system. The feedback from the blue team is an important part of that.

Recall in Chapter 4 the author discussed the classes of events that needed to be detected. The blue team needs to address each detection challenge while planning their defenses, including detection of the unanticipated or exotic activities. The blue team cannot simply rely on detection capabilities but must be on the alert for any unusual activity and be prepared to react as necessary. The red team is motivated to overcome blue team defenses. The blue team needs to be on constant alert.

When the blue team detects red team activities, they need to have plans and procedures on how to mitigate these actions. This implies that the blue team has considered and practiced reacting to various activities. Part of that practice should include command and control of actions. Clean lines of communications with fail-over procedures, if a link in the chain of command is disabled or unavailable, is critical to effective team action. Ad hoc responses may be necessary, particularly in the case of unexpected red team activity, but the command and control of the ad hoc response execution is just as important as it is to preplanned activity execution.

Blue team experiences are just as important to knowledge development as red team experiences. During exercises, all activities should be logged, just like red team activities. Again, this can be done through filming, recording, instrument readings, or through taking notes. Observers should be responsible for collecting data. The blue team should also conduct a hot wash-up of their experiences. Both the team members' individual impressions and observations are extremely valuable to developing the knowledge and capabilities of the enterprise.

Benefits

An obvious benefit of attack/defend exercises is building expertise. Another obvious benefit is testing the target system and developing improvements to its defenses. Less obvious benefits include developing new tactics and techniques, discovering innovations, and creating stronger team relationships.

Operator expertise is an important benefit. While marginal improvements for individuals can be achieved simply by working through a scenario, improving how the team interacts is also a desired outcome. When teamwork improves, sharing of knowledge is an added benefit. Teammates who trust and value each other share information, ideas, and techniques, improving both individuals and the team. Leveraging this synergy during team activity reviews can result in powerful advances. The secret is to truly value each person's contribution; no person should be marginalized or made to be a scapegoat for any real or perceived failure. The value of exercises is in learning and improving. There should be no penalties for making mistakes. Learning occurs from making mistakes and seeing the consequences.

Testing the target system in not simply a matter of seeing what the red team accomplish. The target system is comprised of technologies, people, and organizational constructs. Each of these can potentially be improved and refined through probing, testing, and attacking. Improvements that are discovered through attack/defend exercises can be as trivial as improving the chain of command. Table top exercises in disaster recovery scenarios often reveal that there are few plans for replacing people who fall ill or are unable to communicate with the rest of the team. In fact, one of the best stress tests for organizations is to do a table top exercise of a common operational scenario and then randomly remove players how the organization reacts without them. Simple insights can lead to easy improvements that greatly benefit the organization as a whole. Individual interactions with technologies during exercises can also reveal needed changes. Reviewing all the collected material from the exercise should be done with these types of insights in mind.

While attack/defend scenarios tend to focus on how well the defense is conducted, one of the more powerful outcomes could benefit the offense; the development of new tactics, techniques, and procedures in adversarial situations. Careful recording of red team activities and enabling members to test innovative approaches can be very important benefits to the entirety of operational needs.

Discussion Questions

Test your understanding of the material by thinking through the following questions. Discuss them with other people. Can you think of more questions that would be useful?

1. Devise a table top exercise for a kinetic attack on a surveillance drone. Identify an adversary, select a kinetic attack, and lay out the attack scenario in phases. Who should participate in this exercise? What questions and decisions should be considered at each step in the exercise? Consider all the elements of national power while creating this exercise; diplomacy, information, military, and economy.
2. What kind of intelligence would be needed to set the stage for such a table top exercise? What sources would need to be used?
3. Imagine that you have been asked to command a blue team for a UAS ground station attack/defend exercise. What types of people would you want on your team? How would you organize your team?
4. Imagine instead that you have been asked to command a red team for a UAS ground station attack/defend exercise. What types of people would you want on your team? How would you organize your team?
5. For the attack/defend scenario imagined in challenges 3 and 4, how would you organize the observation aspects? What types of technology would you want to use? How would you collate and curate the information collected? What kinds of reports would you want to develop at the end of the exercise?

Sources for more information

O'Neil, Cathy. (2016) Weapons of Math Destruction: How Big Data Increases Inequality and Threatens Democracy. Crown. ISBN 978-0553418811

Mudd. Philip. (2015) The HEAD Game: High-Efficiency Analytic Decision Making and the Art of Solving Complex Problems Quickly. Liveright. ISBN 978-0871407887

Heuer, Richards J. Jr. (1999) Psychology of Intelligence Analysis. Central Intelligence Agency Center for the Study of Intelligence. Available online at https://www.cia.gov/library/center-for-the-study-of-intelligence/csi-publications/books-and-monographs/psychology-of-intelligence-analysis/PsychofIntelNew.pdf

Red Team Journal. https://redteamjournal.com/about/

Chapter 6: Case Studies in Risk for UAS

This chapter illustrates the material presented in Chapters 4 and 5 using case studies. These cases are drawn from news reporting, with sources provided.

Student learning objectives. After reading this chapter, students should be able to do the following:

– Apply pragmatic risk analysis to operational decisions;

– Understand the benefits of studying the history of how things go wrong; and

– Appreciate the integration of security and intelligence.

Case 1: When the Enemy Hacks Your Data Stream

On February 23, 2016, Israeli authorities arrested a Palestinian on suspicion of hacking into Israeli drones.[1] Details of the case are meager, but it appears that Majd Ouida, 22 at the time of his arrest, had managed to breach protections and obtain access to Israeli surveillance data from the drones from 2011 to 2014.[2] Alternatively, he may simply have just collected the transmissions using appropriate antennas and processing technology; they are broadcast, after all. As David Axe pointed out, "what's really impressive is the fact that Israeli authorities caught him allegedly doing so. That's because there's no straightforward way to know whether someone has intercepted your drone video."[3]

The question that bubbles up to the top is: "how did the Israelis know that their data was being collected by unauthorized people?" This is a classic confidentiality detection problem. When secrets have been stolen in physical form, at least the physical artifact is missing. But when a copy is surreptitiously made, or a conversation is overheard, how can you detect that your secrecy has been compromised? Think back to the example of the competitor's pricing strategy in Chapter 5. If someone snuck into the building and took a photo of the strategy, how would

1. Gross, Judah Ari et al. (2016) Israel charges Islamic Jihad hacker for spying on IDF drones. Times of Israel, 23 March 2016. Available online at https://www.timesofisrael.com/israelcharges-islamic-jihad-hacker-for-spying-on-idf-drones/
2. Ben-Yishai, Ron. (2016) IDF's cyber defense easily breached. Ynet News, 23 March 2016. Available online at https://www.ynetnews.com/articles/0,7340,L-4782445,00.html
3. Axe, David. (2016) How Islamic Jihad Hacked Israel's Drones. The Daily Beast, 25 March 2016. Available online at https://www.thedailybeast.com/how-islamic-jihad-hacked-israels-drones

the competitor know that their secrecy had been violated? If no detection capabilities were in place, such a theft might remain undetected indefinitely.

There are ways to detect violation of confidentiality, but most of them must be engineered into an enterprise before the compromise occurs. These include both technical and procedural solutions. Identifying the potential for secrecy compromise is an important first step. This scenario is where attack/defend exercises are valuable. Challenging a red team to figure out how to access the information collected by a sensor will reveal interesting weaknesses and opportunities to engineer solutions.

Questions to consider:

- What are the tactical implications for the adversary having access to your surveillance data?
- What are the strategic implications of the adversary having access to your surveillance data?
- How would you design an attack/defend exercise to focus on confidentiality compromise?
- How would understanding the intelligence efforts of an adversary help you to understand which weaknesses to focus on?
- What surveillance efforts are available in discovering the compromise of secrets?
- What are the costs and benefits of implementing cryptography for broadcasts from drones?
- Could you use this known compromise as a way of inserting misleading information into the adversary's decision processes?

Case 2: When Your Drone Goes Missing

Unmanned Aerial Systems (UAS) are, by nature, unmanned. This implies the control of the system, such as navigation, is comprised of some combination of remote control and autonomous pre-programmed decision. In the event of loss of remote connectivity, the UAS is typically provided with a set of pre-programmed safe landing sites that conform to its mission profile.

When a UAS is sent into hostile territory, one foreseeable risk is the adversary may attempt to shoot it down. Another foreseeable risk is capturing the UAS, either in whole or in part.

This happened in December 2011, when Iran captured and displayed, to great propaganda fanfare, a surveillance drone operated by the U.S. The drone, a Lockheed Martin RQ-170 Sentinel, had been operating over Afghanistan near the Iranian border. The Iranians claimed that they captured the RQ-170 through cyber warfare means. They later claimed that that they were able to decode all the stored data from the RQ-170 sensor systems. Various claims have been made by both the US and Iran on how the capture was made and under what circumstances. These claims include GPS spoofing, cyber intrusion into the control system, and physical damage. The

US demanded the return of the RQ-170. Iran countered those demands by alleging that Iranian airspace had been violated and that international laws had been violated. By 2016, Iran claimed to have reversed engineered the design of the RQ-170 and created their own version. The event remained in the news for several years.[4]

Questions remain; did the drone fail or was it intentionally brought down? One possibility is the RQ-170 failed in flight and crash landed in an area where Iranians recovered it before friendly forces could. The question of what caused the failure is somewhat tangential. Did the system simply die? Were navigation systems were compromised? Was there some sort of physical damage to the aircraft?

An obvious question that arises is, "why was there no self-destruct capability embedded?" After all, if you expect to fly near or over hostile territory, then there must be a remote risk of capture. Refer to history for risk analysis; The U-2 piloted by Francis Gary Powers, shot down by the Soviet Union in 1960. The U-2 had the hallmarks of stealth (extreme altitude) and risk was considered low. But even so, it was shot down.[5] An auto-destruct system could have been triggered remotely when it was determined that the system was not in friendly control. Alternatively, a destruct mechanism could be logically triggered by lack of signaling, an extended period without authorized contact could be used as a triggering event for auto-destruct.[6]

Since no system is perfect, alternatives to such a capability for self-destruction should be considered as well. In this line of analysis, more questions come to mind. "What security controls were in place to detect the compromise of the sensitive data and equipment?" and "when

4. Sources for this material include the following: Wikipedia. (2018) Iran–U.S. RQ-170 incident. Available online at https://en.wikipedia.org/wiki/Iran–U.S._RQ-170_incident. Last edit date when accessed 1 July 2018; Peterson, Scott. (2011) Downed US drone: How Iran caught the 'beast'. Christian Science Monitory, 9 December 2011. Available online at https://www.csmonitor.com/World/ Middle-East/2011/1209/ Downed-US-drone-How-Iran-caught-the-beast ; Cenciotti, David. (2016) Iran unveils new UCAV modeled on captured U.S. RQ-170 stealth drone. The Aviationist, 2 October 2016. Available online at https://theaviationist.com/ 2016/10/02/iran-unveils-new-ucav-modeled-on-captured-u-s-rq-170-stealthdrone/ ; Opall-Rome, Barbara. (2018) Israel Air Force says seized Iranian drone is a knockoff of US Sentinel. Defense News, 12 February 2018. Available online at https://www.defensenews. com/global/mideast-africa/2018/ 02/12/israel-air-force-says-seized-iraniandrone- is-a-knockoff-of-us-sentinel/
5. Historian, US Department of State. (n.d.) U-2 Overflights and the Capture of Francis Gary Powers, 1960. Office of the Historian of the U.S. Department of State. Available online at https://history.state.gov/milestones/1953-1960/u2-incident
6. Hsu, Jeremy. (2017) Self-Destructing Gadgets Made Not So Mission Impossible. IEEE Spectrum, 9 February 2017. Available online at https://spectrum.ieee.org/tech-talk/consumerelectronics/gadgets/selfdestructing-gadgets-made-not-so-mission-impossible

detected, what reactions were engineered into the system to reduce the risk of data and equipment being exposed?" For example, a localized destruct capability (explosive, acid, etc.) could be built into the protective casing of the systems. The detect mechanism could be integrated into the casing itself through various means, the easiest being as an elemental part of the casing. If the casing were to be opened, the integrity of the case would be destroyed, triggering the reaction. In this case, the result would be the destruction of data and/or equipment. A real-life example of this type of integrated detection/reaction capability can be seen in any museum. Fine grids of wire are integrated into casings, which, when parted, cause an electrical circuit to be broken, which triggers an alarm.[7]

Questions to ponder:

- How could a table top exercise have helped in discovering this risk?
- What information would you need about adversary capabilities to understand the level of risk? How could you obtain that information?
- When should you include minimal risk, but high impact problems, in your engineering analysis?
- How did the capture of this UAS impact the risk to friendly forces?
- How did the capture and analysis of the UAS alter the balance of power in the near and far term? Consider allies of Iran while pondering this question.

Case 3: When Pilots Are Targeted for Assassination

In 2016, it was widely reported that the Islamic State had compiled and published a list of US drone pilots, including home addresses and photographs. The list was posted online with an accompanying message urging followers to attack the pilots by any means available.[8]

Clearly, the knowledge that you personally are a specific target differs from the usual warfare rules of engagement. , where a person in a uniform is a general target while on the battlefield. This type of tactic has occurred at least once previously, in the case of the USS Vincennes shoot-down of an Iranian civilian airliner.[9]

7. Anderson, Ross. (2008) Chapter 16: Physical Tamper Resistance. Security Engineering (2nd edition). Available online at https://www.cl.cam.ac.uk/~rja14/Papers/SEv2-c16.pdf
8. Gadher, Dipesh and Toby Harden. (2016) Islamic State hackers publish hit list of US drone pilots. The Australian, 2 May 2016. Available online at https://www.theaustralian.com.au/news/world/islamic-state-hackers-publish-hit-list-of-us-drone-pilots/
9. Bernstein, Leonard and Richard A. Serrano. (1989) Bomb Blows Up Van Driven by Wife of Vincennes Captain; She Escapes. LA Times, 11 March1989. Available online at http://articles. latimes.com/1989-03-11/news/mn-792_1_pipe-bomb

Questions to consider:

- What types of security controls *could* be put in place to protect the identity of drone pilots?
- What types of security controls *should* be put in place to protect that data?
- How would those security controls increase the complexity of UAS operations?
- What are the risk trade-offs?

Case 4: When Commercial Drones Spy Domestically

In 2017, stories began to emerge alleging that a commercial drone manufactured by DJI, Inc, a Chinese company, was collecting data about U.S. infrastructure and sending it back to China.[10][11][12] The stories stemmed from a memo that was issued in August 2017 from the Los Angeles office of U.S. Immigrations and Customs Enforcement. The memo stated that the office assessed, "with moderate confidence that Chinese-based company DJI Science and Technology is providing U.S. critical infrastructure and law enforcement data to the Chinese government" and "with high confidence the company is selectively targeting government and privately owned entities within these sectors to expand its ability to collect and exploit sensitive U.S. data."[13] The memo, originally classified as Law Enforcement Sensitive, was leaked.[14]

It is not optimal to use equipment that surreptitiously collect data about your facilities and locale. This has been a repeated theme through the years. For example, a Popular Mechanics article from 1997 recounted the story of a camera placed inside a Xerox copying machine.[15]

10. Newman, Lily Hay. (2017) THE ARMY GROUNDS ITS DJI DRONES OVER SECURITY CONCERNS. Wired Magazine, 7 August 2017. Available online at https://www.wired.com/story/army-dji-drone-ban/
11. Mozur, Paul. (2017) Drone Maker D.J.I. May Be Sending Data to China, U.S. Officials Say. New York Times, 29 November 2017. Available online at https://www.nytimes.com/2017/11/29/technology/dji-china-data-drones.html
12. Corfield, Gareth. (2018) Yes, drone biz DJI's Go 4 app does phone home to China – sort of. The Register, 25 Apr. 2018. Available at https://www.theregister.co.uk/2018/04/25/dji_data_security_audit/
13. U.S. Immigration and Customs Enforcement (2017) Da Jiang Innovations (DJI) Likely Providing U.S. Critical Infrastructure and Law Enforcement Data to Chinese Government. ICE-IL-17-0019, 9 August 2017. Available online at https://info.publicintelligence.net/ICE-DJI-China.pdf
14. Smith, Ms. (2017) Leaked DHS memo accuses drone maker DJI of spying for China. CSO Magazine, 3 December 2017. Available online at https://www.csoonline.com/article/3239726/security/leaked-dhs-memo-accuses-drone-maker-dji-of-spying-for-china.html
15. Stover, Dan. (1997) Spies in the Xerox machine: how an engineer helped the CIA snoop on Soviet diplomats.

Another interesting story related to Acoustic Kitty: a cat that was operated on to embed listening electronics inside its body for espionage purposes.[16][17]

Questions to be considered:

- How do you know if you can trust your equipment?
- Assume your equipment is in perfect working order. How would one detect clandestine operations?
- How would you craft a security policy regarding equipment probing?
- What types of operational risks does this type of activity pose, and how can one engineer countermeasures to mitigate it?
- How would one use an attack/defend exercise to test for these activities?
- What would an adversary gain from this type of intelligence activity?
- What risk did DJI expose China to when they implemented these measurements into their equipment? Consider all the elements of national power; Diplomatic, Information, Military, and Economic (DIME).
- What could one gain from discovering and exploiting this intelligence capability without exposing one's knowledge to the adversary?

Case 5: The Drone That Steals Your Wi-Fi Password

Security researchers have discovered substantial activity of small UASs, including spying, data theft, and other nefarious acts. The "Wireless Aerial Surveillance Platform, or WASP, … is equipped with an HD camera, a cigarette-pack sized on-board Linux computer packed with network-hacking tools including the BackTrack testing toolset and a custom-built 340-million-word dictionary for brute-force guessing of passwords, and eleven antennae." As one researcher, describing his invention, said "This is like Black Hat's Greatest Hits. And it flies."[18] As a rare spot of good news, other researchers have developed detection capabilities for this type of activity, "someone using only a laptop and an object that flickers can detect if someone is

Popular Science, 1 January 1997. Available online at https://electricalstrategies. com/about/in-the-news/spies-in-the-xerox-machine/

16. Hillman, Jennifer. (2008) What the CIA Learned from Get Smart. Wired Magazine, 19 June 2008. Available online at https://www.wired.com/2008/06/pl-print-19/
17. National Security Archives. (n.d.) Memorandum for: [deleted], Subject: [deleted] Views on Trained Cats [deleted] for [deleted] Use, March 1967, 2 pp. Available online at https://nsarchive2.gwu.edu/NSAEBB/NSAEBB54/
18. Greenberg, Andy. (2011) Flying Drone Can Crack Wi-Fi Networks, Snoop on Cell Phones. Forbes, 28 July 2011. Available online at https://www.forbes.com/sites/andygreenberg/2011/07/28/flying-drone-can-crack-wifi-networks-snoop-on-cell-phones/

using a drone to spy on them."[19] Meanwhile, UASs are getting smaller and are being implemented in swarms.

These are both security and intelligence issues. The confidentiality issues are paramount; having secrets stolen electronically or using video is a significant problem. The next level of threat to be concerned about is the infiltration of bad data, misleading data, meaconing signals, and jamming signals into SOP.

Questions to ponder:

- What types of intelligence could help detect whether data integrity is being attacked?
- What level of surveillance is required to detect unauthorized UASs operating near or in your sensitive areas?
- What level of operational security training for personnel is required to reduce the potential impact of these types of attacks?
- How can these threats be included in attack/defend scenarios?

Concluding Thoughts

These five cases are real world examples. Unfortunately, the enemy always gets a vote. Keeping up with developments and news from the security and intelligence communities can be time intensive. A management best practice; divvy up the work. Have team members research diverse sources and report back on interesting developments. Make learning about advances an important part of team activities. Learn from your peers. Participate in professional networks. Go to conferences and listen to the experts. There is too much to know for any one person, so you need to develop your personal team as well as participate in a professional team.

References

Anderson, Ross. (2008) Chapter 16: Physical Tamper Resistance. Security Engineering (2nd edition). Available online at https://www.cl.cam.ac.uk/~rja14/Papers/SEv2-c16.pdf

Axe, David. (2016) How Islamic Jihad Hacked Israel's Drones. The Daily Beast, 25 March 2016. Available online at https://www.thedailybeast.com/how-islamic-jihad-hacked-israels-drones

Ben-Yishai, Ron. (2016) IDF's cyber defense easily breached. Ynet News, 23 March 2016. Available online at https://www.ynetnews.com/articles/0,7340,L-4782445,00.html

19. Charlaff, Joe. (2018) Spies in the Sky: Israeli researchers develop a counter-surveillance drone system. The Jerusalem Report, 6 August 2018. Available online at https://aabgu.org/wp-content/uploads/2018/07/JRep-August-6-38-39-Joe-drones.pdf

Bernstein, Leonard and Richard A. Serrano. (1989) Bomb Blows Up Van Driven by Wife of Vincennes Captain; She Escapes. LA Times, 11 March1989. Available online at http://articles.latimes.com/1989-03-11/news/mn-792_1_pipe-bomb

Cenciotti, David. (2016) Iran unveils new UCAV modeled on captured U.S. RQ-170 stealth drone. The Aviationist, 2 October 2016. Available online at https://theaviationist.com/2016/10/02/iran-unveils-new-ucav-modeled-on-captured-u-s-rq-170-stealth-drone/

Charlaff, Joe. (2018) Spies in the Sky: Israeli researchers develop a counter-surveillance drone system. The Jerusalem Report, 6 August 2018. Available online at https://aabgu.org/wp-content/uploads/2018/07/JRep-August-6-38-39-Joe-drones.pdf

Corfield, Gareth. (2018) Yes, drone biz DJI's Go 4 app does phone home to China – sort of. The Register, 25 Apr. 2018. Available at https://www.theregister.co.uk/2018/04/25/dji_data_security_audit/

Gadher, Dipesh and Toby Harden. (2016) Islamic State hackers publish hit list of US drone pilots. The Australian, 2 May 2016. Available online at https://www.theaustralian.com.au/news/world/islamic-state-hackers-publish-hit-list-of-us-drone-pilots/

Greenberg, Andy. (2011) Flying Drone Can Crack Wi-Fi Networks, Snoop on Cell Phones. Forbes, 28 July 2011. Available online at https://www.forbes.com/sites/andygreenberg/2011/07/28/flying-drone-can-crack-wifi-networks-snoop-on-cell-phones/

Gross, Judah Ari et al. (2016) Israel charges Islamic Jihad hacker for spying on IDF drones. Times of Israel, 23 March 2016. Available online at https://www.timesofisrael.com/israel-charges-islamic-jihad-hacker-for-spying-on-idf-drones/

Hillman, Jennifer. (2008) What the CIA Learned from Get Smart. Wired Magazine, 19 June 2008. Available online at https://www.wired.com/2008/06/pl-print-19/

Historian, US Department of State. (n.d.) U-2 Overflights and the Capture of Francis Gary Powers, 1960. Office of the Historian of the U.S. Department of State. Available online at https://history.state.gov/milestones/1953-1960/u2-incident

Hsu, Jeremy. (2017) Self-Destructing Gadgets Made Not So Mission Impossible. IEEE Spectrum, 9 February 2017. Available online at https://spectrum.ieee.org/tech-talk/consumer-electronics/gadgets/selfdestructing-gadgets-made-not-so-mission-impossible

Mozur, Paul. (2017) Drone Maker D.J.I. May Be Sending Data to China, U.S. Officials Say. New York Times, 29 November 2017. Available online at https://www.nytimes.com/2017/11/29/technology/dji-china-data-drones.html

National Security Archives. (n.d.) Memorandum for: [deleted], Subject: [deleted] Views on

Trained Cats [deleted] for [deleted] Use, March 1967, 2 pp. Available online at https://nsarchive2.gwu.edu/NSAEBB/NSAEBB54/

Newman, Lily Hay. (2017) THE ARMY GROUNDS ITS DJI DRONES OVER SECURITY CONCERNS. Wired Magazine, 7 August 2017. Available online at https://www.wired.com/story/army-dji-drone-ban/

Opall-Rome, Barbara. (2018) Israel Air Force says seized Iranian drone is a knockoff of US Sentinel. Defense News, 12 February 2018. Available online at https://www.defensenews.com/global/mideast-africa/2018/02/12/israel-air-force-says-seized-iranian-drone-is-a-knock-off-of-us-sentinel/

Peterson, Scott. (2011) Downed US drone: How Iran caught the 'beast'. Christian Science Monitory, 9 December 2011. Available online at https://www.csmonitor.com/World/Middle-East/2011/1209/Downed-US-drone-How-Iran-caught-the-beast

Smith, Ms. (2017) Leaked DHS memo accuses drone maker DJI of spying for China. CSO Magazine, 3 December 2017. Available online at https://www.csoonline.com/article/3239726/security/leaked-dhs-memo-accuses-drone-maker-dji-of-spying-for-china.html

Stover, Dan. (1997) Spies in the Xerox machine: how an engineer helped the CIA snoop on Soviet diplomats. Popular Science, 1 January 1997. Available online at https://electricalstrategies.com/about/in-the-news/spies-in-the-xerox-machine/

U.S. Immigration and Customs Enforcement (2017) Da Jiang Innovations (DJI) Likely Providing U.S. Critical Infrastructure and Law Enforcement Data to Chinese Government. ICE-IL-17-0019, 9 August 2017. Available online at https://info.publicintelligence.net/ICE-DJI-China.pdf

SECTION III
UAS HEART & SOUL – SENSE AND AVOID (SAA) SYSTEMS / STEALTH

Chapter 7: UAS SAA Methodologies, Conflict Detection

Student Learning Objectives

In Chapter 3: *Understanding Hostile Use and Cyber-Vulnerabilities of* UAS: *Components, Autonomy* Vs. *Automation, Sensors,* SAA, SCADA *and Cyber Attack Taxonomy*, the student was exposed to the Counter UAS *problem* and UAS Vulnerabilities. The heart of any UAS system are its Sense and Avoid Systems (SAA). It is the key to understanding threats and vulnerabilities of the UAS. Understanding SAA methodologies, conflict detection and conflict resolution principles is best practice. In Chapter 7, the student was introduced to SAA sensing practice and a logic case for conflict detection and resolution using the sensing data. Along with Chapter 8: *Designing* UAS *systems for Stealth* and Chapter 9 *Smart Skies Project*, the student should be versed on the dynamics and importance of SAA.

Sense and Avoid (SAA) Function

"The purpose/function of SAA is to remove and replace a human pilot, to detect and resolve certain flight hazard s. Hazards consist of other traffic or ground objects; anything that presents a risk of collision. UAS is unmanned. SAA systems are not necessarily designed to preserve the aircraft. SAA is certainly designed to prevent collisions, ground personnel, and collateral damage to property. SAA must operate for emergency and diversionary events as well as normal operations." (Zeitlin, 2012)

"Human pilots on manned aircraft are required to See and Avoid (also identified as SAA) hazards. They do this with regular visual scans across their Forward Field of View (FFoV) to detect another A/C." Pilot "scanning may be focused toward areas where operations are on-going or informed by controller over radio traffic or by electronic display on an Automated Identification System (AIS) for collision avoidance." (Zeitlin, 2012) Traffic spotted means the pilot must judge "its trajectory relative to their own A/C and determine the risk of collision and whether an A/C maneuver is required. (Zeitlin, 2012) Humans are at disadvantage for the *See and Avoid* process especially in poor weather, when backgrounds are confusing, when visibility is reduced, or pilot workload is excessive.

UAS come in all sizes and capabilities. Those that can be outfitted with robust SaA equipment, can overcome the human limitations. "Some UAS are too small to carry weighty SAA equipment. These A/C can operate within direct radio communication of the pilot and to maintain a Visual Line of Sight (VLoS) between pilot and A/C." (Zeitlin, 2012) Small UAS may have additional restrictions that preclude operating over densely populated areas, near airports, or total weight capacities.

System Configurations and Subsystems

Chapter 3 addressed onboard / off-board SAA configurations, sensors and surveillance volume, sensor capabilities, typical sensors, coordinate systems, ground-based sensing, sensor parameters, self-separation schemes, SAA services and sub-functions with timeline, and SCADA. Additional material to these topics follows.

Sensor Categories

UAS robotic aircraft flights are primarily to collect data. The data collected can be broken down into two broad categories: ***in situ***[1] and **remote sensing.** These two methods are used to collect vast amounts of intelligence, including an enemy's dispositions of forces and probable intentions. (Marshall, 2016)

In situ Sensing

In situ sensing in a UAS means the aircraft is transported to the location where measurements are to be made. There are two methods: brute force on the aircrafts control inputs [such as turning into a storm to measure the vehicles responses] or measure an attribute at the location directly. In the former, the UAS is forced to respond to a stimulus or environmental or state parameter. In the latter, the environmental or state attribute is measured at the location while the aircraft transits the location. Measuring gas composition and temperature changes are examples of the latter, more passive approach. (Marshall, 2016)

Remote Sensing

Remote sensing is the process of measuring an object of interest from a distance. This is done by detecting and measuring the effects of said object, usually in the form of emitted or reflected particles and/or waves. (Marshall, 2016) Remote sensing is not dependent on platform or application. There are three broad categories of remote sensing: terrestrial, airborne, and space-based. It is the Airborne Remote Sensing (ARS) from UAS category that is the topic here. ARS has a broad range of sensor options, from large multiple arrays to single sensor pick-up systems. There are four classes of ARS: framing, push broom, scanner, and receiver systems. All four classes of ARS work on most of the EMS. (Marshall, 2016) ARS work on all UAS platforms. Figure 7-1 (Pater, 2018) shows the wide range of drones with size comparisons. ARS equipment

1. In situ comes from Latin "in place" Merriam Webster's Dictionary defines it as "in the natural or original position or place."

is not limited to large drones. (Marshall, 2016) The sensors performance, function and size may vary, but they can be installed on platforms.

The emitted or reflected particles and/or waves often associated with ARS is sunlight, or EMS Visible Light. (EMSVIS) (Marshall, 2016) Payloads that use Electro-Optical (EO) or digital cameras, the sunlight enters through the lens and strikes a sensor. The light receiving lens on the EO camera is aimed at a target and collects the sunlight that the target reflects. The target is at a distance and the sunlight reflected to the camera is collected and the data converted to useful intelligence.

Figure 7-1 Drone Survival Guide

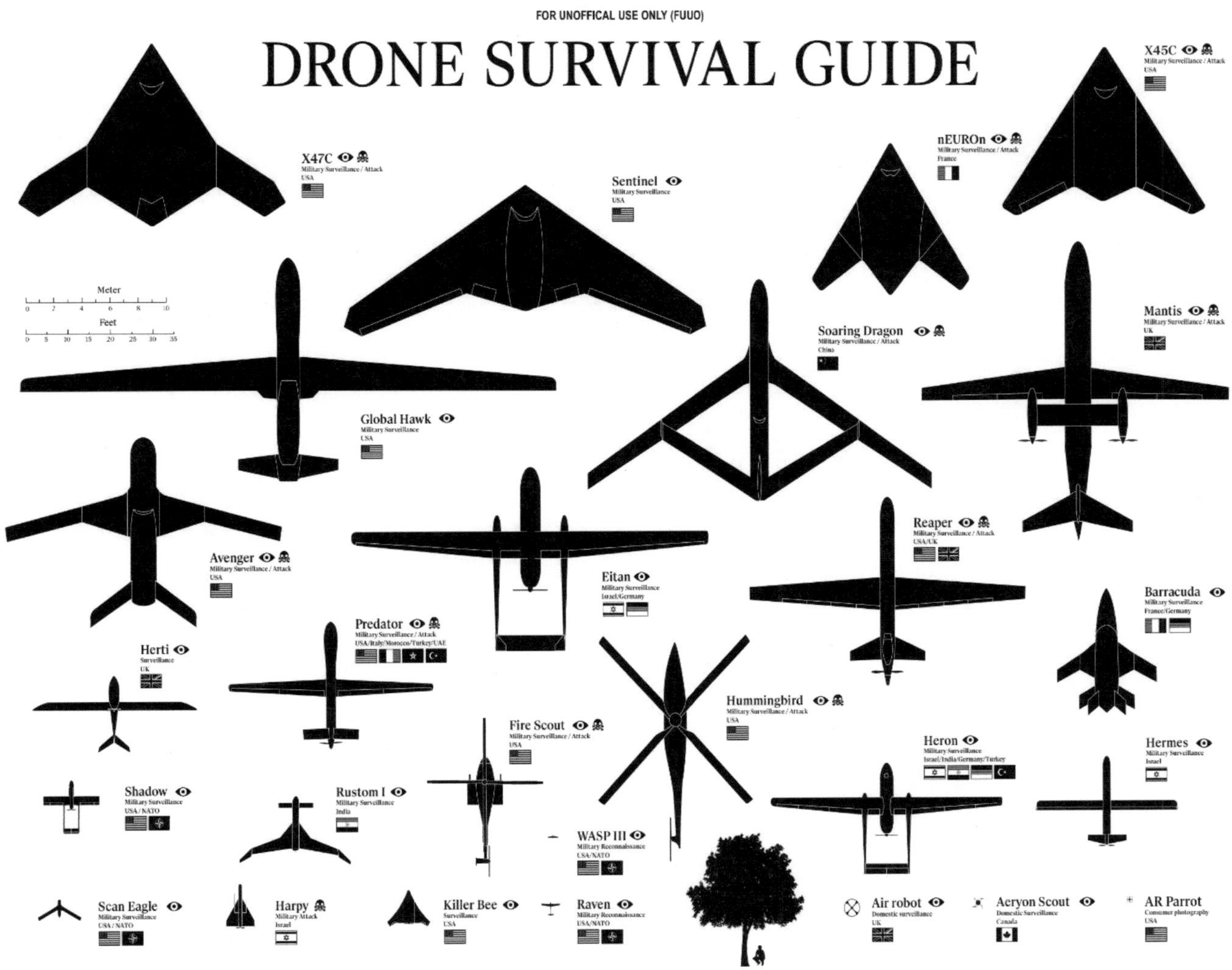

Source: Pater, R. (2018). Drone Survival Guide. Retrieved from http://www.drone.survivalguide.org/DSG.pdf

Remote sensing systems can be divided into two categories; active and passive. Active sensors emit EM radiation, directed at a target and then measure the reflected signal. (Marshall, 2016) Passive sensors do not emit EM radiation, instead measuring what is emitted by other sensors after it is reflected or as the target emits it. In UAS ARS the external target is almost always the sun.

The EMS is broken down into bands. EM bands are segments of energy grouped together by a common property and defined by a specific range of v It helps to understand the terminology: Agency for Science, Technology and Research (ASTRA) defines the following: VNIR = Visible light and near infrared (NIR) 400 – 1400 nm, 0.4 – 1.4 μm wavelength range (MIR); SWIR=Shortwave infrared, 1400-3000 nm, 1.4 -3.0 μm wavelength range; and TIR = Thermal infrared = 8000 – 15000 nm, 8 -15 μm (TIR) vibrational wavelengths. (Marshall, 2016) In ARS, the bands most commonly dealt with are in Table 7-1. (Schowengerdt, 2007)

Table 7-1 2007 Listing Remote Sensing Use of EMS

name	wavelength range	radiation source	surface property of interest
Visible (V)	0.4–0.7μm	solar	reflectance
Near InfraRed (NIR)	0.7–1.1μm	solar	reflectance
Short Wave InfraRed (SWIR)	1.1–1.35μm 1.4–1.8μm 2–2.5μm	solar	reflectance
MidWave InfraRed (MWIR)	3–4μm 4.5–5μm	solar, thermal	reflectance, temperature
Thermal or LongWave InfraRed (TIR or LWIR)	8–9.5μm 10–14μm	thermal	temperature
microwave, radar	1mm–1m	thermal (passive), artificial (active)	temperature (passive), roughness (active)

Source: (Schowengerdt, 2007)

Table 7-1 illustrates that different spectrums have various sources and different surface properties that are measured by wavelength. The imaging sensor responds to specific wavelength of incoming EM radiation. This response can be measured and calibrated. (Marshall, 2016) The calibration process is called radiometric calibration. Once a sensor is calibrated for a subset of the EMS, the sensor response can be linked directly to an individual component wavelength of the EMS being reflected by the target. (Marshall, 2016) Another powerful aspect of ARS is spatial continuity, which is the ability to know what is happening at a target location at every location surrounding the target for quite a distance. (Marshall, 2016)

Units

Since chapter wavelengths are referred to in SI units, Table 7-2 shows the common units used for EMS radiation bands.

Sensor Types

Two general types of ARS: Those that build images (Imaging sensors) and those that do not (spot sensors).

Spot sensors

Spot sensors measure single locations and do not create an image library. They do not facilitate special continuity, rather they opt for simplicity. (Marshall, 2016)

Table 7-2 Common Wavelengths units for Electromagnetic Radiation

Common wavelength units for electromagnetic radiation

Unit	Symbol	Wavelength, (m)	Type of Radiation
Picometer	pm	10^{-12}	Gamma ray
Ångstrom	Å	10^{-10}	X-ray
Nanometer	nm	10^{-9}	X-ray
Micrometer	μm	10^{-6}	Infrared
Millimeter	mm	10^{-3}	Infrared
Centimeter	cm	10^{-2}	Microwave
Meter	m	10^{0}	Radio

Source: (Cummings, 2007)

Imaging Sensors

There are three basic ways to generate images for an ARS: Line scanners (push- brooms) and array sensors. Line scanners move a single or small number of sensors elements back and forth to build up a picture of the target acquisition. It is limited by pivot distance, movement speed, and sensor processing time to build the image. Push-brooms have a fixed array of sensors, which are the width of the final image product across the A/C path, but only one element tall. Array sensors are like digital cameras, in that when triggered the entire array "flashes" simultaneously. (Marshall, 2016)

Camera

The most common UAS remote-sensor is the camera. There are four types of ARS cameras: Visible Spectrum, Near Infrared, Infrared and Hyperspectral. (Marshall, 2016)

Visible Spectrum Cameras (VIS) and Near-Infrared (NIR) Cameras

VIS and NIR cameras operate on the same principles, the only difference being the NIR camera sensor is sensitive to the NIR wavelengths (See Table 7-1). A VIS image consists of three primary colors, red, green, blue (RGB) to create a full color image. NIR images can capture green, red, and NIR reflectance. The human eye cannot see NIR. Therefore, the NIR band of colors is artificially shifted to form human-visible colors. The standard is called Color InfraRed (CIR). (Schowengerdt, 2007)

Long-Wave Infrared Cameras (LWIR)

LWIR sensors respond to heat striking the sensors. The temperature change causes an electronic signal to be generated. Bright spots or high heat flashes can wipe out the thermographic images. (Schowengerdt, 2007)

Infrared sensors beyond 1800 nm of the EM range can not use the same sensor materials as for VIS or NIR cameras. (Marshall, 2016) The concept of influx of radiation into a 2-D array of sensor element is the same. The physical and material properties of the LWIR camera receptors is different. Thermographic images can be created by cryogenic cooling of camera sensors. Another approach is to use room temperature sensors which have individually calibrated sensors reacting to incoming radiation.

Hyperspectral Images

Hyperspectral images use hundreds of color channels per image (not just the single RGB channel). The resulting image is called a *Hyper Cube* and multi-band image. Think of taking many pictures all at the same time and merging them. Problems for UAS use: prohibitive cost, vibration, requirement for stable Position and Orientation System (POS), power consumption, and weight. (Marshall, 2016)

LIDAR

Light Detection and Ranging (LIDAR) is the game changer for ARS. LIDAR uses a laser beam and a receiver to measure the distance to the ground. LIDAR sensor systems can discriminate between multiple return reflections of a single light pulse. (Marshall, 2016) Flash LIDAR is the recent evolution of technology which uses a single point source emission variant of previous versions using separate scanners and emitters. (Marshall, 2016) Research is close to having a Geiger mode LIDAR than can measure the return of a single photon. LIDAR systems require a very accurate measurement of time, sensor pointing angles, and sensor-spatial location. Efficiency decreases with height. Range is a function of UAS capabilities. (Marshall, 2016)

Synthetic Aperture Radar (SAR)

SAR systems are radar systems that map by using a band of EM spectrum in the range of 1 m -> 1mm. (Schowengerdt, 2007) SAR sends out a radar pulse and uses the reflected radar signal and the receivers motion to construct either a 2-D or 3-D image of the returning echoes. "The resolution can be as high as a few centimeters (depending on the wavelength used)." (McGlone, 2004)

Figure 7-2 SAR Imaging Geometry for Strip Mapping Option

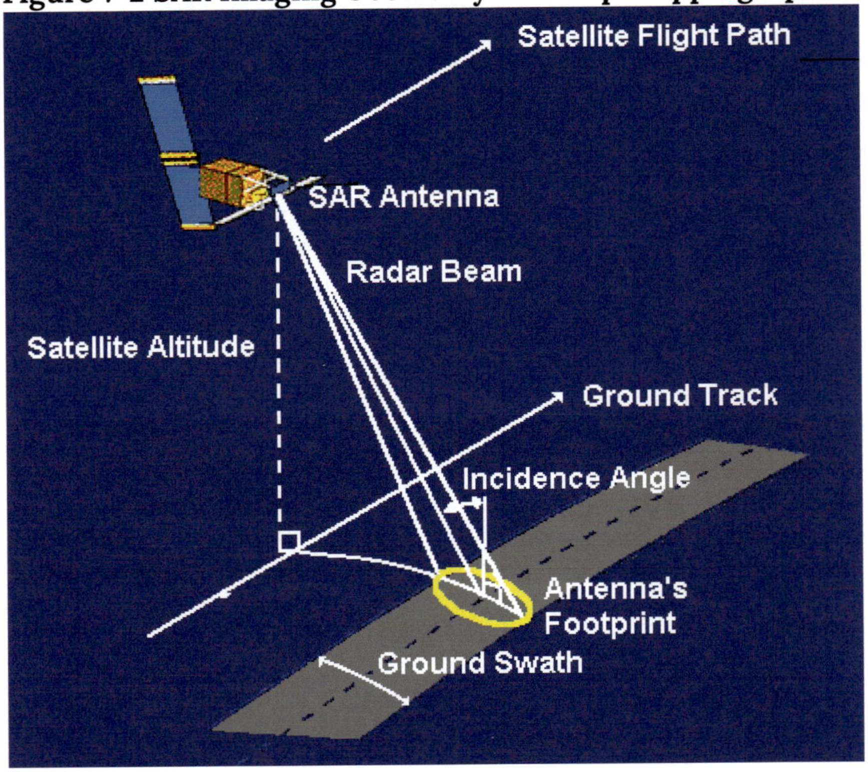

Source: Schowengerdt, R.A. (2007) Remote Sensing (Third edition), Chapter 1: The Nature of Remote Sensing, 2007, Academic Press. 3 Available as 17 pp PDF at: http://www.springer.com/cda/content/document/cda_downloaddocument/9781461419938-c1.pdf

There are two other SAR options: ScanSAR mode and Spotlight. Figure 7-3 shows their relationship to Strip map mode.

Figure 7-3 SAR Modes of Operation

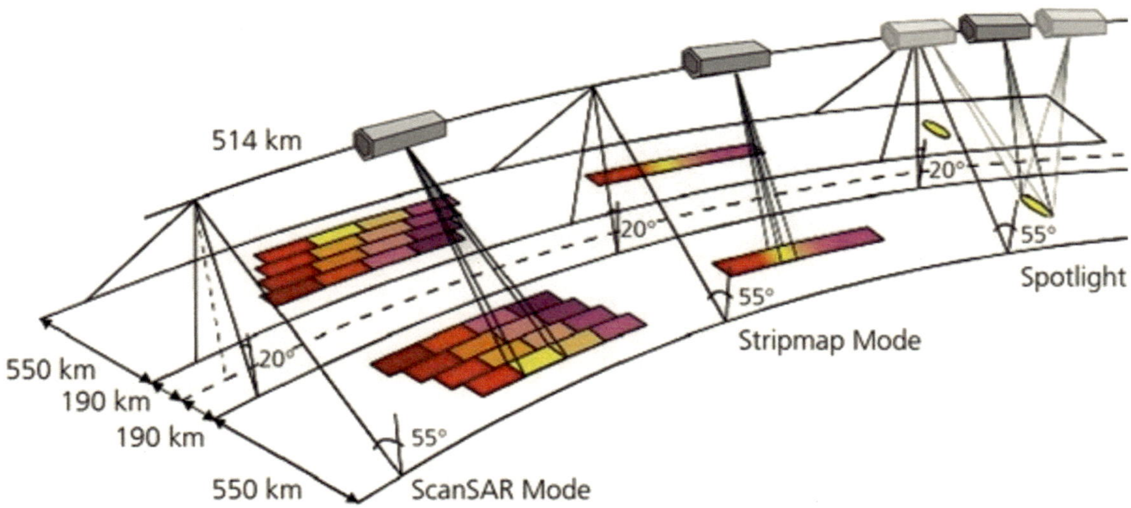

Source: DLR. The three SAR (Synthetic Aperture Radar) modes: Spotlight, Stripmap, and ScanSAR. Viewed September 11, 2018. https://www.dlr.de/dlr/en/desktopdefault.aspx/tabid-10382/570_read-431/

SAR is very computer intensive, uses long wavelengths that can pass through clouds, smoke, and dust in the atmosphere to produce a high spatial resolution that can pick up on surface features, such as waves and textures. (Marshall, 2016) See Figure 7-2 for an example of SAR RADAR geometry for strip mapping option.

Live Video Gimbals for VIS, MWIR and LWIR Cameras

Live streaming images from anywhere via drone cameras use a gimbaled system, which allows operation in a 360-degree arc. This allows continuous observation of a ground target. Many commercially manufactured gimbals are built for military use and ISR operations. Figure 7-4 shows a TASE 500 that works with VIR, MWIR and LWIR cameras.

Figure 7-4 shows a TASE 500 that works with VIR, MWIR and LWIR cameras.

Source: UTC Aerospace Corporation. (summer, 2018). TASE 500 Gimbals fact sheet, http://www.cloudcaptech.com/images/uploads/documents/TASE500_Data_Sheet.pdf

Figure 7-5 Drone Jammer Model KWT-FZQ used for Police interception

Source: GlobalDroneUAV.com (summer, 2018) Police drone jammer effective drone controller. https://globaldroneuav.com/Product/Police-drone-jammer-effective-drone-controller.html

Predicting Conflict

"Non-recreational UAS systems are used for DDD (Dull, Dangerous, and Dirty) missions. UAS deployment requires an ability to navigate in unknown territory, to avoid static obstacles, to avoid moving obstacles like airplanes, birds, intruder UAS, and anything else (missiles or a drone jammer)." (Kopřiva, 2012) Figure 7-5 looks at a Drone Jammer Gun Model KWT-FZQ used for Police interception.

Unless the frequency and range are right, drone jammers are not always successful.[2] UAS systems with special counter jamming modules can maneuver away to safety.[3] Military and commercial UAS systems must do their DDD work in any weather conditions, especially inclement. Military UAS systems must be able to detect "Identification Friend or Foe" (IFF) situations. Of special importance is a UAS near or in commercial controlled airspace. (Marshall, 2016) "The UAS must be able to Sense and Avoid (SAA) potential conflicts regarding air-traffic regulations." (Kopřiva, 2012)

Collision Detection and Resolution (CDR) systems are the next generation automated SAA. (Marshall, 2016)They use many of the sensors described here. CDR systems for UAS were the product of ideas and engineering development "from airport management automated tools like the Traffic Collision Avoidance System (TCAS) (Anonymous, 2018) and the Precision Runway Monitor (PRM)." (Kopřiva, 2012) "Both systems are used to increase safety in the NAS and the mobility of air- traffic." (Kopřiva, 2012) "A separate domain of investigation came from the robotics and AI research facilities." Scientists studying these disciplines were interested in "trajectory planning and obstacle avoidance algorithms for aerial, ground, and maritime systems." (Kopřiva, 2012) CDR has been investigated by several researchers; two of importance are Krozel, who presented a survey of CDR methods, and Albaker who concentrated on CDRs for UAVs.

Conflict Detection and Resolution Principles

Angelov defines "Conflict as an event in which the horizontal or vertical Euclidian distance between two aircrafts breaks the minimal defined separation criterion." (Marshall, 2016) "The criterion varies based on the airspace the UAS is flying in and may be different for different UASs." (Kopřiva, 2012) Recall Figure 3-2 Self -Separation and Collision Volume. (Marshall, 2016)

"The **Safety Zone** (SZ) is defined as the horizontal and vertical separation criteria which form a cylindrical airspace volume around the UAS. The UAS is centered in that volume. Safety Zone (SZ): defined as the horizontal and vertical separation criteria which form a cylindrical airspace volume around the UAS. In figure 3-2 that volume is defined by 1000 ft radius and 200 ft height. It is assumed that initially the UAS is in the center with 100 ft above and below the A/C. Under no circumstances may the SZ be violated by another UAS. For CDR systems different horizontal and vertical criteria may apply." (Kopřiva, 2012)

2. https://www.youtube.com/watch?v=2Bqj0bkSJWE
3. GlobalDroneUAV.com (summer, 2018) Police drone jammer effective drone controller. https://globaldroneuav.com/Product/Police-drone-jammer-effective-drone-controller.html

Figure 3-2 Self -Separation and Collision Volume

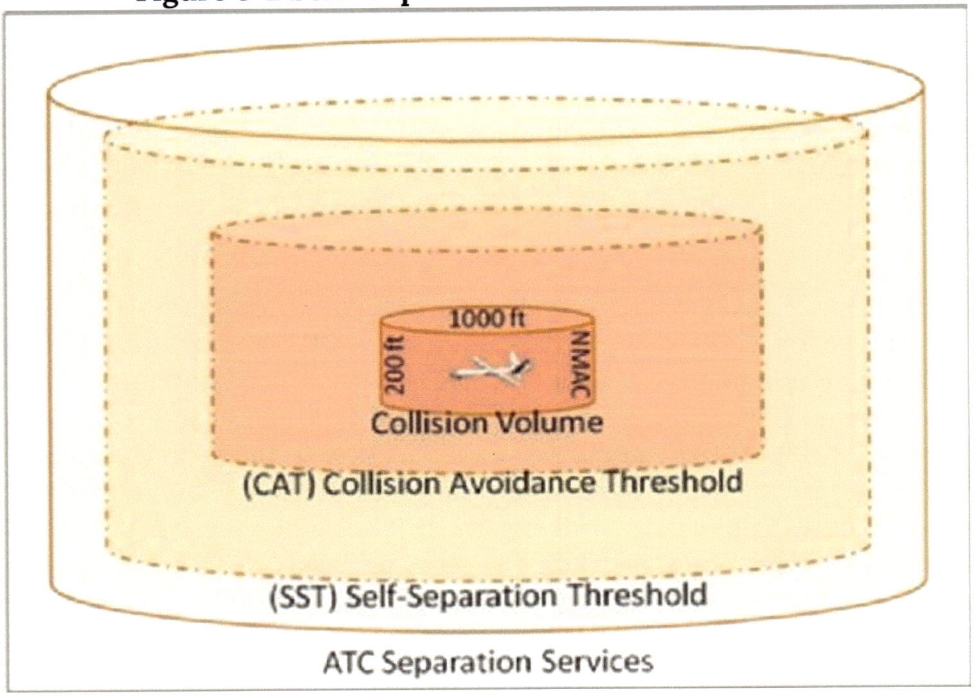

Source: Angelov, P. (2012). *Sense and avoid in UAS research and applications.* Hoboken: NJ.

CDR Architecture

The Conflict Detection and Resolution (CDR) architecture is simple. "It involves five processes:

- *Sensors* – Sensing for both Cooperative and Non-Cooperative A/C
- *State Prediction* – Trajectory predictions and results from Sensing stage
- *Conflict Detection* – Conflict as an event in which the horizontal or vertical Euclidian distance between two aircrafts breaks the minimal defined separation criterion. Conflict as an event in which the horizontal or vertical Euclidian distance between two aircrafts breaks the minimal defined separation criterion.
- *Conflict Resolution* – Determines / generates commands to pilot / autopilot
- *Evasion and Maneuver* – Move in predefined programmed ways including return to Waypoint." (Kopřiva, 2012)

"The function of the CDR system is to detect a potential collision and provide the resolution in terms of an evasion maneuver which will be executed by the UAS's autopilot.

The CDR block logic is simple." (Kopřiva, 2012) The actual mathematics of predictions and models to elucidate these predictions to be carried out in real-time by the UAS autopilot are a totally different story.

Sensing

The subject of sensors and various technologies used was discussed early in this chapter. "Their purpose is to monitor the surrounding environments for both static and dynamic obstacles using the onboard systems. There are two distinct types of sensors of interest; cooperative and non-cooperative." (Kopřiva, 2012)

Cooperative Sensors

"*Cooperative sensors* are those that receive radio signals from another aircraft using on board equipment." (Zeitlin, 2012) "Cooperative sensors provide the ability to sense the environment and to communicate with aircrafts equipped with the same type of sensors by establishing a communications link." (Kopřiva, 2012) The Automatic Dependent Surveillance Broadcast (ADS-B) is an example of a cooperative sensor. (Angelov, 2012) The ADS-B transfers longitude, latitude, altitude, speed, direction over ground, and unique UAS identification. Advanced cooperative sensors can transfer entire flight plan data and weather conditions. (Angelov, 2012)

Non-Cooperative Sensors

"*Non-Cooperative Sensors* sense the environment to gather information about obstacles and other aircraft or UAS." (Kopřiva, 2012) There are no communications links to another aircraft or intruder UAS. "Sensor information needs to be processed to get the correct environment state knowledge." (Kopřiva, 2012)

"Types of non-cooperative sensors include Inertial Measurement Unit (IMU), Laser Range Finders (LRF), Stereo Camera Systems (SCS), Single Moving Camera (SMC) and radar" Active radar is used on the larger UAS. SMC and SCS systems are used on the smaller UAS." (Kopřiva, 2012)

Intruder Aircraft

What is an *intruder* aircraft? As seen in Figure 7-6, Bageshwar shows that an *intruder* aircraft is any aircraft that breaks the self-separation threshold, from any direction, not necessarily in the collision path of the UAS. (Bageshwar, 2015) Note the path of intruder, the potential collision point in space, the decisions and executable commands that must be processed in real time and the maintenance of the Collision Avoidance Threshold (CAT).

Non-cooperative aircraft miss some very good information. Consider the TCAS II. Figure 7-7 shows sample terminology of Traffic Alert (TA), Resolution Advisory (RA), Proximity Intruder Traffic (PIT) and Non-Threat Traffic (NTT) for specified criteria. (Anonymous, 2018) Figure 7-8 shows the TCAS II Conceptual framework. (Anonymous, 2018) Figure 7-9 shows what the pilot sees in his cockpit. Wikipedia has a worthwhile description of TCAS II, its operational modes, maneuver commands, TAs, RAs, configurations, and plenty of references.

Trajectory Prediction

"To detect and resolve a conflict, it is necessary to compare the trajectory of the UAS and the trajectory of the sensed object." (Kopřiva, 2012)

The trajectory is produced by the trajectory computation unit from the raw sensor input data.

There are four basic models for trajectory prediction. "In the *nominal* method, the trajectory is predicted directly from the sensor data without considering any uncertainty of change." (Kopřiva, 2012)

Figure 7-6 Intruder Aircraft and SAA Decisions

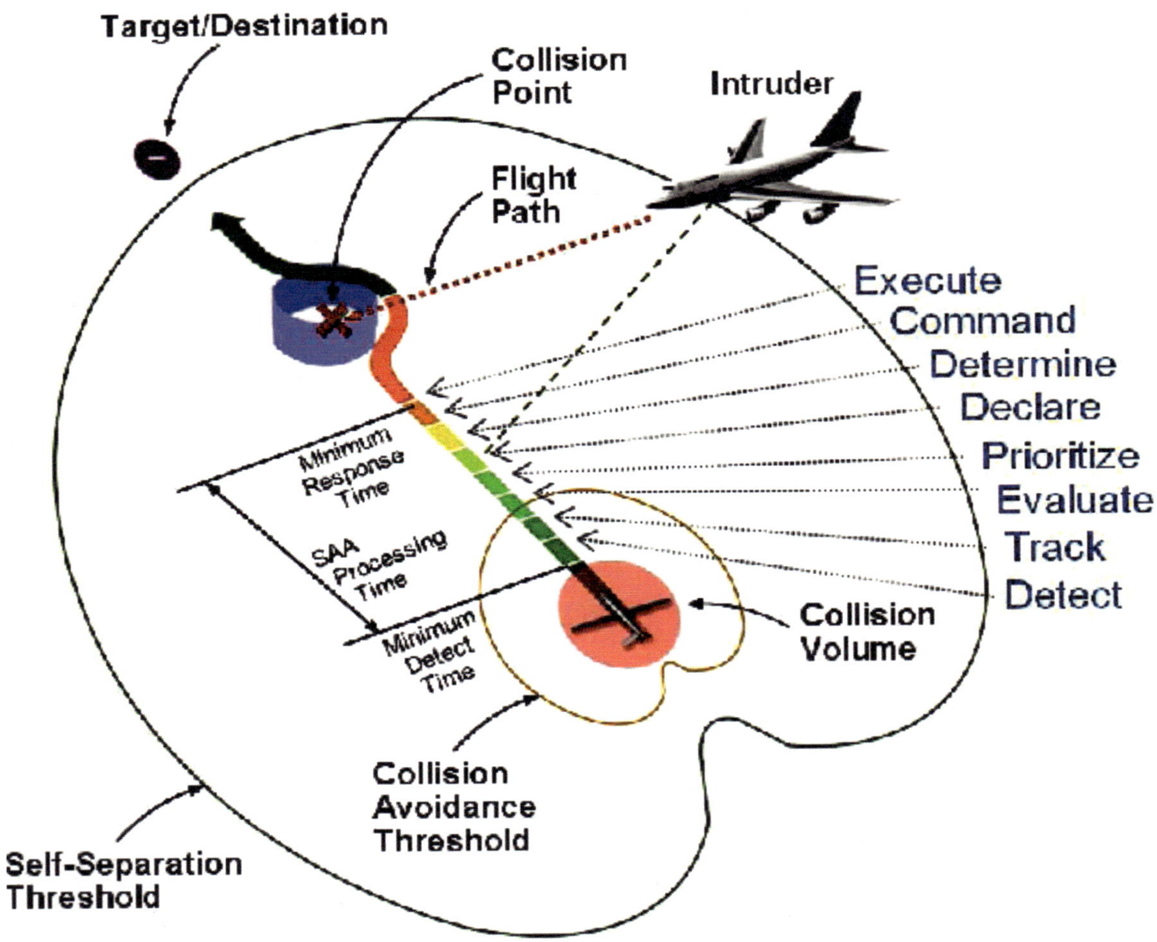

Source: Bageshwar, B. (2015). Multi-intruder aircraft, multi-sensor tracking system. 2015 IEEE/AIAA 34th Digital Avionics Systemes Conference (DASC).

Figure 7-7 TCAS II Terminology

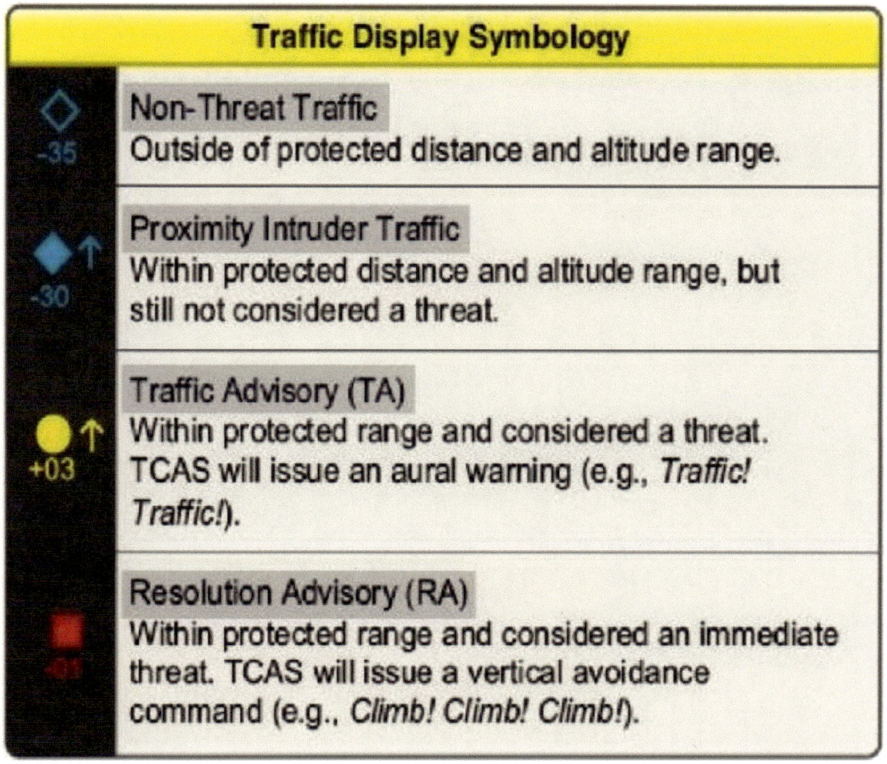

Source: Sanders T. (2017) QUORA, https://www.quora.com/If-a-plane-any-kind-is-not-equipped-with-TCAS-is-it-likely-to-collide-with-other-aircraft-over-oceans-when-there-is-no-radar-penetration

Figure 7-8 TCAS II Conceptual Framework

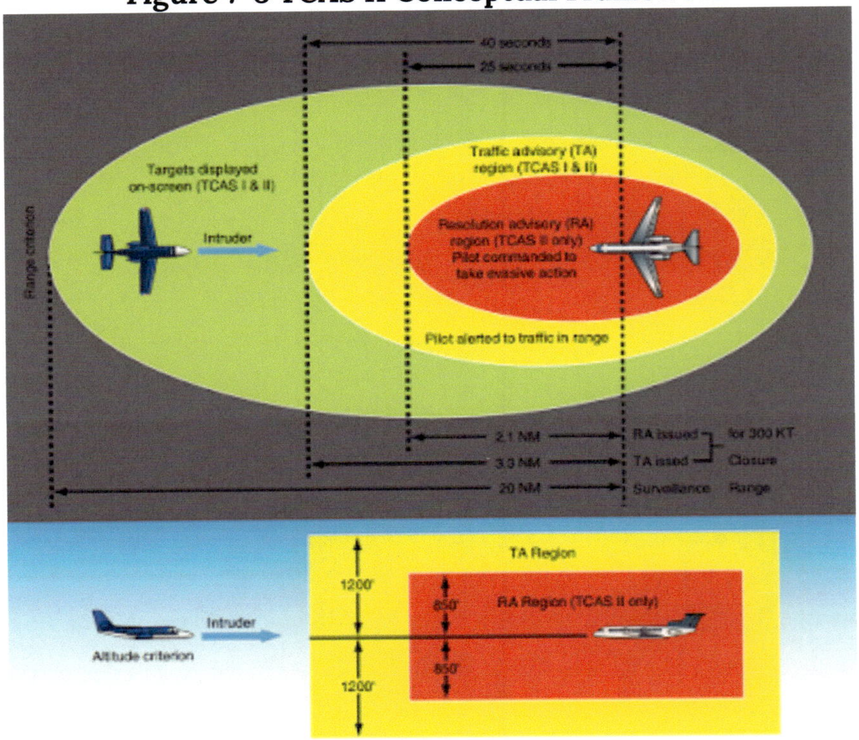

Source: Wijerathne, C.(2017). Collision Avoidance Systems, *Aeronautics Guide*. http://okigihan.blogspot.com/2017/05/collision-avoidance-systems.html

"The output of the nominal trajectory predictor a single trajectory computed from the last sensor scans." (Kopřiva, 2012) This is very similar to boat navigators use of dead reckoning from last known positions determined by landmark or sky information.

Figure 7-9 TCASS II Cockpit View

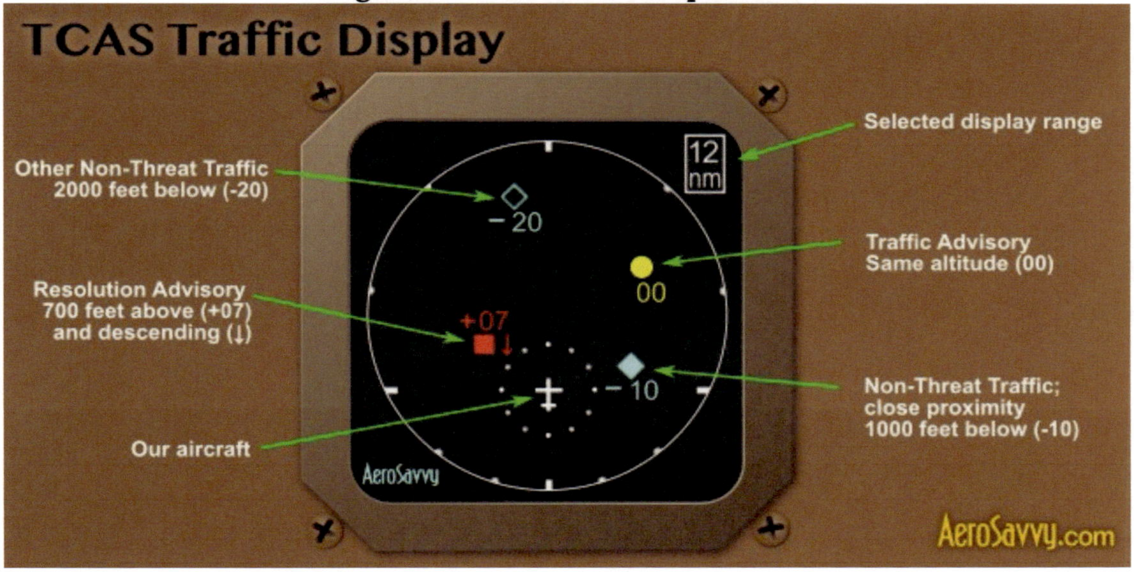

Source: AeroSavvy.com (2015) TCAS: Preventing Mid-Air Collisions. *Aerosavvy*, *Aviation Insight*. https://aerosavvy.com/tcas/

"The trajectory may be computed by various mathematical approaches; linear, non-linear, Taylor series, or Kalman filter. The nominal prediction is good for short- predictions, where probability of change is low. " (Kopřiva, 2012)

"The *worse-cast* prediction covers the full range of maneuvers that an aircraft may perform and using a look-ahead time parameter computes the area where the aircraft may occur. This computed area is the predicted trajectory." (Kopřiva, 2012)

"In the *probabilistic prediction* approach, the uncertainties are used to model the potential variations in the trajectory. All possible trajectories are generated, and each trajectory is evaluated by the probability function. This model is a trade-off between the nominal method and the worse-case method. Decisions are based on the likelihood of the conflict." (Kopřiva, 2012)

"The *flight plan approach* works only with cooperative sensors. Intruder aircraft and home aircraft exchange flight plan data. The exact trajectory is known, and no predictions are required. The plans are exchanged as a set of way-points together with SZ parameters. The advantage of this model is the exact knowledge of the future trajectory. This model requires high bandwidth for the data transmission." (Kopřiva, 2012)

Conflict Detection

"Conflict is detected based on the flight plan representation obtained from the trajectory prediction unit. The unit checks the flight plans (cooperating A/C) of both planes and checks whether the safety zone of any airplane has been violated. If so, then the parameters (position and times of possible conflict) are passed to the conflict resolution unit. Conflict detection can be expressed in 2D Horizontal Plane, 2D Vertical Plane or 3D display." (Kopřiva, 2012)

Conflict Resolution

According to Kopriva, "there are eight methods that the conflict resolution modules use to avoid collisions: rule-based methods (RB), game theory methods (GT), field methods (F), geometric methods (G), numerical optimization methods (NO), combined methods (C), multi-agent methods (MA) and other methods (O)." (Kopřiva, 2012) C and O designations are not considered in this chapter.

1. "*Rule-based methods* use a set of prescribed rules to avoid conflict. The sets of rules are fixed during the system design phase. This is a limited solution because it only works with planes in a shared air-space. This method does not permit changes or integration of further intentions. Little communication between planes is required." (Kopřiva, 2012)
2. "*Game-theory methods* (GT) use game theory to model the differential behavior of the two planes. GT is useful for non-cooperative conflict resolution for the short-term solution. Recall that non-cooperative sensing data which may be used for GT methods are generated from radar, laser range finders, stereo cameras, moving cameras, IR cameras and EO cameras." (Kopřiva, 2012)
3. "*Field methods* (F) treat each airplane as a charged particle and are very close to a reactive mechanism. The field is computed based on current configuration, state, positions of the airplanes, weather and other uncertainties. The UAS based on its own positional data in this field applies control actions depending on the state of the airplane with respect to the field. Evasive maneuvers are generated based on the repulsive forces between the field. Field methods are computer intensive." (Kopřiva, 2012)
4. "*Geometric methods* (G) Geometric methods are really limited to two planes in the CV. They solve an optimization objective function based on trial evasive maneuvers. Multi intruder planes present a complex sub-optimal problem solution, especially when changing parameters of altitude, velocity, and heading are factored in." (Kopřiva, 2012)
5. "*Numerical optimization methods* (NO) use a kinematic model of the aircraft along with a set of constraints and cost metrics to determine the best evasion maneuver." These are formal, elegant solutions to the CR problem. However, the more intruder planes, the NO methods become intractable. "Defining the set of constraints becomes the real challenge for this method." (Kopřiva, 2012)
6. "*Multi-agent methods* (MA) use a multi-agent framework for solution generation. Each aircraft is controlled by one agent. The agents communicate. The agents negotiate the solution using various utility functions." (Kopřiva, 2012)

Evasion Maneuvers

"The last stage in the decision model is evasion and maneuver. The output from the conflict resolution system/module is proposed evasion maneuvers to avoid conflict and collision. UAS have multiple maneuver options: speed-up, slow-down, keep the same speed, turn-left, turn-right, climb, and descend. Combined maneuvers are also possible, i.e. change speed while turning left." (Kopřiva, 2012)

CDR Taxonomy

Angelov proposes a taxonomy for Collision Detection and Resolution System. His Figure 6.5 (Angelov, 2012) illustrates all the subgroups expressed in the previous paragraphs. In his Table 6.1 on the following page, he describes the attributes and abbreviations used for CDR classification broken down by the five classes in his taxonomy. (Angelov, 2012)

Discussion Questions

1. Which one of the five CDR logic systems would be most susceptible to cyber-attacks and why?
2. Propose a method to disrupt the trajectory predictions by 5% and which would report this error to the conflict detection stage.
3. Research the maritime Automated Identification System for Collision Avoidance (AIS) system. Note the similarities to the TCAS II. Note also the GUI and information parameters are superior to TCAS II. Design / report on a CONOP to use the best elements and display of parameters of AIS to upgrade TCAS II to TCAS III

Bibliography

Angelov, P. (2012). *Sense and avoid in UAS research and applications.* Hoboken: NJ.

Bageshwar, B. a. (2015). Multi-intruder aircraft, multi-sensor tracking system. 2015 IEEE/AIAA *34th Digital Avionics Systemes Conference* (DASC).

Cummings, B. (2007). *Wikipedia Wavelength units images.* Retrieved from Wikipedia: https://slideplayer.com/slide/8733356/

FAA *Booklet Introduction to TCAS II Version* 7.1.(2018) Retrieved from FAA: https://www.faa.gov/documentLibrary/media/Advisory_Circular/TCAS%20II%20V7.1%20Intro%20booklet.pdf

Kopřiva, S. Š. (2012). Sense and Avoid Concepts: Vehicle-Based SAA Systems (Vehicle-to-Vehicle). In S. Š. Kopřiva, *Sense and Avoid Concepts: Vehicle-Based SAA Systems (Vehicle-to-Vehicle)* (p. Chapter 6). Wiley Online Library. doi:https://doi.org/10.1002/9781119964049.ch6

Marshall, D. M. (2016). *Introduction to Unmanned Aircraft Systems, 2nd Edition.* New York: CRC Press.

McGlone, J. M. (2004). *Manual of Photogrammetry* (5th ed.). Bethesda, MD: The American Society for Photogrammetry and Remote Sensing.

Pater, R. (2018). *Drone Survival Guide.* Retrieved from Wikipedia: http://www.drone.survival-guide.org/DSG.pdf.)

Schowengerdt, R. (2007). *Remote Sensing (Third edition).* Retrieved from Chapter 1: The Nature of Remote Sensing: http://www.springer.com/cda/content/document/cda_downloaddocument/9781461419938-c1.pdf

Zeitlin, A. (2012, April 11). *Performance Tradeoffs and the Development of Standards.* Retrieved from Wiley Online Library: https://onlinelibrary.wiley.com/doi/abs/10.1002/9781119964049.ch2

Readings

Adamy, D. (2001) EW 101 A *First Course in Electronic Warfare*, Boston: Artech House.

Adamy, D. (2004) EW 102 A *Second Course in Electronic Warfare*, Boston: Artech House.

Adamy, D. (2009) EW 103 *Tactical Battlefield Communications Electronic Warfare*, Boston: Artech House.

Adamy, D. (2015) EW 104 EW *against a New Generation of Threats*, Boston: Artech House.

Adamy, D. (2003) Introduction to Electronic Warfare Modelling and Simulation, Boston: Artech House.

Albaker, B. & Rahim, N. (2009) A Survey of collision approaches for unmanned aerial vehicles, Technical Postgraduates (TECHPOS), 2009 *International Conference for.*

Angelov, P. (2012). *Sense and avoid in UAS research and applications.* Hoboken, N.J.: Wiley.

Austin, R. (2010) UAVS *Design, Development and Deployment*, New York: Wiley.

Barnhart, R.K., Hottman, S.B, Marshall D.M., and Shappee, E. (2012) *Introduction to Unmanned Aircraft Systems.* New York: CRC Press.

Burch, D. (205) RADAR *for Mariners.* New York, McGraw-Hill.

Drone Jammer Model KWT-FZQ: GlobalDroneUAV.com (summer, 2018) Police drone jammer effective drone controller. https://globaldroneuav.com/Product/Police-drone-jammer-effective-drone-controller.html

FAA Booklet: Introduction to TCAS II Version 7.1, (2018) see: https://www.faa.gov/documentLibrary/media/Advisory_Circular/TCAS%20II%20V7.1%20Intro%20booklet.pdf

Gall, D. (2017)QUORA: https://www.quora.com/If-a-plane-any-kind-is-not-equipped-with-TCAS-is-it-likely-to-collide-with-other-aircraft-over-oceans-when-there-is-no-radar-penetration

Gelbart, A., Redman, B.C, Light, R.S., Schwartzlow, C.A., and Griffis, A.J. (2002) Flash LIDAR based on multiple-slit streak tube imaging. LIDAR VII. *Laser Radar Technology and Applications* 4723, pp 9-18.

Hubbard, R. K (1998) Boater's Bowditch, Camden, MA: International Marine.

Krozel, J., Peters, M., & Hunter, G. (April 1997) Conflict detection and resolution for future air transportation, *Technical Report* NASA CR-97-205944.

Monahan, K (2004) *The RADAR Book: Effective Navigation and Collision Avoidance*. Anacortes, WA: Fineedge Publications.

SAR Images: https://crisp.nus.edu.sg/~research/tutorial/mw.htm

SAR: https://www.dlr.de/dlr/en/desktopdefault.aspx/tabid-10382/570_read-431/

Schowengerdt, R.A. (2007) Remote Sensing (Third edition), Chapter 1: The Nature of Remote Sensing, 2007, Academic Press. 3 Available as 17 pp PDF at http://www.springer.com/cda/content/document/cda_downloaddocument/9781461419938-c1.pdf

Staff (2015) Camera. In Merriam Webster's Online Dictionary. http://www.merriam-webster.com/dictionary/camera

Staff (2015) In situ. In Merriam Webster's Online Dictionary. http://www.merriam-webster.com/dictionary/in%20situ

TASE 500 Gimbals: https://www.sypaq.com.au/sensorsandsurveillance/tase-gimbals/

TCAS Cockpit view: AeroSavvy.com (2015) TCAS: Preventing Mid-Air Collisions. *Aerosavvy, Aviation Insight*. https://aerosavvy.com/tcas/

Toomay, J.C. (1982) RADAR *for the Non – Specialist*. London; Lifetime Learning Publications

Traffic CA system, https://en.wikipedia.org/wiki/Traffic_collision_avoidance_system

Wavelength units images, Benjamin -Cummings (2007) and https://slideplayer.com/slide/8733356/

Chapter 8: Designing UAS Systems for Stealth

Student Learning Objectives – The student will be introduced to the design requirements for making a UAS stealthy, i.e. having reduced signatures for sonic, visual, thermal and radar detection.

Designing a UAS for Stealth

Stealth means "to resist detection." Stealth applies to the air vehicle and materials visible to the enemy plus the internal SAA systems that control / create noise, heat, electromagnetic emanations, and changes in light. For ISR platforms and missions, it is essential the UAS systems be undetected in operation. "It is desirable not to alert the enemy (military) or criminals (police) to the ISR operation." It can be assumed that the enemy is using counter-UAV operations and weapons. Stealth design protects the air vehicle from these counter – UAV measures. Stealth in civilian operations results in minimal environmental disturbances. (Austin, 2010)

From a personal privacy standpoint or in civil airspace it is desirable to have the UAV stealth features turned off. [It should be as if we had flicked a switch.] (Austin, 2010)

Detection Signatures

UAS / UAVs are detected by their ***signatures***: noise (acoustic), optical (visible), infrared (thermal) and radar (radio). "These acoustic or electromagnetic emissions occur at the following wavelengths: (Austin, 2010)

A) Noise (acoustic) [16 m-2 cm, or 20 – 16000 Hz]

B) Optical (visible) [0.4 – 0.7 um]

C) Infrared (thermal) [0.75 um – 1 mm]

D) RADAR (radio) [3 mm – 3 cm]" (Austin, 2010)

If the designer is to "reduce the vehicle detectability to an acceptable risk level, it is necessary to reduce the received emissions or reflection of the above wavelengths (expressed as frequencies) below the threshold *signature* value. A good portion of the UAS signatures are a function of the operating height of air vehicle." (Austin, 2010)

A student might look at the answers above and ask what is the significance? What does this really mean? Let's take a short sojourn down EMS lane. Military planners used to think in terms of ground, sea, and air. Space came later. Now there is a "fifth realm" the electromagnetic spectrum (EMS). For EMS, we think in terms of *frequency*. The enhancement in our ability to communicate using the EMS is making significant changes in the way we conduct warfare. (Adamy D. -0., 2015)

Radio communications and wireless transmissions using tuned transmitters and information explosion of the Internet were the heart of a warfare-revolution. (Adamy D. -0., 2015)The certainty of intercept of radio communications and radar signals, and the ability to locate transmitters had a significant impact on military operations. (Adamy D. -0., 2015) Intercept, jamming, emitter location, message security, and transmission security became fundamental to warfare. (Adamy D. -0., 2015) The basic destructive capabilities (energy) employed in warfare have not changed a lot. (fast-moving projectiles, significant overpressure, heat, and sound). However, the ways they are employed have changed significantly through use of the EM Spectrum (EMS). (Adamy D. -0., 2015) Now, we guide the destructive energy of weapons towards their intended targets using the EMS in many ways. Also, the EW specialist uses EMS to prevent those weapons from hitting their intended targets (US), (Adamy D. -0., 2015) Sometimes the destruction of communications capability by an enemy is a goal.

The battlespace, which once had only four dimensions (latitude, longitude, elevation and time [before radio]) now has a fifth dimension: frequency. (Adamy D. -0., 2015) See Table 8-1 Battlespace Dimensions.

Dimension	Function	Action
Latitude	Friendly Force Location	Direction of Weapons
Longitude	Enemy Force Location	Maneuver of Forces
Elevation		
Time	Speed of Maneuver	Timeliness of Attack
	Timing of Weapon Release	Enemy Vulnerability
Frequency	Bandwidth Required	Rate of Information Flow
	Bandwidth	Available Interference
	Frequency of Transmissions	Vulnerability to Jamming
		Vulnerability to Intercept

Table 8-1 Battlespace Dimensions

Source: (Adamy D. -0., 2015)

Bandwidth is defined as the range within a band of wavelengths, frequencies or energy. Think of it as a range of radio frequencies occupied by a modulated carrier wave, assigned to a service over which a device can operate. Bandwidth is also the capacity for data transfer of electrical communications system. Range has a significant impact on radio transmission. Depending on the environment, the strength of a received signal, T, is a function of the square or fourth power of a distance, d, from the transmitter.

A closer transmitter will do a better job of receiving a signal and can usually locate the transmitter more accurately. (Adamy D. -0., 2015) Once we depend on inputs from multiple receivers, the network becomes central to our war making ability. [Think UAS Team collaboration.] We have now entered net-centric warfare. (Adamy D. -0., 2015)

Thinking again about a team or swarm of UAS, the low-hanging fruit target is US communications. We depend on connectivity in everything we do: daily lives, social interactions, business, manufacturing, government, transportation, computers and warfare to name just a few in the extensive list. *Connectivity is any technique for the movement of information from one location or player to another.* Consider the economic impact of having our critical infrastructure (banking, air transportation, etc.) shut down. Damaging the connectivity of system is real damage. We measure connectivity in terms of information flow. In warfare, this is called Information Operations (IO). Fundamental to IO is the frequency at which the information is transmitted or received.

Returning to stealth with respect to UAS design, we note that the intelligence, surveillance, reconnaissance and weapons payload-delivery functions of UAS. These are all IO operations and frequency is at the heart of their success against or denial by the enemy. (Adamy D. -0., 2015)

Electromagnetic Spectrum (EMS)

The German company, Tontechnic-Rechner-Sengpielaudio (TRS) has put together some clever tools for conversions of wavelength to frequency (visa versa) "for Acoustic Waves (sound waves) and Radio Waves and Light waves in a vacuum." (TRS, 2018) Start with Figure 8-1 EMS. Note the inverse relationship between frequency, f and wavelength L (lambda - Greek).

Figure 8-1 EMS

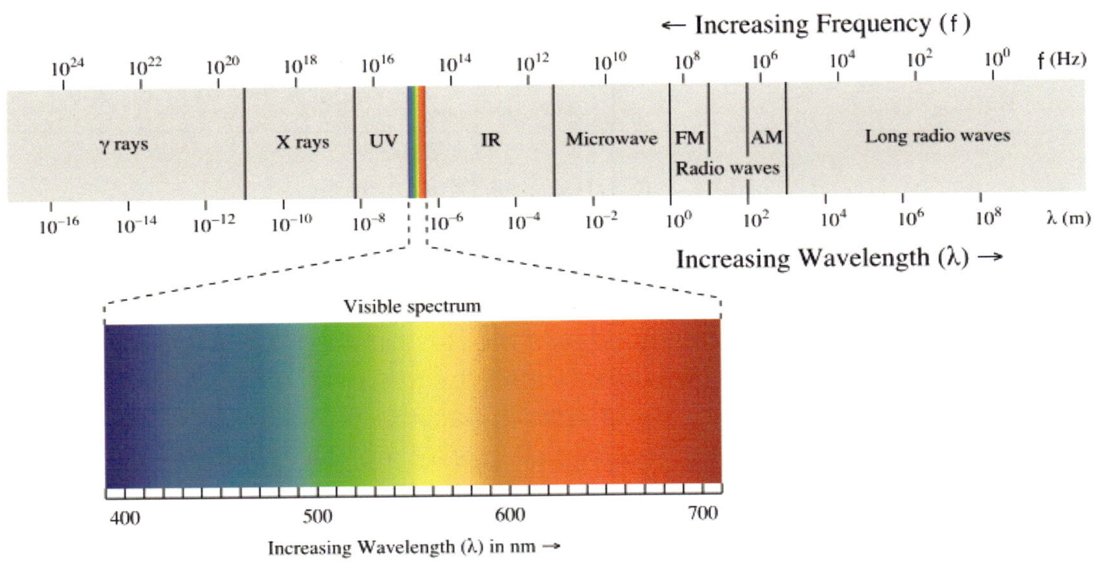

Source: TRS. (2018, July 10). Tontechnic-Rechner-Sengpielaudio. Retrieved from Tontechnic-Rechner-Sengpielaudio Calculator. www.sengspielaudio.com/calculator-wavelength.htm

Note also how small the visible spectrum as part of the enormous EMS. Figure 8-2 shows some of the EMS functions.

Figure 8-2 EMS Functions

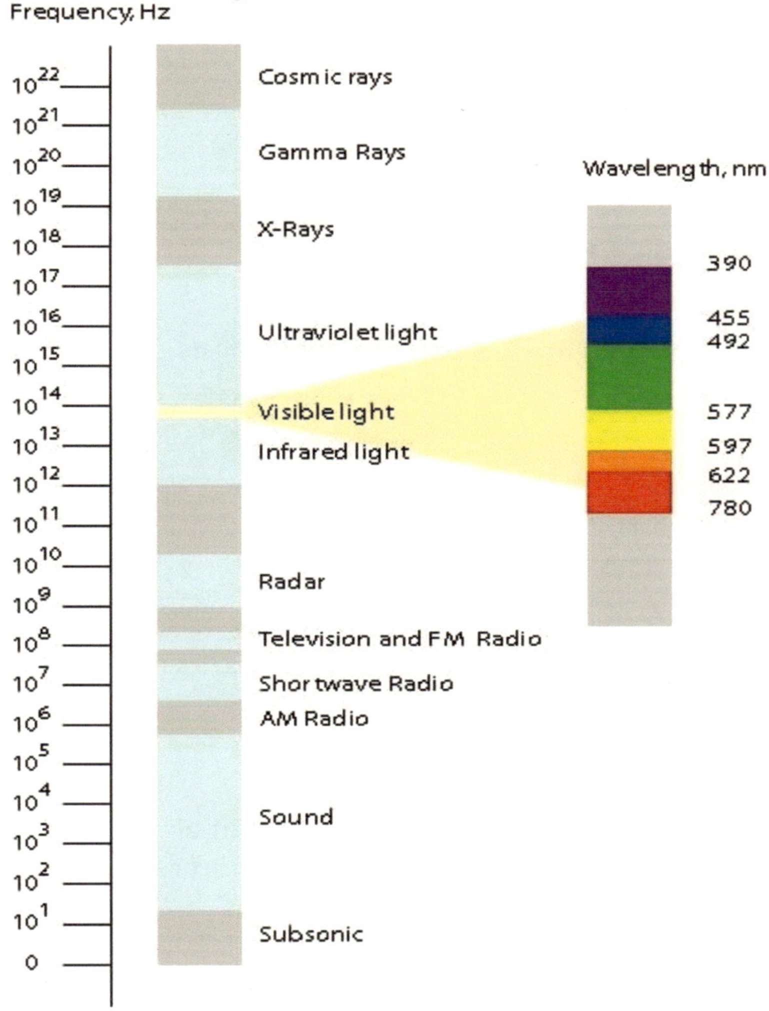

Source: TRS (2018, July 10). Tontechnic-Rechner-Sengpielaudio. Retrieved from Tontechnic-Rechner-Sengpielaudio Calculator, URL: www.sengspielaudio.com/calculator-wavelength.htm

Figure 8-3 shows the conversion for sound and acoustic wave period to frequency and back. (Adamy D. -0., 2015) Figure 8-4 shows the Sound EMS regions (Adamy D. -0., 2015)

Figure 8-3 Conversion for sound and acoustic wave period to frequency and back

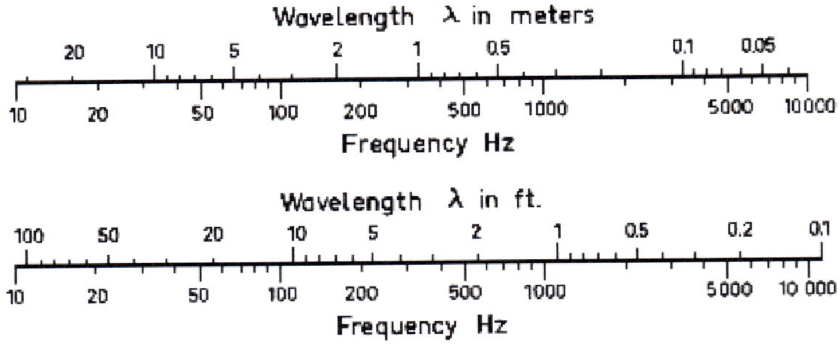

Figure 8-4 Sound EMS Regions

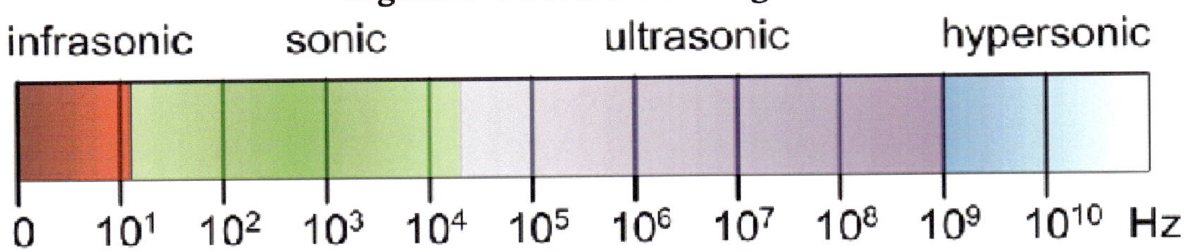

Source for Figures 8-3 &4: TRS (2018, July 10). Tontechnic-Rechner-Sengpielaudio. Retrieved from Tontechnic-Rechner-Sengpielaudio Calculator. www.sengspielaudio.com/calculator-wave-length.htm

"Acoustic waves and Sound Waves in Air

Sound waves are EMS waves which propagate vibrations in air molecules. The 1986 standard speed of sound, **c**, is 331.3 m/s or 1125.33 ft/s at a temperature, T = 0 degrees Celsius." (TRS, 2018)

"The formulas and equations for sound are: **c = L x f; L = c /f = c x T; f=c /L,**

T = time- period or cycle duration and **T = 1/ f** and **f = 1 / T.** The unit for frequency is Hertz = Hz =1/s. the unit for wavelength, L is meters, m. The time-period or cycle duration, T is sec, s. The wave speed or speed of sound, c, is meters/sec, m/s." (TRS, 2018)

Austin states that the design limit for UAS Stealth for acoustic (noise) or sound waves is "[16 m-2 cm, or 20 – 16000 Hz]." (Austin, 2010) Use the bottom of the page converter {Basis: Speed of sound $c = \lambda \times f$ = 343 m/s at 20°C} for 16 m L = 21.4375Hz. This compares to the Austin value of 20 Hz. For the 2 cm = 0.02 m, the resulting valued for f = 17650 Hz. This is above the 16,000 Hz limit from Austin. This might be due to the 20-degree Celsius basis difference. This tells the UAS designer that the upper end of noise – Stealth acceptability 17,150 Hz. **The Stealth range is 20 Hz – 17,150 Hz.**

"Hearing range describes the range of frequencies that can be heard by humans, (aka range of levels). The human range is commonly given as 20 to 20,000 Hz, there is considerable variation between individuals, especially at high frequencies, and a gradual loss of sensitivity to higher frequencies with age is considered normal. Sensitivity also varies with frequency, as shown by equal loudness contours" (Rosen, S, 2011). See Figure 8-5.

"An **equal-loudness contour** is a measure of sound pressure (Db SPL) over the EMS spectrum, for which a listener perceives a constant loudness when presented with pure steady tones. The unit of measurement for loudness levels is the *phon* and is arrived at by reference to equal-loudness contours. Two sine waves of differing frequencies are said to have equal-loudness level measured in phons if they are perceived as equally loud by the average young person without significant hearing impairment." (Staff, 2016) (Fletcher, 1933)

Figure 8-5 Equal Loudness Contours

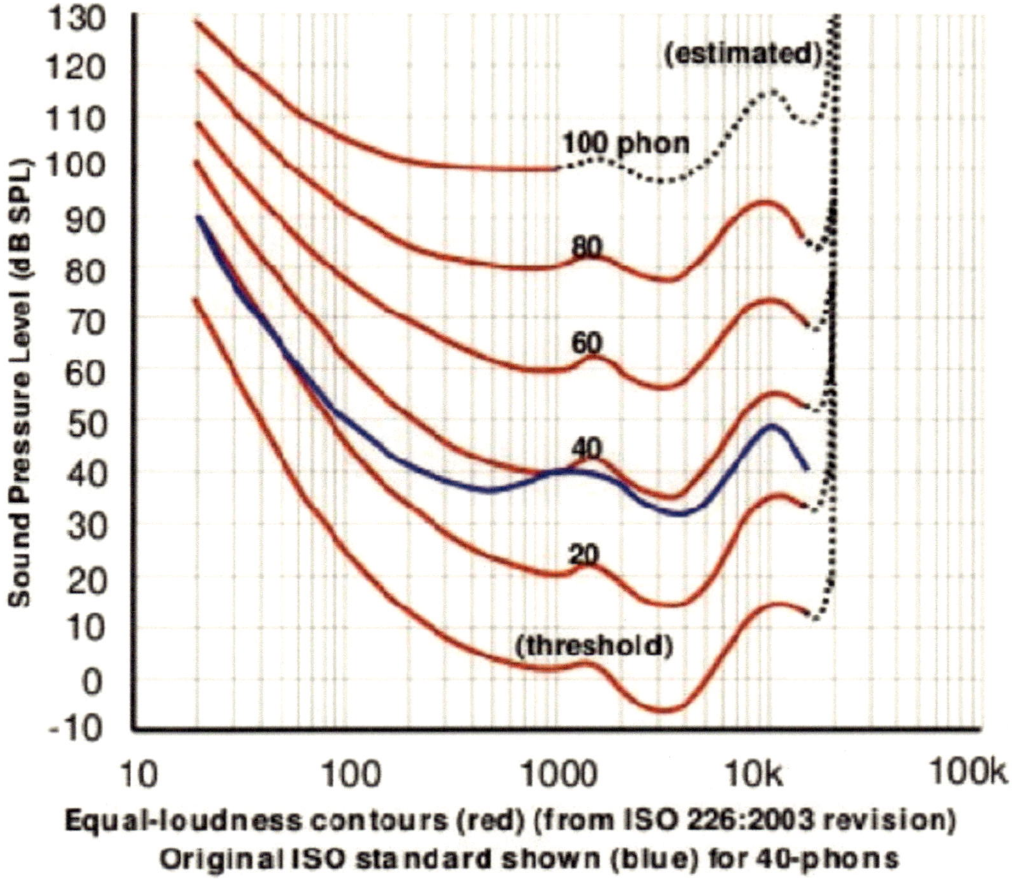

Source: By Lindosland – http://en.wikipedia.org/wiki/Image:Lindos1.svg, Public Domain, https://commons.wikimedia.org/w/index.php?curid=2565926

"Radio Waves and Light Waves in a Vacuum

The formulas and Equations for radio waves and light waves in a vacuum are the same. However, the constant c is different. Lower-case c is the speed of light waves and the speed of radio waves in a vacuum. The speed of light in free space (vacuum) is the speed at which electromagnetic waves propagate, including light waves." (TRS, 2018) "Instead of the speed of sound in air, the speed of light c, is 299,792,458 m/s (or 983,571,056 ft/s.) needs to be used in the formulas as speed of propagation. Wave frequency in Hz = 1/s and wavelength in nm = 10 (**-9) m." (TRS, 2018)

"Radio waves and microwave radiation are both forms of energy known as Electromagnetic Radiation (EMR). Sunlight contains other EMR forms: ultraviolet rays, infrared (heat) waves, as well as visible light waves. These EMW spread in a vacuum at the speed of light ~ 300, 000 km/s as electromagnetic radiation." (TRS, 2018) The propagation speed of electrical signals via optical fiber is about 9/10 of c or ~270, km/s. "Copper as a medium is worse slowing the propagation speed c, to ~200, 000 km/s." (TRS, 2018) Sound is also shown on the EMS chart, but it has no electromagnetic radiation. "Sound pressure is the deviation from local ambient pressure (sound pressure deviation) caused by a sound wave – mainly in air." (TRS, 2018) Wavelength is sometimes given in Angstrom units. 1 A = 10 (**-10) m = 0.1 nm. See Figure 8-6 EMS Reduced.

Figure 8-6 EMS Reduced

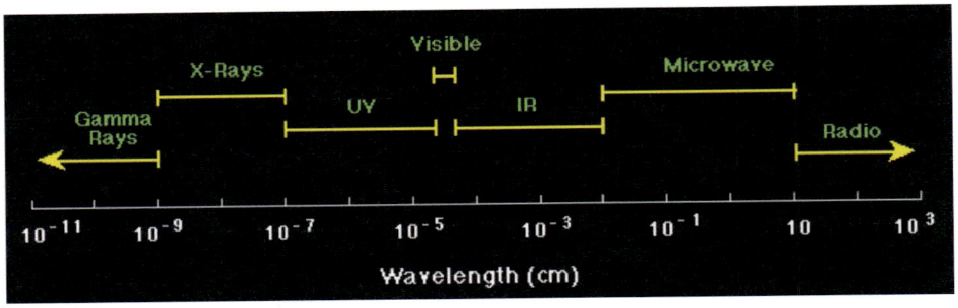

Source: TRS (2018, July 10). Tontechnic-Rechner-Sengpielaudio. Retrieved from Tontechnic-Rechner-Sengpielaudio Calculator, URL: www.sengspielaudio.com/calculator-wavelength.htm

Visible light, gamma rays, microwaves, and radio waves are all part of the EMS. They simply differ by wavelength. (TRS, 2018) Figure 8-7 is a conversion chart for Radio waves and light waves in a vacuum.

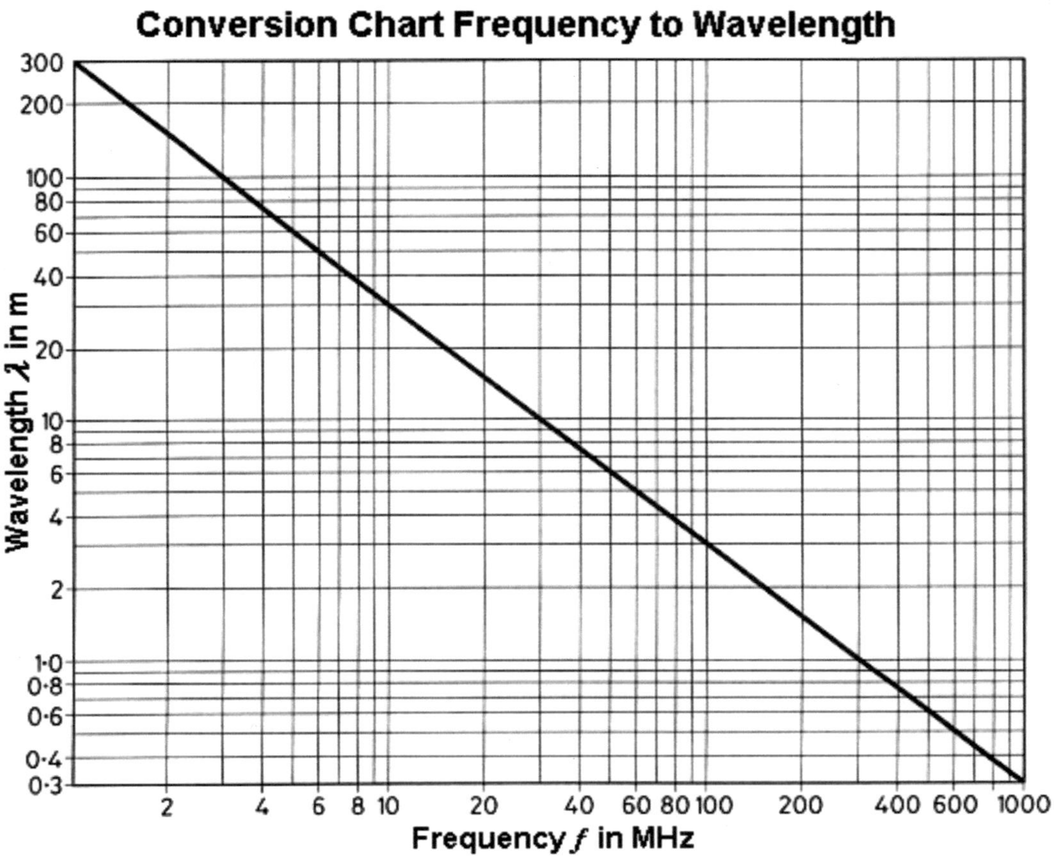

Figure 8-7 Conversion Chart – Frequency to Wavelength Radio and Light Waves in a Vacuum

Source: TRS (2018, July 10). Tontechnic-Rechner-Sengpielaudio. Retrieved from Tontechnic-Rechner-Sengpielaudio Calculator. http://www.sengspielaudio.com/calculator-wavelength.htm

Some useful factors: 1 Terahertz (THz) = 10**3 GHz = 10**6 MHz = 10 **12 Hz; and

1 nm = 10 (**-3) um (micron-meter) = 10 (**-6) mm (millimeter)= 10(**-9) m

1-micron, um = m / 1000000 (1 millionth of a meter).

We have covered Austin's noise, optical and infrared stealth signatures. RADAR is not as simple without another trip down RADAR lane. RADAR is extensively discussed and written about in the 20th century. It is certainly one of the most influential inventions in the last century, arguably more relevant than the cellphone. Our concern is to "paint" or recognize the UAS signature from a distance. If we can "see" the hostile UAS coming, it can be tracked, disabled, destroyed or intercepted and "turned" to a new waypoint or objective.

Figure 8- 8 RADAR Frequency Bands

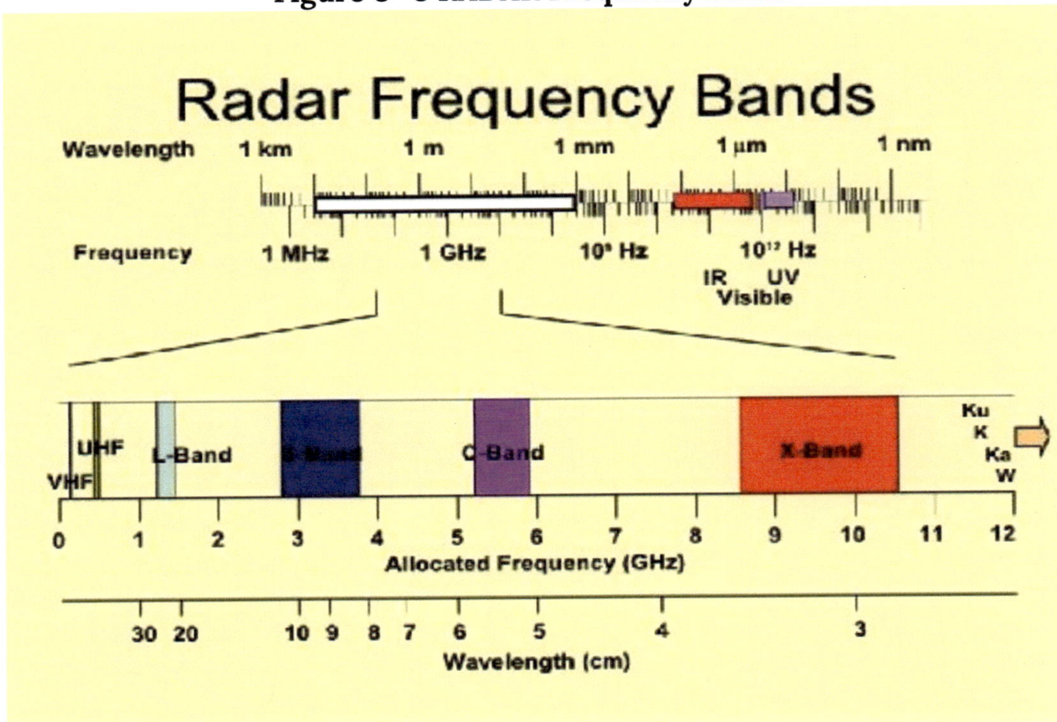

Source: (ITU, 2019)

RADAR / EW / Range Equation

From Austin, we know that the upper frequency for a UAS RADAR signature is 0.03 m = 3 cm. This is approximately 10 GHz frequency. See Figure 8-8. RADAR is usually thought of in terms of frequency Bands. See Figure 8-9 RADAR Bands and their Usage. (ITR , 2019)

Figure 8-9 RADAR Bands

Frequency Range	Wavelength Range	Band Name	Usage
3-30 MHz	10-100 m	HF	Coastal radar systems
30-300 MHz	1-10 m	VHF	Very long range
300-1000 MHz	0.3-1 m	UHF	Very long range
1-2 GHz	15-30 cm	L-band	Long range
2-4 GHz	7.5-15 cm	S-band	Terminal air traffic control, marine radar
4-8 GHz	3.75-7.5 cm	C-band	Satellite transponders, synthetic aperture radar
8-12 GHz	2.5-3.75 cm	X-band	Marine radar, weather, ground surveillance, synthetic aperture radar
12-18 GHz	1.67-2.5 cm	Ku-band	Satellite transponders
18-24 GHz	1.11-1.67 cm	K-band	Satellite transponders, radar guns, weather
24-40 GHz	0.75-1.11 cm	Ka-band	Mapping, surveillance

The key to understanding Electronic Warfare (EW), its role in detecting a UAS approaching a target, is understanding of radio propagation theory. If we understand how radio signals propagate, we can then intercept, jam, or protect in a logical progression. (Adamy D. -0., 2015) Adamy has written five stellar references on EW, use of dB logarithmic mathematics to solve EW equations for strength, gains, losses, radars, interceptors, jamming technologies, current threats, defense systems and more for the reader to research and enjoy. (Adamy D. -0., 2015)

RADAR is Radio Detection and Ranging. It is the use of radio waves and their propagation in the EMS to determine the battlespace elements for an approaching aircraft, UAS, ship, submarine, or any moving vehicle. We are only interested in two equations to understand the RADAR (radio) signature of a UAS. They are the link equation and the RADAR Range Equation, both presented without derivation. "*The operation of every type of RADAR, military communications, signals intelligence, and jamming system can be analyzed in terms of individual communications links.*" (Adamy D. -0., 2015) "A Link includes one radiation source, one receiving device, and all events to the electromagnetic energy as it travels from source to receiver." (Adamy D. -0., 2015)

"Sources and receivers can take on many forms. When a radar pulse reflects off the skin of a UAS or airplane, the reflecting mechanism is a transmitter. It obeys the same laws of that apply to a walky-talky when the transmit button is pushed. Yet there is no power source and no circuitry to the reflection." (Adamy D.-9. , 1998)

One – Way Link Equation

"The basic communication link, known as a one-way link, consists of a transmitter, receiver, transmitting and receiving antennas, and propagation losses between the two antennas along the path." (Adamy D. -0., 2015) See Figure 8-10 Path Through Link.

Figure 8-10 Path Through Link

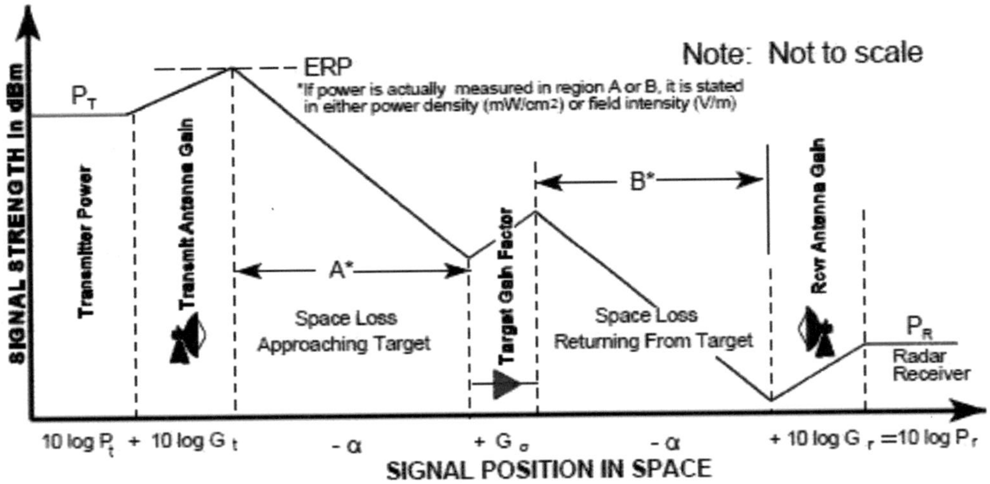

Source: Naval Air Warfare Center Weapons Division, Avionics Department (2013). Electronic Warfare and Radar Systems Engineering Handbook – Two-Way Radar Equation (Monostatic) http://www.rfcafe.com/references/electrical/ew-radar-handbook/two-way-radar-equation.htm

"The diagram shows signal strength in dBm and increases and decreases of signal strength in dB." (Adamy D.-9. , 1998) Figure 8-10 shows the Line-of-Sight link. "The transmitter and receivers can electronically see each other. However, there are interferences / exceptions. The link must not be too close to water or land or in severe weather or asymmetric non-line-of-sight propagation factors. To calculate the received signal level (in dBm), add the transmitting antenna gain (in dB), subtract the link losses (in dB), and add the receiving antenna gain (in dB) to the transmitter power (in dBm)." (Adamy D.-9. , 1998)

"A simple example of the link equation in dB format is:

Transmitter Power (1 Watt) = + 30 dBm
Transmitter Antenna Gain = +10 dB
Spreading loss = 100 dB
Atmospheric loss = 2 dB
Receiving Antenna Gain = +3 dB
Received Power = =30 dBm + 10 dB – 100 dB – 2 dB +3 dB = – 59 dBm" (Adamy D.-9. , 1998)

Figure 8-11 One – Way RADAR Equation

ONE-WAY RADAR EQUATION

Peak Power at Receiver Input, $P_r \text{ (or } S) = P_D A_e = \frac{P_t G_t A_e}{4\pi R^2}$ *and Antenna Gain,* $G = \frac{4\pi A_e}{\lambda^2}$ *or: Equivalent Area,* $A_e = \frac{G\lambda^2}{4\pi}$

So the one-way radar equation is :

$$S \text{ (or } P_r) = \frac{P_t G_t G_r \lambda^2}{(4\pi R)^2} = P_t G_t G_r \left[\frac{c^2}{(4\pi f R)^2}\right]^{*} \qquad (\text{Note: } \lambda = \frac{c}{f})$$

* keep λ, c, and R in the same units

On reducing to log form this becomes:

$$10\log P_r = 10\log P_t + 10\log G_t + 10\log G_r - 20\log fR + 20\log (c/4\pi)$$

or in simplified terms:

$$10\log P_r = 10\log P_t + 10\log G_t + 10\log G_r - \alpha_1 \quad \text{(in dB)}$$

Where: α_1 = one-way free space loss = $20\log (f_1 R) + K_1$ (in dB)
and: $K_1 = 20\log [(4\pi/c)$(Conversion factors if units if not in m/sec, m, and Hz)]

Note: To avoid having to include additional terms for these calculations, always combine any transmission line loss with antenna gain

Values of K_1 (in dB)

Range (units)	f_1 in MHz K_1 =	f_1 in GHz K_1 =
NM	37.8	97.8
km	32.45	92.45
m	-27.55	32.45
yd	-28.33	31.67
ft	-37.87	22.13

Note: Losses due to antenna polarization and atmospheric absorption (Sections 3-2 & 5-1) are not included in any of these equations.

Source: Naval Air Warfare Center Weapons Division, Avionics Department (2013). Electronic Warfare and Radar Systems Engineering Handbook – One-Way Radar Equation. http://www.rfcafe.com/references/electrical/ew-radar-handbook/two-way-radar-equation.htm

Effective Range

The question is what is the maximum range that a RADAR can "see" a UAS in any form: individual, group, team or Swarm? The RADAR range equations can be used to estimate ***maximum distance to detect a UAS***. The smaller the UAS the less reflective area is present to "return "a radar pulse back to its transmitter source. See Figures 8-11 and 8-12 which give the one-way and two-way (return trip) for determining the maximum range of a RADAR unit. At the maximum link range, the received power is equal to receiver sensitivity. "Receiver sensitivity is defined as the smallest signal (lowest power strength) that it can receive and still provide the specified output." (Adamy D. , 2001)

Figure 8-12 Two Way RADAR Equation (Bi-Static)

TWO-WAY RADAR EQUATION (BISTATIC)

Peak power at the radar receiver input is: $P_r = \dfrac{P_t G_t G_r \lambda^2 \sigma}{(4\pi)^3 R_{Tx}^2 R_{Rx}^2} = P_t G_t G_r \left[\dfrac{\sigma c^2}{(4\pi)^3 f^2 R_{Tx}^2 R_{Rx}^2}\right]^*$

Note: $\lambda = c/f$ and σ = RCS
* *keep λ or c, σ, and R in the same units*

On reducing the above equation to log form we have:

$10\log P_r = 10\log P_t + 10\log G_t + 10\log G_r + 10\log \sigma - 20\log f + 20\log c - 30\log 4\pi - 20\log R_{Tx} - 20\log R_{Rx}$

or in simplified terms: $10\log P_r = 10\log P_t + 10\log G_t + 10\log G_r + G_\sigma - \alpha_{Tx} - \alpha_{Rx}$ (in dB)

Where α_{Tx} corresponds to transmitter to target loss and α_{Rx} corresponds to target to receiver loss.

Note: Losses due to antenna polarization and atmospheric absorption (Sections 3-2 and 5-1) are not included in these equations.

Target gain factor, $G_\sigma = 10\log \sigma + 20\log f_1 + K_2$ (in dB)

K_2 Values (dB)

RCS (σ) (units)	f_1 in MHz K_2 =	f_1 in GHz K_2 =
m^2	-38.54	21.46
ft^2	-48.86	11.14

One-way free space loss, $\alpha_{Tx\ or\ Rx} = 20\log (f_1 R_{Tx\ or\ Rx}) + K_1$ (in dB)

K_1 Values (dB)

Range (units)	f_1 in MHz K_1 =	f_1 in GHz K_1 =
NM	37.8	97.8
Km	32.45	92.45
m	-27.55	32.45
yd	-28.33	31.67
ft	-37.87	22.13

Source: Naval Air Warfare Center Weapons Division, Avionics Department (2013). Electronic Warfare and Radar Systems Engineering Handbook – Two-Way Radar Equation (Bi-Static), http://www.rfcafe.com/references/electrical/ew-radar-handbook/two-way-radar-equation.htm

If the received power level is at least equal to the receiver sensitivity, communication takes place over the link. The amount of design signal delta over the minimum receiver sensitivity is called the margin. Both Figures 11 and 12 show the derivations (in normal and dB forms) of the RADAR Ranging Equations for limited environments. Other forms of the basic RADAR Ranging Equation, derivations, definition of terms and examples of radar units for surveillance, tracking and jamming applications can be found in Toomay's simplified reference. (Toomay, 1982) Readers interested in the RADAR units for mariners (picking up a hostile UAS over a ship) can refer to Monahan's (Monahan, 2004) or Burch's references. (Burch, 2015)

Example – Given the operating frequency is 100 MHz, the atmospheric and normal terrestrial loses are minimal. Assume the transmitter output power, Pt = 10 watts. [About double the normal marine VHF set.] The transmitting gain antenna, GT is +10dB, the receiving antenna gain, GR, is +3 dB, and the design receiver sensitivity, Sens = – 65 dBm. {If we find that the received power level (say -59 dBm is at least equal to the sensitivity, then the communication takes place. The margin is this example would be 6 dB higher}. Assume line-of-sight between the two antennas. Calculate the maximum range we can see to the hostile UAS, not using Stealth techniques to reduce the radar visibility. Let PR = received power in dBm. Let d = distance in km. Setting Sens = PR = -65 dBm. Convert to dB math. Plug in the values and solve for 20 log (d). [Logs are base 10 not base e}

Sens = -65 dBm = PR = PT + GT -32.4 -20 log (f) – 20 log (d) + GR
20 log (d) = PT + GT -32.4 -20 log (f) + GR – Sens
And PT = 10 W = +40 dBm, GT = 10 dB, Gr = +3 dB, [20 log (f=100] = +40 dB
20 log (d) = +40 +10 – 32.4 -40 +3 +65 = 45.6
D = antilog (20 log (d)/20) = Antilog (45.6/20) = Antilog 2.28 = 190.54 km = 118.6 miles

We can see the UAS (multiple with bead on leader) at 119 miles from our radar transmitter.

We have come full circle back to the question of designing a UAS for stealth and to get closer to the target.

Closer

In this example, 119 miles gives the US defender some time to respond to his radar set bleep. However, as a hostile actor, suppose the plan is to reach within 5-10 miles of the US target. If a slow UAS is travelling at 60 mph, in the 5 seconds that it takes to spot the UAS on the defender RADAR set, the UAS has travelled about 1 ½ football fields.

{60 mph / 60 min /hour = 1 mile/ per min = 5280 ft /min / 60 sec / min = 88 ft / s x 5 sec = 440 feet / 3 feet /yard = 146.66 yards / 100 yards football field = ~ 1 ½ football fields travelled} Many USAs are designed to travel at over 100 mph.

To reduce the UAS RADAR detectability "(signatures") "to an acceptable level it is necessary to reduce the received emission or reflection of the above frequencies (wavelengths) below a threshold value. This is often just a function of the height the UAS operates at." (Austin, 2010) Hostile UAS fly lower and at greater speeds. The idea is to hit the target with surprise from above. "An unmanned air vehicle has advantages, compared to manned equivalents, in its inherent ability to achieve lower signatures. These advantages are:

- It has dense packaging. No air crew or their support equipment means a dimensionally smaller vehicle. In general, the vehicle mass is less for UAS missions;
- It has shape. UAS shape is not limited to accommodations for crew and need for access or external vision ports. UAS designs consider aerodynamic efficiency and detection – signature reduction." (Austin, 2010)

Acoustic Signature Reductions

"Aircraft noise may be the first warning of its presence; however, it may not immediately be directionally/locatable for detection." "UAS noise emanates predominantly from vortices, tips of wings, rotors, or propellers. Lowering wing span or blade span enhances acoustical stealth." Conventional propulsion systems are a concern because of the noise of combustion. Electric motors develop virtually no noise. "Reducing mass and aerodynamic drag of the UAS reduces noise generation." (Austin, 2010)

The human ear is a problem for the designer. "It is most sensitive to frequencies around 3500 Hz and can hear sound down to a practical threshold of 10 dB. For a given sound pressure level, attenuation of sound with distance in air and insulating material varies as the square of the sound frequency. Low frequency sound presents a greater problem for UAS stealth design." (Austin, 2010)

MALE and HALE systems do not present acoustic issues "noise from their high frequency generators is attenuated by the time it reaches earth. The greater noise problem is posed by smaller UAS using piston engines." (Austin, 2010) Sound comes from their internal combustion and exhaust systems. Sound emanation can be achieved with sound-absorptive materials, silencers and mufflers and directing the intake and exhaust manifolds upward. "Lastly, the level of sound detected depends on the level of background noise for sound contrast." (Austin, 2010)

Visual Signature

"The human eye operates within a small range of the EMS, between 0.4 and 0.7-micron (um) meters peaking at a maximum efficiency approximately 0.55 um. Criteria that determine whether a human can see a UAS in the sky regardless of cloud conditions: a) size and shape of the UAS, b) background contrast and sharpness of edges that contrast; c) atmosphere; d) movement; e) exposure time; f) stability and soberness of observer; and g) glint." (Austin, 2010)

"Size and shape of the UAS combine to determine a threshold of detectability (*sic, at a flying altitude of the* UAS) Contrast C, is defined as a ratio of the difference in luminance between an object and its background to the luminance of its background. So,

$$C= (B - B1) / B1 = \text{delta } B / B1$$

Where B is the luminance of the object and B1 is the luminance of the background, cd/m**2." (Austin, 2010)

Laboratory tests have shown that "there is a 50% probability of detection by the unaided human eye if it subtends an angle of about 0.15 mrad." (Austin, 2010) Other effects of atmosphere, movement, exposure time, stability of observer, and glint require further study.

Thermal Signature

"Infrared (IR) radiation is emitted from a heat source and propagates in the same way as light. The detectability of the radiant body is determined by its contrast with the background." (Austin, 2010) "IR radiation is absorbed well by the atmosphere, particularly water and CO_2 molecules, more so than visible light. The major windows for IR to pass through the atmosphere are in the

3-4 and 8-12 micron -meter wavelength bands. Receivers are designed to operate in one of these two bands." (Austin, 2010)

Design factors include materials to insulate the power plant exhausts and using low-emissivity materials. Sky is cold and offers an excellent contrast for IR detection. However, the technology has not matured, and long-range IR detectors are limited to a 5-degree field of view. "The detection range is a function of temperature (K) to the fourth power. Reducing the target emissions temperatures is an effective evasion tactic for IR detection." (Austin, 2010)

Radio / RADAR Signature

"The radio signature relates to the radio frequency emissions from the UAS and SAA systems and should be minimized." RADAR / Radio signature is reflected frequency "generated by an emitter which is scanning the sky from ground and looking for a return (reflected) pulse from a body entering its sector." (Austin, 2010) "The stealth designer aims to prevent these pulses being reflected to the detector." (Austin, 2010) There are three methods to minimize UAS reflection of pulses to a receptor:

- "Use radar- translucent materials such as Kevlar or glass. Problem is that to protect a UAS sufficiently, the thickness of materials may be prohibitive;"
- "Cover the external surfaces with radar-absorptive materials. Unfortunately, this solution is both material and frequency dependent" and
- "Shape the aircraft externally to reflect radar pulses in a direction away from the transmitter." This is more efficient proposal than a) or b). The goal is to reduce cross-section areas to not be at 90-degree angles to the radiation. The latter yields a clean pulse return. "It is also a design goal to avoid surfaces meeting at 90 degrees as this act as a radar "corner reflector" to give strong returns." (Austin, 2010)

Low flying UAS – Use Navigation Collision Avoidance RADAR

The author has been experimenting with a low flying UAS[1] flying over calm (< 10 knot winds) Chesapeake Bay water at 100 above mean sea level. With a tailwind the Drone can clock upwards of 50-60 mph. The DJI Phantom 4 Pro Plus (PRO+) Drone is not designed for stealth. Experiment results: Can the Drone be detected? Yes. At what distance? Using the marine reference *Boaters Bowditch* equation for RADAR Geographic Range over water, the author calculated the detection range from the pulses of the boat's radar to the UAS.

Distance (nautical miles) = 1.22 X [square root (Height of the Antenna (ft)) + square root (Height of the Target (ft))]. (Hubbard, 1998)

1. DJI Phantom 4 Pro Plus (PRO+) Drone with various payloads.

If the antenna on a cruiser is at 20 ft and the UAS is flying at 100 ft at 55 mph toward the boat on a bow heading, what is the detection range? Answer 17.7 nautical miles.

The question then is how low should an aircraft / UAS fly to avoid radar detection?[2] The answer is "it depends." It depends on the radar and the aircraft itself. Search radars are often usually located on well elevated land-forms to extend the line-of-sight (LOS) distance to the horizon. A conventional (non-stealthy) UAS aircraft flying just above surface level (100 ft up above water could theoretically be seen, as it comes over the horizon line, if there were no obstacles, radio towers, seaplanes, or large freighters to obscure it. The atmosphere attenuates the transmissions from the radar and can distort them.) The atmosphere over land has contaminants that can obscure radar pulses; dust, sand, smoke, rain, bugs, birds, water vapor, fog, clouds etc. These contaminants bend, scatter, and/or absorb transmissions showing up as "noise" on the radar screens. Noise reduces the effective range of the radar. Some noise can be electronically filtered so that the UAS can be distinguished from the noise. Chesapeake Bay has clean atmospheric conditions in fair weather – no contaminant density.

Assume a radar antenna mount on a hill might at 230 ft above ground level. [This is the height of three Radio Towers on land near Annapolis, MD.] If the aircraft is flying 100 ft above the ground, it will be at 30.7 nautical miles distance from the radar as it comes over the horizon. At that distance it will be seen clearly because atmospheric disturbances at that short distance are not an issue. That is what a US capital ship at sea might encounter if looking at a strike fighter or cruise missile coming at 11 nm/m with 3 minutes to impact. Let's now mount the radar on a mountain top (say ski Round Top in Camp Hill, PA). Its antenna is approximately 2600 ft above mean ground level. The UAS aircraft flying at 100 ft will come over the horizon and be detected at ~ 74.4 nautical miles. 100 ft is not realistic unless over the sea on a calm day.

Stealthy UAS aircraft are a problem for radars. Stealth doesn't make them invisible, but it greatly reduces the limit of detection. In the case of our first example, a fast stealth UCAV fighter coming over the horizon at ~18 nm will probably not be seen immediately. The human lag time for the radar operator could be 1-2 minutes, bringing the UCAV within 60 seconds of the radar receiver or ship before discovery – not much time for lethal countermeasures

Discussion Questions

1. Why is RADAR detection of a UAS over water different from in the air?
2. Check the Austin signature limits and track them down on the EMS charts.
3. What stealth measures would be employed for a UAS / drone specific to maritime service – say in the Spratly Islands?

2. The author's CRYPTOWIZ is a nominal 36' cruising yacht with Garman 3205 RADAR with touch screen and Viseo capability, GMR 41. AIS systems are effective to 64 nautical miles.

Bibliography

Adamy, D. -0. (2015). EW 104 EW *against a New Generation of Threats*. Boston: Artech House.

Adamy, D. (2001). *EW 101 A First Course in Electronic Warfare*. Boston, MA: Artech House.

Adamy, D.-9. (1998, Jan). Lesson 4: the basic link for all EW functions. (electronic warfare)(EW Reference & Source Guide). *Journal of Electronic Defense, Jan 1998 Issue.*

Austin, R. (2010). *"Design for Stealth", Unmanned Aircraft Systems UAVS Design Development and Deployment.* New York: John Wiley and Sons.

Burch, D. (2015). *RADAR for Mariners*. New York: McGraw-Hill.

Fletcher, H. a. (1933). Loudness, its definition, measurement and calculation. *Journal of the Acoustical Society of America* , 5, 82-108.

Hubbard, R. K. (1998). *Boaters Bowditch*. Camden, MA: International Marine.

ITU. (2019, July 19). ARTICLE 2 – *Nomenclature – Section I – Frequency and Wavelength Bands.* Retrieved from ITU Radio Communication Edition 2008: https://web.archive.org/web/20111001005059/http://life.itu.int/radioclub/rr/art02.htm

Monahan, K. (2004). *The Radar Book: Effective Navigation and Collision Avoidance.* Anacortes, WA: Fineedge Publications.

Rosen, Stuart (2011). *Signals and Systems for Speech and Hearing* (2nd ed.). BRILL. p. 163. "For auditory signals and human listeners, the accepted range is 20Hz to 20kHz, the limits of human hearing"

Staff. (2016, April 17). *Equal Loudness Contours*. Retrieved from Gutenberg Organization: http://central.gutenberg.org/article/WHEBN0001046687/Equal-loudness%20contour

Toomay, J. (1982). *RADAR for the Non – Specialist. London; Lifetime Learning Publications.* London: Lifetime Learning Publications.

TRS, S. (2018, July 10). *Tontechnic-Rechner-Sengpielaudio.* Retrieved from Tontechnic-Rechner-Sengpielaudio Calculator: www.sengspielaudio.com/calculator-wavelength.htm

Weise, E. (2017, August 23). *could-hackers-behind-u-s-navy-collisions.* Retrieved from USATODAY: https://www.ruidosonews.com/story/tech/news/2017/08/23/could-hackers-behind-u-s-navy-collisions/594107001/

Readings

Adamy, D. (2001) EW 101 A *First Course in Electronic Warfare*, Boston: Artech House.

Adamy, D. (2004) EW 102 A *Second Course in Electronic Warfare*, Boston: Artech House.

Adamy, D. (2009) EW 103 *Tactical Battlefield Communications Electronic Warfare*, Boston: Artech House.

Adamy, D. (2015) EW 104 EW *against a New Generation of Threats*, Boston: Artech House.

Adamy, D. (2003) Introduction to Electronic Warfare Modelling and Simulation, Boston: Artech House.

Austin, R. (2010) UAVS *Design, Development and Deployment*, New York: Wiley.

Burch, D. (2005) RADAR *for Mariners*. New York, McGraw-Hill.

Hubbard, R.K (1998) Boaters Bowditch, Camden, MA: International Marine.

Monahan, K (2004) *The RADAR Book: Effective Navigation and Collision Avoidance.* Anacortes, WA: Fineedge Publications.

Toomay, J.C. (1982) RADAR *for the Non – Specialist*. London; Lifetime Learning Publications

Chapter 9: Case Study Smart Skies Project

Student Learning Objectives

Chapters 7 and 8 covered SAA systems, collision avoidance technologies and designing UAS for stealth. In Chapter 9, the student advances the SAA principles by investigating one of the more interesting published research projects in SAA known as the Smart Skies Project. Smart Skies was a defining project that has spurned many different research papers, funded projects, technology advancements and improved conditions at airports.

Safety

The primary goal of UAS is safe, predictable flying in the NAS. To do this UAS must have SAA "capability that replicates a human function: to ***see and avoid.***" (Wilson, 2012)

This is quite a challenge and has been the delaying factor for wide-spread "integration of UAS into the NAS" other, then limited classes of airspace. Further, the level of safety achievable by UAS designers must be at least equal to that of "manned aviation. Until *Sense and Avoid* (SAA) for UAS reaches equal or better capability to See and Avoid," FAA will continue to restrict operations of UAS in the NAS. (Wilson, 2012)

See and Avoid

Visual Flight Rules (VFR) "require pilots to see and avoid aircraft and other objects while flying visual meteorological conditions." (Wilson, 2012) Current see and avoid technology has clearly prevented collisions, however the state-of-the-art is far from reliable. "Some of the limiting factors are: human visual system, demands of cockpit tasks, and physical and environmental conditions. These combine to make see and avoid an uncertain method of traffic separation." (ICAS, 2012) Wilson (2012) presents several studies that show that pilots do not perform well for unalerted visual searches. Another study showed that pilots "spent more time looking inside the cockpit than outside and that the average scanning performance would leave them vulnerable to late detection of aircraft conflicts" and adequate reaction time. (Wilson, 2012)

Case Study: The Smart Skies Project

"The Smart Skies Project (SSP) was a 3-year AUD $10M joint venture amongst ARCAA, QUT, BR & T, and Insitu Pacific. The Smart Skies Project was, created to enable technologies for UAS operations in non-segregated airspace." (Wilson, 2012) Specific objectives of SSP were to, "develop and demonstrate automated separation management technologies that facilitate use (sic) of the national airspace system by both manned and unmanned aircraft." The second objective was to, "use (sic) the information and experiences gained to support the further development of standards, regulations and safe operating practices for civil and commercial UAS in Australia and overseas (USA)." ARCAA accomplished both goals. Later in the chapter the author will summarize several practical research projects that SSP inspired. "SSP describes "development of an automated separation management system," aircraft-based sense-and-act technologies, and a mobile aircraft tracking system." (Clothier R. R., 2011) More specifically, the enabling technologies that support Class G airspace are MATS, "a network-enabled Mobile Aircraft Tracking System, for detection and tracking" of local air traffic. (Clothier R. R., 2011) MATS has multiple sensors supporting "UAS in non-segregated airspace. MATS uses a cost-effective primary radar and cooperative surveillance systems." (Wilson, 2012) The second enabling technology is the "Vision-based Sense and Act (SAA) – an automated system capable of replicating the See-and-Avoid function of the human pilot." (Clothier R. , 2017) In the SSP environment, SSA "refers to the capability of a system to detect, track, and resolve potential collisions with other aircraft (A/C) or obstacles on the ground. SSP explores the development of automated aircraft-bases SAA "systems capable of avoiding dynamic obstacles (other A/C) or static obstacles (trees, powerlines, or radio towers)." (Clothier R. R., 2011) The automated Static Obstacle Avoidance (SOA) system is particularly effective at low-altitudes aircraft operations. The last enabling technology is the Automated Separation Management System (ASMS) is designed to reduce workload for the Automated Traffic Management by increasing automation. It consists of three systems; the "Automated Dynamic Airspace Controller (ADAC)", the "Common Information Network (CIN),and the Predictive Flight Management System (pFMS)." (Clothier R. R., 2011) All the theory would be useless without flight tests to prove the innovative technologies. "An important objective of SSP was to integrate, demonstrate and validate the performance these enabling technologies through a series of eight integrated flight-test activities." (Wilson, 2012)

Smart Skies Architecture

"The SSP architecture diagram is shown in Figure 9-2 The system includes manned and unmanned aircraft, virtual aircraft, public mobile data and Iridium satellite links," the ADAC and MATS. **What is special about the flight testing in SSP " architecture is that the *manned and unmanned flight tests occurred in Kingaroy, Australia, and the SAA control system, the ADAC was in Palmdale, California*."** (Wilson, 2012) "All the systems in the SSP architecture are linked to the Internet by the Iridium satellite communication network and a 3G public mobile data network. The Iridium satellites provide global coverage at low bandwidth communication. The public mobile data network provided high bandwidth communications, but in a limited geographic coverage." (Wilson, 2012)

Flight Test Capability

The Automated Cessna 172R was used for capture of onboard flight and sensor data. It had a "custom flight management systems and display, with input to the onboard autopilot." (Clothier R. , 2017)

Figure 9-1 SSP Architecture

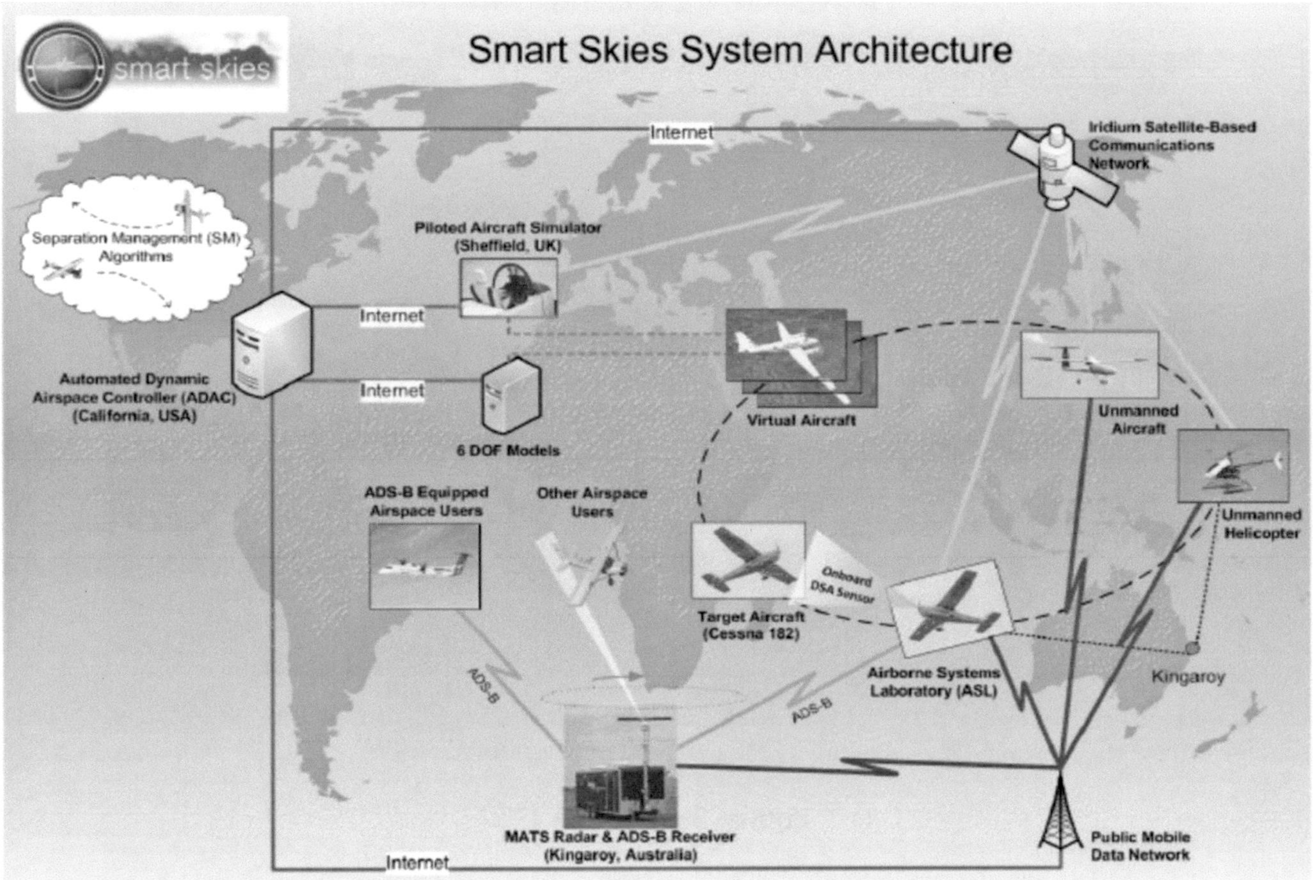

Source: Wilson, M. (2012). The Use of Low-Cost Mobile Radar Systems for Small UAS Sense and Avoid. Sense and Avoid in UAS Research and Applications.

The "Iridium and NextG™ communications facilitated automated telemetry, command and control from anywhere in the world." (Clothier R. , 2017) Figures 9-2 and 9-3 Show the Cessna 172R cockpit and flying views.

Figure 9-2 Cessna 172R Cockpit

Source: Ricci, A. Cessna 172R/S Skyhawk [cockpit view]. Retrieved from https://www.airliners.net/aircraft-data/cessna-172rs-skyhawk/142 (Viewed September 13, 2018)

Figure 9-3 Cessna 172R

Source: Cessna 172R of BSC Flying School at Khon Kaen-KKC, Thailand, 3/12/16 by Alec Wilson from Khon Kaen, Thailand - HS-BSI, CC BY-SA 2.0, https://commons.wikimedia.org/w/index.php?curid=56933619

Also, in the SSP air was the ARCAA Flamingo UAS. See Figure 9-4.

Figure 9-4 ARCAA Flamingo UAS

Source: Robotronica.qut.edu (2015) Australian Research Centre for Aerospace Automation Display. http://www.robotronica.qut.edu.au/demonstrations/arcaa.php

The "ARCAA Flamingo UAS sports a micropilot 2128 autopilot, 20 kg MTOW, 112 feet wingspan, one-hour endurance, Iridium and NextG™ communications facilitated automated telemetry, command and control from anywhere in the world," and seventy plus hours of automated operation. (Clothier R., 2017)

The third flying element of the triad was the ARCAA Heli UAS. See Figure 9-5.

Figure 9-5 ARCAA Heli UAS

Source: Clothier, R., Baumeister, R., Brunig, M., Roberts, J., Duggan, A, & Wilson, M. (June 2011). The SBS -1 ADS-B, see: http://www.kinetic.co.uk/DownloadFiles/Assorted/SBS1-eR-Reference-Manual.pdf

The ARCAA Heli UAS package includes a "custom autopilot, FMS and ground station, 12.kg

MTOW, 1.8m rotor, 45 minutes endurance, scanning laser and stereo vision sensors and Iridium and NextG™ communications facilitated automated telemetry, command and control from anywhere in the world." (Clothier R. , 2017)

The Mobile Aircraft Tracking System (MATS)

"Angelov demonstrates the architecture of the MATS subsystem. MATS interfaces with two external systems, a UGCS and the ADAS. MATS was designed to research and demonstrate a field-deployable surveillance system (see Figure 9-6)." (ICAS, 2012)

Figure 9-6 MATS

Source: Wilson, M. (2010) Boeing Research & Technology Australia, A Mobile Aircraft Tracking System That Supports Unmanned Aircraft Operations, 27th International Congress of The Aeronautical Sciences. http://www.icas.org/icas_archive/icas2010/papers/474.pdf

MATS was "designed to support UAS operations in non-segregated airspace. It uses commercial off-the-shelf (COTS) primary radar and tracking system, Automated Dependent Surveillance Broadcast (ADS-B) and filtering, display and communications systems." (Clothier R. , 2017) MATS also sports a "VHF voice transceiver and a server that performs data fusion and communications management. The UAS flight crew may be located inside the MATS trailer or inside a remote GCS." (Wilson, 2012)

The MATS Radar System

MATS enables two interfaces to work with SAA. "One interface enables MATS to provide information to the UAS pilot. The radar's TCP/IP data networking capability allows tracks and plots to be sent remotely to the CGS." (Wilson, 2012) "This is a pilot/operator in-the-loop SAA system, where MATS provides the sense function and the pilot provides the avoid function by maneuvering the UAS." (Wilson, 2012) MATS uses a second interface to "provide information about cooperative and non-cooperative aircraft to external systems – the ADAC. In this case the avoid function is automated. ADAC accesses the airspace situation and provides updated flight plans to avoid collisions/conflicts." (Wilson, 2012)

"A key part of MATS is the primary surveillance radar system. It has two parts; a front-end COT marine radar and a back-end that performs the detection, tracking and display functions. The Furuno radar operates at three pulse rates; 0.07 us for 10.5 m, 1.2 us for 180 m, and 0.3us for an approximate distance of 45 m." (ICAS, 2012) "The Accipiter® permits multi-target tracking and uses a specially designed Multiple-Hypotheses-Testing (MHT) and Interacting-Multiple-Models tracker (IMM) that locks on to targets that have a Low Radar Cross section (LRC). All radar information is mapped, and multiple layers can be displayed." (ICAS, 2012)

The MATS ADS-B Receiver

"The ADS-B receiver that MATS used is an SBS-1 from Kinetic Avionics Products Limited. The SBS-1 is portable, low-cost 1090 MHz ADS-B receiver. It provided the tracking and logging of information about ADS-B equipped Aircraft." (ICAS, 2012)

MATS Performance and Flight Characterization Testing

In March 2010, MATS was an innovative approach to ground-based tracking. The SSP team wanted MATS to detect and accurately track manned and unmanned airspace users and use that information to maintain separation minima. MATS performance had to be superior to ground observers or observers situated in a chase plane. MATS was used to detail responses to failures of navigation systems and to track recovery of the A/C. MATS was a key component in the ASMS.

MATS strength main component was its Primary Surveillance Radar (PSR). SSP team used flight simulations that were augmented with human spotters to compare against MATS' PSR. The SSP team varied "target cross-sectional area – by using different A/C and target A/C range, velocity, altitude maneuvering and clutter." (Clothier R. R., 2011)

MATS Results

"MATS radar characterization testing verified that the PSR can track a Cessna 172 in a low clutter environment out to a range of 15 nm." (Clothier R. R., 2011) Tracking consistency improved at closer ranges. "MATS also tracked A/C flying on designated air routes at ranges between 16 – 20 nm." (Clothier R. R., 2011) MATS was able to track small fixed-wing UAS targets. MATS data provided additional situational awareness and provided valuable lessons for UAS operators.

Sense-and Act

“Sense-and-Act (SAA) refers to the capability of a system to detect, track, and resolve potential collisions with another A/C or obstacles on the ground.” (Clothier R. R., 2011) SSP explored “development of automated aircraft-based SAA systems capable of avoiding dynamic obstacles (other A/C) or static obstacles (trees, power lines and radio towers).” (Clothier R. R., 2011) The ARCAA C172 was used with full closed-loop control of its navigation systems. SSP performed “controlled experiments involving head-on and over-taking scenarios with a C182. SSP explored range of FoV, image processing and control law configurations.” (Clothier R., 2017)

Dynamic SAA

The challenge of SAA is for the system to meet or exceed human “see-and-avoid” capabilities “provided by a pilot under suitable VFR conditions.” (Clothier R. R., 2011) At all times the SAA must “maintain a visual lookout for other A/C and if necessary, initiate maneuvers to avoid a potential collision scenario.” (Clothier R. R., 2011) Clothier discusses the legal scenarios where SAA must qualify and the limitations of human see-and-avoid abilities. The dynamic Sense-and-Avoid algorithms are discussed in detail in the same paper.

SAA Experiments

Since MATS was designed to “support UAS operations, an interesting place to operate a UAS is a small non-towered airport in Class G space. Some of A/C coming into the airport will carry a VHF. These A/C can easily communicate their position and intent. Unfortunately, a percentage of the A/C will not carry a VHF radio, therefore, coordinating UAS activities will not be possible.” (Wilson, 2012) MATS PSR is the only way to track these A/C. “Non-cooperative aircraft without a VHF represent the most challenging class of airspace user for both manned and unmanned aircraft.” (ICAS, 2012)

Consider a non-cooperative intruder aircraft. If MATS PSR can detect it, “what actions should a UAS pilot take to avoid conflicts with the intruder? So, the scenario that SSP explored was:

- A small UA operating in a non-towered airport,
- UA operations are supported by a PSR whose role is to detect A/C beyond the range of observers at the airport,
- Intruder A/C are non-cooperative and without VHF radio,
- Intruder A/C can arrive from anywhere, anytime and any altitude in class G
- Intruder A/C intentions are unknown (fly-over, sight-seeing, overfly and join a traffic pattern, or land)
- Intruder A/C has a speed advantage over a small UAS. The UAS cannot outrun the Intruder.

The UA may need to move to a new location or change altitude to mitigate the risk of collision.

The airport and USA FAA or Australian Civil Authority sets the AGL rules flying near an airport, city, or populous area." (Wilson, 2012)

UAS Actions

Wilson (2012) describes four geographic areas with locations oriented toward an intruder A/C that UAS may determine its course of action(s). "*Danger* – the UA is between the intruder and the airport but outside the circuit area." The *Circuit* ring is 3 nm radius from the runway. The far end of the danger zone is 6 nm in radius. For the *Transit* area, the "UA is outside the circuit area but is the likely transit of the intruder." It extends to the 6 nm and starts at 3 nm on the same bearing as the intruder. The *Hold* area is where the UA "is positioned laterally to the current flight path of the intruder." "Lastly, the Circuit area is generally set by the speed capabilities of the UAS." (Wilson, 2012) For the Flamingo, the Circuit area was a radius of 3 nm.

> "The UAS actions, when the UAS is in each area are as follows:"
>
> AREA UAS COMMAND
>
> Danger "Descend to less than 400 ft AGL and orbit"
>
> Transit "Track to a hold area and orbit"
>
> Hold "Orbit or maintain speed and heading, whichever is safer"
>
> Circuit "Track to the dead side of the circuit at 400 ft AGL and orbit." (Wilson, 2012)

Three other actions were simulated but discarded generally by the SSP team; land (potential accident on the ground and it becomes a hazard for the intruder trying to land), climb (climb rate is much slower than intruder A/C and fails as avoidance maneuver) and join a traffic pattern (mixing a difficult to see small UAS with larger manned A/C where traffic density may be high.)

SAA Results

The results were a dramatic success. SSP may have been the "first in the world to demonstrate a fully automated real-time onboard collision avoidance using a vision-based SAA system. It used real aircraft, real hardware, and real conditions." (Clothier R. , 2017) More than eighty data sets were collected. In addition the SSP SAA system demonstrated detection ranges more than 6.421 miles to the closest point of approach (CPA). The SSP solved the problems of vibration compensation and prevention. The SSP team found optimal configurations to minimize MDR and "FAR across a range of atmospheric, cloud, and lighting conditions." (Clothier R. , 2017) The SSP team proved their Concept of Operations (CONOP) and laid out a plan to use multi-spectral sensors.

Sense-and Act Systems (Static)

"A second challenge to UAS operations is avoiding ground obstacles. This is a risk for fixed-wing aircraft performing at low altitudes (crop dusting), helicopter operations for SAR power-line inspection, radio tower avoidance, winching operations, taxiways obstructions, and animal detection." (Clothier R. R., 2011) The SSA project called this a Static Obstacle Avoidance system (SOA) and it was "suitable for close range (<30 m) for Rotorcraft UAS operations at low altitudes." (Clothier R. , 2017) ARCAA researches explored several design components; LIDAR and 2D scanning laser and stereo camera sensor.

SSA SOA Results

"SSP team demonstrated that the SOA successes in an automated, beyond visual range of operation, in an unknown obstacle environment. Sixty flights were completed in a range of single and multi-sensor configurations. LIDAR produced the most reliable avoidance capability with an 84% success rate. It detected trees out to 23m, and a microwave tower out to 18m. The wide (270 degrees) horizontal field of view of the laser allowed it to continue sensing obstacles even when flying alongside of them." (Clothier R. , 2017)

Automated Separation Management System (ASMS)

Figure 9-7 shows the ASMS architecture.

Figure 9-7 ARCAA ASMS

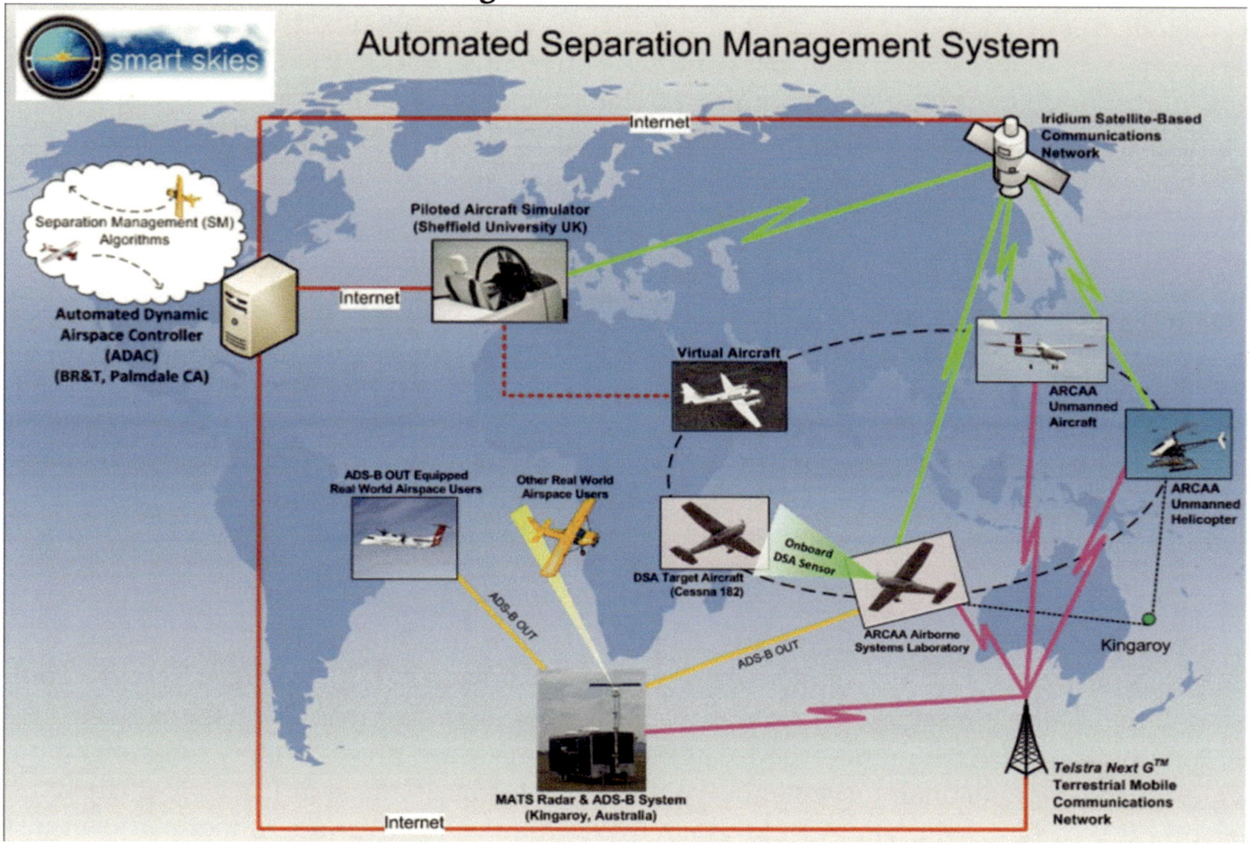

Source: Clothier, R., Baumeister, R., Brunig, M., Roberts, J., Duggan, A, & Wilson, M, S: (June 2011), CT: ASMS, see: http://www.kinetic.co.uk/DownloadFiles/Assorted/SBS1-eR-ReferenceManual.pdf

The ASMS consists of three basic systems; "The Automatic Dynamic Airspace Controller (ADAC), the Common Information Network (CIN), and the Predictive Flight Management Systems (pFMS)." (Clothier R. R., 2011) The ADAC continuously "maintains situational awareness of the state of the airspace system by receiving aircraft tracking information and changes within the airspace environment (weather, dynamic/ temporary airspace activation.) The ADAC also monitors the airspace system state, and if necessary, transmits recommended control information to cooperative airspace users where there is a potential for a Loss of Separation (LOS)." (Clothier R. R., 2011)

The CIN is the computer backbone of the ASMS. It provides communications (multi-channel) "infrastructure to network aircraft and other sensors (surveillance and weather) to the ADAC. CIN can support multiple redundant and geographically distributed ADAC systems." (Clothier R. R., 2011)

"The pFMS system enables communication with the ADAC, but all the participating aircraft required a modified flight management system to include predictive capabilities." pFMS functions include "estimation of the current and future aircraft states (position, attitude, time, and uncertainties), management of multiple communication links within the CIN, receiving, loading and execution of ADAC generated commands, the intelligent management of onboard sensors, and sensor information and display." (Clothier R. R., 2011)

ASMS Results

The AMSM demonstrated unqualified success for fifty trials "for real aircraft, real communications links, real sensors in complex scenarios involving real and simulated aircraft. It was a truly global systems test providing a completely automated separation service from Palmdale, CA to Queensland, Australia. Trials included non-cooperative aircraft detected using the MATS and SAA systems in mixed-mode separation." (Clothier R. , 2017) The ASMS was able to ensure "quality of the separation service under variable communications performance factors such as latencies and drop-out. ARCAA researchers believed that their ASMS was the first in the world to autonomously command and control UAS using a civil mobile cellular network." (Clothier R. , 2017)

The SSP project has spurred hundreds of publications and innovative steps based on the ARCAA team's results. "A few samples are seen here. Sahawneh, L.R, et.al (2017) wrote about a *Ground-Based Sense-and Avoid System for Small Unmanned Aircraft*. The paper describes the development of a small frequency-modulated continuous-wave phased-array radar system that provides three-dimensional surveillance volume. Contarino, V.M. et.al (23 May 2011) wrote a paper on *All Weather Sense and Avoid System* (AWSAS) *for all UAS and manned platforms*. They discuss a new collision avoidance system that would overcome current FAA stated deficiencies so that military and national security agencies can use the NAS in unfettered mode. Lai, J., *Mejias, L & Ford, J.J.* (2011) *wrote about an Airborne Vision-Based Collision-Detection System*. This paper describes the development and evaluation of a real-time, vision-based collision-detection system suitable for fixed-wing aerial robotics." (Wilson, 2012)

Discussion Equations

1. The global market for UAS / UAV is exploding globally. How might terrorist use the UAS as a weapon? What countermeasures would you propose to stop a team or swarm of UAS attacking USA national critical infrastructure?
2. What cyber techniques could be used to change parameters in the SAA systems to take control of an enemy UAS.
3. Research three papers or peer-reviewed articles on the state-of-the-art SAA components. See if you put the dots together for how the SSP might have influenced the chosen works.

Bibliography

Clothier, R. (2017, April 02). *The Smart Skies Project: Enabling Technologies for UAS Operations in Non-segregated Airspace.* Retrieved from QUT ePrints: http://eprints.qut.edu.au/40465/3/40465.pdf

Clothier, R. R. (2011). The Smart Skies project. *IEEE Aerospace and Electronic Systems Magazine.*

ICAS. (2012, December 22). ICAS *Archive* CD1998-2010. Retrieved from ICAS.ORG: http://www.icas.org/ICAS_ARCHIVE_CD1998-2010/ICAS2010/PAPERS/474.PDF

Wilson, M. (2012). The Use of Low-Cost Mobile Radar Systems for Small UAS Sense and Avoid. *Sense and Avoid in UAS Research and Applications.*

Readings

Accipiter® Radar UAS Detection and Tracking, DLA: 07172018, https://www.accipiter-radar.com/products/safety/drone-uav-detection-tracking-alerting/

Adamy, D. (2001) EW 101 A *First Course in Electronic Warfare*, Boston: Artech House.

Adamy, D. (2004) EW 102 A *Second Course in Electronic Warfare*, Boston: Artech House.

Adamy, D. (2009) EW 103 *Tactical Battlefield Communications Electronic Warfare*, Boston: Artech House.

Adamy, D. (2015) EW 104 EW *against a New Generation of Threats*, Boston: Artech House.

Adamy, D. (2003) *Introduction to Electronic Warfare Modelling and Simulation*, Boston: Artech House.

Angelov, P. (2012). *Sense and avoid in UAS research and applications*. Hoboken, N.J.: Wiley.

Austin, R. (2010) UAVS *Design, Development and Deployment*, New York: Wiley.

Burch, D. (2005) *Radar for Mariners*. New York, McGraw-Hill.

Clothier, R. (2011) ARCAA The Smart Skies Project Public presentation

Clothier, R., Baumeister, R., Brunig, M., Roberts, J., Duggan, A, & Wilson, M (June 2011) *The*

SBS -1 ADS-B, see: http://www.kinetic.co.uk/DownloadFiles/Assorted/SBS1-eR-Reference-Manual.pdf

Contarino. M. et.al (23 May 2011) *All Weather sense and avoid system* (AWSAS) *for all* UAS *and manned platforms.* SPIE Defense, Security, and Sensing Conference paper / event, Orlando, FL. SPIEDigitallibrary.org/conference-proceedings-of-spie

Hubbard, R.K (1998) *Boater's Bowditch*, Camden, MA: International Marine.

Mejias, L & Ford, J.J. (2011) *An Airborne Vision-Based Collision-Detection System.* Journal of Field Robotics 28(2), 137-157 Wiley Publications.

Monahan, K (2004) *The Radar Book: Effective Navigation and Collision Avoidance.* Anacortes, WA: Fineedge Publications.

Sahawneh, L.R, et.al (5 December 2017) Ground-Based Sense-and Avoid System for Small Unmanned Aircraft, Journal of Aerospace Information Systems. Published online 24 April 2018 by American Institute of Aeronautics and Astronautics, Inc.

SBS -1 ADS-B, see: http://www.kinetic.co.uk/DownloadFiles/Assorted/SBS1-eR-Reference-Manual.pdf

Smart Skies Project, IEEE A&E Systems Magazine.

Clothier, R (2011) Slide Deck on The Smart Skies Project for ARCAA presentation.

Monahan, K (2004) *The Radar Book: Effective Navigation and Collision Avoidance.* Anacortes, WA: Fineedge Publications.

Toomay, J.C. (1982) *Radar for the Non – Specialist.* London; Lifetime Learning Publications

SECTION IV

UAS WEAPONS & ISR &IO

Chapter 10: UAS Intelligence, Surveillance and Reconnaissance (ISR)

Student Learning Objective

The student will gain knowledge on UAS intelligence, surveillance, and reconnaissance systems. Topics include sensors, missions including history, the current state of industry, and planning and execution considerations for gathering datasets using this technology.

History/Background of UAS ISR

UAS for intelligence, surveillance, and reconnaissance is a well-known application of the technology. “The clear majority of UAVs are used purely for intelligence, surveillance, and reconnaissance (ISR) missions. In current military usage, they range from the Global Hawk, with a wingspan greater than a Boeing 737 airliner, to Nano-helicopters that weigh a few grams, and all points in between” (Lambeth, 2006). FLIR Systems produces the Black Hornet 3 and, “At 32 grams, offers the lowest size, weight, and performance for UAS available…flies 2 kilometers at speeds of over 21 kilometers an hour. The Black Hornet 3 also incorporates sharper imaging processing featuring the FLIR Lepton® thermal micro camera core and a visible sensor to allow greater image fidelity, (with) an improved encrypted military-approved digital datalink, enabling seamless communications and imagery significantly beyond line-of-sight and in closed areas” (“FLIR Launches Next-Generation Black Hornet 3 Nano-UAV,” 2018).

A UAS is simply a “truck” for the sensor system. It is the tool that is used to move, power, and task the sensor in the air. With such a wide variety of “trucks” available today, ISR appears to be limited more by human innovation and imagination than technology itself.

The ability to extend missions, gather previously unknown or unavailable information, and keep human operators out of harm’s way forms the basis of why UAS are more suitable for the dull, dirty, and dangerous missions. UAS do not get bored, tired, lose focus, require minimal training as compared to humans and do not carry the bias of humans, the data gathered is “truth” to the sensor system. “The physical movement of drones is only one aspect of their potential vulnerability. The still image or video cameras routinely fitted to UAVs serve as a live link back to their operators – and enable drones to be used as highly maneuverable real-time eyes in the sky.” (CyberRisk, 2017) “They provide the visual and location information they carry and yield a high-value target for malicious third parties.” (CyberRisk, 2017)

The tactics and techniques that are applied to today’s technology stem from the field of remote sensing. Remote sensing has a long history as it began with humans attempting to see and sense phenomena from a distance; from using pigeons to balloons to aircraft satellites, to UAS. Now UAS are coupled with space-based platforms for information and command and control. This field of study has allowed advances in military movement, attack and defend, as well as civilian surveying, and developing, freedom of movement throughout our world.

During Napoleon's campaigns, the French used observation balloons to monitor enemy activities. In 1831 Napoleon understood the true importance of remote sensing for shaping the battlefield as he stated, "If I always appear prepared, it is because before entering on an undertaking, I have meditated for long and foreseen what may occur" (Napoleon Bonaparte, 1831). This technique continued through the U.S. Civil War.

Remote sensing as defined by Lillestand and Keifer is, "the science and art of bating information around an object, area or phenomenon through the analysis of state acquired by a device that is not in contact with the object, area or phenomenon under investigation" (Lillesand, Kiefer, & Chipman, 2014). In simple terms, it is the ability to study different point of interest from afar without having to physically sample the item.

History of Photography

The history of photography is one that started with the early Greeks and pinhole cameras. These were very basic, and the image was upside down and was not permanent (Bellis, 2017). It was not until the 1800s that an image was able to be permanently stored on media. Cameras and film were progressing at different speeds. They finally came together in the 1830s. Nicephorus Niepce and Louis Daguerre, both of France, created the capability to produce permanent images that could be replicated. Nicephorus Niepce died before France recognized the achievement, and because of that, the first photographs were known as daguerreotypes. One of the problems that continued to plague the early photographers was the amount of time it took to develop the film. Prior to 1829, it was not unusual for a film to take eight hours to develop the film.

Figure 10-1 Boston Harbor, 1860, James Wallace Black

Source: Schultz, C. (2013). This Picture of Boston, circa 1860, is the World's Oldest Surviving Aerial Photo. Retrieved from: https://www.smithsonianmag.com/smart-news/this-picture-of-boston-circa-1860-is-the-worlds-oldest-surviving-aerial-photo-14756301/#RohlhYJZRcJzyVy7.99

Prior to 1829, it was not unusual for a film to take eight hours to develop. Daguerre discovered a method that took less than thirty minutes (Lucibella, 2013). Once the photographic and film development methodology was published, aerial photographs were taken with a wide variety of objects – most typically kites and balloons. The first American to take aerial photographs was James Wallace Black, who took photographs of the Boston harbor in 1860, see Figure 10-1. The 1906 San Francisco earthquake photographs were taken from a kite (King, 2012).

Remote Sensing

Remote sensing systems continue to change the way wars are fought. Information is gathered and analyzed that allows leaders to "see more, understand better and decide quicker." Observing enemy activity using high ground to survey the battlefield and then relay this information to military leaders goes back to ancient times. Military leaders understand the importance of high ground. With remote sensing systems, this high ground can be extended far beyond humans' normal observation abilities.

The introduction of the airplane by World War I enhanced the world's military capability to gather intelligence beyond enemy lines; gathering aerial photographs that would provide vital intelligence before, during and after a battle.

Limitations on airplanes, photography systems, and the time required to process intelligence lead to ever-increasing research and development by militaries around the world.

The US recognized the limitations of using standard aircraft for its intelligence gathering and began research on new remote sensing systems to improve its intelligence gathering capability. These platforms would be designed to fly undetected at higher altitudes, faster airspeeds, and with advanced camera systems. This capability assisted in tracking enemy equipment, movements, as well as their research and development programs. Early systems would offer the ability for the first time in the US to have an "eyes on" ability for treaty verification, something that the Soviet Union was not prepared for, nor did it want the US to have.

"The U-2 project was undertaken in the early 1950s; first flight occurred in August 1955. This project was created by the Central Intelligence Agency (CIA) because they required better intelligence gathering systems to amass intelligence on the Soviet Union. Collection efforts against the Soviet Union with modified bomber aircraft had taken place. However, those missions were vulnerable to counterintelligence observation, anti-aircraft fire, and fighter intercepts. It was thought a high-altitude aircraft such as the U-2 would be hard to detect and almost impossible to shoot down. Understanding the principles of high ground, the CIA began working on platforms that could be used specifically for reconnaissance operations." (Malesky, 2002)The U-2 gained public attention, "during the U-2 Crisis when pilot Francis Gary Powers was shot down over Soviet territory on May 1, 1960." (Malesky, 2002)

"The U-2 aircraft that was funded by the CIA and built by the U.S. starting in the 1950s. It became the subject of many "incidents," or diplomatic confrontations, with the Soviet Union during the Cold War. The most famous of these run-ins is referred to as *the* U-2 incident began on May 1, 1960, and what was to have been the twenty-fourth U-2 overflight of the Soviet Union." (FAQS, 2018) At this time, Gary Powers was one of the most experienced U-2 pilots in the world and "the overflights had become routine to him." (Malesky, 2002) The CIA issued its U-2 pilot's cyanide capsules and a poison needle hidden in a bisected silver dollar. Powers never took them along. Even more alarming was that, "Powers had become more and more cavalier. The morning of departure for Peshawar, he even packed his wallet with an assortment of American German and Turkish money for his layover in Norway; his wife packed him a lunch for the shuttle flight to Pakistan" (Barnes, 2005).

"Gary Powers took off from a U.S. air base at Rawalpindi, Pakistan. The mission profile on this flight codenamed Grand Slam was to be the most ambitious U-2 flight yet. Powers flight plan would take him from Turkey to Soviet nuclear-weapons facilities in the Ural Mountains, then over various railroads, then to intercontinental ballistic missile sites in Siberia, then back across northern Russia, finally to several shipyards before leaving Soviet airspace above the Arctic Circle and landing in Bodo, Norway." (Malesky, 2002)

"Powers was detected by Soviet radar while still fifteen miles from the Afghan-Soviet border. The radar detection was not unusual. In fact, all previous U-2 flights over the Soviet Union had been detected at some point. The previous U-2 flights had not counted on stealth to save them, but on the fact that the Soviets had no fighter jets or, for the first few years, surface-to-air missiles that could fly high enough to shoot it down. However, the newly developed surface-to-air missile, designated as the SA-2 had improved capabilities" (Malesky, 2002).

The U-2 program would again be in the spotlight on October 14, 1962, when it, "photographed the Soviet military installing nuclear warhead missiles in Cuba, precipitating the Cuban Missile Crisis. However, later in the Cuban missile crisis, another U-2 was shot down, killing the pilot, Major Rudolph Anderson." (Burr, 2012).

The CIA armed with the knowledge that the Soviets had developed Surface to Air Missiles (SAMS) that could intercept the U-2, began development of a faster, higher-flying reconnaissance aircraft, even before the U-2 became operational. Lockheed Martin proposed an aircraft codename OXCART that later became the famous USAF SR-71, unofficially known as the "Blackbird." Although the intent of the system was not only to enhance the US intelligence capability, it was more survivable. The speed of the SR-71 would prove to be its survivability capability. "The SR-71 remained the world's fastest and highest-flying operational manned aircraft throughout its career. From an altitude of 80,000 ft (24 km), it could survey 100,000 square miles per hour (72 square kilometers per second) of the Earth's surface. On July 28, 1976, an SR-71 broke the world record for its class – an absolute speed record of 2,193.1669 mph (3,529.56 km/h), and a US "absolute altitude record" of 85,068.997 feet (25,929 m)." (Haynes, 1996). The SR-71 had been fired upon many times, the standard countermeasure was simply to accelerate. Twelve aircraft are known to have been lost, all through non-combat.

Original capabilities for the SR-71 included Optical/Infrared Imagery systems, (the infrared systems were discontinued in the later years of the program) Side Looking Radar (SLR), later Synthetic Aperture Radar System (ASARS-1) an Electronic Intelligence (ELINT) gathering system. Onboard systems recorded sensor information and maintenance data (Malesky, 2002).

UAS now usher in a new era of capabilities and vulnerabilities. "Protocols implemented on the ground station applications enabling communications with the UAVs (and permitting users to pilot them via wireless remote control) were found to be unsecured. This allowed hackers to install malware on the systems running the ground stations. In addition, the telemetry feeds used in monitoring the vehicles and facilitating information transfer through wireless transmission were vulnerable to interception, malicious data injection, and the alteration of pre-set flight paths," (CyberRisk, 2017) Forward progress is not without tradeoffs, as depicted in the chart below, each ISR platform has limiting factors and tradeoffs. See Table 10-1.

Table 10-1 ISR Platform Tradeoffs: Space

SPACE	UAVs	MANNED
LIMITED # ASSETS	VULNERABILITY	VULNERABILITY DRIVEN CONOPS
TRADES BETWEEN QUALITY/REVISIT/SAMPLE RATE	DUTY CYCLE	LIMITED FOOTPRINT
LARGE DATA SOURCE	SMALL FOOTPRINT	DUTY CYCLE
DISSEMINATION & INTEGRATION	RELIABILITY	PREDICTABLE
PREDICATABLE	ATTRITION COSTS	HIGH ACQUISITION & SUSTAINMENT COST
HIGH ACQUISITION COST	Cyber Risk	REGIONAL BASING

Source: (Snyder, 2003)

UAS ISR-*Purpose/Market Sector/Product/Economic Opportunity*

Purpose of Technology: Intelligence, Surveillance, Reconnaissance for Dual Use-Military-Law Enforcement and Civilian Sectors

Markets Sector Serviced:

Military/Government Applications (Law Enforcement-counter drugs/terrorism etc.)

- Border Patrol / Monitoring

- Military monitoring of ports and inland activity for national security
- Guardian Angel for ground troops
- Node and Network Discovery
- Monitor Shipping/Pipeline Monitoring
- Damage assessment
- Prevent Movement
- Re-supply
- Radio Relay / Translator
- Chem./Bio Attack Rapid Response

Commercial applications

- Precision Agriculture (Imagery/Crop Spraying)
- Humanitarian Assistance and Disaster Response
- Delivery of Healthcare and Food Supplies
- Wildlife protection and research
- Environmental Monitoring/Research
- Energy/Mining Infrastructure Protection
- Sports/Media Entertainment and Journalism

Product/ Economic Opportunity:

Information is the product that ISR creates, and unmanned vehicles can provide unprecedented amounts of it. However, there is a need to manage the flood of data. "Data is the new oil," Intel Corp. Chief Executive Officer Brian Krzanich, he cited a growing competitive "separation" between companies that collect and understand their data and those that do not. A single autonomous car can generate the same data trove as 3,000 people surfing the Internet, while a small drone fleet could easily create 150 terabytes of data per day. (Tyson, 2016). A singe UAS in a single mission can create even more data than a single autonomous car.

Hacking drones is now becoming a market sector all on its own. "Military technology companies from around the world are rushing to design, build, and sell drones that hack and track, while others want to own the business of hacking of the drones themselves. The burgeoning market is foreshadowing battles that could play out in the skies and, for some companies, bring significant profits" (O'Neil, 2018).

Power by the hour offers the ability to sell the information once to a specific customer or to offer it to several customers, allowing for a greater rate of return. A turn-key UAS offering may include the information as a product with training and support of the equipment with a host country owning/controlling/maintaining and operating system or a combination of product sales and service support contracts.

Mission Drives the Sensor Requirements

The mission drives the payload choices, and the payload should drive the platform. In reality this tends to be the reverse. The platform creates compromised trade-offs between a higher quality sensor that weighs more and consumes power versus a lower quality sensor that is less expensive, lighter, but will require more passes to get useable information or will provide a less clean data set. "As military UASs continue to evolve and shrink-think of swarms of tiny drones-their resulting payload footprints pose numerous SWaP (Size, Weight and Power) design space constraints and tradeoffs, together with sensor processing, data link bandwith and security issues as well" (Cole, 2016). Keeping in mind that the more passes to collect data can increase the cyber footprint and create needless cyber vulnerabilities. "You can get information much faster if you do the processing in near-real time onboard the aircraft and then get the data down the link" (Cole, 2016).

Standard ISR Camera Sensors

The average payload on a UAS is a passive sensor that does not emit any signals or energy, it sends and or records this data entirely in passive mode. Standard ISR payloads consist of a sensor, normally a camera such as the popular Go-Pro cameras shown below, that offer many user-friendly features while still acquiring high-quality data that can be processed onboard the aircraft, sent down on a live downlink for processing on the ground, or a combination of these techniques. "Military UAS users are seeking actionable intelligence from their sensors in real time-whether the sensor is part of a radar, electronic warfare or ISR sensor chain" (Cole, 2016)

Go Pro Camera Comparison. See Figure 10-2.

Feature	Hero5 Camera	Hero6 Camera	Fusion Camera

Photo	12MP / 30 fps Burst	12MP / 30 fps Burst	18MP / 30 fps
Video	4K30	4K60	5.2K30
Spherical Capture	No	No	Yes
Waterproof	33ft (10m)	33ft (10m)	16ft (5m)
Voice Control	Yes	Yes	Yes
Video Stabilization	Yes	Advanced	Advanced
Quick Stories	Yes	Yes	No
HDR Photo Capture	No	Yes	No
Touch Zoom	No	Yes	No
Auto Low Light	Yes	Yes	No
Exposure Control	Yes	Yes	No
Advanced Wind Noise Reduction	3-mic processing	3-mic processing	4-mic processing
360 Audio	No	No	Yes
GPS	Yes	Yes	Yes
Wi-Fi +Bluetooth ®	Yes	Yes	Yes
5GHz Wi-Fi for Offload to Phone	No	Yes	Yes

Figure 10-2 GoPro Camera Comparisons

Source: (GoPro, 2018). The Ultimate GoPro. In F. t. R. G. F. You (Ed.), GoPro: GoPro.

Newer designs include the modularity of cameras including a camera that is a, "drone that comes with a 4K camera and has autonomous features... (and is able to) transform the drone into something that can fly, a home security camera, a wearable, or a camera mounted on a stick. It's a versatile product packed into a single device" (McKalin, 2018).

Multispectral and Hyperspectral Sensors

To gain additional insight into target areas, the use of multispectral and hyperspectral sensors allows for extended data collection. "The main difference between multispectral and hyperspectral is the number of bands and the bandwidths. Multispectral imagery generally refers to three to ten bands. To be clear, each band is obtained using a remote sensing radiometer. A hyperspectral image could have hundreds or thousands of bands" These sensors tend to need copious amounts of power to operate and create pairing issues for datalinks that can handle the amount of data being generated. This is a consideration if the data is going to be encrypted as the data rates will be derogated and real-time monitoring may not be possible. "Multispectral and hyperspectral imagery gives the power to see as humans (red, green and blue), goldfish (infrared) and bumble bees (ultraviolet). We can see even more than this as reflected EM radiation to the sensor"("Multispectral vs Hyperspectral Imagery Explained," 2018). Pictured in Figure 10-3 is Raytheon's Multi-Spectral Targeting System (MTS) for Predator MQ-1.

Figure 10-3 Raytheon's Multi-Spectral Targeting System (MTS) for Predator MQ-1

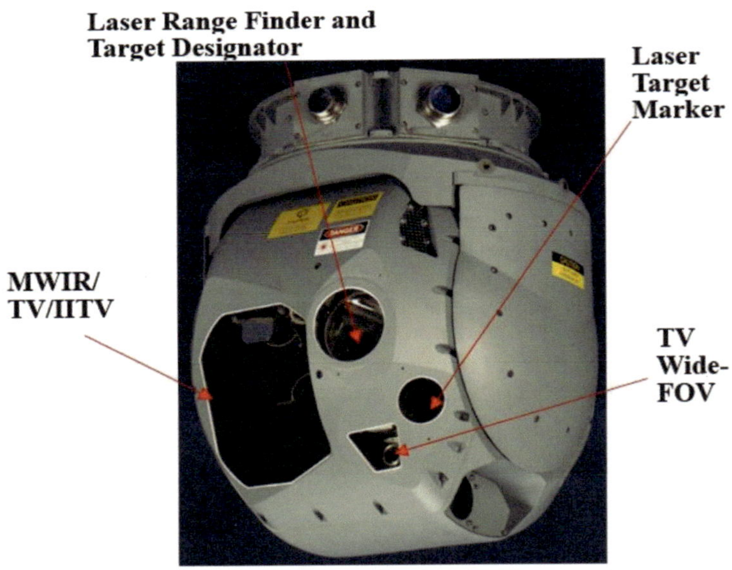

Source: (Snyder, 2003) Snyder, J. (2003). The Latest Tools, Techniques and Opportunities in UAV Design. In. Arlington, VA: Mongo Industries, LLC.

A UAS is an excellent pairing for a multispectral or hyperspectral sensor due to the proximity the UAS allows to the target area and the increase in data collection available in relation to the proximity of the sensor to target.

> "Hyperspectral imaging combines taking pictures of a scene or object, with a spectral view at each point of resolution in the scene. The result is a 3-dimensional data set that can be sliced to view multiple images at separate wavelengths or sliced to show how the spectra vary along different spatial positions across the image in one direction. If the acquisition system or the object is moving, the 4th dimension of time is added." ("Hyperspectral Imaging," 2018).

Below is a US Government photograph of Arizona mining operations captured with the hyperspectral imaging. See Figure 10-4

Figure 10-4 Hyperspectral Imaging

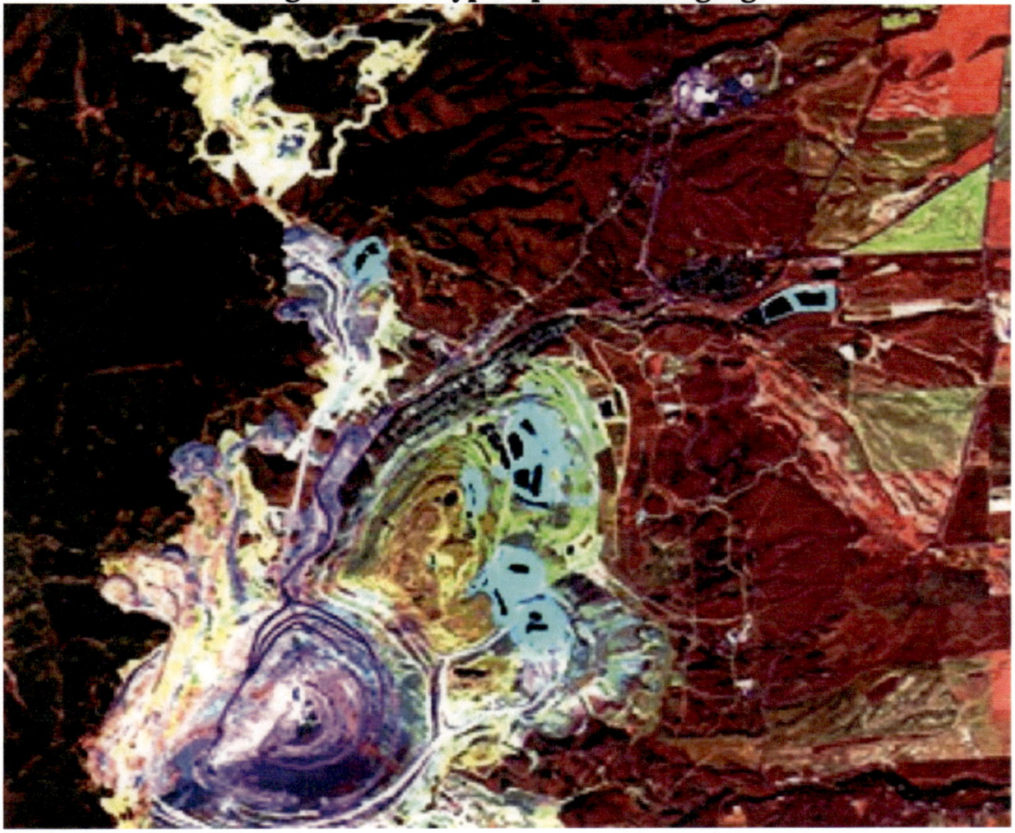

Source: Hyperspectral Imaging. (2018). Retrieved from http://www.sensorsinc.com/applications/military/hyperspectral-imaging/ Multispectral vs Hyperspectral Imagery Explained. (2018). Retrieved from https://gisgeography.com/multispectral-vs-hyperspectral-imagery-explained/

A Changing World Creates a Changing Target Set and Sensor Requirement- SWIR

Since 9-11, the requirements for ISR capabilities quickly moved from the ability to photograph a military installation, or large piece of equipment such as a tank, to the difficult requirement to locate individuals or small groups. "Past reconnaissance needs were more strategic in nature, today's needs are highly tactical, demanding an elevated level of persistence and the ability, in many cases, to identify individual humans in the field of interest. Many approaches have been employed and proposed, but in recent years, the exceptional capabilities of shortwave infrared (SWIR) technology have made SWIR the "Next Generation" of imaging technology for ground, airborne and space technology." ("Using SWIR in Intelligence, Surveillance, and Reconnaissance (ISR) Military and Security Systems," 2018)

Figure 10-5 Shortwave InfraRed (SWIR) bands

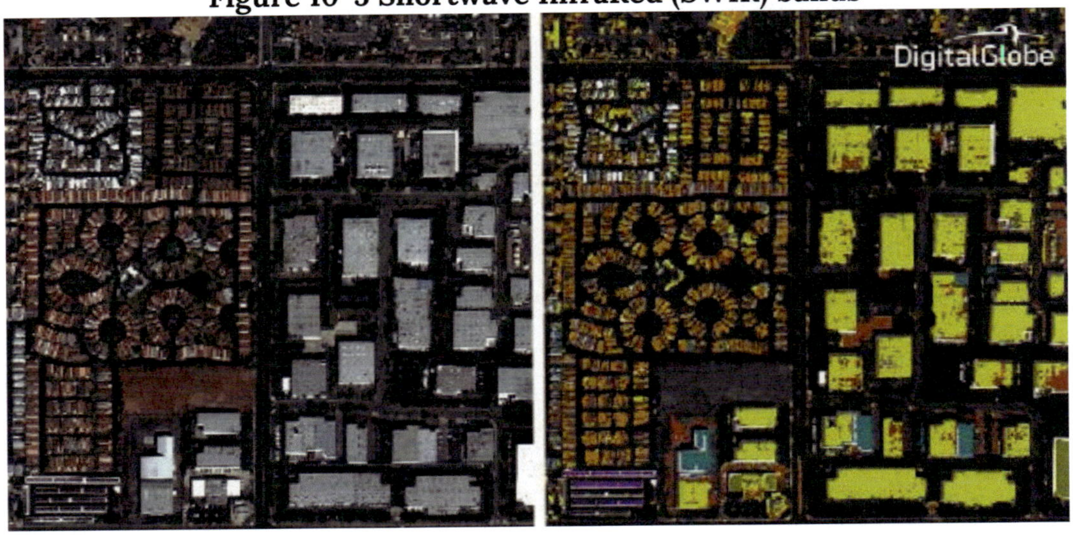

Source: Young, D. (2018). Our eyes can play tricks on us but shortwave infrared (SWIR) imagery reveals all – part 1 of 2. Retrieved from URL: http://blog.digitalglobe.com/technologies/our-eyes-can-play-tricks-on-us-but-shortwave-infrared-swir-imagery-reveals-all-part-1-of-2/

Pictured here is image (Figure 10-5) with eight Shortwave InfraRed (SWIR) bands which are ideal for material identification and mapping. Visible imagery is shown on the left and classification using SWIR imagery is shown on the right. Note that different roof top materials that look the same in the visible imagery are clearly differentiated in the SWIR imagery (Young, 2018).

SWIR cameras and sensors can see reflected light in the shorter wavelengths. Small targets such as humans become distinguishable, with the "typical difference being that all hair shows as white due to the lack of moisture in hair. Conversely, skin shows darker, due to its high moisture content. It is said that long and medium wave sensors provide detection, while SWIR and visible sensors provide recognition." ("Using SWIR in Intelligence, Surveillance, and Reconnaissance (ISR) Military and Security Systems," 2018). This spectrum is illustrated below, and it is easy to see why using SWIR sensors on a UAS is an advantage over the old ISR sensors. See Figure 10-6.

Figure 10-6 SWIR advantage over old ISR sensors

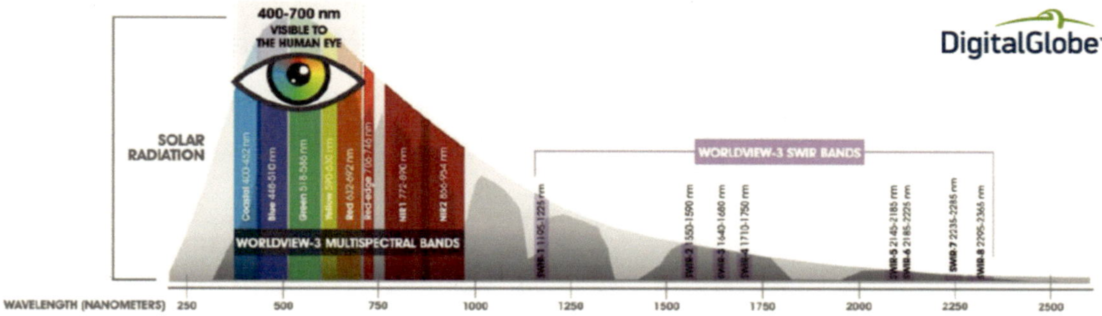

Source: Young, D., (2018). Our eyes can play tricks on us but shortwave infrared (SWIR) imagery reveals all – part 1 of 2. Retrieved from: http://blog.digitalglobe.com/technologies/our-eyes-can-play-tricks-on-us-but-shortwave-infrared-swir-imagery-reveals-all-part-1-of-2/

Bomb-Sniffing Drone Technology

UAS can be outfitted with sensor suites that can detect improvised explosive devices and other active landmines from past wars. This not only offers direct security, but it also offers a safer way to detect landmines. "Sensors look for gamma rays or other particles with the signatures of specific materials, such as explosives or a nuclear device. It is the same technology used at security checkpoints to scan luggage and shipping containers in airports, but the breakthrough for the UW-Madison scientists was making the radiation source small enough to mount on a drone" (R. Schultz, 2016). Drug and bomb-sniffing drones can detect dangerous chemicals from 1.8 MILES away." Unmanned drug-sniffing drones have been introduced in the Netherlands. They fly over houses (video), sniff for weed and scan for grow lights. Police say they are not breaking the law because the samples can be taken without entering the building." (Zenpus, 2009). See Figure 10-5 for bomb sniffing logic for UAS Figure 10-5 for bomb sniffing logic for UAS.

Cave Mapping

Unique missions' development will mean the UAS uses are only limited to the innovation of the operator and limits of physics. One such application is the mapping of underground mines, normally a dangerous job attempted by humans, it is one that UAS are well suited to accomplish. UAS, "will be able to carry out safety checks by monitoring the build-up of water and checking the extent of roof collapses, and search for valuable mineral deposits that may have been missed" (Hambling, 2017).

A standard quad-copter UAS is outfitted with, "powerful LED lights, cameras, and sonar. Initially, they tried flying it using the drone's on-board camera to guide them, an approach known as First Person View (FPV) piloting" (Hambling, 2017).

Figure 10-7 for bomb sniffing logic for UAS

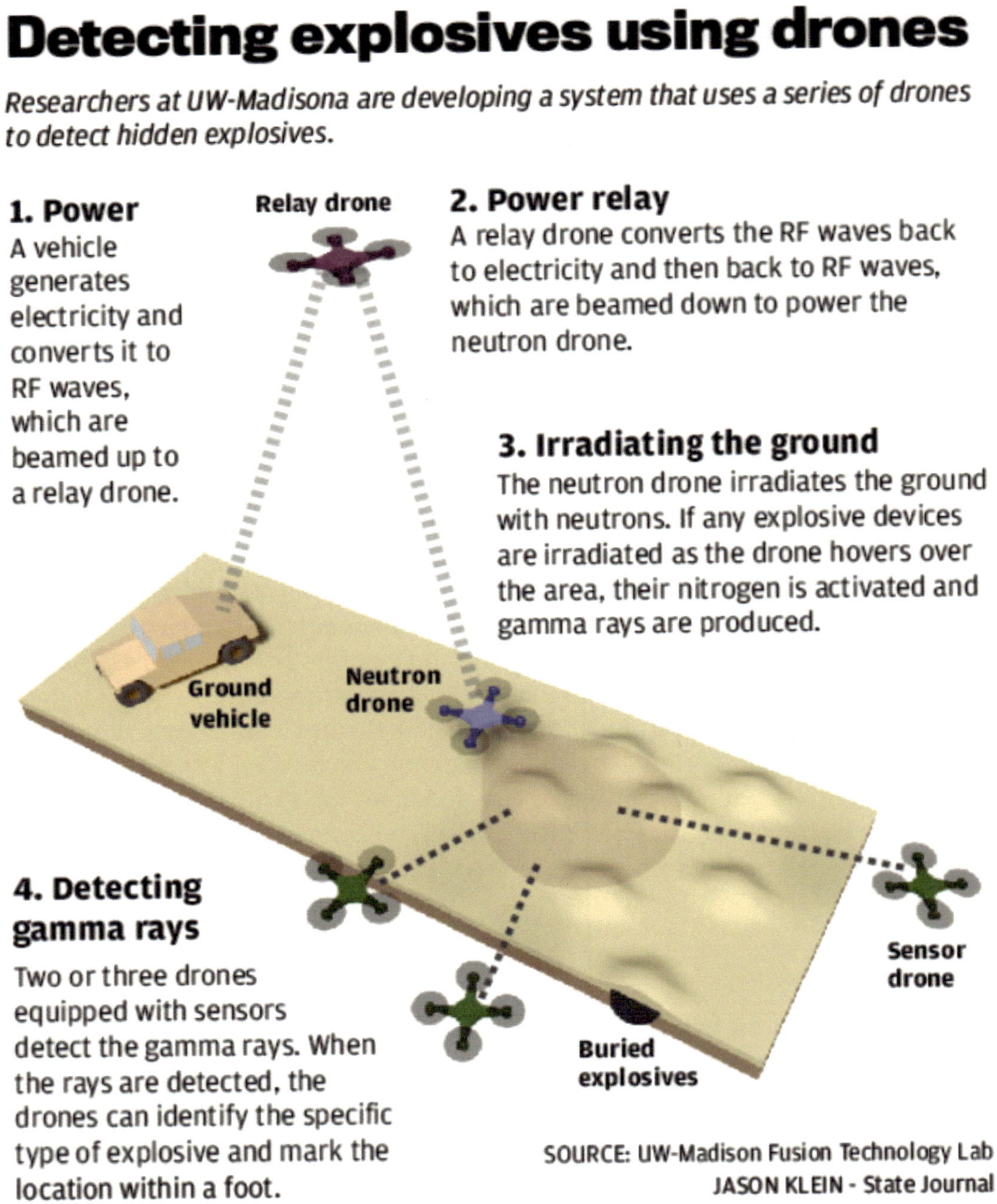

Source: Schultz, R. (2016). Bomb-Sniffing Drone Technology. Retrieved from: https://www.uasvision.com/2016/04/29/bomb-sniffing-drone-technology/

Using a multi-mission sensor suite, the variety of data that can be collected and stored in a brief period is far superior to human lead collections. "Sonar sensors, which use sound waves to detect objects, produce less data than video cameras. This means they can be used to create a 3D model more quickly, possibly even in real time...the sonar model is less accurate; a yet-to-be-published paper shows that it provides effective navigation. Given that it uses fewer data and therefore less processing power, the mapping could potentially be done onboard the drone" (Hambling, 2017).

UAS offers the ability to create modular designs for sensor suites, allowing many options for data acquisition to be tested in a short amount of time. This offers the ability to determine the best mix of sensors, platforms, and data acquisition capabilities, "we are also experimenting with lidar, which maps using lasers. Also, they are running tests with data from X-ray fluorescence analyses, which detect different elements, to train machine learning to identify minerals in rock walls." (Hambling, 2017).

Mission and Sensor Planning and Considerations

Several issues must be considered in planning a successful UAS mission in order to acquire the required data in a useable format, in a timely and secure manner as "the next front in the cyber-war is literally above your head" (O'Neil, 2018). Information can be stored on the vehicle for later retrieval, or it can be real time or near real time sent back to the ground for processing. Several factors will determine the optimal method to obtain and secure the information, including encryption threat environment, timeliness, and the consequences (if any) if the information is intercepted or compromised. "Maldrone backdoor malware kit has been developed as a universal hack, applicable to all makes and models of UAV. Maldrone silently interacts with a drone's device drivers and sensors, allowing the user to hijack and control the UAV remotely" (CyberRisk, 2017). The below graphics offer some sense of planning that must be done in looking at flight paths and sensor limitations, including security of the information.

One technique for securing information that must be sent using non-secure or hackable methods is to obfuscate the true data that is being collected. This can be done by increasing the number of passes and collecting a larger dataset than the mission requires, effectively flooding the sensor with data so an adversary would not be able to discern the truly valuable data from the noise. "Poorly secured or unsecured wireless networks are particularly vulnerable, with attack scenarios envisaged where compromised or purpose-bought UAVs could be flown or discreetly landed near a hot spot and used to stage Man in the Middle (MIM), data injection, and similar attacks over guest and short-range Wi-Fi, Bluetooth, and other wireless connections. The success of such attacks might be bolstered by the fact that traditional security measures operate on the assumption that no-one could get close enough to such short-range wireless connections to pose a serious threat". (CyberRisk, 2017)

"As depicted in Table 10-2 below a mission that might only need one or two passes at a lower altitude, instead might be flown at multiple altitudes and additional passes over the target allowing for the obfuscation of the true targeted data. This technique also allows for additional data analysis on an area, however, due to storage, data links, timing and threat environments this technique is not always employed." (Bosak, 2014)

"One easily can deduce $72km^2$ map surface in a single direction straight flight.

However, assuming returning flight and only 60% overlap, one gets (100%+40%) *720m=1008m strip width and only one hour of flying in one direction.

This yields only $50.4km^2$ map surface with returning flight." (Bosak, 2014)

Table 10-2 "Surface, map shape, and flight altitude" (Bosak, 2014)

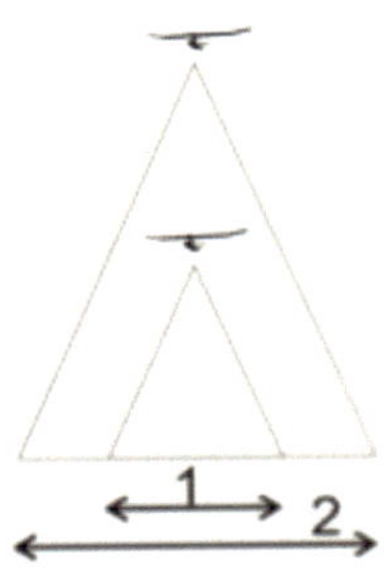	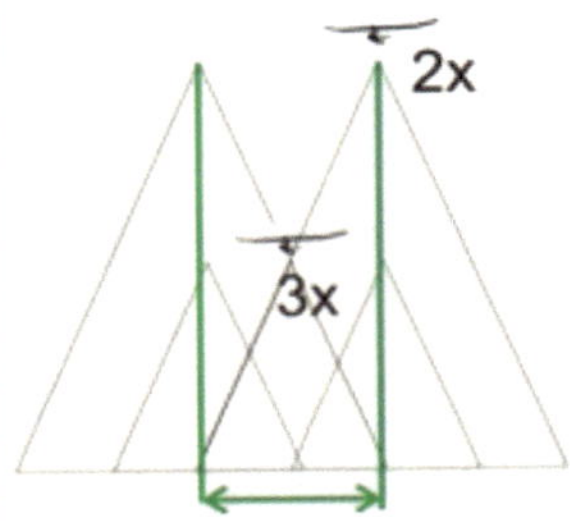	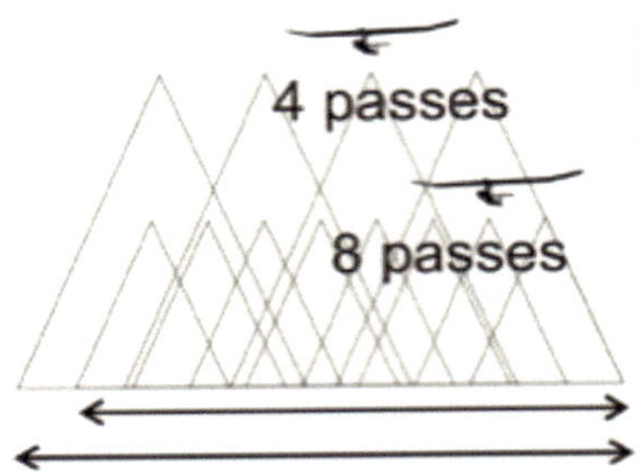
"If you are mapping a linear object, doubling the altitude, you are doubling surface coverage" (Bosak, 2014)	"Choosing to make two passes may improve geometry matching, because of overlap requirement, flying high you are reducing flight time only by about 1/3, because outside regions have valid bitmap but poor geometry." (Bosak, 2014)	"With regular map shape and multiple passes, flying two times higher requires two fewer passes and flight time and provides an extra surface at map edges. However, the area should be flat as there will be no multi-angle information allowing to orthonormalize high objects along the edges." (Bosak, 2014)

	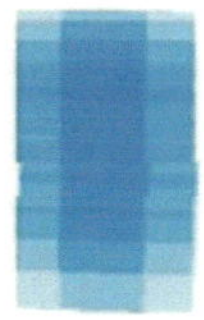	
"A single-leg mission. Typical along-overlap is high, around 75-90%." (Bosak, 2014)	"Two-leg mission. Typical 60% side-overlap shown. Along-overlap 75-90%." (Bosak, 2014)	"Three-leg mission. Typical 60% side-overlap shown." (Bosak, 2014)

Source: (Bosak, 2014)

Importance of stabilized head

"Since increasing overlap is so costly in mission time and so important in urban areas, UAS can use roll-stabilized head and has aerodynamic design successfully attenuating oscillations in pitch and yaw axes. The use or roll-stabilized head increases useful surface during turns and increases processing success rate thanks to overall more predictable photo properties. See Figure 10-8." (Bosak, 2014)

Figure 10-8 Importance of stabilized head

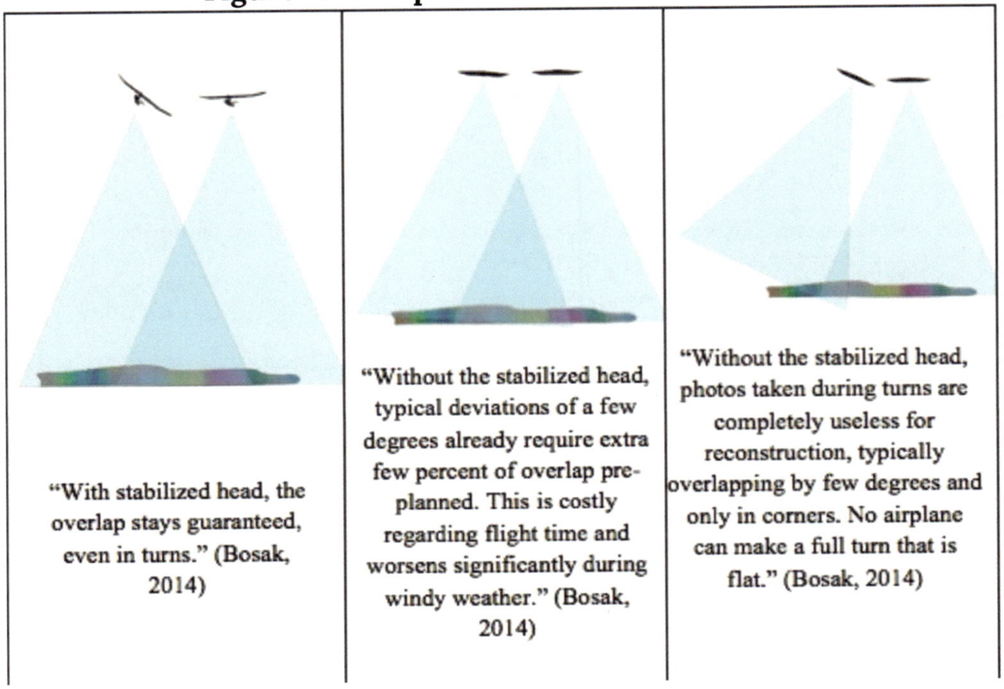

Source: Bosak. (2014). Secrets of UAV Photomapping. Retrieved from: http://ww1.aerialrobotics.eu/pteryx/pteryx-mapping-secrets.pdf

"Both small UAV like flying wings and even large UAS with several meters wingspan tend to respond for navigation with changing 0-5 deg roll both directions even when ordered to 'fly straight' over the ground. The reason is, while the ground path is straight, the wind blows in any direction, usually as much as 45 deg different at altitude than at ground level, with little direction change but much more wind speed variation. This means the UAS has to bank left and right all the time in order to stay on its path"

Protecting the Systems from the Cyber Threat

Sensors, datalinks, platforms and power supplies tend to be built independently without cyber protection standards built in leaving the systems vulnerable. The very nature of "plug and play" tends to create incompatibility in cyber protection with virtually no data standards.

"Analysis of the configuration and flight controllers/microprocessors of several popular UAV models having multiple rotors revealed weaknesses associated with both the telemetry links streaming data to and from a drone via serial port connections (in which information could be captured, modified, or injected), and the UAVs' connections to their ground station interface (whose data link could be spoofed, enabling hackers to assume complete control of the vehicle)" (CyberRisk, 2017).

The dangers of weak security protocols is becoming more apparent as reported by CNN on December 17, 2009 as, "Insurgents were able to use a mass-market software program to view live feeds from U.S. military Predator drones monitoring targets in Iraq. There also is evidence that UAV feeds also have been hacked in Afghanistan, but there was no evidence the militants were able to take control of the remote aircrafts' systems in either country. The inexpensive software, created by a Russian company called SkyGrabber, is downloadable off the Internet. It allows users to take advantage of unprotected communication links in some of the UAVs" (Mount & Quijano, 2009).

An attacker can have many objectives in launching an attack. Possible goals include the loss of data integrity, the link to the vehicle, or the payload. An example of an attack tree that demonstrates different branches of attacks based on a MIM attack is shown below in Figure 10-7.

"The loss of integrity (Level 1 of the attack tree) is achieved by manipulating the communication stream between the client and the server creating cyber-physical impacts. The branches of the tree represent the methods attackers can achieve their goal. The arrows in the attack tree indicate the sequence through which each attack proceeds" (Ahmad Yousef et al., 2018).

Encryption offers some level of protection.

Figure 10-9 MIM Attack Effects

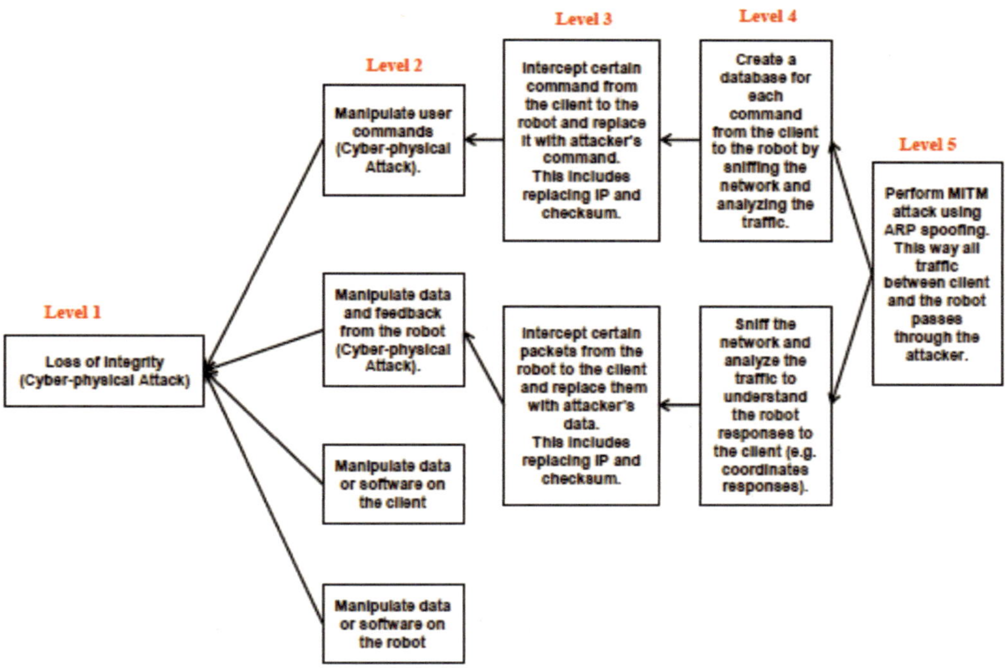

Source: Ahmad Yousef, K. M., AlMajali, A., Ghalyon, S. A., Dweik, W., & Mohd, B. J. (2018) Analyzing

Cyber-Physical Threats on Robotic Platforms. Sensors (Basel, Switzerland), 18(5), 11-12. doi:10.3390/s18051643

However, compromises occur due to mission requirements. Encryption can overload data links, take up valuable bandwidth and cause undesirable flight and control characteristics if a real time data link and control is required for the mission. Not using encryption is a trade-off that a Pentagon official said, "that many of the UAV feeds need to be sent out live to numerous people at one time, and encryption was found to slow the real-time link. The encryption therefore was removed from many feeds. Removing the encryption, however, allowed outsiders with the correct tools to gain unauthorized access to these feeds" (Mount & Quijano, 2009).

"Is there a perfect solution to the security issues involved in UAS operations? No. There is always a degree of risk to be mitigated." (Nichols, 2002) However, what if a the UAS payload or flight control systems is lost to a hacker or an enemy nation it would be advisable to build a way of remotely disabling or destroying the information, sensor and or vehicle? Over the years many concepts and ideas have come forward including the idea of a "zero out" chip that would automatically delete all information on a UAS upon a compromised status. Cornell University and Honeywell Aerospace in laboratory testing is developing a method of vaporizing electronic circuits, without laying a hand on the actual device.

The design method is, "When the shell is exposed to a certain frequency of radio waves, tiny graphene-on-nitride valves between the cavities open, allowing the chemicals to mix and react. Along with applications such as data protection, it is hoped that the technology might also find use in things like environmental sensors that can be remotely vaporized once they're no longer needed" (Coxworth, 2018).

Conclusions

Although ISR payloads have increased in capability and the wide range of data collections, the goals and mission of remote sensing platforms remain constant. The ISR platforms and the sensors that they carry stared off as balloons, and went into airplanes, then into satellites, and are now being put into unmanned aerial systems with modular payloads, allowing vast amounts of information to be collected.

The wide variety of sensor payloads should drive the missions and data collects, however in the real world it is often the platform that dictates the sensors choices, which can impact the ability to collect the data required. The weight and power requirements of the sensor must be considered when planning the flights. The mission profile (number of passes and angles) are all considerations when pairing a sensor with a platform including all mission data collection parameters that drive the ISR requirements. Sensor data security and the threat of attacks within the cyber domain must be addressed. Mission planning will require tradeoffs between access to the target area, sensor capability and availability, information time dominance and cyber/data security requirements.

Students are encouraged to continue researching and learning about the new sensors systems, data links and cybersecurity techniques for use on UAS. As the market expands, more sensors will be built and optimized specifically for UAS, leading to more cyber attacks and the vulnerability of the data, as well as the UAS vehicles before during and after missions.

Discussion questions

1. How could you use physical security and UAS flight characteristics to supplement a UAS cybersecurity plan?
2. How has UAS changed the nature of ISR after the attacks on 9-11?
3. What sensors would you use to map the health of agricultural crops?
4. What is the best way to secure the data collected by UAS, collecting it and storing onboard or using data links to process the information on the ground? Please explain your answer.
5. Name three unique missions that UAS payloads are configured for today that in the past humans would have been assigned to do. Research three additional missions' payloads and discuss how these payloads are being integrated on UAS.

Bibliography

Bosak, K. (2014). *Secrets of UAV Photomapping.* Retrieved from http://ww1.aerialrobotics.eu/pteryx/pteryx-mapping-secrets.pdf

CyberRisk. (2017, March 16). *he Usage of Drones in Cyber Attacks – Both as Targets for Attack and as Potential Attack Vectors.* Retrieved from CyberRisk Blog: https://www.cyberisk.biz/the-usage-of-drones-in-cyber-attacks/

Malesky, L. A. (2002). Just one more U-2 overflight of the Soviet Union was one too many for Francis Gary Powers. *Military History*, pp. 19(5), 26. .

Nichols, R. K. (2002). *Wireless Security: Models, Threats, and Solutions.* New York: McGraw Hill.

Readings

Ahmad Yousef, K. M., AlMajali, A., Ghalyon, S. A., Dweik, W., & Mohd, B. J. (2018). Analyzing Cyber-Physical Threats on Robotic Platforms. *Sensors (Basel, Switzerland)*, 18(5), 11-12. doi:10.3390/s18051643

Barnes, T. (2005). Mayday for the U-2. Retrieved from http://area51specialprojects.com/u2_mayday.html

Bellis, M. (2017). The History of Photography: Pinholes and Polaroids to Digital Images. Retrieved from https://www.thoughtco.com/history-of-photography-and-the-camera-1992331

Burr, W. (2012). *Cuban Missile Crisis Day by Day: From the Pentagon's "Sensitive Records"*. Washington, DC: National Security Archive Electronic Retrieved from https://nsarchive2.gwu.edu/NSAEBB/NSAEBB398/.

Cole, S. (2016). Small UAS payloads pose Swap and bandwidth challenges. Retrieved from http://mil-embedded.com/articles/small-pose-swap-bandwidth-challenges/

Coxworth, B. (2018). Circuits self-destruct in response to radio waves. Retrieved from https://newatlas.com/radio-waves-vaporize-electronics/53134/

FLIR Launches Next-Generation Black Hornet 3 Nano-UAV. (2018). Retrieved from https://www.businesswire.com/news/home/20180605005630/en/FLIR-Launches-Next-Generation-Black-Hornet-3-Nano-UAV

GoPro. (2018). The Ultimate GoPro. In F. t. R. G. F. You (Ed.), *GoPro*: GoPro.

Hambling, D. (2017). Drone maps mines to explore unsafe caverns and seek out minerals. Retrieved from https://www.newscientist.com/article/2127123-drone-maps-mines-to-explore-unsafe-caverns-and-seek-out-minerals/

Haynes, L. (1996). SR-71 World Record Speed and Altitude Flights. Retrieved from http://www.wvi.com/~sr71webmaster/spd_run001.html

Hyperspectral Imaging. (2018). Retrieved from http://www.sensorsinc.com/applications/military/hyperspectral-imaging/

King, F. (2012). Kite Photo of Post-Quake San Francisco (1906). Retrieved from http://www.bigmapblog.com/2012/kite-photo-of-post-quake-san-francisco-1906/

Lambeth, B. S. (2006). Air Power Against Terror: America's Conduct of Operation Enduring Freedom. In. Santa Monica, CA: RAND Corporation.

Lillesand, T., Kiefer, R. W., & Chipman, J. (2014). *Remote Sensing and Image Interpretation*: Wiley.

Lucibella, M. (2013). January 2, 1839: First Daguerreotype of the Moon. Retrieved from https://www.aps.org/publications/apsnews/201301/physicshistory.cfm

McKalin, V. (2018). PITTA Is a Modular Camera That Can Transform into a Drone. Retrieved from https://sanvada.com/2018/01/02/pitta-modular-camera-can-transform-drone/

Mount, M., & Quijano, E. (2009). Iraqi insurgents hacked Predator drone feeds, U.S. official indicates. Retrieved from http://www.cnn.com/2009/US/12/17/drone.video.hacked/index.html

Multispectral vs Hyperspectral Imagery Explained. (2018). Retrieved from https://gisgeography.com/multispectral-vs-hyperspectral-imagery-explained/

O'Neil, P. (2018). Drones emerge as new dimension in cyberwar. Retrieved from https://www.cyberscoop.com/apolloshield-septier-drones-uav-cyberwar-hacking/

Schultz, C. (2013). This Picture of Boston, circa 1860, Is the World's Oldest Surviving Aerial Photo. Retrieved from https://www.smithsonianmag.com/smart-news/this-picture-of-boston-circa-1860-is-the-worlds-oldest-surviving-aerial-photo-14756301/#RohlhYJZRcJzyVy7.99

Schultz, R. (2016). Bomb-Sniffing Drone Technology. Retrieved from https://www.uasvision.com/2016/04/29/bomb-sniffing-drone-technology/

Snyder, J. (2003). The Latest Tools, Techniques and Opportunities in UAV Design. In. Arlington, VA: Mongo Industries, LLC.

Tyson, M. (2016). "Data is the new oil" declares Intel CEO Brian Krzanich. Retrieved from http://hexus.net/ce/news/automotive/99277-data-new-oil-declares-intel-ceo-brian-krzanich/

The Usage of Drones in Cyber Attacks – Both as Targets for Attack and as Potential Attack Vectors. (2017). Retrieved from https://www.cyberisk.biz/the-usage-of-drones-in-cyber-attacks/

Using SWIR in Intelligence, Surveillance, and Reconnaissance (ISR) Military and Security Systems. (2018). Retrieved from http://www.sensorsinc.com/applications/military/swir-for-isr

Young, D. (2018). Our eyes can play tricks on us but shortwave infrared (SWIR) imagery reveals all – part 1 of 2. Retrieved from http://blog.digitalglobe.com/technologies/our-eyes-can-play-tricks-on-us-but-shortwave-infrared-swir-imagery-reveals-all-part-1-of-2/

Zenpus, S. (2009). Drug-Sniffing Drones Take to the Skies in the Netherlands. Retrieved from https://hardware.slashdot.org/story/09/04/30/1629253/drug-sniffing-drones-take-to-the-skies-in-the-netherlands

Chapter 11: UAS Weapons

Student Learning Objective

The student will gain knowledge of UAS weapons, both lethal and non-lethal options, and the current state of industry and planning and execution considerations for weapons employment. The student will examine how world events are driving the UAS weapons industry as well as a glimpse at the future of weapons for UAS platforms.

History

Unmanned vehicles have been around for quite some time. Some were even weapons in and of themselves. The first consistent use of a UAS, or RPV as they were called then, was in Vietnam. The Lightning Bug program using the BQM-34 Firebee was equipped with both gravity weapons and missiles. "The Navy, meanwhile, canceled the most extensive armed UAV (unmanned aerial vehicle) program in U.S. history the same year the armed Firebee was tested, retiring its QH-50 DASH drone helicopter, which carried torpedoes and even nuclear depth bombs that were never used in combat." (Whittle, 2015).

The Firebee underwent weapons testing late in the war. The only "weapon" used during the war was the leaflet bombs. The Firebee weapon set included MK-82 varieties that included early laser-guided bombs, early maverick missiles, early cluster bombs, TV guided weapons and unattended sensors. The BQM-34 series with various designations flew over 3,500 missions (Northrup Grumman Brochure) in Vietnam. See Figure 11-1. Most were reconnaissance, but a considerable number were decoy missions to address the enemy's surface to air missile systems. "Despite the drone successfully destroying various mock enemies on the ground, the Air Force lost interest in the project. For one, after the Vietnam War formally ended in 1973, Washington slashed defense spending across the board." (Joseph, 2016).

Figure 11- 1 BGM-34b

Source: Parsch, A. (2003). Teledyne Ryan AQM/BQM/MQM-34 Firebee. Retrieved From http://www.designation-systems.net/dusrm/m-34.html#_BGM (Photo By Bud Wolford)

Desert Storm: Though UASs were used to excellent effect during Desert Storm, other than cruise missiles, and armed UAS were not used.

Events in 2000

"The first predator flight over Afghanistan was on September 7th, 2000." (Kaplan, 2016) "An unmanned spy plane called the Predator begins flying over Afghanistan, showing incomparably detailed real-time video and photographs of the movements of what appears to be Bin Laden and his aides. It flies successfully over Afghanistan sixteen times" (9-11 *Commission Report*, 2004). "The Predator has been used in the Balkans and Iraq since 1996. President Clinton is impressed by a two-minute video of Bin Laden crossing a street heading toward a mosque inside his Tarnak Farms complex. Bin Laden is surrounded by a team of a dozen armed men creating a professional forward security perimeter. One Predator crashes on takeoff and another is chased by a fighter, but it apparently identifies Bin Laden on three occasions. Its use is halted in Afghanistan after a few trials, as seasonal winds are picking up. It is agreed to resume the flights in the spring, but the Predator fails to fly over Afghanistan again until after 9/11" (Derek, 2000). (Kaplan, 2016)

"The USAF BIG SAFARI program office managed the Predator program and was given direction on 21 June 2000 to explore options to arm the aircraft. This led to it being fitted with reinforced wings and stores pylons to carry munitions, as well as a laser designator. The RQ-1 conducted its first firing of a Hellfire anti-tank missile on 16 February 2001, over a bombing range near Indian Springs Air Force Station north of Las Vegas, Nevada. An inert AGM-114C successfully hit a tank target. This led to a series of tests on 21 February 2001 in which the Predator fired three Hellfire missiles, scoring hits on a stationary tank with all three missiles. Following the February tests, the decision was made to move immediately to increment two of the testing phase, which involved more complex tests to hunt for simulated moving targets from greater altitudes with the more advanced AGM-114K version." (Derek, 2000).

Figure 11-2 MQ-1 In The Smithsonian

Source: Bowden, M. (Nov. 2013) In Smithsonian Magazine, https://www.smithsonianmag.com/history/how-the-predator-drone-changed-the-character-of-war-3794671

Post 9/11/2001

War on Terror: In a much-accelerated program, the Predator UAS was pressed into service with weapons. The designation for the weaponized version became the MQ-1. See Figure 11-2. Predators had been flying surveillance missions up until that point. Predators were armed with Hellfire missiles. Soon they were flying over Afghanistan. Dull, dirty, and dangerous are often cited as the reasons for unmanned vehicles, especially armed UAS. However, the driving force behind the proliferation of armed UAS in the war on terror has been shortening the kill chain and persistence in the target area. "Subsequent to 9/11, approval was quickly granted to ship the missiles, and the Predator aircraft and missiles reached their overseas location on 16 September 2001. The first mission was flown over Kabul and Kandahar on 18 September without a weapons payload. Subsequent host nation approval was granted on 7 October, and the first armed mission was flown on the same day." (Derek, 2000). "The one pictured here was deployed in Afghanistan, where it became the first drone to fire Hellfire's in combat. In all, it flew 261 sorties in Afghanistan, totaling more than 2,700 hours, before the Air Force donated it to the Air and Space Museum in 2003." (Bodwden, 2013).

The combination of persistence and weapons capability turned out to be very effective. "And yet the most important breakthrough was still to come. The original drones broadcast a view only to operators on the ground. As the United States continued to fight in Afghanistan and Iraq, the drones' cameras and sensors were linked to the global telecommunications system. Now a drone could be piloted—and its live feed viewed and its missiles aimed—from anywhere in the world. The pilots could be insulated from the risks of combat" (Bodwden, 2013).

The U.S. military quickly mounted "caps," or permanent observation platforms, over large areas. Using computers to analyze data feeding continuously from drones, military and spy agencies isolated and tracked targets night and day. Whole enemy networks could be mapped simply by following a target's moves and contacts over time, tying together visual imagery with other kinds of intelligence; intercepted phone calls, e-mails, and text messages. Munitions could be fired at the time and place of a drone operator's choosing (Bodwden, 2013).

"An Iraqi MiG-25 shot down a Predator performing reconnaissance over the no-fly zone in Iraq on 23 December 2002. This was the first time in history a conventional aircraft and a drone had engaged each other in combat. Predators had been armed with AIM-92 Stinger air-to-air missiles and were being used to "bait" Iraqi fighters, then run. In this incident, the Predator did not run but instead fired one of its Stingers. The Stinger's heat-seeker became "distracted" by the MiG's missile and missed the MiG. The Predator was hit by the MiG's missile and destroyed." (Knights, 2005). This became the first air-to-air engagement by an armed UAS.

Figure 11-3 MQ-9 Reaper

Source: Twisted Sifter. (2010) The World's Deadliest Drone: MQ-9 Reaper. Retrieved From https://twistedsifter.com/2010/05/worlds-deadliest-drone-mq-9-reaper/

In 2007 the Predator was joined in Afghanistan by the Predator B which later was renamed as the Reaper and designated as the MQ-9. "The MQ-9 is fitted with six stores pylons. The inner stores pylons can carry a maximum of 1,500 pounds (680 kg) each and allow carriage of external fuel tanks. The mid-wing stores pylons can carry a maximum of 600 pounds (270 kg) each, while the outer stores pylons can carry a maximum of 200 pounds (91 kg) each. An MQ-9 with two 1,000 pound (450 kg) external fuel tanks and one thousand pound of munitions, has an endurance of 42 hours. The Reaper has an endurance of fourteen hours when fully loaded with munitions. The MQ-9 carries a variety of weapons including the GBU-12 Paveway II laser-guided bomb, the AGM-114 Hellfire II air-to-ground missiles, the AIM-9 Sidewinder[16], and the GBU-38 JDAM (Joint Direct Attack Munition)." (Grier, 2009). The JDAM weapon was added to the list in 2017.

"The pilots first conducted combat missions in Iraq and Afghanistan in the summer of 2007. On 28 October 2007, the Air Force Times reported an MQ-9 had achieved its first "kill," successfully firing a Hellfire missile against Afghanistan insurgents in the Deh Rawood region of the mountainous Oruzgan province. See Figure 11-3. By 6 March 2008, according to USAF Lieutenant General Gary North, the Reaper had attacked sixteen targets in Afghanistan using 500 lb. (230 kg) bombs and Hellfire missiles (Tirpak, 2000).

The Reaper can carry as many as sixteen Hellfire missiles, but the typical load is as pictured with four Hellfires and two GBU-12 or GBU-38. "On 13 September 2009, positive control of an MQ-9 was lost during a combat mission over Afghanistan, after which the control-less drone started flying towards the Afghan border with Tajikistan. An F-15E Strike Eagle fired an AIM-9 missile at the drone, successfully destroying its engine. Before the drone impacted the ground, contact was reestablished with the drone, and it was flown into a mountain to destroy it. It was the first US drone to be destroyed intentionally by allied forces." (Reichhardt, 2009). See Figure 11-4.

"On 27 June 2014, the Pentagon confirmed that a number of armed Predators had been sent to Iraq along with U.S. Special Forces following advances by the Islamic State of Iraq and the Levant. The Predators were flying thirty to forty missions a day in and around Baghdad with government permission, and intelligence was shared with Iraqi forces." ("Unmanned aircraft Predator armed with Hellfire missiles used in Iraq to protected U.S. advisers," 2014). Combat operations have continued into 2018 in Iraq and Syria to combat ISIS.

"During his eight years in office, Mr. Obama authorized roughly 550 drone strikes in Pakistan, Yemen, Somalia, and other nations in which the U.S. was not explicitly at war.

In just his first twelve months, Mr. Trump green-lighted at least eighty strikes in those countries and "is on pace to surpass the strike tempo of both of his predecessors, which perhaps signals a great willingness to use lethal force," the survey says." (Wolfgang, 2018). Operations in and around Afghanistan have also increased the number of strikes.

Figure 11-4 Typical Reaper Load Out with the Hellfire Missiles on the right and a GBU-12 on the Left

Source: Twisted Sifter (2010). The World's Deadliest Drone: MQ-9 Reaper. Retrieved From https://twistedsifter.com/2010/05/worlds-deadliest-drone-mq-9-reaper/

Weapons Systems (Lethal)

Hellfire: The combination of the Predator/Reaper with the Hellfire missile brought the armed UAS into its own. There are several varieties of Hellfire missiles, but the weapon of choice has been the AGM 114K with the semi-active laser guidance. It is simple, accurate, and useful in a variety of environments. It is not without its limitations, which will be discussed later. "AGM-114K. One of the newest missiles in the Hellfire family, this missile features dual warheads, electro-optical countermeasure immunity, and a programmable guidance section for trajectory shaping/seeker logic changes. This missile is referred to as the Hellfire II missile. The missile is sixty four inches long and weighs one hundred pounds." (AH-64D *Longbow Hellfire Modular Missle System*, 2009).

As an active emitting missile that is inertially guided and radar-assisted, "the RF Hellfire missile uses an active RF signal to detect and track targets. It emits RF energy and homes-in on the reflected RF energy. The missile is sixty nine inches long and weighs one hundred eight pounds" (AH-64D *Longbow Hellfire Modular Missle System*, 2009). The RF version would allow all-weather operations including launching through an under cast, but this weapon is not yet used by any UAS. It is the primary weapon of the Longbow helicopter.

"The AM-114M version was originally developed for the Navy. Its warhead is solely blast fragmentation, which is effective against boats and lightly armored vehicles.

The AGM-114N variant uses a thermobaric ("metal augmented charge") warhead that can suck the air out of a cave, collapse a building, or produce an astoundingly large blast radius out in the open" (AH-64D *Longbow Hellfire Modular Missle System*, 2009).

A new AGM-114R "multi-purpose" Hellfire II is scheduled to go into production, with some guidance and navigation improvements and, "goes one step further than the K-A variant: it's intended to work well against all three target types: armored vehicles, fortified positions, or soft/open targets. The Romeo is the mainstay of the future Hellfire fleet, used from helicopters and UAVs, until and unless Hellfire itself is supplanted by the JAGM program" ("US Hellfire Missile Orders, FY 2011-2017," 2017).

"There are multiple kinds of HELLFIRE warheads to include a High-Explosive Anti-Tank, or HEAT, weapon and a Blast-Fragmentation explosive along with several others.

- The HEAT round uses what is called a tandem warhead with both a smaller and larger shaped charge. The idea is to achieve the initial requisite effect before detonating a larger explosion to maximize damage to the target.
- The Blast-Frag warhead is a laser-guided penetrator weapon with a hardened steel casing, incendiary pellets designed for enemy ships, bunkers, patrol boats and things like communications infrastructure, according to Army documents.
- The Metal Augmented Charge warhead improves upon the Blast-Frag weapon by adding metal fuel to the missile designed to increase the blast overpressure inside bunkers, ships, and multi-room targets.
- The Metal Augmented Charge is penetrating, laser-guided and used for attacks on bridges, air defenses, and oil rigs. The missile uses blast effects, fragmentation and overpressure to destroy targets." (Osborn, 2017).

The HEAT warhead is primarily designed for armored targets. It can penetrate the defenses on all current tanks, including reactive armor. The K version uses a shaped charge warhead with a scored steel fragmentation sleeve for counter personnel and light skinned vehicle purposes. The frag pattern is significantly affected by the impact angle of the weapon. The steeper, the better.

"There are five basic considerations for using laser spot trackers (LSTs) or laser-guided weapons (LGWs):

- LOS must exist between the designator and the target and between the target and the LST/LGW.
- Pulse repetition frequency (PRF) codes of the laser designator and the LST/LGW must be compatible.
- The direction of attack must allow the LST/LGW to sense enough reflected laser energy from the target for the seeker to acquire and lock onto the target.
- The laser target designator must designate the target at the correct time, and for the correct length of time. If the length of time is insufficient, the seeker head could break the lock, and the flight pattern of the LGW becomes unpredictable.
- The delivery system must release the LGW within the specific LGW delivery envelope to ensure the weapon can physically reach the target.

There is an increased hazard to friendly forces when aircrews release weapons behind friendly positions. The final decision to release standoff LGWs from behind friendly positions in a CAS environment rests with the ground commander." (*Close Air Support-Joint Publication* 3-09.3, 2014).

Figure 11-5 Hellfire Weapon Engagement Zone

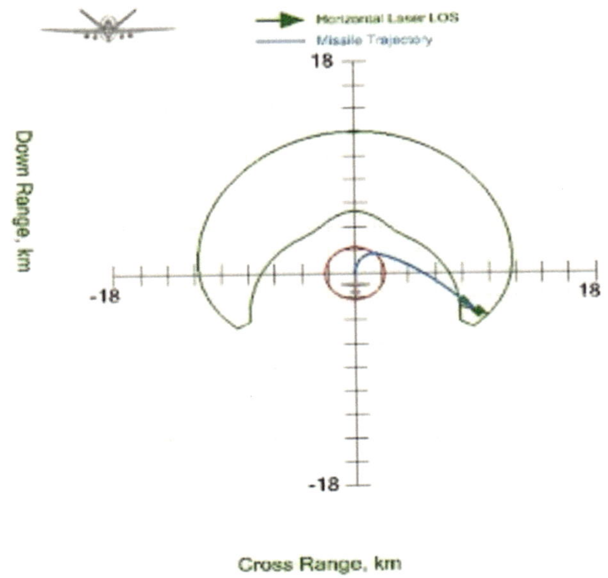

Source: Wayne, S. (2018). The Future Of Unmanned Systems. Retrieved From http://rpasarethefuture.blogspot.com/2015/11/rpa-autopilot-time-dependent-behavior.html

In open terrain this weapon, with its semi-active laser seeker, is relatively easy to employ from altitudes flown by the Reaper (Predators have been retired). However, when operations move to the more urban environment, laser operations have additional challenges. Chief among them is the ability to keep the laser spot on the target throughout the entire weapon flight time. This includes the entire slant range due to the altitude of the Reaper. Example approximate times of flight: “1 km- 3 s / 2 km- 7 s / 3 km- 10 s / 4 km- 14 s / 5 km- 19 s / 6 km- 24 s / 7 km- 29 s / 8 km- 36 s” (“Hellfire,” 2018). The aircraft’s motion during this time could place a building in the line of sight or cause part of the laser beam to reflect off a closer object or elevated terrain (hill, bridge, vehicle, etc.) The weapon guides to the first reflected laser energy. The taller the urban canyon, the more difficult it becomes.

A Hellfire launched from a Reaper at altitude has a significant WEZ (Weapon Engagement Zone) that can extend out to twelve kilometers. The weapon can also literally hit a target that is behind the wing line of the aircraft; although that zone is quite small. In Figure 11-5, the bounded green area is the projected successful WEZ. The red area reflects the weapon’s minimum range limitations, i.e., arming. The gap from the minimum range to the WEZ is a function of the altitude that the aircraft is flying. Airspeed and altitude both change the shape of the WEZ.

Although the Hellfire missile has a relatively small warhead, it still can produce a significant amount of collateral damage. Often a significant amount of time is spent waiting for non-combatants to leave the lethal zone of the missile. Sometimes the entire attack is thwarted for this reason. The typical blast/frag radius is about 15 to 20 meters.

Figure 11-6 GBU-12 Loaded On The Reaper UAS

Source: Defense Talk. (2007). USAF MQ-9 Reaper GBU-12 Paveway Laser Guided Bomb, Retrieved from https://www.defencetalk.com/military/images/usaf-mq-9-reaper-gbu-12-paveway-laser-guided-bomb.32757/

a. **GBU-12**: "The Guided Bomb Unit-12 (GBU-12) utilizes a Mk82 500-pound general purpose warhead. The operator illuminates a target with a laser designator and then the munition guides to a spot of laser energy reflected from the target. See Figure 11-6. The GBU-12 is a member of the Paveway II series of laser-guided bombs (LGBs)" ("Guided Bomb Unit-12 (GBU-12) Paveway II," 2017). Laser guided bombs made their debut in the Vietnam War, but rose to prominence in Operation Desert Storm. The guidance system uses a bang-bang methodology. When the weapon is no longer pointed directly at the laser, the guidance fins go to full deflection to correct the error. This generates a sinusoidal type of path about the direct line of sight. The reason this is important is twofold. First it uses up the bombs energy which starts out low due to the Reaper's low speed at launch. Second, the bomb tends to spend more time on the low side of the direct line of sight, causing the bomb to fall short of the target in some scenarios. Techniques to overcome these effects include raising the spot location slightly on the target when possible and/or using a release point that ensures the bombs dive angle to the target is greater than 45 degrees. At that point the bomb gains energy throughout the engagement.

The GBU-12 provides a significant degree of both blast and fragmentation. See Figure 11-7.

Figure 11-7 Sample GBU-12 Delivery Envelope

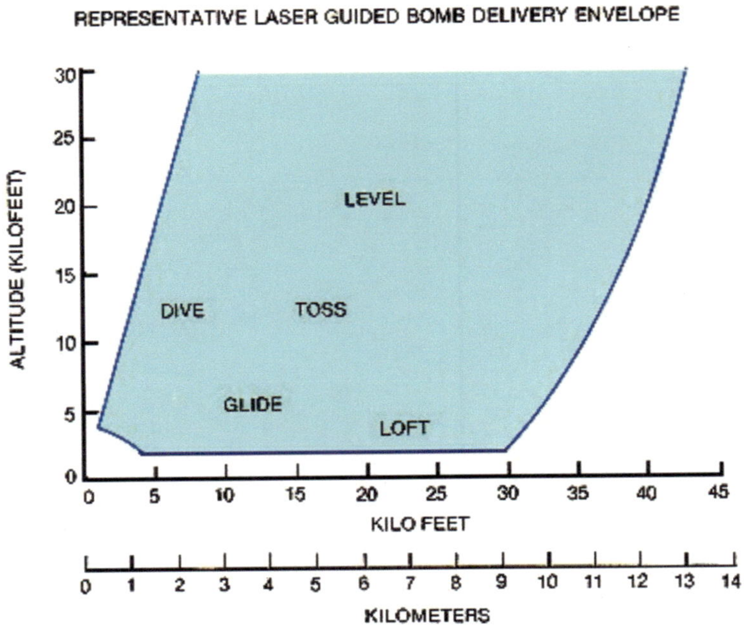

Source: Global Security. (2017) Guided Bomb Unit-12 (GBU-12) Paveway II. Retrieved from https://www.globalsecurity.org/military/systems/munitions/gbu-12.htm

The weapon can destroy a wide variety of targets. Certain targets, like tanks, require a direct hit or a coupling effect from the blast to be effective. A miss distance of ten feet or less can produce the desired coupling effect. The GBU-12 is affected by the same laser limitations in urban terrain as the Hellfire, but more pronounced. Because of the collateral damage area of the 500 lb. warhead, it is not normally the weapon of choice in populated areas for a war on terror scenario. The collateral damage radius can extend to about 2,500 ft. This effect makes the weapon ideal for groups of combatants in the open or in a building. When targeting combatants or equipment inside of a structure, a small delay fuse is incorporated into the overall fuse to ensure detonation occurs inside the structure for maximum effect. Above is an example of a Paveway II bomb delivery envelope. Actual ranges for a Reaper will generally be shorter due to the lower launch speeds of the aircraft ("Guided Bomb Unit-12 (GBU-12) Paveway II," 2017).

Figure 11-8 GBU-38 Left And GBU-54 Right

Source: Trevithick, J. S: (2017). USAF Reaper Drones Can Finally Drop GPS Guided Joint Direct Attack Munitions. Retrieved from http://www.thedrive.com/the-war-zone/10046/usaf-reaper-drones-can-finally-drop-gps-guided-joint-direct-attack-munitions

b. **GBU-38/GBU-54**: "The Air Force has added the Joint Direct Attack Munition (GBU-38) a GPS-guided bomb, to the Reaper drone force, dropping the first one in a combat strike in Operation Inherent Resolve..." (Clark, 2017). The weapon has the JDAM GPS/INS tail unit and a nose laser guidance unit. Operational use was started with just the tail unit and the weapon being dropped on pre-programmed coordinates. "It also has a selectable fuse...." "I can target the third floor of an apartment building, or we could target an enemy vehicle, or we could target enemy personnel in the open all with the same weapon because I know I can adjust the fuse with this weapon and I just don't have that option with the GBU-12." The GBU-38 is "also more useful in cloudy weather" because it uses coordinates and is not laser guided" (Clark, 2017). See Figures 11-8 and 11-9.

The GBU-54 is expected to replace the GBU-38. It also includes the JDAM tail unit and a laser guidance nose unit. Although the laser guidance unit is relatively new, it is currently scheduled to be incorporated into the overall guidance package. The GBU-54 makes it easier and more effective to take out moving targets.

Figure 11-9 MQ-9 Reaper With GBU-38 Jdams Loaded

Source: Clark, C. (2017). Air Force drops first GPS bomb from Reaper: GBU-38 JDAM. Breaking defense. Retrieved from https://breakingdefense.com/2017/05/air-force-drops-first-gps-bomb-from-reaper-gbu-38-jdam/

GBU-39B/B: The Air Force has awarded a contract to General Atomics to integrate this weapon capability into its MQ-9 Reaper. Work is expected to be completed by 2021 but is already operational with the Special Operations Command. The GBU-39B/B is a laser-guided version of the small diameter bomb. The wings that deploy on release will give the Reaper a 40 nm stand-off weapons capability. The GBU-39B/B (Figure 11-10) also has INS/GPS precision guidance.

Figure 11-10 GBU-39B/B

Source: Eshel, T. (2017). Drones Double Weapon Loadout with Laser-SDB. Retrieved from https://defense-update.com/20171128_lsdb.html

The choice of laser guidance or GPS guidance is a function of several considerations such as weather, laser spot size at longer ranges and whether the target is moving or stationary. "The Laser, Small Diameter Bomb variant, planned for the Reaper integrates the JDAM's semi-active laser, enabling the bomb to hit targets moving at up to 80 km/h (50 mph) and has been fielded by the U.S. Special Operations Command since 2014" (Fergus, 2017). Because the Laser Small Diameter Bomb (LSDB) is approximately half the weight of the GBU-12/38/54 type bombs, the Reaper can double the number of precision bombs that it carries. The LSDB also has a significantly smaller collateral damage zone. Thus, making them more user-friendly in urban and near-urban environments. Since all of these weapons have a laser guidance capability, it is key that the designating aircraft or ground unit has the matching code for the weapon about to be employed. Otherwise, the weapon will not guide. Also, the different codes allow multiple weapons to be guided to separate targets by multiple aircraft or ground units.

Pulse Repetition Frequency (PRF) Codes

"Laser coding permits the simultaneous use of multiple laser designators and laser-guided weapons/seekers. Laser designators and seekers use a PRF coding system to ensure that a specific seeker and designator combination work in harmony. By setting the same code in both the designator and the seeker, the seeker tracks only the target that is designated with that code.

Code Description

The system uses either a three digit or four-digit numeral system, depending on the type of laser equipment. Three-digit settings range from 111 to 788, while four-digit settings range from 1,111 to 1,788. All three and four-digit designator/seekers are compatible. Lower numbered PRF codes provide higher quality designation due to faster pulse repetition.

Code Allocation and Assignment

Laser guided weapons system codes must be controlled and coordinated". (JPJT, 2018) (Globalsecurity, 2016)

c. **Future weapons** for UAS that are smaller than the Reaper system: Several variant weapons are in design and testing phases that provide a smaller UAS with a precision weapons capability and a weapon with a much smaller collateral damage radius. Below is a chart of the current group of smaller weapons. See Table 11-1.

Table 11-1 Future Weapons (Snyder, 2018)

Company	Weapon Name	Weight	Notes
MBDA	SABER	10 lb. or 30 lb.	Glide or Rocket, GPS/INS/Laser
ATK	Hatchet	6 lb.	Glide, GPS/Laser
ATK	Hammer	16 lb.	Glide, Laser
Lockheed Martin	Shadow Hawk	11 lb.	Glide, Laser
Raytheon	Griffin	33 lb. or 45 lb.	Gravity or Rocket, GPS/INS/Laser
Raytheon	Pyros	12 lb.	Gravity, GPS/INS/Laser
DRS	Spike (F2M2)	5 lb.	Rocket, EO/Laser/INS
IAI	LAHAT	28 lb.	Rocket, Laser
Thales	LMM	28 lb.	Rocket, Laser/IR
Aerovironment	Switchblade	2.2 lb.	Electric motor, EO/Data Link
Thales	Starstreak II	31 lb.	Laser, 3.5 Mach
SAAB	BILL 2	23 lb.	Rocket, Wire guided
Lockheed Martin	Scorpion	35 lb.	Glide, GPS/INS/Laser
Lockheed Martin	MHTK	5 lb.	Semi-active radar
MBDA	Viper Strike	42 lb.	Glide, GPS/Laser (Used on MQ-5)

Weapons (Non-lethal)

Law enforcement use of UAS is still in its initial stages. However, several pilot programs and policy efforts are underway to add non-lethal weapons into the mix. Tasers and tear gas canisters are early contenders. "While no state in the US allows the use of deadly force by drones, North Dakota allows officers to equip them with stun guns and other non-lethal weapons" (Kozlowska, 2017). North Dakota police are authorized to use drones with tasers and pepper spray. The Israelis are examining stun grenades, rubber pellet grenades, coloring agent grenades, and sticky gel grenades. "The U.S. Army is seeking non-lethal warheads to be deployed on tiny UAVs; the U.S. Army describes the possible uses of the non-lethal UAV. "Potential commercial applications might include, but are not limited to: crowd control for local law enforcement; border protection for Homeland Security; or temporary incapacitation of non-violent criminals for local SWAT teams and/or law enforcement" ("U.S. military seeking non-lethal UAVs," 2012).

The military can use these types of non-lethal weapons in certain circumstances. The battlefield, however, lends itself to addition non-lethal weapons. Also, the increased size of the unmanned military systems provides other options. Non-lethal can be further delineated by the type of target; either anti-personnel or anti-material.

Figure 11-11 Active Denial System

Source: Hambling, D. (2009). Pain Beam to Get Tougher, Smaller, More Powerful. Wired Hellfire. Retrieved from https://quizlet.com/26596148/hellfire-flash-cards/

Anti-personnel non-lethal weapons include chemical agents, optical weapons, acoustics, directed energy, and restraining mechanisms. Anti-material non-lethal weapons include chemical, biological, directed energy, and restraining mechanisms. See Figure 11-11.

Anti-Personnel:

Chemical Agents: "Non-lethal chemical capabilities generally include agents that induce sleep or produce irritation (calmative, neural inhibitors, irritants, and odor producing chemicals)" (Siniscalchi, 1998). These types of weapons lend themselves to small areas and favorable weather conditions

Optical Weapons: Low energy lasers directed against the eyes or night vision goggles. Flash weapons are also effective directly against personnel. Examples include flash-bang grenades and high-power strobes.

Acoustics: The two main types of acoustical weapons are the loud hailer and the low-frequency wave. The loud hailer uses directed high decibel sound to disorient, discourage and potentially disabling a target. The low-frequency system (below 50 Hz) is used to disorient or even cause nausea.

Directed Energy (High Powered Microwaves): These weapons can be dialed up or down to produce the desired effect. Additionally, certain frequencies generate specific effects. Examples include pain rays (active denial system), systems designed for nausea and disorientation or even incapacitation. At the highest settings, some of these weapons can cause permanent damage or death.

Restraining mechanisms: These are typically used to restrict operations in certain areas or protect a location, building or personnel. These include things as simple as net guns to sticky foam or super-slick liquids.

Anti-Material

Chemical/Biological: These are typically used as a more clandestine method of attacking vehicles, supplies, or infrastructure. This category includes "Supercaustic agents, derived from chemical, biological, or biological enzymes, can rapidly deteriorate rubber, plastics, or spoil petroleum supplies. These are claimed to be millions of times more caustic than hydrofluoric acid and can be delivered as a liquid or aerosol. Liquid metal embrittlement agents can alter the molecular structure of metals making them weak and susceptible to structural failure. The embrittlement agents are normally formulated for a specific metal or alloy which may complicate the flexibility that is needed for combat employment. Polymer agents are extremely strong adhesives. Polymers, called stick-ems, can be applied as a liquid or foam to deny the mobility of equipment and personnel. Alternatively, super-lubricants (slick-ems) are being developed as an anti-traction capability that could disrupt the movement of vehicles" (Siniscalchi, 1998). The target area for these kinds of weapons are typically finite but can have broader effect; such as railroad tracks, pipelines, truck parks, or supply depots. Plants are another potential target. The use of genetically engineered blights on rice, cocoa, poppies, etc. could have a very outsized effect on the conflict. Defoliants have been used with success but with very harmful side effects. New defoliants without the side effects can make a significant impact in certain regions of the world.

Directed Energy/Electromagnetic Pulse (DE /EP):

"This technology offers a significant capability against modern electronic equipment susceptible to damage by transient power surges. This weapon generates a very short, intense energy pulse producing a transient surge of thousands of volts that kills semiconductor devices. See Figure 11-12.

Figure 11-12 Counter-electronics High-powered Microwave Advanced Missile

Source: Stoker, L. (2012) Electromagnetic pulse weaponry: Boeing CHAMP video and jammer grenades. Retrieved from https://www.army-technology.com/features/featureelectromagnetic-pulse-weaponry-boeing-champ-jammer-grenades

The conventional EMP and HMP weapons can disable non-shielded electronic devices including practically any modern electronic device within the effective range of the weapon. The effectiveness of an EMP device is determined by the power generated and the characteristic of the pulse. See Figure 11-13.

Figure 11-13 Portable Jammer

Source: Stoker, L. (2012). Electromagnetic pulse weaponry: Boeing CHAMP video and jammer

grenades. Retrieved from https://www.army-technology.com/features/featureelectromagnetic-pulse-weaponry-boeing-champ-jammer-grenades

The shorter pulse waveforms, such as microwaves, are far more effective against electronic equipment and more difficult to harden against. Current efforts focus on converting the energy from an explosive munitions to supply the electromagnetic pulse. This method produces significant levels of directionally focused electromagnetic energy" (Stoker, 2012). These conventional weapons are designed to engage a finite area unlike the potential widespread effect of a high altitude nuclear weapon. Weapons in this category could target a single UAS or take down a whole electrical grid.

Restraining mechanisms: Many of these weapons are designed to stop a vehicle. The police use many of these technologies to assist with ending a pursuit without further endangering the lives of the fleeing criminals. These include Spike Strips, Grapplers, and a variety of wheel nets. Grapplers are the only ones attached to a pursuing vehicle and used to capture the wheel of the vehicle in front. Some wheel nets are designed to capture the wheels of the vehicle while anchored firmly to the ground to stop the vehicle. Other nets combine the net effect with the spike strip effect. With the proliferation of small UAS comes the advent of counter UAS weapons. A couple of those weapons use nets to capture and secure the UAS. See Figure 11-14.

Figure 11-14 Drone launches a net to capture another drone

Source: Titcomb, J. (2016). This drone catches other drones by shooting nets at them. Retrieved from https://www.telegraph.co.uk/technology/news/12093204/This-drone-catches-other-drones-by-shooting-nets-at-them.html

Protecting the Weaponized Systems from the Cyber Threat/Response

- Threat: Susceptible to the enemy taking control of the UAS via the data link.
- Response: Encrypt all data links. User identifier codes.
- Threat: Susceptible to GPS denial attacks or GPS signal manipulation.
- Response: Military encrypted and directional GPS receivers such as SAASM (Selective Availability Anti-Spoofing Module).
- Threat: Susceptible to the enemy using high powered microwaves against the UAS or the weapon.
- Response: Electronic hardening and shielding efforts.
- Threat: Susceptible to the enemy manipulating the UAS head flux compass.
- Response: Low drift inertial navigation units and/or differential GPS signals.
- Threat: Susceptible to the enemy jamming the data link signal to and/or from the UAS.
- Response: Frequency hopping and other counter jamming techniques. The addition of the appropriate level of artificial intelligence (AI). A local controller that can overpower the remote jammer.
- Threat: Susceptible to the enemy waging a cyber-attack on one or more of the satellites being used to control the UAS from a remote location.
- Response: Military hardened commercial satellites.
- Threat: Susceptible to the enemy waging a cyber-attack on the controlling facility.
- Response: UAS control isolation from other information assets in the facility. Securing possible attack entry points to the control station, including power.

Conclusions

Armed UAS came into their own during the long-running war on terror, forever changing the nature of warfare. Their persistence in the target area and the ability to immediately prosecute the target has given the U.S. and its allies a significant advantage in an area of the world that historically proved very difficult for outsider militaries. This is true especially in the rugged areas of Afghanistan and surrounding countries. Unmanned aircraft are not limited by terrain. What was the enemy's advantage in this arena has become part of their weakness. The Predator/Reaper has been the focus so far for the U.S. armed UAS effort. Initially, their only weapon was the Hellfire Missile. The advent of the Reaper UAS with its larger payload capability added the 500 lb. Laser guided bomb to the equation. The continued success of the Reaper in the war has brought it several new versions of the 500 lb. weapon. The latest version is the GBU-54, with both INS/GPS and laser guidance, handle stationary and moving targets in a variety of weather conditions. Still, the Reaper can only carry two of these precision bombs along with four Hellfire missiles. Thus, a new weapon is being added to the mix; the GBU-39B/B. At approximately half the weight of the GBU-54, the Reaper can double its bomb load to four. The GBU-39B/B also gives the Reaper a 40nm standoff range. The next major effort is the arming of smaller UAS. A significant variety of weapons have emerged as contenders for this role. All are precision weapons. Some are rocket powered. Some are gravity weapons. All produce a much smaller collateral damage radius, which is ideal for both urban and rural operations, where non-combatants could be a factor. In parallel with this development program is the advent of combining unmanned aircraft, especially smaller ones with non-lethal weapons. There is a significant amount of overlap in non-lethal weapon types between the military and law enforcement. Law enforcement has taken some of the early steps in advancing their non-lethal capabilities. Their early weapons of choice have been a type of Taser, pepper balls, tear gas and flash/bang grenades. This will give law enforcement and the military a whole new set of tools for a wide variety of mission types. One of the most significant tools for the military is likely to be the use of directed energy weapons. Though non-lethal, they can have a major impact on both personnel and material targets. Weaponized unmanned vehicles are still in its early stages, but due to their significant success so far, expect to see many more varieties in the future. Protecting these assets against the current and future cyber-attack challenges will play a major role in their continued success.

Questions

1. How do world events drive the weaponization of UAS?
2. What cyber factors should be considered when pairing a UAS with a specific weapons system?
3. What non-lethal options could the military and lawn enforcement choose instead of lethal action in a major city?
4. Do you think the armed UAS will replace manned combat and support aircraft?
5. What dangers do you see in using an all armed UAS air force?

Bibliography

Globalsecurity. (2016, March 16). *Laser Employment*. Retrieved from Global Security Org: https://www.globalsecurity.org/military/library/policy/usmc/mcwp/3-16/fdraft_appk.pdf

JPJT. (2018, August 26). *Appendix K Laser Employment*. Retrieved from JPJT Army Manual: http://docplayer.net/40451907-appendix-k-laser-employment.html

Kaplan, F. (2016, September 9). *a_history_of_the_armed_drone*. Retrieved from Slate News: http://www.slate.com/articles/news_and_politics/the_next_20/2016/09/a_history_of_the_armed_drone.html

References

9-11 *Commission Report*. (2004). Retrieved from Washington DC: https://www.9-11commission.gov/report/911Report.pdf

AH-64D *Longbow Hellfire Modular Missle System.* (2009). Fort Rucker, Alabama: US Army Aviation Center of Excellence Retrieved from http://gomotherrucker.com/mem/fdrgifhsnu4/ah64/studenthandouts/0923LBHMMSLOT11.pdf.

Bodwden, M. (2013). How the Predator Drone Changed the Character of War. Retrieved from https://www.smithsonianmag.com/history/how-the-predator-drone-changed-the-character-of-war-3794671/#oPiV9wmC9GGYevbF.99

Clark, C. (2017). Air Force Drops First GPS Bomb From Reaper: GBU-38 JDAM. *Breaking Defense.* Retrieved from https://breakingdefense.com/2017/05/air-force-drops-first-gps-bomb-from-reaper-gbu-38-jdam/

Close Air Support-Joint Publication 3-09.3. (2014). Washington DC US Department of Defense Retrieved from https://fas.org/irp/doddir/dod/jp3_09_3.pdf.

Derek, M. (2000). Complete 911 Timeline US Drone Use in Pakistan or Afghanistan. Retrieved from http://www.historycommons.org/timeline.jsp?timeline=complete_911_timeline&complete_911_timeline__war_on_terrorism__outside_iraq=complete_911_timeline_pakistan___afghanistan_drone_stikes

Eshel, T. (2017). Drones Double Weapon Loadout with Laser-SDB. Retrieved from https://defense-update.com/20171128_lsdb.html

Fergus, K. (2017). General Atomics to integrate precision-guided small diameter bomb onto Reaper drone. Retrieved from https://thedefensepost.com/2017/11/28/reaper-drone-small-diameter-bomb-integration/

Grier, P. (2009). Drone aircraft in a stepped-up war in Afghanistan and Pakistan. Retrieved from https://www.csmonitor.com/USA/Military/2009/1211/Drone-aircraft-in-a-stepped-up-war-in-Afghanistan-and-Pakistan

Guided Bomb Unit-12 (GBU-12) Paveway II. (2017). Retrieved from https://www.globalsecurity.org/military/systems/munitions/gbu-12.htm

Hambling, D. (2009). Pain Beam to Get Tougher, Smaller, More Powerful. *Wired*

Hellfire. (2018). Retrieved from https://quizlet.com/26596148/hellfire-flash-cards/

Joseph, T. (2016). Vintage Videos Show the US Air Force's First Armed Drone Dropping Bombs. Retrieved from https://motherboard.vice.com/en_us/article/53dkeq/vintage-videos-show-the-us-air-forces-first-armed-drone-dropping-bombs

Knights, M. (2005). *Cradle of Conflict: Iraq and the Birth of Modern US Military Power.* Annapolis, MD: US Naval Institute Press.

Kozlowska. (2017). Is the US ready for police drones outfitted with tasers, tear gas, and lethal weapons? Retrieved from https://qz.com/947368/police-in-connecticut-could-be-first-in-us-to-fly-drones-outfitted-with-lethal-weapons/

Osborn, K. (2017). The US is running out of Hellfire missiles because of ISIS. *Business Insider*.

Reichhardt, T. (2009). Robot airplane goes AWOL, gets shot down. *Air & Space Magazine*.

Siniscalchi, J. (1998). *Non-Lethal Technologies: Implications for Military Strategy*. Maxwell Air Force Base, Alabama: US Air Force Center for Strategy and Technology Retrieved from https://fas.org/man/dod-101/sys/land/docs/occppr03.htm.

Snyder, J. (2018). The Latest Tools, Techniques and Opportunities in UAV Design. In. Arlington, VA: Mongo Industries, LLC.

Stoker, L. (2012). Electromagnetic pulse weaponry: Boeing CHAMP video and jammer grenades. Retrieved from https://www.army-technology.com/features/featureelectromagnetic-pulse-weaponry-boeing-champ-jammer-grenades

Tirpak, J. A. (2000). Find, Fix, Track, Target, Engage, Assess. Retrieved from http://www.air-force-magazine.com/MagazineArchive/Pages/2000/July%202000/0700find.aspx

Titcomb, J. (2016). This drone catches other drones by shooting nets at them. Retrieved from https://www.telegraph.co.uk/technology/news/12093204/This-drone-catches-other-drones-by-shooting-nets-at-them.html

Trevithick, J. (2017). UASF Reaper Drones Can Finally Drop GPS Guided Joint Direct Attack Munitions. Retrieved from http://www.thedrive.com/the-war-zone/10046/usaf-reaper-drones-can-finally-drop-gps-guided-joint-direct-attack-munitions

U.S. military seeking non-lethal UAVs. (2012). Retrieved from http://www.homelandsecuritynewswire.com/dr20120429-u-s-military-seeking-nonlethal-uavs

Unmanned aircraft Predator armed with Hellfire missiles used in Iraq to protected U.S. advisers. (2014). Retrieved from http://www.armyrecognition.com/june_2014_global_defense_security_news_uk/unmanned_aircraft_predator_armed_with_hellfire_missiles_used_in_iraq_to_protected_u.s._advisers_2806.html

US Hellfire Missile Orders, FY 2011-2017. (2017). Retrieved from https://www.defenseindustrydaily.com/us-hellfire-missile-orders-fy-2011-2014-07019/

USAF MQ-9 Reaper GBU-12 Paveway Laser Guided Bomb. (2007). Retrieved from https://www.defencetalk.com/military/images/usaf-mq-9-reaper-gbu-12-paveway-laser-guided-bomb.32757/

Wayne, S. (2018). The Future of Unmanned Systems. Retrieved from http://rpasarethefuture.blogspot.com/2015/11/rpa-autopilot-time-dependent-behavior.html

Whittle, R. (2015). Hellfire Meets Predator. Retrieved from https://www.airspacemag.com/flight-today/hellfire-meets-predator-180953940/#bBdb2HLUkHjsZ5fK.99

Wolfgang, B. (2018). Trump outpacing Obama in drone strikes; 80 in the first year. Retrieved from https://www.washingtontimes.com/news/2018/jun/7/donald-trump-outpacing-barack-obama-drone-strikes-/

The World's Deadliest Drone: MQ-9 REAPER. (2010). Retrieved from https://twistedsifter.com/2010/05/worlds-deadliest-drone-mq-9-reaper/

Chapter 12: UAS System Deployment and Information Dominance (ID)

Student Learning Objectives – The student will be introduced to the concepts of Information Dominance (ID), Information Superiority (IS), Offensive Information Operations (OIO), Network-Centric Operations (NCO) and deployment objectives for the US Coast Guard in areas of operation using unmanned air and sea vehicles. This chapter will hone in on the *Information Dominance (ID) goal* that UAS/UAV systems present opportunities for excellence in Offensive Information Operations (OIO), and Network – Centric Operations (NCO).

UAS in Military and Commercial Service

At first glance most developed UASs are applicable to both military and civilian deployment. The military forces use UAS in dull, dirty, dangerous (DDD) roles in which human pilots would be at risk. These roles have expanded far beyond the DDD boundary. (Austin, Unmanned Aircraft Systems: UAVS Design, Development and Deployment, 2010) The reason civilian UASs have not outnumbered military UAS (in the CONUS) is because of the restrictions in the civilian market.

Civilian uses require operations in open airspace, rather than on a battlefield or within a military enclosure. Regulating authorities have not yet accepted their general operation. (Austin, Unmanned Aircraft Systems: UAVS Design, Development and Deployment, 2010) FAA is in charge of protecting the National Airspace (NAS). The FAA regulations regarding UAS seem to center on preventing injury to persons and damage to property due to failures of the UAS and preventing injury or damage caused by collisions between UAS and other airborne vehicles. (Austin, Unmanned Aircraft Systems: UAVS Design, Development and Deployment, 2010) In the former case, FAA has made effective regulations to assure the airworthiness of the systems and meeting these requirements to insure protection of persons and property. If there are drawbacks, they are cost and bureaucracy.

UAS collisions in the NAS is another thing entirely. Authorities are still searching for a completely reliable method of sensing the presence of another airplane vehicle and avoiding collision with it. (Austin, Unmanned Aircraft Systems: UAVS Design, Development and Deployment, 2010) This requirement is true in open airspace and in dedicated UAS space. Cost of the SAA and product support may also be inhibiting the civilian UAS market. Both UAS markets are subject to the principles of Information Dominance (ID) because UAS / UAV systems are an essential component of the collection technologies used. As an intermediate step to full commercialization, FAA may require a Part 107 waiver process in controlled airspace. (FAA, 2018)

Information Dominance (ID)

"Information warfare (IW) -based technologies are categorized by their information operations roles and by three distinct levels of technology maturity:

- *Core Technologies* – current state-of-the-art, essential technologies necessary to sustain the present level of information.
- *Enabling technologies* –concerned with the next generation of IW capabilities. They represent a significant enhancement in operations. [Tactical level changes]
- *Emerging technologies* – far on the horizon applications where feasibility is demonstrated. These are so call Black Swan events which involve radical improvements in capability and approach to information operation." [Strategic level changes] (Waltz, 1998) 1972 RSA changes to public –key cryptography represented a Black Swan (Taleb, 2010) event in the crypto – world. (Rivest, 1978)

Numerous DOD technology studies have evaluated the potential developments that may impact Information Warfare with respect to UAS / UAV / UUV systems. A few samples:

Unmanned Systems Integrated Roadmap Fy2011-2036 covers interoperability, autonomy, airspace integration, and manned-unmanned teaming (MUM) (Army, 2013)

The Navy Unmanned Undersea Vehicle: (UUV) *Master Plan* covers the important mission categories for use of UUVs. These include ISR, mine countermeasures (MCM), Anti-submarine warfare (ASW), Inspection and Identification (ID), Oceanography, communications / navigation network node (CN3), payload delivery, information operations (IO) time critical strike (TCS), barrio patrol and sea base support. (Navy, 2004)

Joint Publication JP 6-01, *Joint Electromagnetic Spectrum Management Operations* covers the entire spectrum of offensive and defensive electromagnetic spectrum operations for all levels of defense activities. (Army-M, 2012)

US *department of Homeland Security Cybersecurity Strategy*- a fascinating document that covers risk identification, vulnerability reduction, threat reduction, consequence mitigation and enabling cybersecurity outcomes. (DHS, 2018)

The U.S. Navy's Plan for Information Dominance – a broad ranging document about US Navy policies to gain ID and to use information as a weapon. (NavyID, 2010)

The Military Critical Technologies List Part II: Weapons of Mass Destruction Technologies. An older document that enumerates critical technologies for directed energy weapons (DEW) and information warfare. (DoD-IW, 1998)

"*Information Dominance* as defined by the US Navy is the operational advantage gained from fully integrating the Navy's information functions, capabilities and resources to optimize decision making and maximize warfighting effects." (Google, 2018) There are other definitions:

"*Information Dominance* – the degree of information superiority that allows the possessor to use information systems and capabilities to achieve an operational advantage in a conflict or to control the situation in operations other than war while denying those capabilities to the adversary." (Army6, 1996)

"*Information Dominance* – A condition that results from the use of offensive and defensive information operations to build a comprehensive knowledge advantage at a time, place, and on decision issues critical to mission success." (Griffith, 1997)

"*Information Superiority* – the capability to collect, process, and disseminate an uninterrupted flow of information while exploiting or denying an adversary's ability to do the same." (Chiefs, 2014)

The definitions above have conceptual similarities. The author actual prefers the Information Superiority approach which involves technologies for collecting data, processing that data into knowledge and disseminating that knowledge to live respondents. Table 12-1 is a condensed form of (Waltz, 1998) Table 10.1 pages 362-363 specific to IW/ID. Table 12-1 is essentially a top-level matrix for Information Warfare based on enabling and emerging core technologies. Recognize that the prime goal of the broader purview of Information Warfare is Information Dominance and by inclusion Information Superiority (IS)

Table 12-1 General Technology Categories for Information Warfare

Information Warfare

Core – Sustainable

Attack

- Electronic attack based on brute force
- And precision jamming, deception
- Semi-automated network attack
- Dynamic Malicious codes

Defend

- Robust cryptography, trusted computers,
- Network security authentication protocols
- Electromagnetic hardening

Enabling –Technology Jump

Attack

- Semi /fully-automated network attack
    - Dynamic adjustment
- Tactical electronic attack with DEW
- High energy chemical lasers
- Dynamic and autonomous malware/ logic
- Precision directed DEW

Defend

- Trusted network challenge response, duel /
- Authentication and response
- Multi – type authentication
- Network intrusion detection systems
- Steganography – use of SPAM
- Anti-DEW weapons
- Blockchain
- All optical networks

Information-Based Warfare

Core-Sustainable

Collect

- Airborne Reconnaissance (manned and MAME UAS)
- Space surveillance
- HUMINT
- GPS / GNSS
- EO, IR, SAR multi-spectral sensing

Process

- Data warehouse
- Data fusion
- Automatic target recognition (ATR)
- Data mining
- Text-based databases, hyperlinking

Disseminate

- Push-pull dissemination
- Data compression
- Global broadcast
- 3-D visualization

Enabling – Technology Jump

Collect

- High altitude UAS
- Unattended intelligent ground sensors
- Commercial high – resolution imaging satellites
- Integrated ground to space sensors
- Protected GNSS / tracking
- Hyperspectral, integrated aperture sensing
- Barrier penetrating sensors
- Micro UAVs
- Intelligent Level 5 SAA

Process

- High-bandwidth global broadcast
- Medium-bandwidth global communication satellite network
- Global cellular and microcellular wireless voice and data
- Virtual reality visualizations
- DNA and molecular computing storage

Disseminate

- High-bandwidth broadcast, multicast, point-cast networking
- Networks of sensors in space, air, surface operating together
- Global real-time tailored knowledge delivery

As seen in Table 12-1, information dominance technologies fall into "three general areas: collection, processing and dissemination. *Collection* includes methods of sensing physical phenomena and platforms that allow sensors to carry out their missions. These include direct and remote sensing devices along with the relays of data to user." (Waltz, 1998)

Processing (power) refers to the numbers of operations per second, information storage capacity (in bits). Technologies to increase processing power are many, subtle, overt, heterogeneous, involving hardware, software, networks, machine / human interface, autonomous, knowledge-management, indexing and beyond the scope of this work.

Dissemination technologies are communications technologies to increase bandwidth and effective use of bandwidth (compression techniques, data / knowledge organization). Enhancements to these technologies include increased storage and shortened latency times.

Unmanned aircraft systems play a significant role in the collection technologies. Collection technologies include the advanced platforms and sensors to acquire a depth of data. (Waltz, 1998)

High-Altitude Endurance (HAE) and Medium – Altitude Endurance Unmanned Air Vehicles (UAVs)

"The Global Hawk (HAE) and Predator models (MAE) introduced penetrating airborne surveillance with a broad area search capability. HAE UAVs provide long dwell times over target areas. Both the Global Hawk and the Predator series complement short – and close-range UAVs, which do not have deep penetration capability, and satellite surveillance, which does not have revisit rates." (Waltz, 1998) Close-range UAVs support small unit operations with ranges to 30km and short-range UAVs have medium-altitude endurance (30-50 hours with 150 to 300 km range)

Both the Global Hawk and Predator sport communication relay capabilities, precision SIGINT, local precision navigation capabilities, and operate a sensor network with autonomous and cooperative behavior in hostile space. (Waltz, 1998)

Offensive Information Operations (OIO)

The names and definitions have changed since Waltz's signature work in 1998, on the subject of Information Warfare, but the operations remain the same – but more sophisticated, networked and complexity-rich. (Waltz, 1998) Our objective is to see where UAS /UAV systems fit into his various taxonomies of IW / ID. Coverage of Waltz's work, in this chapter, is only "helicopter -view" and compressed. The reader is encouraged to explore further the wealth of literature on this subject.

"Offensive operations are uninvited, unwelcome, unauthorized and detrimental to the target; therefore, the term *Attack* to refer to all of these operations." (Waltz, 1998)

"Offensive information operations are malevolent acts conducted to meet strategic, operational, or tactical objectives of authorized government bodies; legal, criminal or terrorist organizations; corporations; or individuals. The operations may be performed covertly, without notice to the target, or they may be intrusive, disruptive, and destructive. The effects on information may bring physical results that are lethal on humans." (Waltz, 1998)

President Obama used drones to great success tracking down terrorist High Value Targets (HVTs) and assassinating them. Because of his elevated use of UAVs for this purpose, he left somewhat of a legal mess in his wake. (Zenco, 2016) However, the information dominance side of the operations were an unqualified success.

Offensive information attacks have two basic functions: to capture or to affect information. Information here refers to data /information / knowledge content. ID is measured in terms of:

- "Functions – broken down into <u>offensive measures</u> of *capture and affect* used to effectively gain a desired degree of control of a target's information resources. Capturing information is an act of theft of a resource if captured illegally, or technical exploitation if the means are not illicit. Affecting information is an act of intrusion with intent to cause unauthorized effects, usually harmful to the information owner." (Waltz, 1998) Both capture and affect by UAS are collection processes.
- Tactics – <u>Attack tactics</u> -the operational processes employed to plan, sequence, and control the countermeasures of an attack. These tactics consider objectives, desired effects [covertness, denial, disruption of service; destruction, modification, or theft of information], degree of effects; and target vulnerabilities." (Waltz, 1998)
- Understanding attack mechanisms helps information security designers to prepare for defense. In the information business, the common standard of information security (INFOSEC) is CIA, which means confidentiality, integrity and availability. (Nichols R. K., 2002) Reviewing the attack tactics – factors above, brings to mind Parker's brilliant expansion of the CIA basis for securing information. (Parker, 2015)
- "Parker expanded the traditional INFOSEC framework. Traditionally, users were concerned with preservation of: confidentiality, integrity and availability (CIA) information from disclosure, modification, destruction, or use; by prevention, detection, recovery; to *reduce loss or reduce risk of loss*. Parker was ahead of his time. He saw INFOSEC as preservation of six elements: availability, utility, integrity, authenticity, confidentiality, possession of information; from accidental or intentional destruction, interference, use of false data, modification or replacement, misrepresentation or repudiation, misuse or failure to use, access, observation or disclosure, copying, stealing or endangerment. This was done by: avoidance, deterrence, prevention, detection, mitigation, transference, sanction, recovery, or correction to meet a standard of due care,

  Avoid loss, reduce loss, or / and eliminate loss." (Parker, 2015) Parker even envisioned the means to accomplish his information security framework. Two controls were to be robustly instituted: government controls to include: employee clearances, the principle of need-to-know, mandatory access control, classification of information, and cryptography and business controls include: need-to-withhold, discretionary access control, copyright and patent, and digital signatures. (Parker, 2015)

- “Techniques –the technical means of capturing and affecting information of humans – their computers, communications, and supporting infrastructures.” (Waltz, 1998)
- Motive – varied but most common are: ideological, revenge, greed, hatred, malice, challenge, theft.
- “Invasiveness – Attacks may be active or passive. Active attacks invade and penetrate the information target. Passive attacks sit on the line and observe behaviors, information flows, timing, and energy.” (Waltz, 1998)
- “Effects –may vary from small- harassment to theft, from narrow, surgical modification of information to large-scale cascading of destructive information that brings down critical infrastructure.” (Waltz, 1998) The Stuxnet attack on the Iranian centrifuges in in 2015 was a brilliant example of large-scale effects on critical infrastructure. (Holloway, 2015)
- Ethics and legality- Traditional intelligence activities are allowed in peacetime (capture information by UAS) but information attacks that affect information are not covered adequately by law. Unlike real property, information is a property that may be shared, abused, copied, and stolen without evidence or the knowledge of the legitimate owner.

A taxonomy of attack countermeasures may be viewed in a two-dimensional attack matrix:

Rows are labelled perceptual, information, or physical. Columns are headed by attack category: capture or affect. From a UAS standpoint, we are only interested in characterizing the information infrastructure level of the attack. Before we extract the Figure 8.1 Attack Matrix row in (Waltz, 1998) page 255, two more avenues of approach are available to the attacker:

- “Direct, or internal, penetration attacks this involves penetrating a communication link, computer, or database to capture and exploit internal information, or to modify, add, delete, insert, or install a malicious process.” (Waltz, 1998)
- “Indirect, or external, sensor attacks – perfect for UAS / UAVs flying above the targets. The attacker presents open phenomena to the systems sensors or information to sources, media, Internet, satellite, third parties, to achieve counter information objectives. These attacks include insertion of information, spoofing of information (GPS), to sensors or observation of behavior of sensors or links interconnecting fusion nodes.” (Waltz, 1998)

Extracting just one information row from Waltz’s Attack Matrix (Waltz, 1998) we have:

Table 12-2 Extracted Information Infrastructure Row from Waltz Attack Categories (Waltz, 1998)

- “Object: Capture
- Level of Attack: Information Infrastructure – Capture Information Resource
- Security Property Attacked: Privacy is breached
- Avenue: Indirect (Observe, Model Infer)

    - Passive intercept of message traffic
    - Non-intrusive mapping of network topology
    - Cryptographic analysis" (Waltz, 1998)
    - "EMS spectral analysis and categorization" (Nichols R. K., 2002)
    - Direct: Penetrate and Observe
        - Network attack and penetrate to secure unauthorized access to data
        - Trojan horse program
        - Install sniffer
        - Install spoofing software
- "Object: Affect
- Level of Attack: Information Infrastructure – Affect Information Resource
- Security Property Attacked: Integrity of data is invalidated; Availability of services degraded
- Avenue: Indirect: Cause effects through sensors or over the open network without penetration of target
    - Deceive: issue deceptive e-mail (phishing) message or conduct deceptive network behavior
    - Disrupt, Deny or destroy: Deny network data collection service by DDOD or Syn flood attacks that disrupt access to public or private sources. Insert an open message traffic and data that diverts attention and processing resources. Insert sensor data that upsets guidance or control process" (Waltz, 1998)[UAS perfect].
- Direct: Penetrate and affect targeted infrastructure and affect
    - Deceive: Insert Trojan horse with deception action. Modify, corrupt data by viral agent.
    - Disrupt, Deny, Destroy: Insert malicious code to deny or disrupt service ibn single host computer or entire network

Modern military activities are focused on network- centric operations. They are concerned with the primary threats to networks and especially the messaging and interconnecting links. The explosion of wireless and IoT devices has ramped up INFOSEC concerns. Some system designers are now trying to mitigate the treats to their networks based on Blockchain technology. Blockchain is harped as a revolution in cryptographic protection. Started with Bitcoin, proponents would use Blockchain technologies to protect business, money and military networks. (Tapscott, 2016) "The managing author disagrees with this approach on many grounds including reviewed security holes, privacy issues, single-source point of failure and the fifty –one percent emersion/ dominance attack. (Jay, 2018) Lastly, the data is held in the dark web where the worst of malevolent actors anonymously play." (Nichols R. K., 2018)

Going back to first principles of INFOSEC, primary active threats to networks and network messaging (includes communication links, EMS vectors, interconnected nodes, wired / wireless hardware, and access points, frankly everything that talks to anything in the networks). The messages may have different formats, however, they are known. UAS systems are not just collectors of information or signals in the sky. They can convert, modulate, attenuate, insert, delete information as programmed to do so, and can initiate a network attack! (Nichols R. e., 2016)

There are five types of network intrusive actions: Access, Denial, Inter-message or spoofed message, Intra-message, and data storage. The functional objective of access is for an invalid user to gain access to the system, and the unauthorized user elevates his/her access at a higher level than authorized. The functional objective of denial is to deny service requests and to disrupt the flow of messages in the system, rendering it completely inoperable or reduced in operating capacity to some degree. There are four inter-message intrusive actions: spoofing (like GPS in naval situations), modification, replay, and leakage. All these are compromises of identity, authentication, message, or content in transit. Intra-message violations are either repudiation (Didn't order the book) or security content (breach of firewall or security device / rule). Data storage intrusions include message pre-plays or direct corruption of sources or integrity while in storage (sending random bits into backup disks is an example). (Waltz, 1998)

A simple NCO attack strategy can be initiated from any source including UAS over the target network. (Nichols R. e., 2016) Phase 1 is Reconnaissance begins by searching for and collecting passwords or cryptographic password files to be computer-brute-forced; gaining access, finding unused accounts, and establishing covert access. Phase two is penetrate and act which involves gaining entry, check for surveillance, gain system control, attack by searching directories, acquiring useful data, searching for evidence and destroying both evidence and audit trails, and surveillance if possible, then replacing control and logging off as if the attacker was never there. (Nichols R. e., 2016)

Practically every known computer system is vulnerable to attacks. Clark wrote the Red Team Field Manual of software attacks on every modern system and structure on the market, including *NIX, Windows, networking, web, databases, programming and wireless. (Clark, 2013) To be fair, White and Clark also wrote a Blue Team Field Manual which covered countermeasures such as scanning, vulnerability analysis, network discovery, service disabling, firewalls, detection (visibility) PCAP tools, NETCAT tools, respond and analysis, remediation, tactics, incident management, and security incident identification (SCHEMA). (White, 2017) One of the best books on practical countermeasures for network security is by (Nichols R. R., 2000) entitled *Defending your Digital Assets against Hackers, Crackers, Spies and Thieves.*

Network-centric Operations (NCO)

Figure 12 -1 UAS Surveillance Network

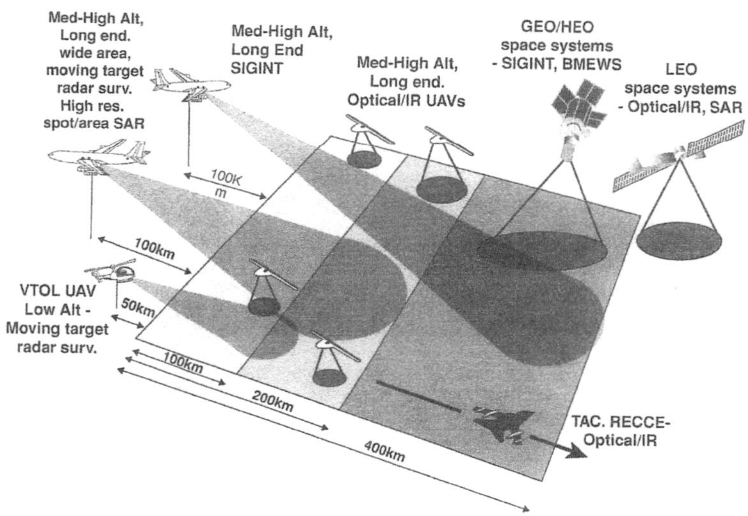

Source: Austin, R, S: (2010), CT: Unmanned Aircraft Systems: UAVS Design, Development and Deployment. London: Wiley Aerospace Series.

UAS systems – especially military ones – rarely operate in a vacuum. They receive and process information from many sources, which include satellites, manned aircraft, naval vessels and ground-based systems. UAVs may be the top supplier of information to a network. Figure 12-1 illustrates a surveillance network using airborne systems. (Austin, Unmanned Aircraft Systems: UAVS Design, Development and Deployment, 2010) Every theater of war and all their military activities may be coordinated through a network. This is called network-centric operations (NCO). An "NCO has four characteristics using robust technologies:

- Secure communications between systems via multiple links,
- Standardization of interfaces between key systems,
- Adaptable and user-friendly interfaces with human operators,
- Coordinated radio frequency bandwidth." (Austin, Unmanned Aircraft Systems: UAVS Design, Development and Deployment, 2010)

The strength of a UAS NCO is that the UAS leader is not limited to surveillance. It is not just the collection agent / technology. The HALE UAS may disseminate information which it has self-acquired or received. The UAS is one of the reasons for information dominance achieved via an NCO. UAS systems have come a long way in just a brief time. See Figure 12-2 UAV evolution since 1995. A range of activities, air-, sea-and land-borne, covering reconnaissance, surveillance, support, defensive and attack operations may be coordinated through a UAS lead NCO. (Austin, Unmanned Aircraft Systems: UAVS Design, Development and Deployment, 2010)

However, NCOs do have weaknesses. If a major system fails, like GPS, then the whole network may be subject to catastrophic failure. Most NCOs have both fail-safe and fall back options.

Figure 12-2 UAV Evolution

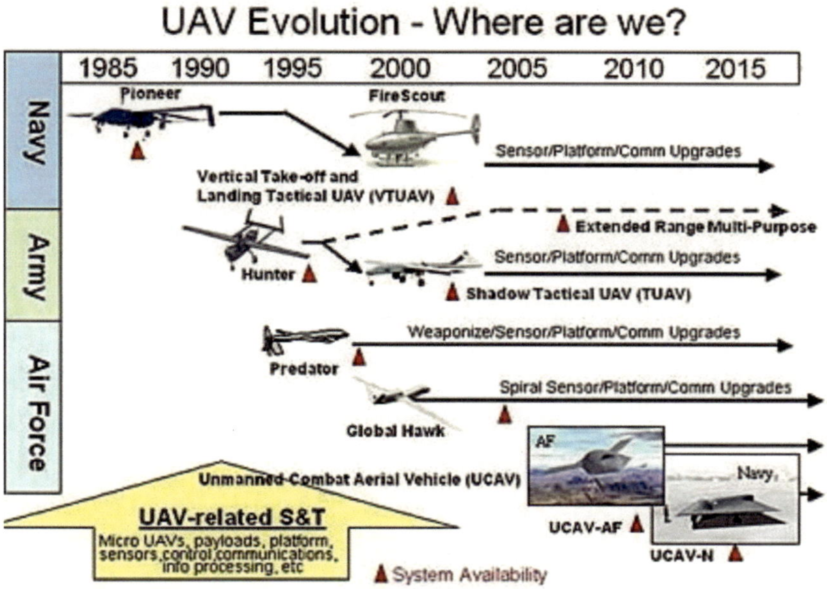

Source: Library, L. (2018, August 23). Government Resources: Defense, Military, and Security: Drones (Military). Retrieved from https://library.louisville.edu/ekstrom/gov_defense/dronesmil

Coast Guard Roles

The U.S. Coast Guard (USCG) plays a vital role in *Information Dominance (ID)*. The National Fleet Policy in 2014, established the partnership of the USCG and USN to enhance both branches capabilities and identify emerging threats. With the combination of Department of Defense and Homeland Security in the maritime infrastructure the partnership brings the U.S. ability to gather intelligence to the next level. (See Figure 12-3) However, this is not the first time the USCG took part in the world of intelligence. The CG-210, a 75-foot Coast Guard patrol boat, was the first boat in U.S. history to become a signal-intercept ship. (Bennett, 2016) During the 1920's the CG-210 employed counterintelligence to stop illegal rum runners. Over 12,000 rum-runner messages were decrypted in a three-year span by Elizabeth Friedman. (Bennett, 2016)

Figure 12-3 United States Coast Guard and Navy

Source: By U.S. Navy photo by Photographer's Mate Airman Apprentice Patrick Gearhiser [Public domain], via Wikimedia Commons https://commons.wikimedia.org/wiki/File:US_Navy_060517-N-4014G-130_The_Pre-Commissioning_Unit_Texas_(SSN_775)_sails_past_the_Coast_Guard_cutter_Sea_Horse_(WPB-87361).jpg

As of today, the Coast Guard "sees a clear opportunity to perform many of its missions faster, cheaper and more safely through the use of short-range unmanned aircraft systems," said Lt. Cmdr. Ryan Lampe, short-range UAS platform manager in the Office of Aviation Forces. (Haring, 2018) The sUAS of choice, (Host, 2018), is used to gather intelligence, surveillance, reconnaissance and provide real time imagery. Real time imagery includes taking photos of the ocean surface, checking for anomalies, and alerting the aircraft's operator for further investigation. The ScanEagle can provide VHF/UHF communications relay and Target illumination. However, after seven years of use the ScanEagle will be challenged by the USCG acquisition of the Aerosonde. The Aerosonde (Figure 12-4) is larger sUAS that has the capabilities to fly over 150,000 flight hours in temperature extremes. The sUAS also has communications relay, but the stand out feature is the available day and night full-motion video.

Figure 12-4 sUAS Aerosonde

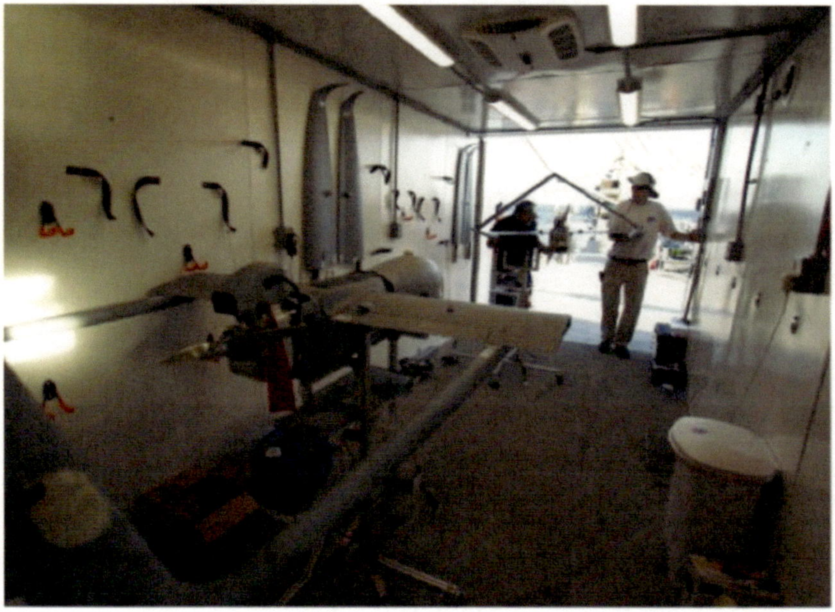

Source: Textron (2018, August 23). Aerosonde Data Sheet. Retrieved from Textron Systems, https://www.textronsystems.com/sites/default/files/resource-files/TS%20US%20Aerosonde%20Datasheet.pdf

There is one other sUAS the USCG is exploring to add to their inventory. The Puma (See Figure 12-5) is the USCG hand-launched sUAS that is current being tested. The Robotic Aircraft Sensor Program for the Maritime Environment (RASP-M), under the DHS, has allowed for the testing of Puma in Mississippi and Connecticut. The advantage of the hand-launched sUAS is the ease of launch and the ability to carry a payload, such as a high definition camera. This provides intelligence ranging from if an approaching vessel has weapons to data of a maritime environmental incident.

Figure 12-5 sUAS Puma

Source: Textron (2018, August 23). Aerosonde Data Sheet. Retrieved from Textron Systems https://www.textronsystems.com/sites/default/files/resource-files/TS%20US%20Aerosonde%20Datasheet.pdf

The Coast Guard uses National Security Cutters (NCS) or commonly referred to "go-fast boats". (Biesecker, 2018) for sUAS. The NCS are technology advanced, capable of launching small boats and serve as a flight deck. NCS perform homeland Security and defense operations in the maritime space.

"Unmanned aircraft systems have the potential of being major force multipliers for the Coast Guard," said Cmdr. Dan Broadhurst, UAS Division Chief for the Office of Aviation Forces (CG-711). (Haring, 2018) "They can provide persistent, tactical wide-area surveillance, detection, classification and identification functions that we currently do not have access to." (Haring, 2018) (See Figure 12-6)

Figure 12-6 United States Coast Guard UAS Concept

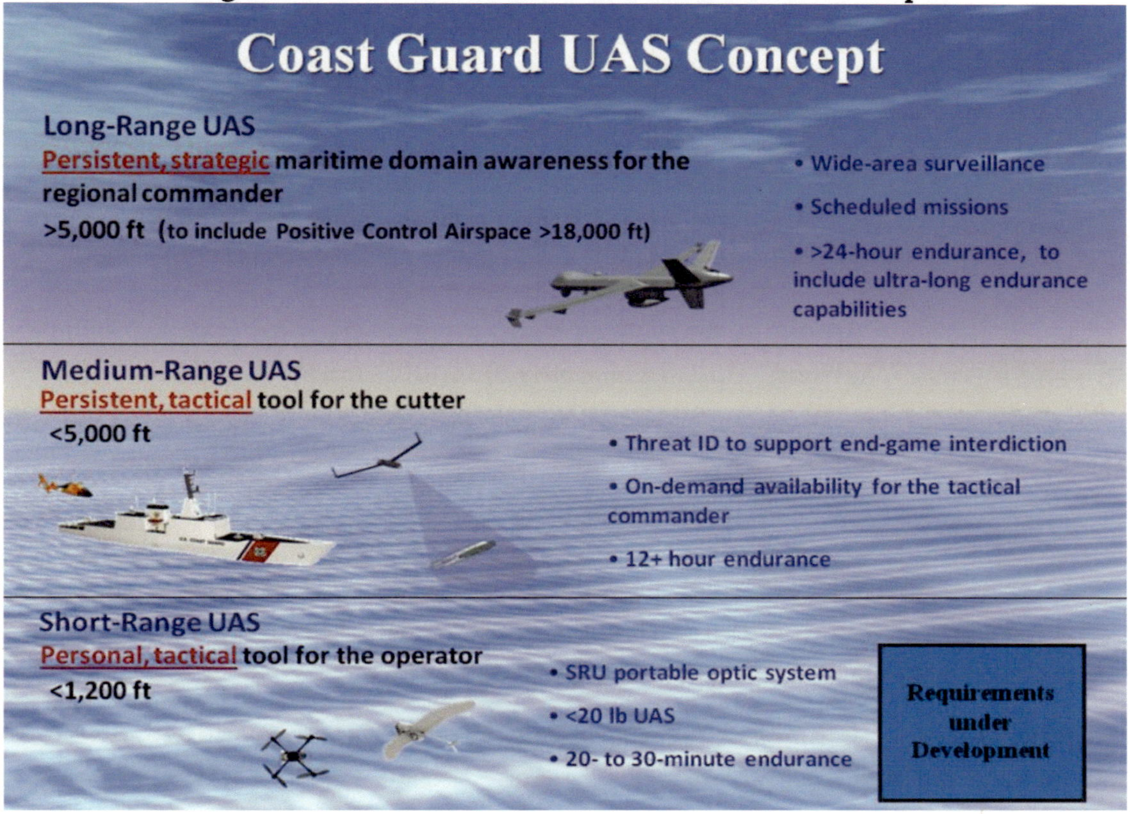

Source: Haring, L, (2018, January 19). Research, Development, Test and Evaluation Spotlight: Long-Range, Ultra-Long Endurance Unmanned Aircraft System. Retrieved from http://coastguard.dodlive.mil/2018/01/rdte-spotlight-long-range-ultra-long-endurance-unmanned-aircraft-system/

USCG is currently researching to acquire long-range, 24-hour endurance, UAS. The long-range drone would be used for intelligence, surveillance and reconnaissance missions. Additional USCG requirements include the UAS to conduct operations at a 15,000-foot mean above sea level and various maritime sensors, including electro-optic and infrared full-motion video, surveillance radar, radio frequency and direction finding, and tactical communications radio and datalink. (Biesecker, 2018)

Often referenced as the forgotten service, the USCG is far from being retired. The USCG has a long history with their intelligence and counterintelligence skills. The USCG is evolving with technology, not just by using UAS functionality. They have used UAS for the past eight years. The Coast Guard has ongoing research that allows for the branch to identify and obtain the best in industry UAS for intelligence and reconnaissance missions. They will continue to be a valuable partner to the USN in defense at sea and protecting the homeland.

Discussion Questions

1. UAS systems are both collection agents and directive information agents. Enumerate the points in the network- centric model that are most vulnerable to UAS surveillance or intrusion.
2. How vital is the UAS platform in terms of Information Dominance on the battlefield?
3. The USCG seems to be a silent service, with a huge mission with limited personnel active over all our water and coastlines. Research and report on five areas where they use UAS as effectively as any of military services.
4. Do the same research but focus on Civilian components / uses for UAS surveillance, collection and intrusion actions – specifically on a computer network.

Bibliography

Army, U. (2013). *Unmanned Systems Integrated Roadmap* FY2011 – 2036. Washington: Createspace Independent Publishing Platform.

Army6, U. (1996, August 27). *Information Operations*. Retrieved from FM 100-6: https://fas.org/irp/doddir/army/fm100-6/index.html

Army-M, U. (2012). *Joint Publication JP 6-01, Joint Electromagnetic Spectrum Management Operations*. Washington: US Government.

Austin, R. (2010). *Unmanned Aircraft Systems: UAVS Design, Development and Deployment*. London: Wiley Aerospace Series.

Bennett, M. (2016, August 8). *The TOP SECRET story of Coast Guard code breaking*. . Retrieved from coastguard.dodlive.mil: http://coastguard.dodlive.mil/2016/08/the-top-secret-story-of-coast-guard-code-breaking/

Biesecker, C. (2018, April). *Long-Range Coast Guard Drone to Undergo Tech Demo – Avionics*. Retrieved from Aviation Today: https://www.aviationtoday.com/2018/04/12/long-range-coast-guard-drone-undergo-tech-demo

Brothers, E. (2018, June 18). *Insitu to provide UAS services to US Coast Guard – Aerospace Manufacturing and Design*. Retrieved from Insitu – Aerospace Manufacturing and Design. http://www.aerospacemanufacturinganddesign.com/article/insitu-uas-services-us-coast-guard-061818/

Chiefs, J. (2014, November 20). *Information Operations.* Retrieved from JP 3-13 Change 1: http://www.jcs.mil/Portals/36/Documents/Doctrine/pubs/jp3_13.pdf

Clark, B. (2013). *Red team Field Manual* (RTFM). New York: NP.

DHS. (2018, May 15). *DHS-Cybersecurity-Strategy.* Retrieved from DHS: https://www.dhs.gov/sites/default/files/publications/DHS-Cybersecurity-Strategy_1.pdf

DoD-IW. (1998, February). *Military Critical Technologies List Part II.* Retrieved from FAS: https://fas.org/irp/threat/mctl98-2/mctl98-2.pdf

FAA. (2018, August 26). *Request a Part 107 Waiver or Operation in Controlled Airspace.* Retrieved from FAA: https://www.faa.gov/uas/request_waiver/

Griffith, J. a. (1997, January). *Information Dominance v Information Superiority.* Retrieved from IWAR: http://www.iwar.org.uk/iwar/resources/info-dominance/issue-paper.htm

Haring, L. (2018, January 19). *Research, Development, Test and Evaluation Spotlight: Long-Range, Ultra-Long Endurance Unmanned Aircraft System.* Retrieved from http://coast-guard.dodlive.mil/2018/01/rdte-spotlight-long-range-ultra-long-endurance-unmanned-air

Holloway, M. (2015, July 16). *Stuxnet Worm Attack on Iranian Nuclear Facilities.* Retrieved from Stanford.edu/courses/2015: http://large.stanford.edu/courses/2015/ph241/holloway1/

Host, P. (2018, January 31). *US Coast Guard evaluating ScanEagle ocean surface anomaly detector payload.* Retrieved from https://www.janes.com/article/77496/us-coast-guard-evaluating-scaneagle-ocean-surface-anomaly-detector-payload

Information Dominance definition. VAdm Card, K.L & VAdm Rogers, M.S. (2013) *Navy Strategies for Achieving Information Dominance 2-13-2-17: Optimizing Navy's Primacy in the Maritime and Informational Domains. Washington, US Navy*: Retrieved 10/12/2018 from https://www.pub-lic.navy.mil/fcc- c10f/Strategies/Navy_Strategy_for_Achieving_Information_Dominance.pdf

Jay, J. (2018, June 1). *Blockchain-platform-eos-found-containing-critical-security-vulnerabilities.* Retrieved from SC Media – SCmagazine UK: https://www.scmagazineuk.com/blockchain-plat-form-eos-found-containing-critical-security-vulnerabilities/article/1472602

Library, L. (2018, August 23). *Government Resources: Defense, Military, And Security: Drones (Military).* Retrieved from https://library.louisville.edu/ekstrom/gov_defense/dronesmil

Mighty-Team. (2018, August 23). 5 *differences between the Navy and Coast Guard.* Retrieved from We are the mighty: Team Mighty. (2018, April 2). 5 differences between the Navy and Coast Guard. Retrieved from https://www.wearethemighty.com/articles/5-differences-between-the-navy-and-the-coast-guard

Navy, U. (2004). *The Navy Unmanned Undersea Vehicle (UUV) Master Plan.* Washington.

NavyID, U. (2010, May). *US Navy's Vision for Information Dominance.* Retrieved from DoD Publications: http://edocs.nps.edu/dodpubs/topic/vision/vision2010.pdf

Nichols, R. e. (2016). Drone Wars: Threats, Vulnerabilities and Hostile Use of UAS. INFOWARCON16 *Proceedings.* Nashville, KY: INFOWARCON.

Nichols, R. K. (2002). *Wireless Security: Models, Threats, Solutions. New York: McGraw-Hill.* New York: McGraw-Hill.

Nichols, R. K. (2018, May 2). A Primer on Cryptocurrency & Blockchain. KSU *Invited Presentation before Lions Club* . Salina, KS, USA: KSUP.

Nichols, R. R. (2000). *Defending your Digital Assets against Hackers, Crackers, Spies and Thieves.* New York: McGraw-Hill RSA Press # 1.

Parker, D. B. (2015, September 12). *Toward a New Framework for Information Security?* Retrieved from Wiley Online: https://onlinelibrary.wiley.com/doi/abs/10.1002/9781118851678.ch3

Rivest, R. S. (1978, February 1). "A Method for Obtaining Digital Signatures and Public-Key Cryptosystems" (PDF). *Communications of the* ACM, 21 (2), pp. 120–126. doi:10.1145/359340.359342.

Taleb, N. N. (2010). *The Black Swan: the impact of the highly improbable* (2nd ed.). London: Penguin.

Tapscott, D. a. (2016). *Blockchain Revolution: How the technology behind Bitcoin is changing money, business and the world.* New York: Penguin Random House.

Textron. (2018, August 23). *Aerosonde Data Sheet.* Retrieved from Textron Systems: https://www.textronsystems.com/sites/default/files/resource-files/TS%20US%20Aerosonde%20Datasheet.pdf

Waltz, E. (1998). *Information Warfare: Principles and Operations.* Boston: Artech House.

White, A. a. (2017). *The Blue Team Field Manual.* New York: NP.

Zenco, M. (2016, January 12). *Reflecting-on-obamas-presidency/obamas-embrace-of-drone-strikes-will-be-a-lasting-legacy.* Retrieved from NY Times: https://www.nytimes.com/roomfordebate/2016/01/12/reflecting-on-obamas-presidency/obamas-embrace-of-drone-strikes-will-be-a-lasting-legacy

SECTION V

COMPUTER APPLICATIONS & DATA LINKS – EXPOSING UAS VULNERABILITIES VIA ELECTRONIC WARFARE (EW) & COUNTERING WITH LOW PROBABILITY INTERCEPT SIGNALS (LPI)

Chapter 13: Data Links Functions, Attributes and Latency

Student Learning Objectives

The student will learn about the data-link function of the UAS which allows for bi-directional communication and data transmissions between UAV and its ground station. The Data Link is a vital component of an UAS. The student will learn the respective functions of each component part and considerations are necessary and how to evaluate their importance when developing a UAS. While the focus of the lesson will be on military applications, the considerations will be equally important for those designing and deploying UAS for civilian purposes. While the design of the Datalink must have requisite attributes that allow the system to function as intended in various environments globally, it must be able to do so securely and effectively. Issues such as Data Link security, interception, deception and signal latency are all attributes that must be balanced to achieve fast and secure data communications between the components of the UAS.

What are the Types of UAV's and how are they Categorized?

UAV's are most often divided into four categories based upon their mission duration and operational radius.

- High Altitude, Long Endurance ("HALE") most often deployed for reconnaissance, interception or attack;
- Medium altitude, moderate range most often used for reconnaissance and combat effect assessment;
- Low cost, short range small UAV's. (See Figure 13-1.)

Components of the UAS Data-Link and their functions

The UAV and Ground Control Station

There are four essential communication and data processing operations the UAS must be able to efficiently and effectively carry out. These functions are vital to the ability of the remote operator or autonomous operation system (auto-pilot) to immediately issue a command, have it processed and executed and send feedback to the operator or auto-pilot confirming its execution by the UAV.

- UAV Base System. Bi-Directional communication between the UAV and the ground station, providing sensor data to the operator confirming that the command was carried out.

- UAV Sensor System, which in military applications would likely include cameras, INS GPS and radar, where data is collected and processed by the sensors and base system and thereafter communicated to operator or autonomous operation system (auto-pilot).
- UAV Avionic System. Which coverts control commands from the operator or auto pilot to the UAV avionic functions, including engine operation, flight surfaces such as flaps, rudders, spoilers and stabilizers as well as responsive feedback to the operator after the command is executed.
- In-flight communication, always wirelessly between the ground stations via line of site or indirect communication via satellite (Hartman K. &., 2013).

Figure 13-1 Mini Drones

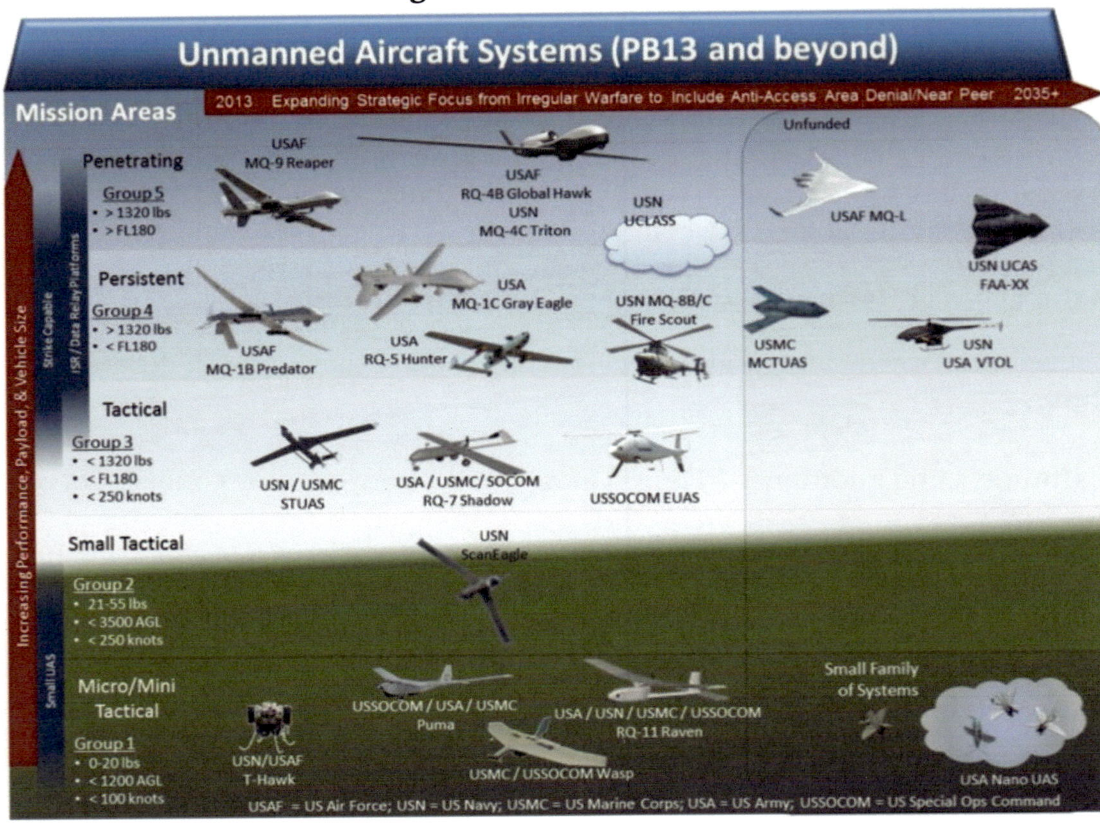

Source: Jang, C. (2017). Taking Drones to The Next Level – Cooperative Distributed Unmanned – Aerial- Vehicular Networks for Small Drones and Mini Drones. IEEE *Vehicular Technology Magazine*, Volume 12, Issue 3, pp. 73-82.

The Datalink – Essential Operations, Functions and Capabilities

The Data-Link is essentially the neurological system for the UAS transmitting data from the eyes and ears of the operator, which is processed in their mind and then communicated through the Data-Link to the UAV either directly via line of sight, radio communication from the ground station, indirectly via satellite or via cloud-based multi-UAV networks. (See Figure 13-2 for example of a remote-controlled attack UAS, MQ-9. Think about the data-links required for this UAS.)

No matter the category of UAV, the UAS must at a minimum have the capability of the remote operator having the ability to communicate data commands to the UAV wirelessly and the UAV in turn, must be able to receive data, process commands and transmit sensor, avionic and performance data which then must be transmitted back to the ground station all of which must be safely and securely transmitted via wireless radio frequency communication.

When designing, developing and deploying UAS, there are many considerations that are vital to the successful development and robust deployment for a given application. At its core an unmanned system is designed to operate remotely in theatre by a pilot located a few feet or, as in military applications thousands of miles away.

The Data-Link is the pathway by which the UAV communicates with the ground station and operator as well as how the operator send commands to the UAV to control the its mission, evaluate changing threat vectors, navigate, respond to terrain and atmospheric condition during the mission and respond thereto as well as control intelligence gathering and in military applications, payload delivery.

The challenges presented to the UAS designer, especially when it comes to the Data-Link is how to maximize the strength of the Data-Link signal while maintaining the security of the data transmitted while protecting against possible countermeasures and threats including, but not limited to Data Jamming ("DJ), Data or Signal Deception and Anti-Radiation Munitions ("ARM")

Figure 13-2 Remote-Controlled Attack UAS, MQ-9

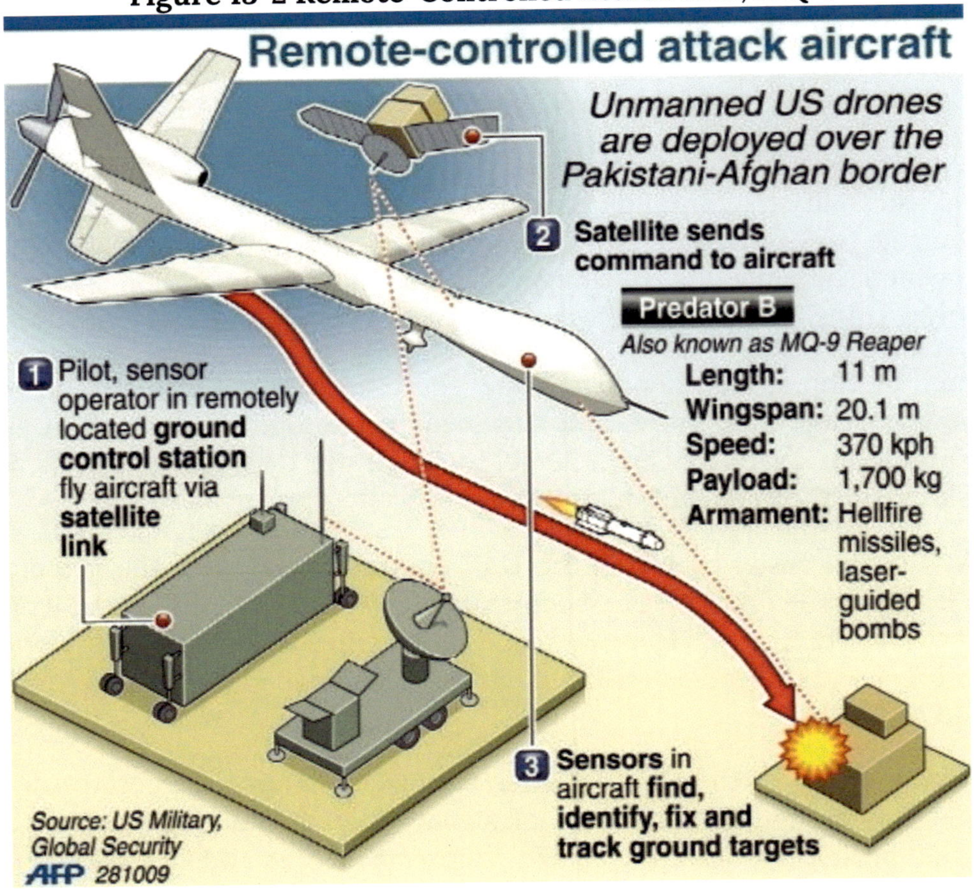

Source: Deadliest Unmanned Killing Machines in USA Arsenal. (January 01, 2011). Retrieved from https://tarwa.blogspot.com/2011/01/deadliest-unmanned-killing-machines-in.htm

All essential functions required of the UAS Data- Link communication system are:

- A ground station from which the operator can communicate via radio uplink permitting the operator to control the UAV, in terms of navigation and payload deployment.
- A downlink which is used to relay sensor, payload, and avionics data from the UAV back to the ground station and to the operator of the UAS.
- Bi-directional communication regarding distance and azimuth to the UAV to aid in precise navigation and targeting accuracy (Fahlstrom, 2012).

In addition, the Data-Link must be designed to seamlessly interface with systems onboard the UAV through the Air Data Terminal (ADT) and associated antenna arrays needed to receive radio signals from the Ground Station or satellite relayed signals in UAS that operate in beyond line of sight UAS designs.

Once received by the ADT the data must then be processed, sometimes compressed, and then instantly transmitted to the appropriate subsystems onboard the UAV such as navigation, flight surface, engine operation and targeting. Similarly, the Ground Station must also have the same abilities to receive downlink data from the UAV directly from the satellite relay. Once the downlink data is received it to must be processed, possibly compressed or converted, and then forwarded to the appropriate systems, sensors, displays or databases at the ground station.

No matter the method of transmission of the uplink or by Radio Frequency ("RF") signals of varying frequencies are usually secured by spread spectrum techniques. (Kakar, 2017). See Figure 13-3 for a simple Data-Link.

Figure 13-3 Data Links Overlay: Ground Station, Satellite, UAS

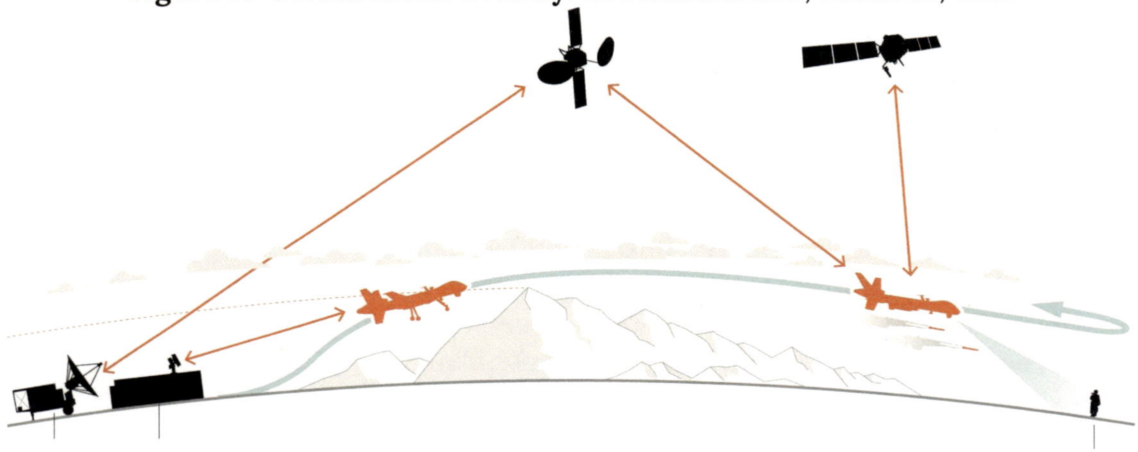

Source: Whitlock, C. (June 20, 2014). When drones fall from the sky. https://www.washingtonpost.com/sf/investigative/2014/06/20/when-drones-fall-from-the-sky/?noredirect=on&utm_term=.09b5d3e895bd

Attributes to consider in the design of the Data Link

"It is important to understand that the concept of data security in terms of the transmission, processing, storage and interpretation. It is not so much a static process but a fluid one. Given the speed with which modern technology is developed or hacked, the UAS designer should view security as a temporary" (Schneier, 2000).

> "If you think technology can solve your security problems, then you don't understand the problems and you don't understand the technology." (Schneier, 2000).

It is a certainty that the security of the Data Link will never be a static secure condition, it must have a robust and reactive architecture that allows it to function in geographically diverse and distant theatres globally. What may be an effective Data Link in a Nebraska field may not be in the mountains of Afghanistan.

The current consensus is that there are seven "must have" attributes of the Data Link for a UAS (Fahlstrom, 2012).

Globally available secure frequency with sufficient bandwidth and assignability. Absent this Data Link will be unable to support regular global training, testing and immediate deployment should the need arise.

- **Resistance to unintentional interference.** This attribute is a vital considering a myriad of other RF activities or sources may be operating simultaneously in theatre or nearby.
- **Low Probability of detection and interception.** The Data Link must be difficult to detect to reduce the likelihood of interception. If the Data Link signal is intercepted it becomes significantly easier for an adversary to use direction finding technology to locate and disable or destroy UAS components.
- **Signal Encryption and Security.** If an adversary is successful in detecting the Data Link it should not be able to locate any component of the UAS. Accordingly, Data Link encryption and decryption must be engineered into ground station, satellite and UAV.
- **Anti-Deception Capability.** The Data Link must securely transmit, receive and relay legitimate transmissions. Data and commands between the ground station, satellite relay and UAV must be capable of differentiating legitimate from counterfeit signals.
- **ARM Resistant Capability.** This attribute is especially important to the Ground Station / Operator as an Anti-Radiation Missiles ("ARM") pose a significant threat since the Ground Station transmissions to the satellite relay and UAV are especially susceptible to this type of munition.
- **Anti – Jam Capability.** Similarly, to the Anti-Deception and Signal Security attributes the Anti-Jam capability requires the Data Link to maintain its functionality and efficacy, especially considering the proliferation of new and more effective Jamming technologies presently available and in development.

There is a significant amount of crossover between technologies and design elements of one attribute which may also influence another. We will discuss those "tradeoffs" in the next chapter.

Global Radio Frequency Functionality and Adaptability

What is Radio Frequency? Broadly defined in the context of UAS Data Link development it is electromagnetic radiation (EMR) being used to transfer energy and information by radio waves. Most Data Links currently use some form of wireless RF communication. Consult Figure 13-4 EMS. Certain frequencies or bands may not be available in disparate regions of the globe where conflicting RF frequencies may be in use or may not be available. Failure to account for this likelihood can result in decreased range, quality or overall efficacy of the Data Link. One of the simplest and most overlooked design attributes is limiting the use of known conflicting or unavailable frequencies, thereby enhancing robust peacetime training and actual combat deployment.

Figure 13-4 Partial EMS

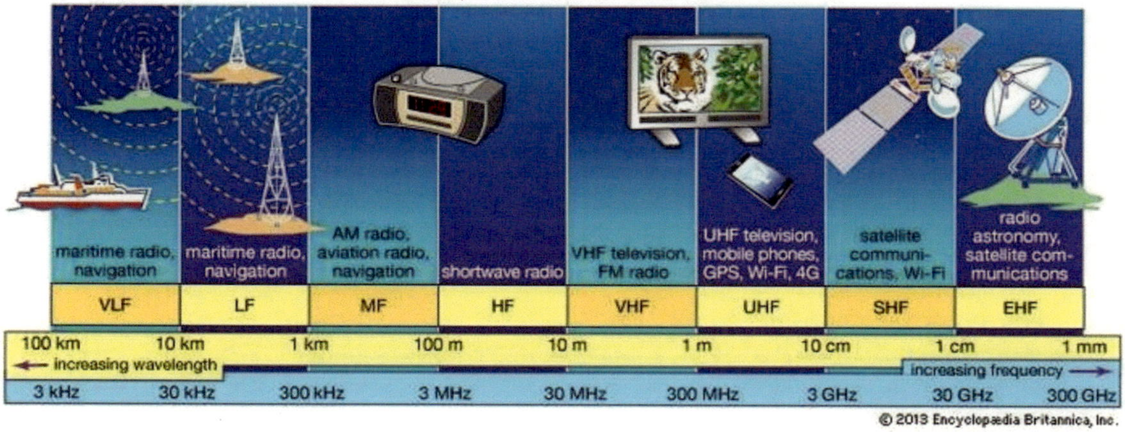

Source: VHF Communications. (2017). In *Encyclopedia Britannica*. Retrieved from https://www.britannica.com/technology/VHF

Without designing a UAS with global availability, the likelihood of training and testing blackout zones increases, thereby compromising rapid global deployment. The best practice is an ample consideration and allocation of frequencies available to meet known as well as unexpected contingencies. Considering that by some estimates in 2035 commercial UAS usage significantly exceed military, and governmental UAS deployments (Kakar, 2017).

Table 13-1 demonstrates the airwaves are populated with a wide array of civilian, commercial, and military Radio Frequency applications. Table 13-2 shows the RF band designations. Considering future technologies, such as global Wi-Fi, the availability of bandwidth will likely contract, while the risk of unintentional interference expands. Given the crowded low frequency airwaves, the design trend is towards VHF and even L band Data Links. This may have a positive effect upon the UAS since using higher frequency bands allows more data to be transmitted faster (Jain, 2017).

Table 13-1 Standard Definitions of Radio Spectrum Segments

TABLE 1: STANDARD DEFINITIONS OF RADIO SPECTRUM SEGMENTS		
Name	**Frequency range**	**Applications**
Low frequency (LF)	30 to 300 kHz	Navigation, time standards
Medium frequency (MF)	300 kHz to 3 MHz	Marine/aircraft navigation, AM broadcast
High frequency (HF)	3 to 30 MHz	AM broadcasting, mobile radio, amateur radio, shortwave broadcasting.
Very high frequency (VHF)	30 to 300 MHz	Land mobile, FM/TV broadcast, amateur radio
Ultra high frequency (UHF)	300 MHz to 3 GHz	Cellular phones, mobile radio, wireless LAN, PAN
Super high frequency (SHF), millimeter-wave range	3 to 30 GHz	Satellite, radar, backhaul, TV, WLAN, 5G cellular
Extremely high frequency (EHF)	30 to 300 GHz	Satellite, radar, backhaul, experimental, 5G cellular
Terahertz , tremendously high fequency (THF) or far infrared (FIR)	300 GHz to IR	R & D, experimental

Source: Spectrumeffect.com (2018)

Resistance to Unintentional Interference

Many of us who are old enough to remember the analog age will remember the heyday of AM (amplitude modulation) radio broadcasting from the 1950's to mid – 1970's. Although not capable of providing high quality sound, AM radio was an excellent broadcasting technology for long distances where terrain and structures may cause line of sight challenges. The downside of AM broadcasting is that is it subject to significant interference from other electromagnetic interference such as powerlines, fluorescent lights, environmental conditions and competing broadcasts. The result can be static, inconsistent signals, or even signal echo on the same frequency.

Table 13-2 shows the RF band designations.

TABLE 2: MICROWAVE LETTER BAND DESIGNATIONS

Band	Frequency range	Applications
L	1 to 2 GHz	Satellite, navigation (GPS, etc.), cellular phones
S	2 to 4 GHz	Satellite, SiriusXM radio, unlicensed (Wi-Fi, Bluetooth, etc.), cellular phones
C	4 to 8 GHz	Satellite, microwave relay, Wi-Fi, DSRC
X	8 to 12 GHz	Radar
K_u	12 to 18 GHz	Satellite TV, police radar
K	18 to 26.5 GHz	Microwave backhaul
K_a	26.5 to 40 GHz	Microwave backhaul, 5G cellular
Q	30 to 50 GHz	Microwave backhaul, 5G cellular
U	40 to 60 GHz	Experimental, radar
V	50 to 75 GHz	New WLAN, 802.11ad/WiGig
E	60 to 90 GHz	Microwave backhaul
W	75 to 110 GHz	Automotive radar
F	90 to 140 GHz	Experimental, radar
D	110 to 170 GHz	Experimental, radar

Source: Spectrumeffect.com (2018)

The availability and allocation of overused civilian frequencies may result in new spectrums which may be more robust, powerful and secure. Incorporating new communication technologies into the UAS will help to ameliorate the challenge of Data Link interference from unintentional or environmental sources (Jain, 2017).

Examples of recently implemented tools to mitigate interference include using cellular technology and power control framework (Yajnanarayana, 2018). Encryption of RF signals may provide additional interference protection as will new cable, connector and power supply shielding (Cannon Corporation, 2017). Lastly novel approaches are currently being developed including a self-interference cancellation solution where the UAS Data Link will self-detect and correct interference by situationally moving to other RF bands and frequencies (Chen, 2014).

Low Probability of Intercept ("LPI")

LPI is a vital attribute for the Datalink if for no other reason than it is the most likely source of human casualties. In most instances it is desirable to have the ground station in proximity to the operator or pilot of the UAV. Distance matters in UAS design since the greater the distance data must travel, the greater the latency.

The threat of interception of the RF emissions from the ground station an essential attribute of the uplink design. Less likely, though still a design concern is in certain instances the downlink can be targeted for interception. One scenario where interception risk to the downlink increases is when a UAV is hovering for extended periods of time. Not only will this put the UAV at risk of attack but also the risk that the downlink communications may be intercepted increases and consequentially the target of the surveillance may be alerted to its presence, causing them to camouflage or cease activities.

Electronic Support Measurement (ESM) systems allow adversaries to locate the source of RF emissions in an area. This technology is especially effective in locating the ground station since it is usually stationary for extended periods of time. The longer a source emits radiation, the greater the chance of interception. Since it is not feasible to constantly move the ground station increasing LPI can seem to be daunting task. Another relatively simple, yet effective methods to minimize the risk of intercept is by increasing armaments and location of the ground station. Shielding vital components and pilots from an attack can help ensure UAS and personnel survivability in the case of a successful interception attack. Finally, designers should not be comfortable in the assumption that LPI is inherent higher in a moving UAV. Innovative technologies capable of detecting and intercepting signal emanations from moving objects are becoming more effective and readily available.

Effective enhancement of LPI can be accomplished by some of the following tactics and technologies:

- Ground Station Handover Method. (GSHM)
- Direct-sequence spread spectrum (DSSS)
- Frequency-hopping spread spectrum (FHSS).
- Dynamic power management. (DPM)
- Directional transmission. (DT)
- Low duty cycle methods (LDCM) (Okcu, 2016). See Figure 13-5 Overlay.

Figure 13-5 LDCM Method Overlay

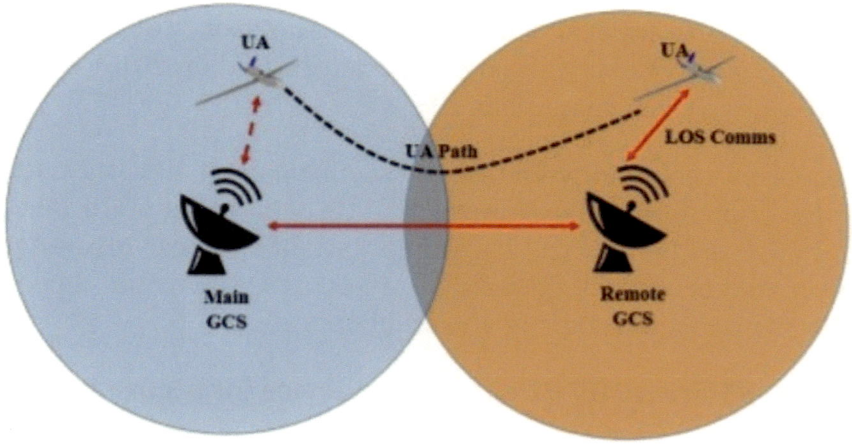

Source: Okcu, H. (2016). Operational Requirements of Unmanned Aircraft Systems. *Journal of Advances in Computer Networks*, Vol. 4, No. 1, 28-30.

Signal Encryption and Security

Best practice in all areas of information security is constantly enhancing inaccessibility of the data and components of a network or system. This seemingly simple concept is no less relevant when designing a UAS Data Link. The wireless RF Data Link is an enticing target one which if successfully attacked can disable or damage multiple systems connected to the Data Link. Here is a list of some recent Data Link hacks upon civilian, commercial and military UAS.

1. **Maldrone**, where malware is injected into critical areas of the UAS operation system through security flaws in the Datalink.
2. **GPS Spoofing** is a hack which essentially can alter or delay UAV commands via GPS and accordingly can cause collisions, faulty guidance and theoretically virtual UAV hijacking whereby a civilian UAV can be turned into an attack vector against military UAV's even though military GPS systems are well protected. This was used by the Iranian military to capture a United States military drone in 2011.
3. **Zigbee and Killerbee** which are essentially sniffing and penetration tools which when successful can cause a major threat to UAS by Denial of Service attacks (Rodday, 2015).

Data encryption is a vital tool to create a secure Data Link. Just as in wired computer networks, the wireless UAS employs CPU's and operating systems to perform functions involving massive amounts of data which must be immediately processed and transmitted internally and externally.

An exciting new protocol is the Commercial Solutions for Classified Program ("CSfC"), developed United States National Security Agency – Central Security Service. This is a layered approach to data security for UAS where two or more commercially available encryption and cybersecurity protocols provide enhanced Data Links security. The layered approach is beneficial since it reduces development time and expense while leveraging crossover of cybersecurity threat similarities across civilian, commercial or military applications (Keller, 2016). See Figure 13-6 Harris KGV-72 encryption device for secure messages.

Whatever the protocol or technology Data Link security must not only address the threats known today but also capable of adjusting to new threats as they are discovered. Constant evaluation, testing, and even “white hat” hacking is one of the best insurance policies against attacks. It is far better to discover a vulnerability or security flaw while the UAS is being tested and used for training as opposed to during an actual operation.

Figure 13-6 Harris KGV-72 encryption device for secure messages

The Harris Corp. RF Communications segment in Rochester, N.Y., designs and manufactures the KGV-72 encryption device, shown above, which provides the ability to process classified messaging traffic.

Source: Harris Corporation. (2009). Kgv-72 Type-1 Programmable Encryption Device. http://jproc.ca/crypto/kgv72.pdf

There is no security without physical security. It has been estimated that up to 95% of all security incidents in 2014 were the result of human error (Howarth, 2014). The best Data Link security requires constant technological examination as well as robust physical security. Physical security and access controls must be implemented and enforced for all who operate, meet, provide third party support or services to the UAS design and operations team.

Resistance to Deception

A closely related, yet separate attribute of Data Link design is avoidance of deception by an adversary. Spoofing an apparently authentic command or GPS direction data can cause a UAV to become uncontrollable or even crash. Deception resistance is currently a focal concern with respect to the data uplink between the ground station and UAV. Just one deceptive command can cause a UAV to crash, be captured, hijacked or even attack a friendly target.

Not only can deception cause these types of undesirable consequences, a successful effort can also jeopardize a wide array of secrets and information. Although it remains unclear whether jamming or deception was used, it is believed that Iran has re-engineered a US built RQ-170 Sentinel Drone which was captured by some form of deception or jamming in December, 2011(Opall-Rome, 2018). See Figure 13-7 Enemy Captured RQ-170.

Currently there are multiple methods to achieve an acceptable level of resistance to this form of attack. Many methods that help protect the Data Link from other security threats will also provide protection against deception attacks. Those include, Spread Spectrum Data Link transmissions using secure authentication codes. These codes can be a software embed in the ground station transmission to the satellite relay or UAV. Both the UAV and ground station will have encoding and decoding software to authenticate commands without direct modification to the uplink. (Fahlstrom, 2012) If a satellite relay is implemented in the UAS it can also have authentication software thereby establishing end-to-end data security. Anecdotally, AJ and LPI can also be enhanced by encoding Data Link transmissions.

Figure 13-7 Enemy Captured RQ-170

Source: Opall-Rome, B. (February 12, 2018). Israel Air Force says seized Iranian drone is a knockoff of US Sentinel: https://www.defensenews.com/global/mideast-africa/2018/02/12/israel-air-force-says-seized-iranian-drone-is-a-knockoff-of-us-sentinel/

One particularly promising GPS spoofing detection systems was hypothesized by a team at Cornell University in Ithaca, New York. This spoofing detection system essentially "spoofs' the Spoofer. Below is a figure of the proposed system architecture for this yet potential anti-deception system.

As depicted above, three Global Navigation Satellite System (GNSS) satellites whose signals would be tracked in the non-spoofed case: satellites j-1, j, and j+1. It also shows the potential location of a Spoofer that could send false versions of the signals from these same satellites. The Spoofer has a single transmission antenna. Satellites j-1, j, and j+1 are visible to the receiver antenna, but the Spoofer could "hijack" the receiver's tracking loops for these signals so that only the false spoofed versions of these signals would be tracked by the receiver" (Psiaki, 2013). See Figure 13-8 Spoofing the Spoofer.

Figure 13-8 Spoofing the Spoofer

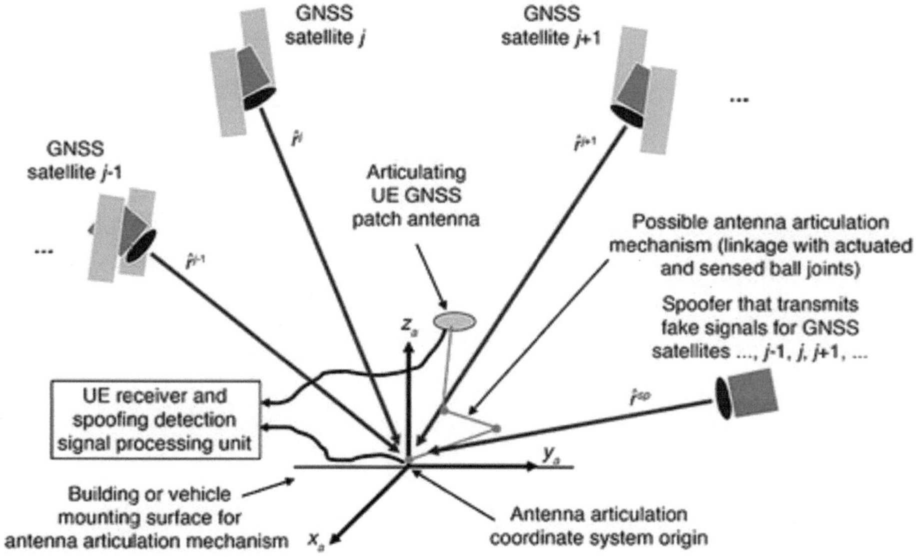

Spoofing detection antenna articulation system geometry relative to base mount, GNSS satellites, and potential spoofer.

Source: Kakar, J. M. (2017). Waveform and Spectrum Management for Unmanned Ariel Systems Beyond 2025. Ithaca, New York: arXIv.org, Cornell University.

This and other recent technologies which can aid in securing the UAS Data Links of civilian, commercial and military UAS applications is an important reminder of the fluidity of the discipline. As the ancient Chinese Philosopher Sun Tzu wrote of 2500 years ago in "The Art of War",

> "The whole secret lies in confusing the enemy, so that he cannot fathom our real intent." Words matter when it comes to avoiding detection, but perhaps more importantly understand that the unexpected should always be expected when (Giles, 2013)

When it comes the securing a UAS Datalink designers would be wise to live by these it comes to technological warfare and cyber -attacks against military technology.

Anti-ARM

Anti-Radar Missile (ARM's) sense and target sources of RF signals radiation to provide an attack vector to destroy the emission source. Since a UAS Data Link, especially the uplink, emits RF radiation from the ground station transmission antenna, it is susceptible to being attacked by ARM weaponry.

ARM threat only exists when RF radiation is emitted. Limiting RF transmissions to instances when commands are being sent is a simple yet effective Anti-ARM defense. In addition, various decoy technologies exist to reduce the ARM threat as well as placing the transmission antenna farther away from the Ground Station to minimize ARM damage. Finally, various signal spectrum spreading techniques, not to mention physical armor for the Ground Station itself will all increase the Anti- ARM characteristics of the UAS Datalink. (Fahlstrom, 2012)

Other proposed Anti – ARM capabilities involve hardening the Ground Station and particularly the antenna array. On such proposal is to quickly identify and then react to the signature of an incoming ARM and rapidly recoil a more heavily armored antenna before an ARM which successfully penetrates other Anti-Arm defenses. This proposal is an important example of physical security, even the age of technology can still be an excellent defense. When one considers the millions of dollars involved in design, development and deployment of UAS, it makes sense to harden its battlefield armor. It most certainly better to be able to have a UAS survive an ARM attack and be re-deployed to re-engage the enemy. (Czeszejko, 2013) See Figure 13-9 ARM Processes.

Anti-Jam (AJ) Capabilities

If one considers ARM threats an attack vector based upon the source of the Data Link communications, then jamming can be considered UAS countermeasure designed to address signals containing commands that are transmitted by the ground station. Jamming is a countermeasure used to inhibit the ability of a UAV to successfully communicate with its operator by directing powerful electromagnetic radiation ("noise") at the Data Link in order to "drown out" communications and data transfer. Similarly, global navigation satellite system jamming can impede the pilot's ability to fly the UAV increasing risk of a catastrophic failure or crash (Droneshield , 2017).

Jamming the GNSS data flow between the UAV and pilot takes away the eyes and ears of the UAS. Without the ability to guide itself or be flown by a pilot using GNSS data, exponentially increase the risk of crash, mission failure, loss of life and investment.

Figure 13-9 ARM Processes

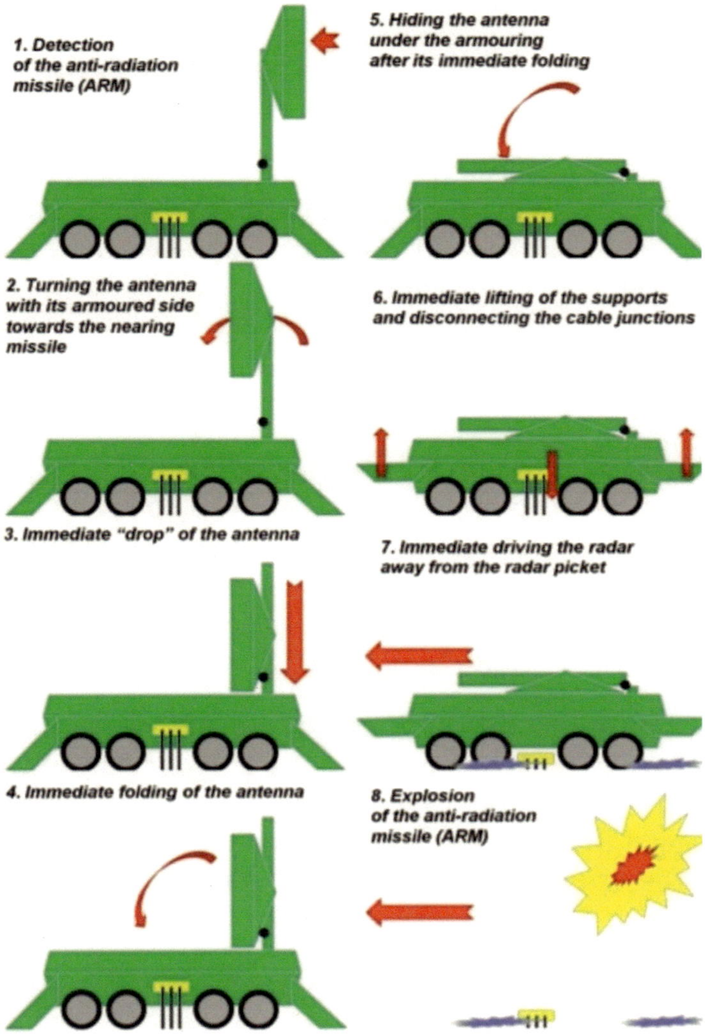

Source: Czeszejko, S. (2013). Anti – Radiation Missiles vs. Radars. In *International Journal of Electronics and Telecommunications*, 59(3), 285-291.

Jamming is an attack similar in nature to a *brute force* attack upon a network. Instead of using technology to randomly generate massive amounts of passwords or passphrases, jamming is intended to overwhelm the Data Link with RF noise or static (remember the AM radio example). Recent history demonstrates the efficacy of successful jamming. During the Russian incursion into Crimea and Ukraine separatist conflict in 2014 Russian jamming effectively kept the eyes and ears of the world from observing their activities (Hudson, 2016).

Increasing the AJ of a UAS can be achieves in multiple ways. First necessary to determine the amount of jamming radiation (noise) the Data Link can withstand before its ability to function falls below minimal acceptable levels. This is referred to as the AJ margin and is usually measured in decibels (dB) (Fahlstrom, 2012).

One technique for increasing the AJ margin U is implementing Spread Spectrum Communication.

"Spread spectrum is a means of transmission in which the signal occupies a bandwidth in excess of the minimum necessary to send the information; the band spread is accomplished by means of a code which is independent of the data, and a synchronized reception with the code at the receiver is used for dispreading and subsequent data recovery." (Pickholtz, 1982)

Other techniques to increase AJ Margin include:

- Adaptive Filtering (AF)
- Time – Frequency Domain Filtering (FDF)
- Adaptive Antennas (AA) (Iqbal, 1991).

One technique, Bilinear Signal Representations ("BSR"), also seems to hold great promise. Essentially BSR identifies jamming signals, separates them and then re-synthesizes legitimate Datalink signals (Kandangath, 2003) (Collins, 2013). See Figure 13-10 and Figure 13-11 for BSR Representations.

It is also important to remember that jamming is solely a concern in relation to the uplink, in fact preprogrammed flight instruction can still allow a successful mission when jamming successfully drowns out signal from the Ground Station to the UAV. However, if the downlink is jammed the ability for the operator to receive real time data can be diminished or disrupted thereby eliminating the flexibility for the controller to make on-the-fly adjustments or changes to the mission (Fahlstrom, 2012)

Figure 13-10 BSR Representation

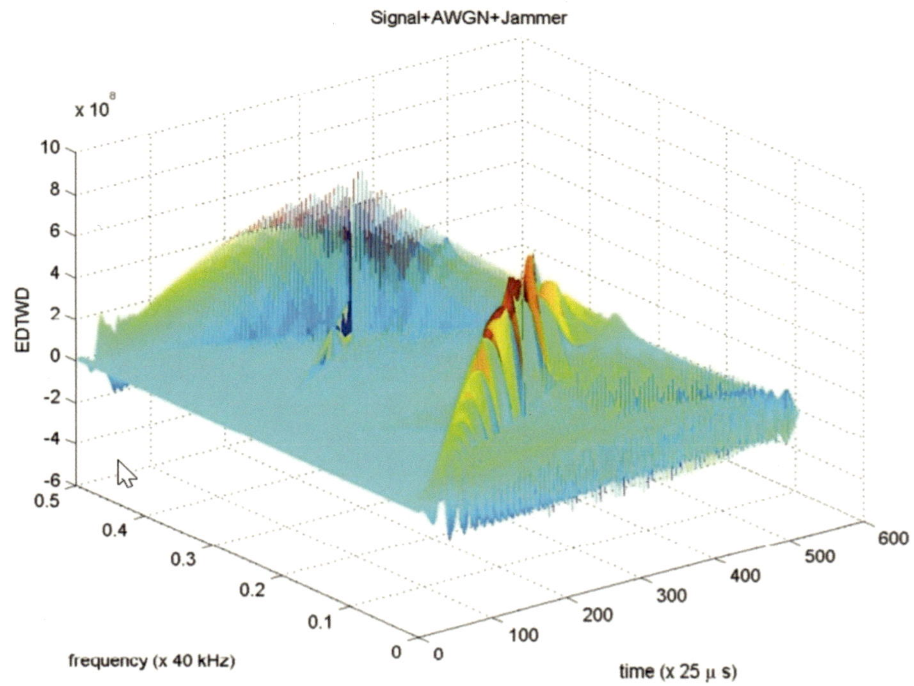

Time-Frequency representation for the received contaminated signal. The GPS signals have low power and are spread all over the region, while the narrowband jammer signals are localized and have strong peaks which makes it easy to extract the jammer.

Source: Kandangath, A. (2003). Jamming Mitigation Techniques for Spread Spectrum Communication Systems. Tempe, AZ: University of Arizona, Tech. Rep., 2003.

Figure 13-11 BSR Representation (alt)

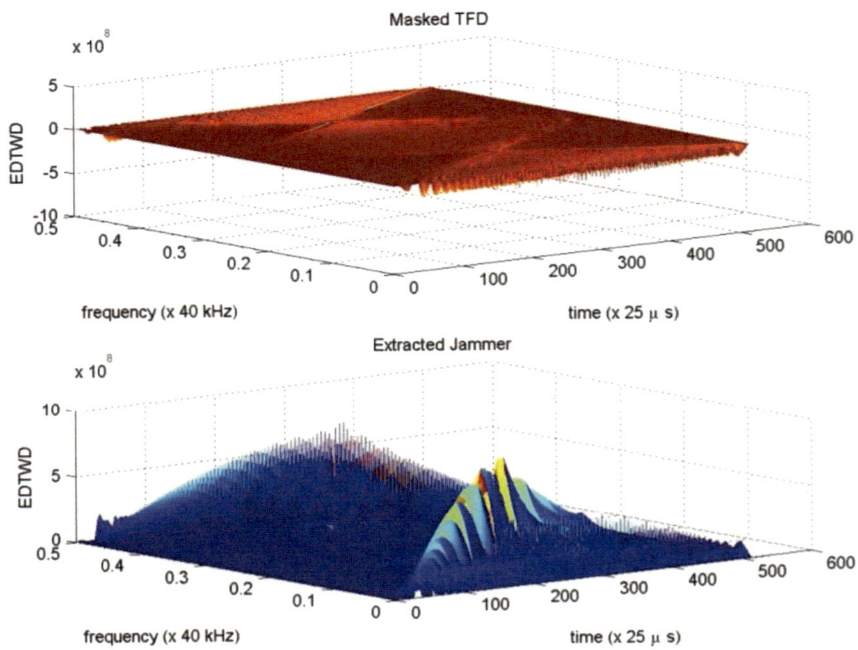

Time-Frequency representation for the masked signal and the extracted jammer. The jammer can be extracted at different threshold levels which depend on the value of α. In this example α = 3.0.

Source: Kandangath, A. (2003). Jamming Mitigation Techniques for Spread Spectrum Communication Systems. Tempe, AZ: University of Arizona, Tech. Rep., 2003

Additional Considerations

Digital vs Analog

Analog signals are a method of data transmission data which vary with time, they are inherently low-latency because they travel at the speed of light (Reid, 2017). The evolution of wireless Data Links favors digital modulation. This makes sense since most data processing activity throughout the UAS requires digital data or analog data converted to digital. Higher interference margin, ease of interfacing between components and systems support conclusion that to the extent practical, Data Link communication should favor digital data transmission(United States Marine Corps, 2015) (Fahlstrom, 2012).

One caveat to the favored use of digital data transmission. Digital transmission handles far more data much faster than analog. If successfully intercepted and decrypted an adversary can extract massive amounts of data, intelligence, intellectual property and even top-secret information. An example of such a threat occurred in 2017 when the United States Army banned the use of DJI drones, a Chinese manufacturer. (Mortimer, 2017) See Figure 13-12 US Army Warning Letter.

System Interface Considerations

Maximizing the efficacy of the Data Link requires that transmission be secure and resilient while capable of swiftly delivering accurate data. The Data Link payload must be capable of interfacing and supporting four types of critical UAS functions:

- Aircraft Control, everything but payloads and weapons
- Payload, product and control
- Weapons, kinetic and electronic
- Situational Awareness (United States Department of Defense, 2005).

Figure 13-12 US Army Warning Letter

FOR OFFICIAL USE ONLY

DEPARTMENT OF THE ARMY
OFFICE OF THE DEPUTY CHIEF OF STAFF, G-3/5/7
400 ARMY PENTAGON
WASHINGTON, DC 20310-0400

DAMO-AV 2 August 2017

MEMORANDUM FOR RECORD

SUBJECT: Discontinue Use of Dajiang Innovation (DJI) Corporation Unmanned Aircraft Systems

1. References:

a. Army Research Laboratory (ARL) report, "DJI UAS Technology Threat and User Vulnerabilities," dated 25 May 2017 (Classified).

b. Navy memorandum, "Operational Risks with Regards to DJI Family of Products," dated 24 May 2017.

2. Background: DJI Unmanned Aircraft Systems (UAS) products are the most widely used non-program of record commercial off-the-shelf UAS employed by the Army. The Army Aviation Engineering Directorate has issued over 300 separate Airworthiness Releases for DJI products in support of multiple organizations with a variety of mission sets. Due to increased awareness of cyber vulnerabilities associated with DJI products, it is directed that the U.S. Army halt use of all DJI products. This guidance applies to all DJI UAS and any system that employs DJI electrical components or software including, but not limited to, flight computers, cameras, radios, batteries, speed controllers, GPS units, handheld control stations, or devices with DJI software applications installed.

3. Direction: Cease all use, uninstall all DJI applications, remove all batteries/storage media from devices, and secure equipment for follow on direction.

4. Point of Contact: Headquarters, Department of the Army G-3/5/7 Aviation Directorate, 703-693-3552.

Source: Scott, A. (2017, August 4). U.S. Army halts use of Chinese-made drones over cyber concerns. Reuters. AND Mortimer, S. (2017, August 4). US Army calls for units to discontinue use of DJI equipment. sUAS News.

Data Link delivery of secure, high-speed data allows electronic systems to interface seamlessly with mechanical system must also be capable of concurrent self-monitoring and agile reaction to attacks. This attribute allows accurate communication between the ADT, ground station and entire UAS network (Fahlstrom, 2012).

Lastly a cautionary note. Meeting data requirements may negatively impact AJ margin and deception resistance which can be far more difficult and expensive to remedy after the original design process (Saeedipour, 2005). Simply because UAS design deals largely with the cyber realm, should not be taken as license to ignore brick and mortar reality. Specifically, Newton's Third Law, "for every action, there is an equal and opposite reaction" takes on new significance when considering which Data Link attribute should be emphasized and which may be subject to trade-off (Sunil, 2008).

Data-Rate

Data rate is a vital since the pilot sends commands through the uplink to the UAV the result of the command execution are confirmed by onboard sensors, video or other indicators to confirm execution. If the data rate of the downlink is not sufficiently robust latency comes into play. In NLOS configurations, the execution data is sent by the ADT to a satellite and then relayed via downlink to reach the operators eyes. Inadequate data rate can cause many unwanted consequences from duplicate commands, expired intelligence, total failure of the mission or loss of the UAV.

The data lifecycle of a UAS underscores the imperative of data rates capable of handling vast amounts of data during a UAV mission. As the figure below depicts just a UAS functions demonstrate the importance of rapid data transmission, as well as high speed onboard and ground-based data processing. See Figure 13-12 Data Lifecycle of UAS.

Figure 13-13 Data Lifecycle of UAS

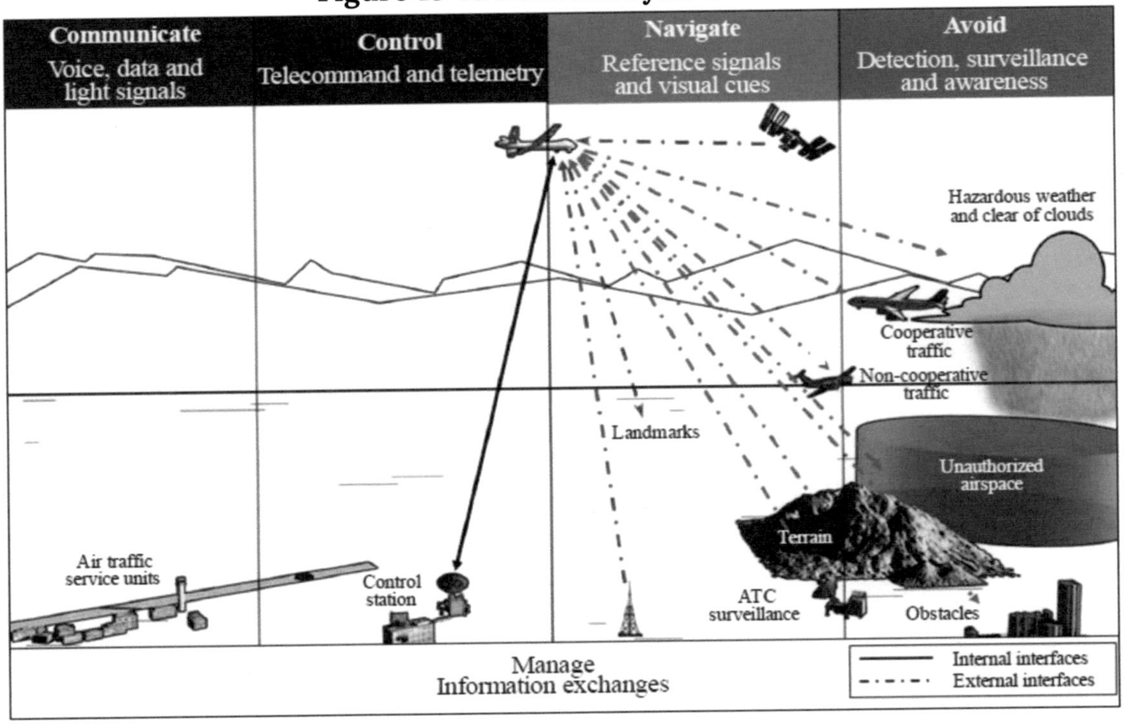

Source: International Telecommunication Union. (2009). Characteristics of unmanned aircraft systems and spectrum requirements to support their safe operation in non-segregated airspace. Geneva, Switzerland: International Telecommunication Union.

It is important to note the figure above only depicts the en route portion of a mission. In actuality the bi-directional data flow commences with pre-flight communication and instruction, continues through departure, en route, arrival and post-flight. Coupled with payload, handover, contingency, time stamping and emergency contingency planning to the data flow to these phases it becomes readily apparent that data rate is a seminal issue in the UAS design process. (International Telecommunication Union, 2009)

Finally the exponential growth of UAV's in service globally, consideration of the finite supply of bandwidth and how to best account for the available bandwidth will most certainly involve attribute trade-offs. Difficult choices will need to be made regarding bandwidth availability as it relates to specific missions, geography and available wireless technology. Accordingly, instead of designing regional as opposed to globally adaptable UAS, best practice as well sound budgetary policy dictates that interchangeability and interoperability be prioritized.

Closed Loop Control

Certain aspects of UAS design can benefit from employ using loop control between a UAV and the same or another location. While it is possible to pre-program the UAV recovery process, human intervention in this phase of the mission will ensure an agility in adapting to changing physical, environmental or threat conditions in the landing zone.

Given the highly variable conditions in the landing and retrieval zone, it is often the case that a separate closed loop Data Link, between the UAV and pilot can enhance safely and effectively retrieving the UAV. Closed loop Data Links require a additional reception and processing capability onboard both the UAV and ground station.

The threats to closed loop control systems are no different than in the main Data Link. Latency, restricted bandwidth, AJ, deception, and line of sight are important considerations. Although discussed latency later in the chapter when it comes to closed loop control, especially video transmission of approach and recovery, the margin of error becomes smaller with declining altitude, limited runway length, environmental and physical conditions in the landing zone. (Fahlstrom, 2012)

Imagine operating a video game or flight simulator on a desktop or gaming console and the visual representation presented on the screen was actually delayed by 2 seconds (ie. Aircraft appears to be 1 mile away from target when it is actually .6 miles away from the target). A command to reduce power and altitude on a one mile glidepath subject to a 2 second delay becomes far more challenging. Suffice it to say the game would not last very long, nor would it be very successful. See Figure 13-14 Flight Simulation Game.

Figure 13-14 Flight Simulation Game

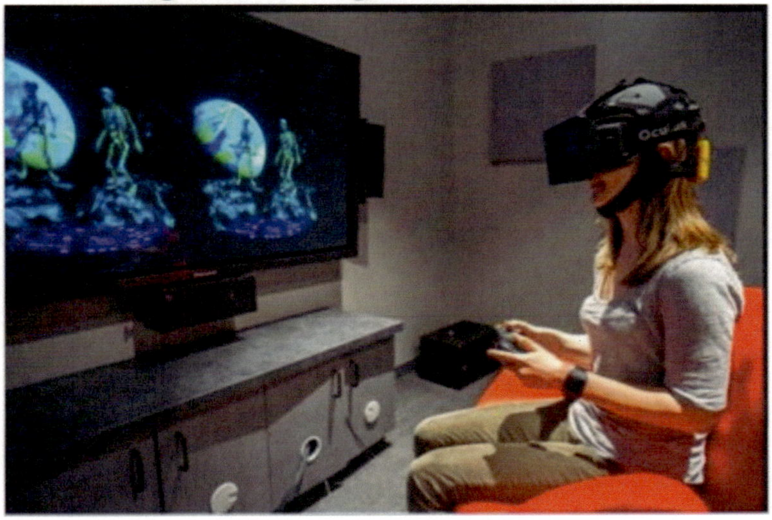

Source: Neuroscape. (Summer, 2018). Technology: Bridging the gap between neuroscience and technology. https://neuroscape.ucsf.edu/technology/

Interchangability, Interoperability and Standarization

At first glance it appears Interchangeability and Interoperability (I & I) are the same thing. In reality they are distinct concepts which require separate analysis and consideration of the UAS design process (Yu, 2016).

Some, such as Fahlstrom consider I & I separate and distinct attributes, writing "interoperability would mean that an ADT from one data link could communicate with a Ground Data Terminal ("GDT") of another and vice versa." (Fahlstrom, 2012) p. 201-202. While others such as Yu, seem to blend I & I into one concept citing the 1999 Joint Chiefs of Staff definition is, "The ability of systems, units or forces to provide services from other systems, units or forces and use the services so exchanged to enable them to operate effectively together." (United States Department of Defense, Joint Chiefs of Staff, 2016) (Yu, 2016).

Finally, standardization must be considered from the moment of initial design of the UAS. If the system is to have a sufficient service life thereby justifying the expense of its development, standardization with the myriad of other systems is essential.

Standardization is a set of "requirements, specifications, guidelines, or characteristics that can be used consistently to ensure that materials, products, processes and services are fit for their purpose." (Standardization, 2017) In 1998 NATO began to consider methods by which technology can be built with uniform standards and interoperability. As unmanned warfare accelerated towards the end of the last century, NATO began to develop protocols to enhance and ensure standardization between member nations and their technology.

STANAG 4856 – Standard Interfaces of UAV Control System (UCS) for NATO UAV interoperability was created in order to maximize communication between, "different UAV's and their payloads, as well as different Command, Control, Communication, Computers and Intelligence systems. The integration of components from different sources as well as the interoperability of legacy systems." (Marques, 2017) Designers must not just consider the reality that interoperability between allied UAV's, they must also consider the challenge of maximizing interoperability with other assets. Figure 13-15 JTIDS view below demonstrates just some of the myriad of systems which must to be able to operate and communicate, in real time, as seamlessly as possible.

Figure 13-15 JTIDS View

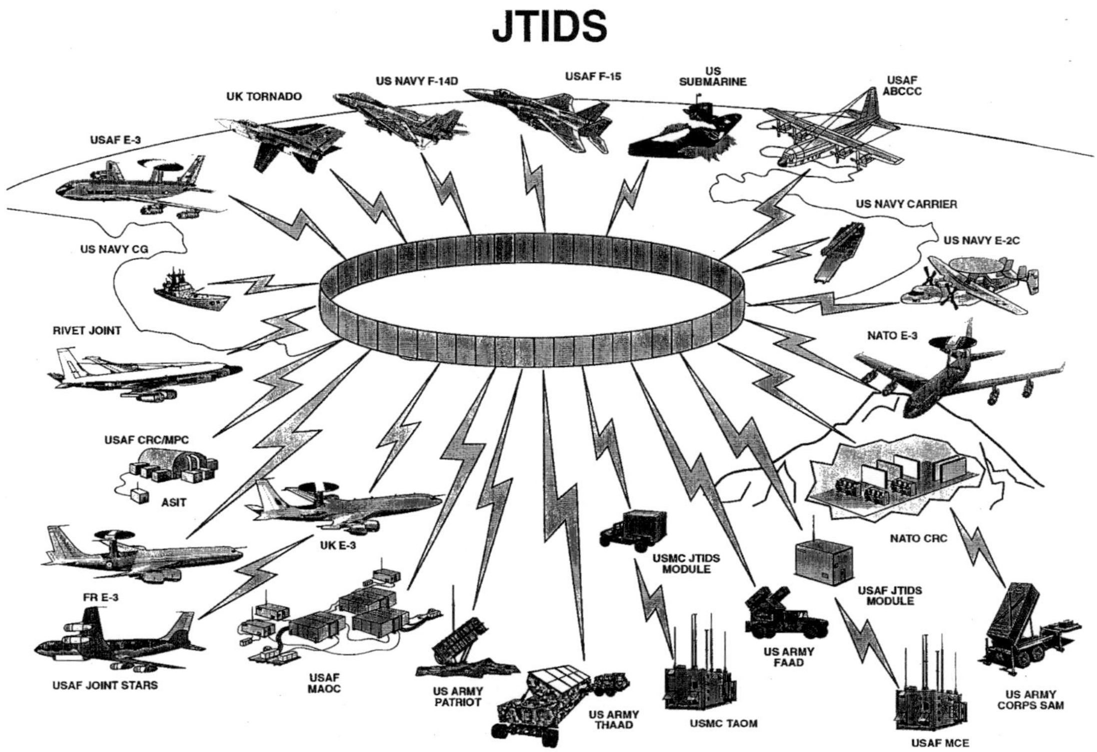

Source: GlobalSecurity.org. (Summer, 2018).Joint Tactical Information Distribution System (JTIDS). https://www.globalsecurity.org/military/systems/ground/jtids.htm

The challenge to designing Interoperability and Standardization (I&S) in UAS design is multifold with the battlefield attributes of central command, communication and coordination taking center stage. The ability of allies to communicate is vital for cohesive, coordinated and effective operations.

Recognizing the importance of I&S, the Department of Defense is developing a Joint Architecture for Unmanned Systems, ("JAUS"), National Information Exchange Model ("NIEM") and most recently Future Airborne Capability Environment (FACE). The objective of FACE is to develop a "Technical Standard for a software capability designed to promote portability, and create software product lines across the military aviation community." (Blais, 2016).

Datalink Latency

The Current Environment

Latency is defined as the interval between the time when data is processed and a signal is transmitted and when the signal is received and then processed in order to be displayed and interpreted by the operator. To help understand latency ask yourself why do we see lightning before we hear thunder? See Figure 13-16 Lightning Strike and Latency.

Light is an electromagnetic wave which needs no through which to travel. It moves at 300,000,000 meters per second, so a lighting strike in your general vicinity is seen by the naked eye almost immediately. On the other hand thunder is sound, a mechanical wave which needs to travel through a medium, in this case air molecules. Sound travel through air at a speed of 340 meters per second which is approximately one million times slower than the speed of light. Hence lightning will be seen before thunder is heard due the latency of the transmission of sound as opposed to light (Park, 1997).

Figure 13-16 Lightning Strike and Latency

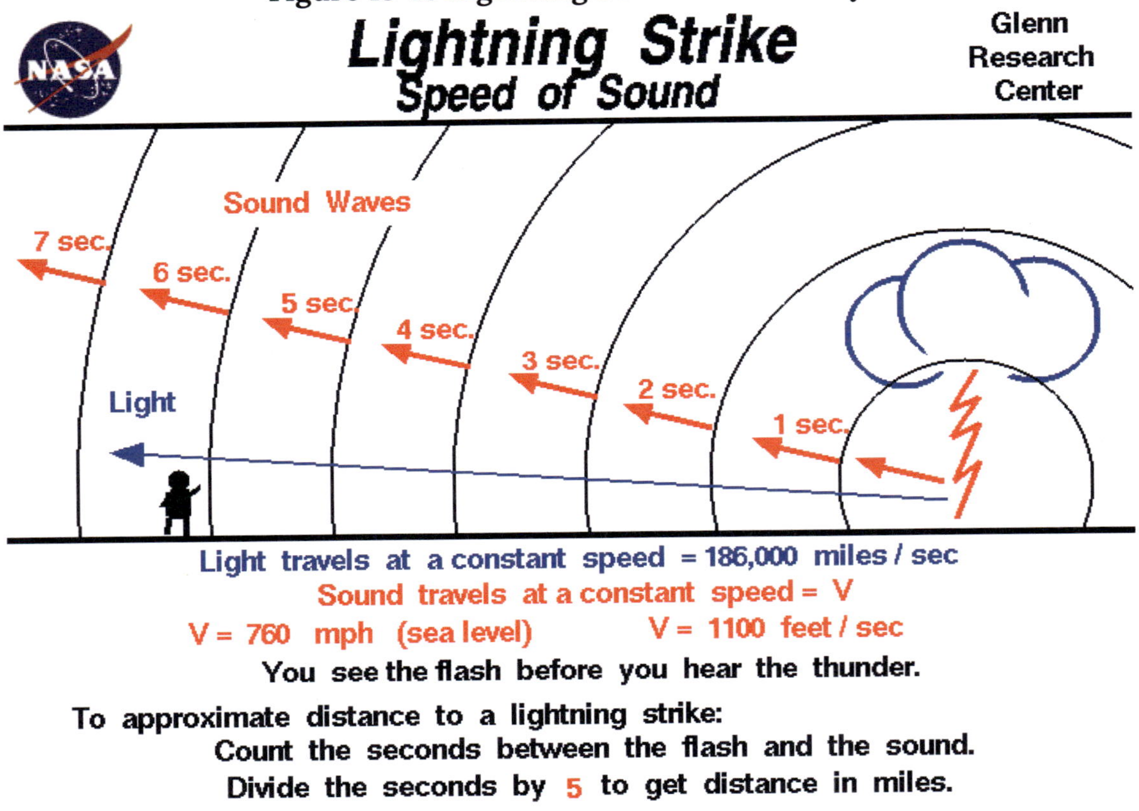

Source: NASA, Glenn Research Center. (Summer, 2018). Speed of Sound. https://www.grc.nasa.gov/www/k-12/airplane/sound.html

Returning to the UAS datalink, line of sight visual transmission and reception is ordinarily nearly instantaneous because of the speed of light. Unfortunately line of sight, environment and distance significantly limit UAV operational radius. Even LOS operation of a UAV can be hindered by weather, vegetation, time of day and topography. Air and ground traffic density become more problematic with greater latency (International Telecommunication Union, 2009).

According to some estimates domestic airspace must be able to accommodate up to 10 million combined air vehicles a day by 2035. When you consider that in 2015 that in US airspace was traversed 50,000 times per day such an exponential increase will be problematic (Atkinson, 2015).

Flight Control Technology

Presently there are 3 classes LOS UAV operation:

- Low Endurance
- Medium Endurance; and
- High Endurance (Valavanis, 2013).

Low Endurance:

Operate almost exclusively in line of sight with a minimum of automated onboard flight control technology. Usually LOS UAS employ C Band frequency with low frequency of between 3.7-4.2 GHz for the downlink and 5.9-6.4 for the uplink. One of the main reasons for C Band datalinks in Low Endurance UAS is that low frequency signals are less susceptible to weather related degradation.

Medium Endurance

Operate in primarily in LOS applications however some do have Beyond Line of Sight ("BLOS") capability. To the extent they are operating in LOS missions lower frequency C Band is used, if the Medium Endurance UAS is deployed BLOS then they usually will usually be operating on Ultra High Frequency ("UHF") (300 MHz) to Ku Band (15 GHz). The downlink is between 11.7 – 12.7 GHz and the uplink between 14-14.5 GHz.

High Endurance

High Endurance UAS deployed BLOS operate on Ultra High Frequency ("UHF") (300 MHz) to Ku Band (15 GHz). Downlink frequency is between 11.7 – 12.7 GHz and uplink between 14-14.5 GHz. High Endurance UAS may also employ Common Data Link ("CDL") technology on either I-band satellite communication ("SATCOM") or KU band between 14.5 and 15.38 GHz. To minimize SATCOM latency autopilot is often favored in LOS RF operations (Valavanis, 2013).

The trade-off between security and latency is a difficult balance to achieve. US Air Force Major General James Poss. (Ret.) put it best.

"Even the Air Force struggles with this problem because the more secure they make their links, the more control latency and potential for link loss they introduce. At least with today's technology. Air Force Predator pilots routinely deal with a five to eight second delay on their controls when flying BLOS due to encryption overhead and the time it takes to relay commands via satellites" (Poss, 2017). Consult Figure 13-17 Security – Latency Trade-off.

While there are options to minimize latency while balancing the attributes needed for a robust and effective UAS Data Link, latency is a constant in any transmission travelling any distance. The greater the distance, the more time to travel. The greater the data payload the more time to transmit.

Innovative technologies designed to address latency in UAS Data Links are presently being studied. One that may hold much promise is the introduction and continued development Artificial Intelligence in UAS design. To be sure the less data that needs to be sent to the UAV on the uplink or back to the ground station on the downlink, the less latency in executing commands and near-real time UAV control. (Bennis, 2018) AI holds much promise but must also be considered a risk. Just as a UAV pilot could become incapacitated or go rogue, the same risks exist with AI implementation in UAS.

Figure 13-17 Security – Latency Trade-off

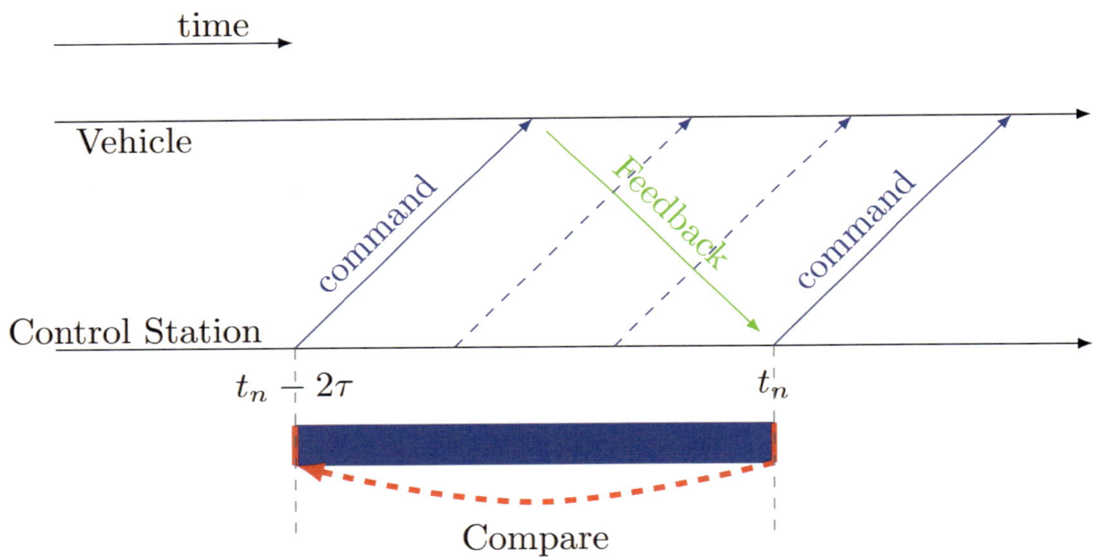

Source: UAV Research Lab at the University of Sydney. (2018). Adapting UAV Control for Latency. UAV – Lab.

Discussion Questions

1. Since so many attributes of a UAS have distinct levels of importance depending on environment, mission and payload, how can the designer create a system that is agile, and globally deployable, thereby increasing service life and efficacy while still being sensitive to the cost of development?

2. How would you rate each of the attributes discussed in this chapter and are there other attributes you think need consideration?
3. If UAS countermeasures continue to evolve, what is the best method of ensuring that current attributes and functions remain effective?
4. Bearing question 3 in mind, at what point does cost of development exceed the value service life of a UAS? How would such an analysis be made and what factors would you consider relevant?

References

Atkinson, N. (2015, September 7). Designing a way to keep increasingly crowded airspace safe. *phys.org*.

Bennis, M. D. (2018). *Ultra- Reliable and Low-Latency Wireless Communication: Tail, Risk, Scale.* arXil.org.

Blais, C. L. (2016). *Unmanned systems interoperability standards.* Monterey CA: The Naval Postgraduate School.

Cannon Corporation. (2017). Shielding against electromagnetic and RF interference for safety and mission success. *Military and Aerospace Electronics*, 1.

Chen, J. (2014). MIMO *Enhancements for Air-to-Ground Wireless Communications.* Los Angeles: UCLA Electronic Theses and Dissertations.

Collins, T. F. (2013). *Implementation and Analysis of Spectral Subtraction and Signal.* Worcester, MA: Worcester Polytechnic Institute.

Congressional ResearchService. (2018). *Artificial Intelligence and National Security.* Washington, DC: Congressional Research Service.

Czeszejko, S. (2013). Anti – Radiation Missiles vs. Radars. *International Journal of Electronics and Telecommunications*, 285-291.

Droneshield . (2017). Drone Defence: Jammers 101. 1.

Fahlstrom, P. G. (2012). Data – Link Functions and Attributes. In P. G. Fahlstrom, *Introduction to UAV Systems, Fourth Edition* (p. 193). John Wiley & Sons, Ltd.

Giles. (2013). *Sun Tzu On The Art of War.* Abingdon, Oxon: Routledge.

Hartman, K. &. (2013). The Vulnerability of UAV's to Cyber Attacks – An Approach to the Risk Assessment. *5th International Conference on Cyber Conflict.* Tallin: NATO CCD COE Publications.

Howarth, F. (2014). *The Role of Human Error in Successful Security Attacks.* Armonk, NY: IBM – Security Intelligence.

Hudson, J. (2016, October 28). International Monitor Quietly Drops Drone Surveillance of Ukraine War. *Foreign Policy.*

International Telecommunication Union. (2009). *Characteristics of unmanned aircraft systems and spectrum requirements to support their safe operation in non-segregated airspace.* Geneva, Switzerland: International Telecommunicaton Union.

Iqbal, R. B. (1991). Performance Analysis of Interference Rejection Techniques in Spread Spectrum Communication. *TENCON 91' IEEE Region 10 International Conference on EC3-Energy, Computer, Communication and Control Systems*, Vol. 3. IEEE.

Jain, R. T. (2017). Wireless Datalink for Unmanned Aircraft Systems: Requirements, Challenges and Design Ideas. *American Institute of Aeronautics and Astronautics*, 2.

Jang, C. e. (2017). Taking Drones To The Next Level – Cooperative Distributed Unmanned – Aerial- Vehicular Networks for Small Drones and Mini Drones. *IEEE Vehicular Technology Magazine, Volume 12, Issue 3*, pp. 73-82.

Kakar, J. M. (2017). *Wavefrom and Spectrum Management for Unmanned Ariel Systems Beyond* 2025. Ithaca, New York: arXIv.org, Cornell University.

Kandangath, A. (2003). *Jamming Mitigation Techniques for Spread Spectrum Communication Systems.* Tempe, AZ: University of Arizona, Tech. Rep., 2003.

Keller, J. (2016, August 1). Cybersecurity and encryption for the masses. *Military and Aerospace Electronics.*

Marques, M. M. (2017). STANAG 4586 – *Standard Interfaces of* UAV *Control System* (UCS) *for* NATO UAV *Interoperability.* Brussels, Belgium: North Atlantic Treaty Organization .

Mortimer, G. (2017, August 4). US Army calls for units to discontinue use of DJI equipment. sUAS *News.*

Nuriev, N. G. (2017). Physical modeling of electromagnetic interferences in the unmanned aerial vehicle in the case of high-voltage transmission line impact. *Russian Aeronautics*, 292-293.

Okcu, H. (2016). Operational Requirements of Unmanned Aircraft Systems. *Journal of Advances in Computer Networks*, Vol. 4, No. 1, 28-30.

Opall-Rome, B. (2018, February 12). Israel Air Force says seized Iranian drone is a knockoff of US Sentinel. *Defense News.*

Park, C. C. (1997). *The Environment: Principles and Applications.* London, UK: Routledge.

Pickholtz, R. L. (1982, May). Theory of spread spectrum communications – a tutorial. *IEEE Transactions on Communications*, Vol. 30 No. 5, pp. 855-884.

Poss, M. G. (2017, February 22). It's the Data Link Stupid. *inside Unmanned Systems.*

Psiaki, M. L. (2013, June 1). Innovation: GNSS Spoofing Detection. *GPS World.*

Reid, J. (2017, November 30). The Difference Between Analog and HD (Digital) Transmission. *Rotor Drone Magazine.*

Rodday, n. (2015). *Exploring Security Vulnerabilities of Unmanned Aerial Vehicles.* Amsterdam: University of Twente.

Saeedipour, H. R. (2005). Data Link Functions and Attributes of an Unmanned Aerial Vehicle IUAV) System Using Both Ground Station And Small Satellite. *5th IAA Symposium on Samm Satellites for Earth Observation.* Berlin, Germany: International Association of Astronautics.

Schneier, B. (2000). *Secrets and Lies, Digital Security in a Networked World.* Hoboken, NJ: John Wiley and Sos, Ltd.

Scott, A. (2017, August 4). U.S. Army halts use of Chinese-made drones over cyber concerns. *Reuters.*

Standardization, I. O. (2017). "Standard". http://www.iso.org/iso/home/standards.htm.

Sunil, K. S. (2008). *Newton's third law of motion.* Houston, TX: Rice University.

UAV Research Lab at the University of Sydney. (2018). Adapting UAV Control for Latency. *UAV – Lab.*

United States Department of Defense. (2005). *Unmanned Aircraft Systems Roadmap 2005-2030.* Washington, DC: Office of The Secretary of Defense.

United States Department of Defense, Joint Chiefs of Staff. (2016). *Department of Defense Dictionary of Military and Associated Terms – Joint Publication 1-02.* Washington, DC: United States Department of Defense.

United States Marine Corps. (2015). *Unmanned Aircraft Systems Operations.* Washington, DC: Department Of The Navy.

Valavanis, K. P. (2013). *Unmanned Aircraft Systems: the Current State-of-the-Art.* New York: Springer.

Yajnanarayana, V. W.-P. (2018). *Interference Mitigation Methods for Unmanned Aerial Vehicles Served by Cellular Networks.* https://arxiv.org/pdf/1802.00223.pdf

Yu, A. M. (2016). *Unmanned Aircraft Systems.* Hoboken, New Jersey: John Wiley and Sons.

Chapter 14: Exposing UAS Vulnerabilities via Electronic Warfare (EW) and Countering with Low Probability Intercept Signals (LPI)

Student Learning Objectives

In previous chapters, we have studied the EMS, Data-links and cyber-vulnerabilities of UAS. This chapter introduces electronic warfare as a method of overwhelming, destroying, or controlling the information, transmitted by communication data-links, to alter the mission of the UAS deployment.

Modern Communication Threats to UAS

Unmanned Aerial Systems (UAS) are in widespread use for reconnaissance, EW, and weapons delivery. **They are extremely dependent on interconnection with ground stations by command and data links.** (Adamy D. , 2001) **The increased use of Low Probability Intercept (LPI)** [Introduced in Chapter 13 *Data – Links Functions, Attributes, and Latency*] **has become a significant challenge to electronic warfare (EW) communication links.** (Adamy D., 2001) This chapter explores LPI and Jamming, after a circuitous route through Intelligence / Information Operations (IO), followed by EW. The student should then have enough background to understand the criticality of LPI and Jamming of UAS communication links. Air defense missiles and associated radars make significant use of interconnecting links. (Adamy D. , 2001) SUAS sometimes use cellphones to command and control the UAVs. Cell phones are widely used for command and control function in nonsymmetrical warfare situations. (Adamy D. , 2001) ISIS and other terrorist groups use cell phones to trigger improvised explosive devices.

Cybersecurity attacks on data communications links are highly classified. Similarly, modern radar threats to hostile installations are also generally classified. So, in this chapter, description of EW techniques is generalized; no classified information. This way if the student must apply EW in real-world situations, they can plug in the parameters learned from classified sources. Before examining LPI and communications signals/link- jamming, we first review the EW environment specific to UAS.

Definitions

Nichols (2000) defines Cybersecurity in terms of cyber-conflict. (Nichols, 2008) Alford (2000) authored effective definitions for the DoD. These will illustrate the bigger picture of *Information Operations* (IO) and the subset known as Electronic Warfare (EW).

Cybersecurity (in the context of Cyber conflict) is defined as, "**the broad *tree* of investigation and practice devoted to *cybercrimes*, *Computer Forensics* (CF), *Information Assurance* (IA), *Information Security* (INFOSEC), *Communications Security* (COMSEC), *and especially Cyber Counter Intelligence* (CCI).**" (Nichols, 2008)[1]

"**Cyber Warfare (CW / CyW).** Any act intended to compel an opponent to fulfill our national will, executed against the software controlling processes within an opponent's system. CyW includes the following modes of cyber-attack; cyber infiltration, cyber manipulation, Cyber assault, and cyber raid." (DAU, 2018) (DAU, 2018)

"**Cyber Infiltration (CI / CyI).** Penetration of the defenses of a software-controlled system such that the system can be compromised, disabled, manipulated, assaulted, or raided." (DAU, 2018) (DoD, 2018)

"**Cyber Manipulation (CM / CyM).** Following infiltration, the control of a system via its software which leaves the system intact, then uses the capabilities of the system to do damage.

For example, using an electric utility's software to turn off power." (DAU, 2018) (DoD, 2018)

"**Cyber Assault (CA / CyA).** Following infiltration, the destruction of software and data in the system, or attack that compromises system capabilities." (Alford, 2000) Includes viruses and system overloads via e-mail (e-mail overflow)." (DoD, 2018; DoD, 2018)

"**Cyber Raid (CR / CyR).** Following infiltration, the manipulation or acquisition of data within

1. Intelligence involves computer-based sensitive information, or information operations for three distinct sciences operating in the cyber realm: Cyber Counter Sabotage (CCS), Cyber Counter Terrorism (CCT), and Cyber Counter Espionage (CCE). In the UAS – Cybersecurity Certificate offered at KSU, Cybersecurity focuses on the prior three investigation areas. Additional concerns are 1) protection of UAS/ UAV/ Drones from cyber-attacks by negligent and hostile means and 2) teaching cybersecurity risk assessment principles to practitioners involved with UAS operations on land, sea, air, or satellite platforms. The impact of Loss of Signal (LOS) or intentional interference in UAS communications or navigation systems cannot be overstated. At the lowest end of the scale is the risk of a downed vehicle, mid-range risk is collision and failure to sense and avoid other vehicles in commercial/military traffic, and at the top of the risk scale is the hostile takeover of a payload to be used against the U.S. or U.S. interests. It is not "good enough" to operate, fly or support UASs; professionals must also be concerned with protection of their charges.

the system, which leaves the system intact, results in transfer, destruction, or alteration of data. For example, stealing e-mail or taking password lists from a mail server." (DAU, 2018) (DoD, 2018)

Cyber-Attack. See CyI, CyM, CyA, or CyR.

Cybercrime (CC / CyC). Cyber-attacks without the intent to affect national security or to further operations against national security." (Alford, 2000)

"C^4ISR. The concept of Command, Control, Communications, Computers, Intelligence, Surveillance and Reconnaissance." (DoD, 2018) (Kaye, 2001) See Figure 14-XXX (C4ISystems, 2013)

Electronic Warfare (EW) is defined as the art and science of preserving the use of the **Electromagnetic Spectrum (EMS)** for friendly use, while denying its use by the enemy. (Adamy D. , 2001)

"Information Assurance (IA). Measures that protect and defend information and information systems by ensuring their availability, integrity, authentication, confidentiality, and non-repudiation. These measures include providing for restoration of information systems by incorporating protection, detection, and reaction capabilities." (Barker, 2003) (Kaye, 2001)

"Information Operations (IO). The integrated employment of the core capabilities of electronic warfare, computer network operations, psychological operations, military deception, and operations security, in concert with specified supporting and related capabilities, to influence, disrupt, corrupt, or usurp adversarial human and automated decision-making process, information, and information systems while protecting our own." (Barker, 2003) (Kaye, 2001)

"Information Superiority (IS). The capability to collect, process, and *disseminate an uninterrupted flow of information while exploiting or denying an adversary's ability to do the same.* A *newer form of this is that: degree of dominance in the information domain which permits the conduct of operations without effective opposition.*" (Alford, 2000) (Kaye, 2001)

"Information Warfare (IW). Information operations conducted during time of crisis or conflict to achieve or promote specific objectives over a specific adversary. IW is any action to Deny, Exploit, Corrupt or Destroy the enemy's information and its functions, protecting those actions and exploiting our own military information functions." (Alford, 2000) (Kaye, 2001)

"Intentional Cyber Warfare Attack (ICWA). any attack through cyber-means to intentionally affect national security (cyber warfare) or to further operations against national security.

Includes cyber-attacks by unintentional actors prompted by intentional actors. (Also see "unintentional cyber warfare attack.") IA can be equated to warfare; it is national policy at the level of warfare. Unintentional Attack(UA) is basically crime. UA may be committed by a bungling hacker or a professional cybercriminal, but the intent is self-serving and not to further a national objective. This does not mean unintentional attacks cannot affect policy or have devastating effects.

Intentional Cyber Actors (I-actors). Individuals intentionally prosecuting cyber warfare (cyber operators, cyber troops, cyber warriors, cyber forces)." (Alford, 2000)

"Network Centric Operations (NCO). NCO involves the development and employment of mission critical packages that are the embodiment of the tenets of Network Centric Warfare (NCW) in operations across the full mission spectrum. These tenets state that a robustly networked force improves information sharing and collaboration, which enhances the quality of information, the quality of awareness, and improves shared situational awareness. This results in enhanced collaboration and enables self-synchronization improving sustainability and increasing speed of command, which ultimately result in dramatically increased mission effectiveness. (Kaye, 2001)" (MORS, 2018) (Kaye, 2001)

OPSEC. (Operations Security) (DoD-01, 2018) "Determining what information is publicly available in the normal course of operations that can be used by a competitor or enemy to its advantage. OPSEC is a common military practice that is also applied to civilian projects such as the development of new products and technologies.

OPSEC – The Official Definition

(From JP 1-02, Department of Defense Dictionary of Military and Associated Terms, www.dtic.mil/doctrine/new_pubs/jp1_02.pdf.) OPerations SECurity (OPSEC) is a process of identifying critical information and subsequently analyzing friendly actions attendant to military operations and other activities to:

1. Identify those operations that can be observed by adversary intelligence systems,
2. Determine what indicators adversary intelligence systems might obtain that could be interpreted or pieced together to derive critical information in time to be useful to adversaries, and
3. Select and execute measures that eliminate or reduce to an acceptable level the vulnerabilities of friendly actions to adversary exploitation." (DoD-01, 2018)

"Psychological Operations (PO) Planned operations to convey selected information and indicators to foreign audiences to influence their emotions, motives, objective reasoning, and ultimately the behavior of foreign entities." (Alford, 2000) (Kaye, 2001)

"Psychological Warfare (PW / PSYWAR) The planned use of propaganda and other psychological actions to influence the opinions, emotions, attitudes and behavior of hostile foreign groups." (Kaye, 2001)

"Unintentional Cyber Actors (U-actors). Individuals who unintentionally attack, but affect national security and are largely unaware of the international ramifications of their actions. Unintentional actors may be influenced by I-actors, but are unaware they are being manipulated to participate in cyber operations. U-actors include anyone who commits CyI, CyM, CyA, and CyR without the intent to affect national security, or to further operations against national security. This group also includes individuals involved in CyC, journalists, and industrial spies. The threat of journalists and industrial spies against systems including unintentional attacks caused by their CyI efforts should be considered high.

Unintentional Cyber Warfare Attack (UCWA/ UA). Any attack through cyber-means, without the intent to affect national security (cybercrime)." (Alford, 2000)

Information Operations (IO) and the part EW plays

Figure 14-1 shows the global view of Information operations. Note how nicely all the prior definitions fit into the puzzle? Note that EW is a key component of IO, but not the singular dominant puzzle piece. EW "boxes" will be discussed under the EW section below.

Figure 14-2 shows the DoDs JOPES (Joint Operation Planning and Execution System) plan for 2025. In this plan, Information Operations (IO) are integrated across the military commands and equipment / knowledge / software is supplied by both the military and commercial businesses. (DoD-02, 2018)

Figure 14-3 is a look at the DoD Vision of the future of the Internet where IO and IW are *gamed* into a fully integrated global network system. (Merrick, 2016) All three figures above show how information operations encompasses all the information collection and action activities of the intelligence, defense, military, computer technology, and civilian organizations for common missions of cybersecurity and information superiority. What is not so obvious is that UAS systems are an integral part of IO planning and actions. Further, global investment in the UAS market are growing exponentially and the leader in development is China, not the USA. Much of China's investment billions are for PLAN which is hostile to USA interests. In Chapters 15 and 16, we address China's "Grand Strategy" in Africa (a UAS testing ground for several powers) and the Spratly Islands (Intelligence center in the South China Seas).

Figure 14-1 Information Operations

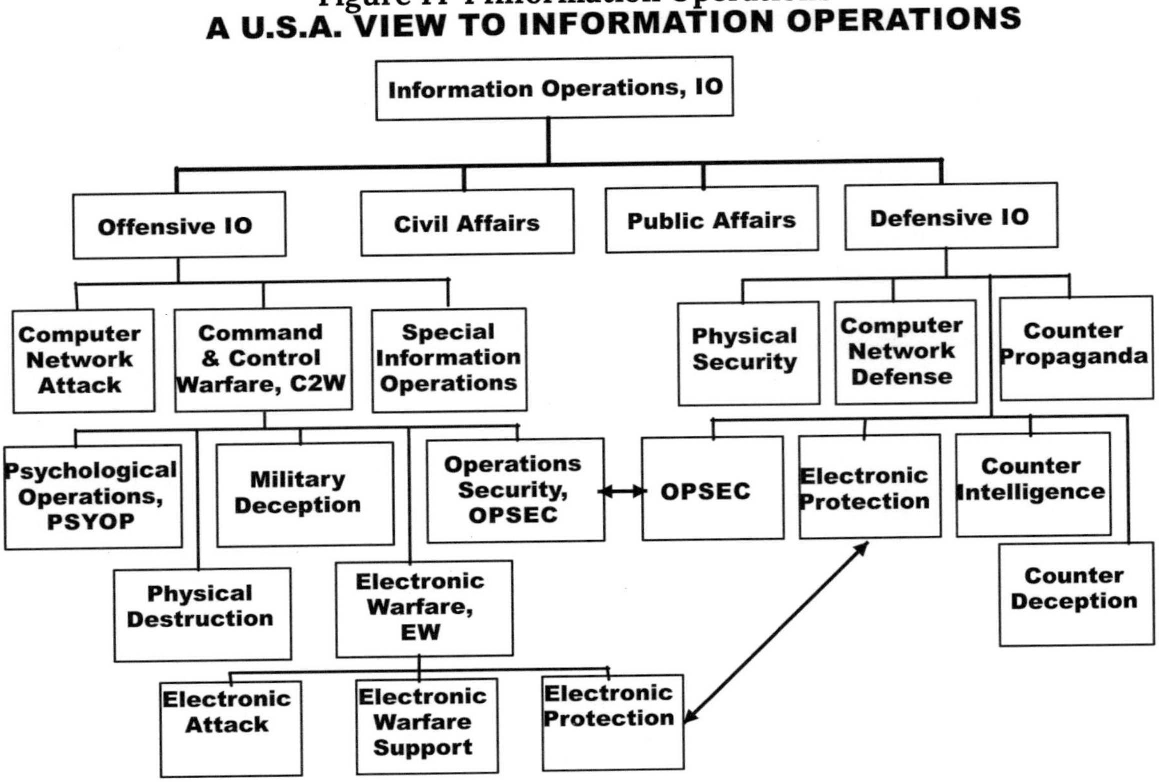

Source: C4ISystems. (2013). Basics-of-information-operations. Retrieved from http://c4isys.blogspot.com/2013/11/basics-of-information-operations-24.html

Figure 14- 2 DoD JOPES

INFORMATION OPERATIONS INTEGRATION INTO JOINT OPERATIONS (NOTIONAL)

Core, Supporting, Related Information Activities	Activities	Audience/ Target	Objective	Information Quality	Primary Planning/ Integration Process	Who does it?
Electronic Warfare	Electronic Attack	Physical, Informational	Destroy, Disrupt, Delay	Usability	Joint Operation Planning and Execution System (JOPES)/ Targeting Process	Individuals, Governments, Militaries
	Electronic Protection	Physical	Protect the Use of Electro-magnetic Spectrum	Security	JOPES/Defense Planning	Individuals, Businesses, Governments, Militaries
	Electronic Warfare Support	Physical	Identify and Locate Threats	Usability	Joint Intelligence Preparation of the Battlespace(JIPB)/SIGINT Collection	Militaries
Computer Network Operations	Computer Network Attack	Physical, Informational	Destroy, Disrupt, Delay	Security	JIPB/JOPES/Targeting Process	Individuals, Governments, Militaries
	Computer Network Defense	Physical, Informational	Protect Computer Networks	Security	JOPES/J-6 Vulnerability Analysis	Individuals, Businesses, Governments, Militaries
	Computer Network Exploitation	Informational	Gain Information From and About Computers and Computer Networks	Security	JIPB/Targeting Process	Individuals, Governments, Militaries
Psychological Operations	Psychological Operations	Cognitive	Influence	Relevance	JOPES/Joint Operation Planning	Businesses, Governments, Militaries
Military Deception	Military Deception	Cognitive	Mislead	Accuracy	JOPES/Joint Operation Planning	Militaries
Operations Security	Operations Security	Cognitive	Deny	Security	JOPES/Joint Operation Planning	Businesses, Governments, Militaries
Supporting Capabilities	Information Assurance	Informational	Protect Information and Informatin Systems	Security	JOPES/J-6 Vulnerability Analysis	Businesses, Governments, Militaries
	Physical Security	Physical	Secure Information and Information Infrastructure	Usability	JOPES/Defense Planning	Businesses, Governments, Militaries
	Physical Attack	Physical	Destroy, Disrupt	Usability	JOPES/Joint Operation Planning	Governments, Militaries
	Counterintelligence	Cognitive	Mislead	Accuracy	JIPB/Human Intelligence Collection	Governments, Militaries
	Combat Camera	Physical	Inform/Document	Usability, Accuracy	JOPES/Joint Operation Planning	Governments, Militaries
Related Capabilities	Civil Military Operations	Cognitive	Influence	Accuracy	JOPES/Joint Operation Planning	Governments, Militaries
	Public Affairs	Cognitive	Inform	Accuracy	JOPES/Joint Operation Planning	Businesses, Governments, Militaries
	Public Diplomacy	Cognitive	Inform	Accuracy	Interagency Coordination	Govemments

Information Operations Integration into Joint Operations (Notional)

Source: U.S. Department of Defense. Information Operations Integration into Joint Operations (Notional). Viewed on September 1, 2018. Retrieved from https://en.wikipedia.org/wiki/File:IO_Integration_into_Joint_Operations_-_Notional.jpg

Figure 14-3 DoD Vision: Future of Internet with IW / IO integration

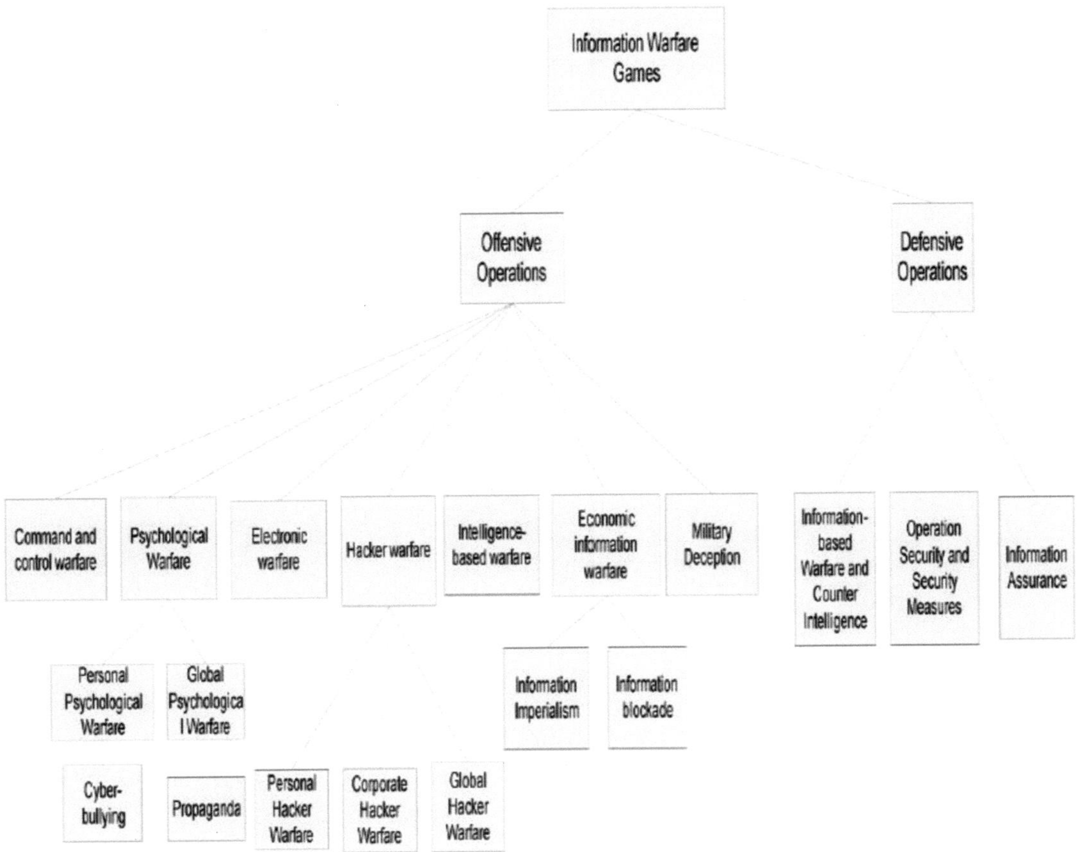

Source: Merrick, K. (2016). Future Internet. 10.3390/fi8030034 Review, 8(3), p. 34.

Electronic Warfare (EW) Purview

Adamy sets the standards for EW instruction. Moir summarizes the topic with respect to military operations, UAS, and military avionics systems. (Moir, 2006) (Toomay, 1982) and (Burch, 2015) bring Radar to the non-specialist reader. A Google search on the key = RADAR yields 296,000,000 results (0.49 seconds). There is substantial material on the subject. The challenge is determining the UAS applicability.

Warfare is conducted by adversaries who go to great pains to understand their enemy's intentions, strengths, weaknesses, and to minimize the threats to their own forces and territory.

The detection and interception of messages/data, combined with ground observations, provide an ability to observe troop movements and facilitate counter-actions by opposing forces.

As communication developed (satellite, MW, IR, Cell, Radars) so did the methods of interception.

Radar started as a simple detection mechanism and grew into a means of surveillance and guidance. (Moir, 2006)

Communication Links for UAS are critical and must be secured

Modern warfare is conducted in a rich electromagnetic environment with radio communications and radar signals from many sources. ***Unmanned aircraft systems (UAS) / UAV / UUV / Drones are an integral part of modern warfare. UAS communications networks and links to ground stations are critical to the successful military use of UAS.*** Securing UAS links from EW attacks is a fundamental concern to military planners and civilian authorities. The USMC takes UAS contributions to its Close Air Support (CAS) military planning. Figure 14-4 shows USMC PCAS Vision; precise, digital, portable air-ground strike coordination.[2] This scenario is part of the USMC Guardian Angel program.[3] In September 2016, the Commandant of the Marine Corps, stated the importance of UAS in the USMC battle plans of the future:

> "Marine Corps must: develop layers of persistent, armed, multi-spectral, and beyond-line-of-sight (BLOS) UAS above our units to produce responsive intelligence and targeting information, extend our command and control (C2) across a shifting battlespace, and deliver non-kinetic and kinetic fires in support of MAGTF operations."[4]

UAS BLOS communications require stable communications. Disrupting these communications links is a goal of hostile forces.

2. Several definitions are necessary to discover the power of Figure 14-4. PCAS = Persistent close air support; CAS = Common situational awareness; FRAGO = Fragmentary Order – to send timely changes of existing orders to a subordinate; MAGTF = Marine air-ground task force; MEB = Marine expeditionary brigade (14,500 marines and sailors); MALE-T =Medium altitude long endurance – tactical UAS; UUNs / DUNSs = Urgent / deliberate universal needs statements; HUD = Heads-up display; JTAC = Joint Terminal Attack Controller; JFO =Joint fires observer; JDAM = Joint direct attack munitions; Maverick Missile = The AGM-65 Maverick is an air-to-surface missile (AGM) designed for close air support. It is the most widely produced precision-guided missile in the Western world, and is effective against a wide range of tactical targets, including armor, air defenses, ships, ground transportation and fuel storage facilities; APKWS = Advanced precision kill weapon system and PGM = Precision guided missile.
3. Cuomo, S. Maj., (September 2017) USMC Vision Guardian Angel: https://www.mcamarines. org/gazette/guardian-angel-uas
4. Headquarters Marine Corps, Marine Corps Operating Concept: How an Expeditionary Force Fights and Wins in the 21st Century, (Washington, DC: September 2016)

Figure 14-4 PCAS Vision: precise, digital, portable air-ground strike coordination

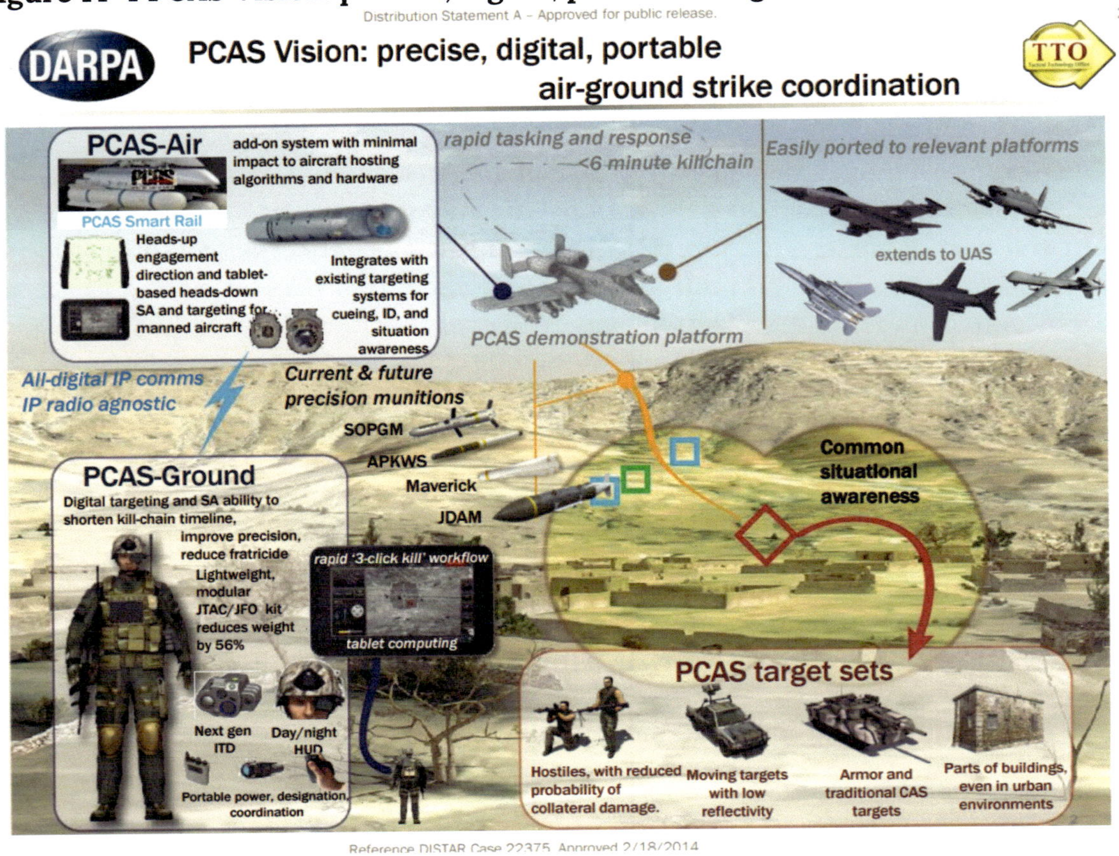

Source: Cuomo, S. (September 2017). "Guardian Angel" UAS, *Marine Corps Gazette*, 101(9). https://www.mca-marines.org/gazette/guardian-angel-uas

DoD has also made plans to build UAS into its mission.

Figure 14-5 shows a complex operational concept to be executed over the next twenty years. (DoD-03, 2015) Think of the designers' struggles to meet required future communication links to the UAS.

Figure 14-5 High -Level C4 Operational Concept Incorporating UAS

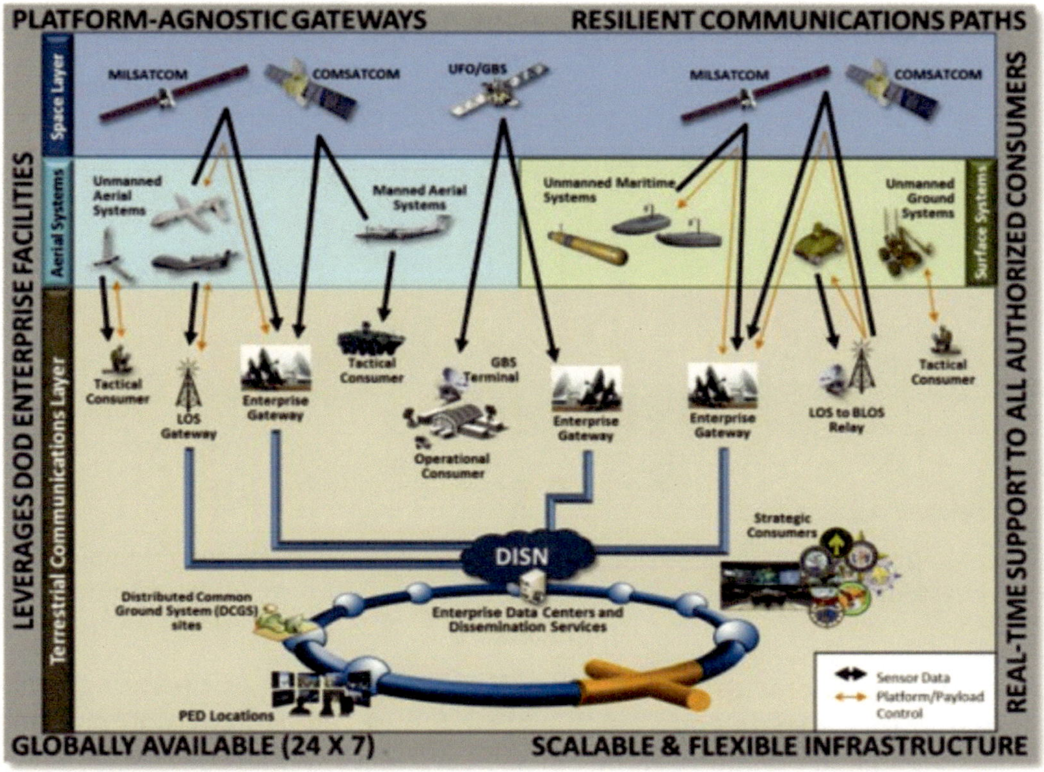

High-Level C4 Infrastructure Operational Concept Graphic (OV-1)

Source: (DoD-03, 2015) DoD-03. (2015). Unmanned Systems Roadmap 2013 to 2038. Retrieved from DTIC: http://www.dtic.mil/dtic/tr/fulltext/u2/a592015.pdf

Figure 14-6 shows NASA's Unmanned Aircraft Systems (UAS) Integration in the National Airspace System (NAS) Project (2018). All communications are in real-time. (NASA, 2018)

Figure 14-6 NASA's Unmanned Aircraft Systems (UAS) Integration in the National Airspace

System (NAS) Project (2018)

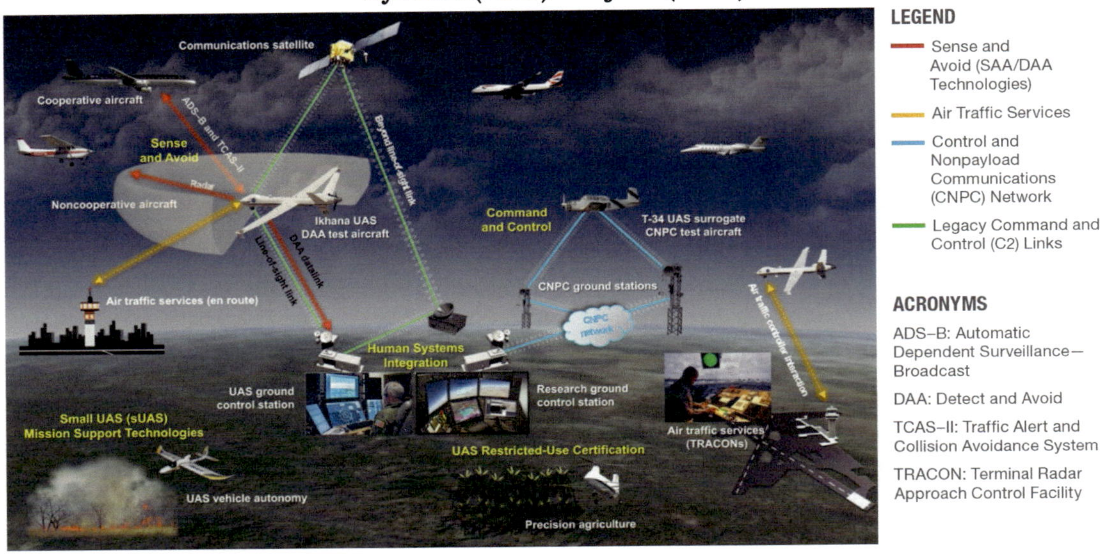

Source: NASA. (July 28, 2018). NASA Unmanned Aircraft Systems (UAS) Integration in the National Airspace System (NAS) Project. https://www.nasa.gov/feature/autonomous-systems DLA 07282018

In a theater-wide scenario, with combined land, sea, and air forces operating against enemy territory, which in turn, are defended with similar forces, we can envision multiple roles for UAS. These include military planning, air defenses, air superiority aircraft, defense suppression, HVT assassination,[5] maritime operations, offensive operations, naval forces, land forces, cyber forces, TV reporters, mobile telephone traffic, and sound satellite links. (Moir, 2006) The radio frequency spectrum (and EMS) covered by emitters used by the forces is broad as was illustrated in Figures 8-1, 8-3, 13-4 and Table 13-1. No single transmitter /receiver can cover this range for transmission or reception. Most communications and radar systems are designed for use in specific bands, designated by international convention. (Moir, 2006) See Figure 8-8 Radio frequency bands and Figure 8-9 Radar frequency bands.

The key role of EW is to search these radio-frequency bands to cull information that can be used for intelligence analysis or by front-line operators. (Moir, 2006) The information gathered may affect a tactical advantage on the battlefield, or in any stage before or after. (Moir, 2006)

Intelligence Cycle

5. HVT = High Value Target – cutting the head off the snake

In the CIA, intelligence is viewed as a continuous cycle of activities to keep ahead of adversaries always (peace, transitions, war). Figure 14-7 shows the Intelligence Cycle.

Figure 14-7 Intelligence Cycle

The intelligence cycle

Source: Count upon Security. (August 15, 2015). The Five Steps of the Intelligence Cycle. https://countuponsecurity.com/2015/08/15/the-5-steps-of-the-intelligence-cycle/ DLA 07282018

The cycle begins with a user requesting intelligence be gathered on a scenario. The analysts will develop a set of requirements for the collection stage. Intelligence is then gathered via surveillance platforms; espionage, co-operative exchange, or open source documents. The intelligence is sent to a processing group whereby the data is analyzed, collated, validated, and securely disseminated to the users or to trusted allies. They are read, evaluated, and acted on based on the intelligence. Or the user may rethink the needs/parameters of the project, returning the direction back to a modified intelligence requirements document, which is then transferred over to the collection group, analysis, reporting, and so forth.

EW Generalities

Time to fill in the EW boxes in 14-1. **Electronic warfare (EW)** is defined as the art and science of preserving the use of the **electromagnetic spectrum (EMS)** for friendly use while denying its use by the enemy. (Adamy D., 2001) The EMS is from DC to light and beyond. See Figures 8-1 to 8-4 EMS Spectrum views. EW covers the full radio frequency spectrum, the infrared spectrum, and the ultraviolet spectrum.

Legacy EW definitions

EW was classically divided into: (Adamy D., 2001)

- **ESM** – Electromagnetic Support Measures – the receiving part of EW;
- **ECM** – Electromagnetic Countermeasures – jamming, chaff, flares used to interfere with operations of radars, military communications and heat-seeking weapons;
- **ECCM** -Electronic Counter-Counter Measures – measures taken in design or operation of radars or communications systems to counter the effects of ECM.[6]

Not included in the EW definitions were Anti-radiation Weapons (ARW) and Directed Energy Weapons (DEW).

USA and NATO have updated these categories:

- **ES** – Electronic warfare Support (old ESM)
- **EA** – Electronic Attack – which is the old ECM but also includes ASW and DE weapons;
- **EP** – Electronic Protection – (old ECCM) (Adamy D. , 2001)

ES is different from Signal Intelligence **(SIGINT).** SIGINT is made up of Communications Intelligence **(COMINT)** and Electronic Intelligence **(ELINT).** All these fields involve the receiving of enemy transmissions. (Adamy D. , 2001)

COMINT receives enemy communications signals to extract intelligence.

ELINT uses enemy non-communications signals for determining the enemy's EMS signature so that countermeasures can be developed. ELINT systems collect substantial data over large periods to support detailed analysis.

ES/ESM collects enemy signals, either communication or non-communication, with the object to do something immediately about those signals or the weapons associated with those signals. The received signals might be jammed or the information sent to a lethal responder. Received signals can be used to type and locate the enemy's transmitter, locate enemy forces, weapons, distribution, and electronic capability. (Adamy D. , 2001)

Adamy (2001) is correct when he suggests that the, "*key to understanding EW principles (particularly the RF) part is to understand radio propagation theory. Understanding propagation leads logically to understanding how they are intercepted, jammed or protected.*" (Adamy D. , 2001) Adamy has written five books and published numerous papers on the subject. The student can explore

6. ECCM was considered TS classified with most secret protocols and design algorithms. TS = Top Secret

his works or use (Moir, 2006) or (Toomay, 1982) (light reading and dated text to dive into RF propagation.) Petti's book on ECCM is simplistic but has legacy views of the EP function. (Pettit, 1982) One of the classics that should read is by Lt. Col. USAF Fitts, *The Strategy of Electromagnetic Conflict* (Fitts, 1980) Lastly, a decent text on ELINT beginnings is authored by Wiley. (Wiley, 1993)

Main Contention

It is the author's contention that UAS communication links are vulnerable and must be evaluated to protect US Unmanned Aircraft in the cyber or electronic domain. Further, those links may be electronically jammed, cyber-spoofed (especially navigational), or made ineffective with electronic or cyber interference.[7]

Cyber-spoofing is conjectured in Chapter 16 in the GPS cyber-spoof discussion of the Spratly Island complex, Red Drones, and US Capital ships colliding with commercial vessels. The remainder of this chapter will focus on the electronic jamming, LPI countermeasures and UAS vulnerabilities to electronic interference.

Communications Jamming

The purpose of communication is to move information from one location to another. All the following types of transmitted signals are communications:

- "Voice or non-voice communications (video or digital format)";
- "Command signals to control remotely located assets;"
- "Data returned from remotely located equipment";
- "Location and motion of friendly or enemy assets (land, sea, or air);"
- UAS communications links from it ground station for control of the aircraft;
- UAS communications links from another aircraft or satellite affecting its flying characteristics;
- UAS communication signals (from any source) that affect the SAA / navigation / payload / waypoints;
- Computer-to-computer communications;
- Data links;
- Weapon-firing links;
- ISR data links;
- Cell phones. (Adamy D., 2009)

7. This a main theme of this book. In addition, this chapter started off with the answer – Low Probability of Intercept (LPI) as a countermeasure to reduce risk of EA to the UAS mission.

"The purpose of communications jamming is to prevent the transfer of information. Communications jamming requirements depend on the signal modulation (strength), the geometry of the link, and the transmitted power." (Adamy D., 2009) *Another way to think of jamming is a method to "interfere with the enemy's use of the electromagnetic spectrum. Use of* EMS *involves the transmission of information from one point to another*". (Adamy D., 2009)

"The basic technique of jamming is to add an interfering signal," along with the desired signal, into an enemy's receiver. "Jamming becomes effective when the interfering signal is strong enough to overwhelm the desired signal." This prevents the enemy from recovering the information from the desired signal. (Adamy D., 2009) There are two possible methods for a successful jam: either the jamming signal is stronger than the desired "signal or the combined signals received have characteristics that prevented the processor from properly extracting the desired information." (Adamy D., 2009) A simple case of jamming unintentionally is when your AM news station (listening in the car) becomes overwhelmed by junk music. You can hear the beginning of the interference as noise, then the junk signal is strong, then as the car moves out of the area, the AM news station regains its status. (Adamy D., 2009)

The cardinal rule of jamming is that you jam the receiver, NOT the transmitter. (Adamy D., 2001)

Figure 14-8 shows the communication jamming geometry. (Adamy D., 2009)

"The primary difference between radar and communication jamming is in the geometry. Whereas a typical radar has both the transmitter and the associated receiver at the same location, a communication link, because its job is to take information from one location to another, always has its receiver in a different location from that of the transmitter." (Adamy D. L., 2004)

In Figure 14-8, the "communications-jamming geometry has one-way links from the desired transmitter and the jammer to the receiver." (Adamy D. L., 2004)

Figure 14-8 shows the communication jamming geometry

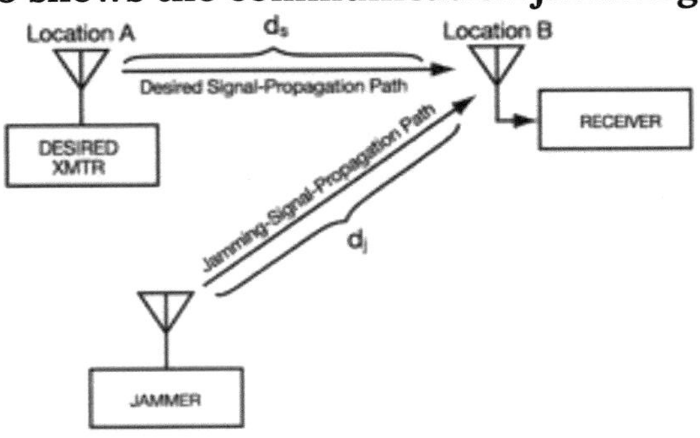

Source: Adamy, D. (2009). EW 103 Tactical Battlefield Communications Electronic Warfare. Boston, MA: Artech House.

"Communication is often done using transceivers (each including both transmitter and receiver), but only the receiver at location B in the figure is jammed. If transceivers are in use and one desires to jam the link in the other direction, the jamming signal must reach location A." (Adamy D. L., 2004)

"There are some important communications instances in which transceivers are not used. For example, in UAV links as shown in Figure 14-9, this figure shows the data link (or "downlink") being jammed. Again, the receiver is the objective to be jammed." (Adamy D., 2009)

Figure 14- 9 UAV Link Jamming Geometry

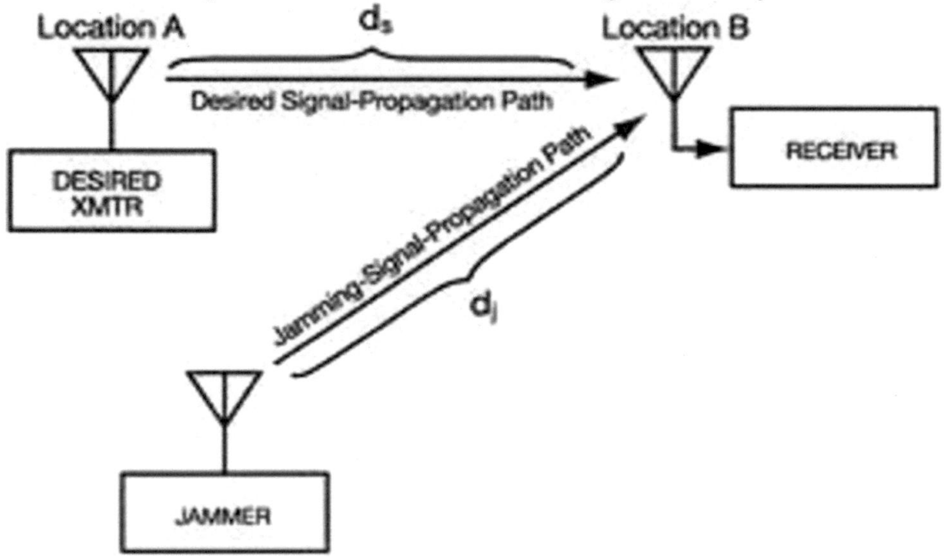

Source: Adamy, D. L. (2004). EW 102 A Second Course in Electronic Warfare. Boston: Artech House. Also see: https://www.globalspec.com/reference/61136/203279/5-8-communications-jamming

"Figure 14-9 shows that a jammer operating against a UAV data-link must target the receiver at the ground station.

Another difference of radar jamming is that the radar signal makes a round trip to the target, so the received signal power is below the transmitted power by the fourth power of the distance (often stated as 40 log range). Since the jammer power is transmitted one way, it is only reduced by the square of distance." (Adamy D. L., 2004)

Table 14-1 shows the Types of Jamming. (Adamy D., 2001)

Type of Jamming	Purpose
Communications jamming	"Interferes with enemy ability to pass information over a communication link"
Radar jamming	"Causes radar to fail to acquire its target, to stop tracking target, or to output false information"
Cpver jamming	"Reduces the quality of the desired signal so that it cannot be properly processed, or the info is lost / unrecoverable"
Deceptive jamming	"Causes radar to improperly process its return signal to indicate the correct range or angle to target"
Decoy	"Looks like the target more than the actual target; causes a guided weapon to attack the decoy rather than intended target."

Table 14-1 Types of Jamming

Source: (Adamy D., 2001)

To be effective, the jammer must get its signal into the enemy's receiver – through the associated antenna, input filters, and processing gates. This depends on the signal strength the jammer transmits in the direction of the receiver and the distance and propagation conditions between the jammer and the receiver. (Adamy D., 2009)

Jammer-to-Signal Ratio

The real test of jammer effectiveness is the effectiveness with which information flow is stopped. "A jammer interferes with communication by injecting an undesired signal into the target, receiver along with any desired signals that are being received." (Adamy D. , 2009) "The obstructing signal must be strong enough that the receiver cannot recover the required information from the desired signals." The ratio of the jamming signal to the desired signal is known as the jamming-to-signal ratio **(J/S),** stated in dB.[8] Effective J/S depends on the transmitted modulation, but the Adamy formula works in general. (Adamy D., 2001)

Refer to Figure 14-9 UAV Link Jamming Geometry. The jammer-to-receiving link and the desired transmitter-to-receiver UAV link in Figure 14-9 can employ any propagation model. (Adamy D., 2009)

"The formula for communication J/S is:

Eq. 14-1

$$J / S = ERP_J - ERP_S - L_J + L_S + G_{RJ} - G_R$$

Where: **J/S** = the ratio of the jammer power to the desired signal power at the input to the receiver being jammed in dB

ERP_J the effective radiated power of the jammer in dBm
ERP_S the effective radiated power of the desired signal transmitter, in dBm
L_J the propagation loss from jammer to receiver, in dBi[9]
L_S The propagation loss from the desired signal transmitter, in dBm
G_{RJ} the receiving antenna gain in the direction of the jammer, in dBi
G_R The receiving antenna gain in the direction of the desired signal transmitter, in dBi." (Adamy D., 2001)

Many UAS (especially UAV or sUAS) have a target receiving antenna with a 360-degree azimuth coverage. They use whips or monopoles. They are inexpensive. With a 360-degree antenna, the communications J/S equation simplifies to:

Eq. 14-2

$$J / S = ERP_J - ERP_S - L_J + L_S$$

8. Any number expressed in dB is logarithmic base 10. dB mathematical concepts with examples may be found in Chapter 2 of Adamy, D., (2001) EW 101. A value expressed in dB is a ratio converted to logarithmic form. A linear number is converted to dB form by the formula: N(dB) = 10 log (base 10) [N]. dB values are converted back to linear format by the formula N = 10 **N (dB/10). dB numbers are usually reference to some standard with constant value. A common example is signal strength expressed in dBm = dB value of Power / 1 milliwatt, used to describe signal strength. For example, 4 watts power level = 4000 mw. Divide by 1 mw standard then convert 4000 to dB = 10 log (4000) = 36.02 dBm. dB forms are used because of the wide range of numbers and orders of magnitude for the EMS.
9. dBi = dB value of antenna gain relative to the gain of an isotropic antenna (perfect antenna). 0 dBi is the gain of an omnidirectional (isotropic) antenna.

The receiving antenna has the same gain toward the jammer and the desired signal transmitter. The two gain terms cancel out. (Adamy D., 2009)

Figure 14-9 shows a simple example where the J /S calculation would indicate a successful jam (desired signal fully compromised). (Adamy D., 2001) The terminology is slightly different for the power terms (removing the "effective radiated" and using "power total" instead). The principle is still the same. (Adamy D., 2009)

US Army Field Manual FM 34-40-7 (23 Nov 1992) *Communications Jamming Handbook*, presents three alternative methods for calculating the jamming power required and distance to target. For the designer of an anti-UAS Drone gun, which transmits a jammer signal to a UAS to overwhelm the desired ground station command signals, one needs the know the power and height of the drone. Since the drone is moving the jammer signal must radiate in such a manner that it covers a volume of space until target "UAS lock." Recall Figure 7-5 Drone Jammer Model KWT-FZQ. Table 14-2 shows the details for this anti-drone gun.[10]

10. Tri-band Anti Drone Rifle KWT-FZQ/DG10-A. Manufacturer: Globaldroneuav.com https://globaldroneuav.com/Product/Police-drone-jammer-effective-drone-controller.html

Figure 14- 10 J/S Calculation Example

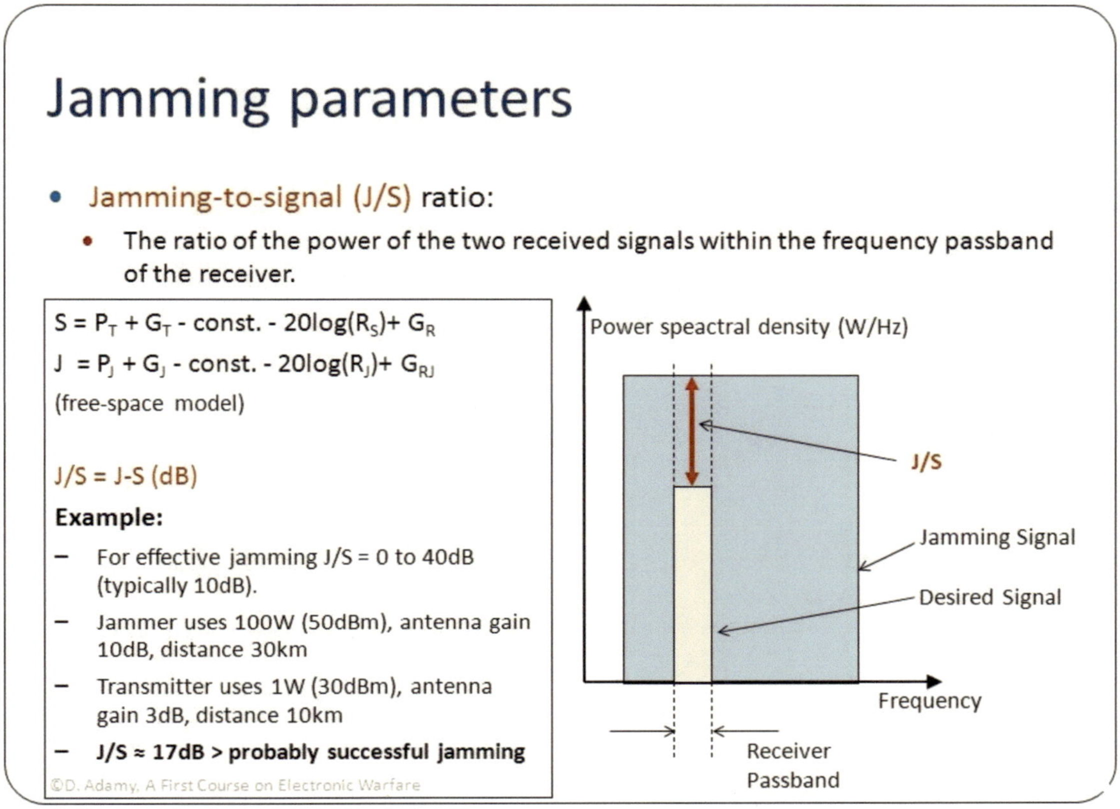

Source: Cagalj, M. (2014). FELK 19: Security of Wireless Networks: University of Spain PPTX Presentation.

Quote from manufacturer Globaldroneuav.com: (Adamy D. , 2009)

> "Anti-drone rifle is multiband device which sends a high-power electromagnetic wave that is used to interfere with UAV flight and force the remote low-altitude drone to drop, fly to home place or to land stably. By triggering the rifle, it disturbs the GPS location and breaks the video transmission control link of UAV instantly to achieve the effective control of "Unapproved fly." It's a portable gun all-in-one design with robust strength and fast use features make its easy start by one triggering after battery assembly. Multi antennas and multi-bands frequency works simultaneous in the front end of gun, which enables the effective control range more than 1000 meters. The tripod makes longtime watching much easier, which facilitates it popular in such no-fly zones as airport, prison and military areas and so on. What is more, it is very convenient to guarantee the activities going safely and orderly in the important security areas and urban management sites."

Table 14-2 Tri-band Anti Drone Rifle KWT-FZQ/DG10-A[11]

Functions and features

1. **Full range cover within three frequency section and high-power transmission helps to achieve the ideal effects.**
2. Fast trigger, easy use and daughter switch design make control more ease and comfort.
3. Dual lithium batteries for power supply last work time longer.
4. The strong internal line connector and external fuses port make the whole vehicle and component parts fastened securely.
5. All aluminum alloy case body design and glass fiber material for antenna cover make its appearance lighter and faster.

11. Tri-band Anti Drone Rifle KWT-FZQ/DG10-A

Technical parameters			
SN	**Parameter name：**		**Parameter index record：**
1	Power supply	Work voltage V	DC13V~16.8V
2		Work current A	≤9A@DC14.8V
3		Work time	≥1.5h
4	Radio Frequency	**Work frequency range MHz**	**(1550±5) MHz～(1620±5)MHz** **(2400±5) MHz～(2483±5)MHz** **(5725±5) MHz～(5852±5)MHz**
5			40dBm@1550～1620MHz（±1dB）
6		Output power dBm	37dBm@2400～2483MHz（±1dB）
7			37dBm@5725～5852MHz（±1dB）
8		Out of band rejection	＜-36dBm@30～1000MHz ＜-30dBm@≥1GHz
9	Specification & environment	Weight	4.8kg±0.2kg(Mainframe +battery） 0.6kg±0.1kg（sighting telescope）
10		Dimension	1323mm×403mm×341 mm, with battery and antenna
11		Work environment humidity	≥95%
12		Work temperature	-25℃～55℃
13		Storage temperature	-40℃～70℃

Appearance dimension ：（**mm**）**L×W×H**；
1323mm×403mm×341 mm
Weight（Kg）**:4.7kg±0.2**kg（mainframe + battery）
0.6kg±0.1kg（sighting telescope）

Source: Tri-band Anti Drone Rifle KWT-FZQ/DG10-A Manufacturer: Globaldroneuav.com https://globaldroneuav.com/Product/Police-drone-jammer-effective-drone-controller.html

Alternative: A Chinese firm makes an anti-drone gun that costs about $35,000 USD and operates on 5.8 GHz and 2.4 GHz.[12] 80% of consumer drones operate on these frequencies. "The gun tricks the drone into thinking it has lost connection with its controller." "RC signal lost" is flashed on drone screen – aircraft returning to home point." The drone can be recovered intact. This gun has an operational limit of about 700 meters (0.43496 miles).

Calculating the minimum of "amount of jammer power output required in watts" for this easy drone capture would be of interest. (Army, 1992) FM 34-40-7 Appendix gives a slightly different version of the Adamy equation 14-2:

Eq. 14-3

$$P_j = P_t \times K \times (H_t / H_j)^2 \times (D_j / D_t)^N$$

Where:

"P_j = Minimum amount of jammer power output required , in watts
P_t = Power output of the enemy drone, in watts
H_j = Elevation of the jammer location above sea level, feet
H_t = Elevation of enemy transmitter location above sea level, in feet
D_j = Jammer location – to-target receiver location distance, in km
D_t = Enemy transmitter location -to- target receiver location, in km
K = 2 for jamming frequency modulated receivers (jamming tuner accuracy)
N = Terrain and ground conductivity factors

5 = very rough terrain with poor ground conductivity
4 = Moderately rough terrain with fair to good ground conductivity
3 = Farmland terrain with good ground conductivity
2 = Level terrain with good ground conductivity

F = Frequency in MHz" (Army, 1992)

"Note: The elevation of the jammer location and the enemy transmitter location does not

12. Video Report, Quote by Amy Hu. Data Expert Technology LTD, https://www.youtube.com/watch?v=o057LmNGsJA DLA 07312018

include the height or length of the antenna above the ground. (Army, 1992) It is the location deviation above sea level.

Given the following parameters:

"P_j = Minimum amount of jammer power output required , in watts = (SOLVE)
P_t = Power output of the enemy transmitter -to drone, in watts = 5 watts
H_j = Elevation of the jammer location above sea level, feet, use 385m =.385 km
H_t = Elevation of enemy transmitter location above sea level, in feet use 386m =.386 km
D_j = Jammer location – to-target receiver location distance, in km = 700 m = 0.700 km
D_t = Enemy transmitter location -to- target receiver location, in km = 372m = 0.372 km
K = 2 for jamming frequency modulated receivers (jamming tuner accuracy) = 2
N = Terrain and ground conductivity factor = Use 4 for moderate terrain with fair to good ground conductivity" (Army, 1992)
F = Frequency in MHz, use 37.5 MHz in the band

Parameters were chosen so that the height ratio would drop-out and the distance would induce some ground conductivity effects consistent with the FM 34-40-7 examples.

Plugging the numbers and solving for P

$$\mathbf{P_j = 5 \times 2 \times (1)^2 \times (0.7 / 0.372)^4 = 10 \times (1.88)^4 = 10 \times 12.46 = 125\ watts}$$

So, under these <u>hypothetical conditions</u> the jammer gun requires 125 watts (2 60-watt light bulbs) to take down the drone. Theoretically, if the jammer was using a log periodic array (LPA) the power could be cut in half to 62.5 watts (1 bulb). Now if this calculation is reasonable, the buyer is spending $35,000 USD to take down a small irritating drone (invasion of privacy) using a 60-watt bulb. A double-aught shotgun shell with a 12-gauge Remington and yellow shooter sunglasses will have the same effect (might even be more satisfying) for 1/100 the cost. The medium size drones present a more interesting case. More power is needed to lock on to the higher altitude UAS. The term of interest in the jamming equation from FM 34-40 -7 is the ratio of the distances to the fourth power (or second power for perfect terrain). That can have a major impact on jammer output power. (Army, 1992)

Radar Range Equation

Equation 14-3 <u>is not the only place we see a term taken to the 4th power</u>. The famous "Radar Range Equation *is dominated by the* R^4 *factor in the denominator*. There is no corresponding function in the numerator of equation 14-4, with an exponent greater than unity. (Toomay, 1982) There is no magic bullet to achieve a high-performance system. If low cross section targets are to be engaged, a combination of high-power, high gain, large aperture, and low-noise needs to be dictated." (Toomay, 1982)

See Figure 14-13 Simple Radar Block Diagram for discussion, after a quick history lesson.

Radar is an excellent technology to "see the UAS" coming into defensive space. RADAR is an acronym for "Radio Detection and Ranging." "Radio waves occupy a small portion of the EMS from frequencies of a few kHz, a few thousand cycles per second, to a few million MHz, over 10^{12} cycles per second." (Toomay, 1982) We can thank Heinrich Hertz, who in 1886, conducted "experiments showing that radio waves reflected, refracted, were polarized, interfered with each other, and travelled at high velocity". The Maxwell equations were confirmed and the properties of "reradiation and of known velocity portended the future of radar." (Toomay, 1982)

Radio waves were first used for communications, generate by spark gaps generating short intense pulses of current to achieve the needed electromagnetic radiation. The real breakthrough in radar was by Lee Deforest in 1906. He invented vacuum tubes. Sinusoidal oscillators came next in evolution. [Yes, transistors and mini-transistors came down the line for another revolution, but who cares?] All the ingredients were in place (Valid theory, experiments, and practical transmitters and receivers), but it was Robert Watson Watt in 1930's, who fielded a radar system in the field. (Toomay, 1982) The English Air Ministry were glad he did so! The WWII air Battle of Briton was won by using Watson's radar to spot and warn the English fighters as to the German bombers and fighter escorts location, altitude, distribution, and number of planes in the attack.

The Chain-Mall Radar stations (codename Chain Home) were strategically located all along the English Coast. See Figure 14-11. Figure 14-12 shows an example of one of these primitive radar stations. The German High command discounted the radar theory until it was too late in 1943.

Figure 14-11 Chain Home Radar Stations Defending England

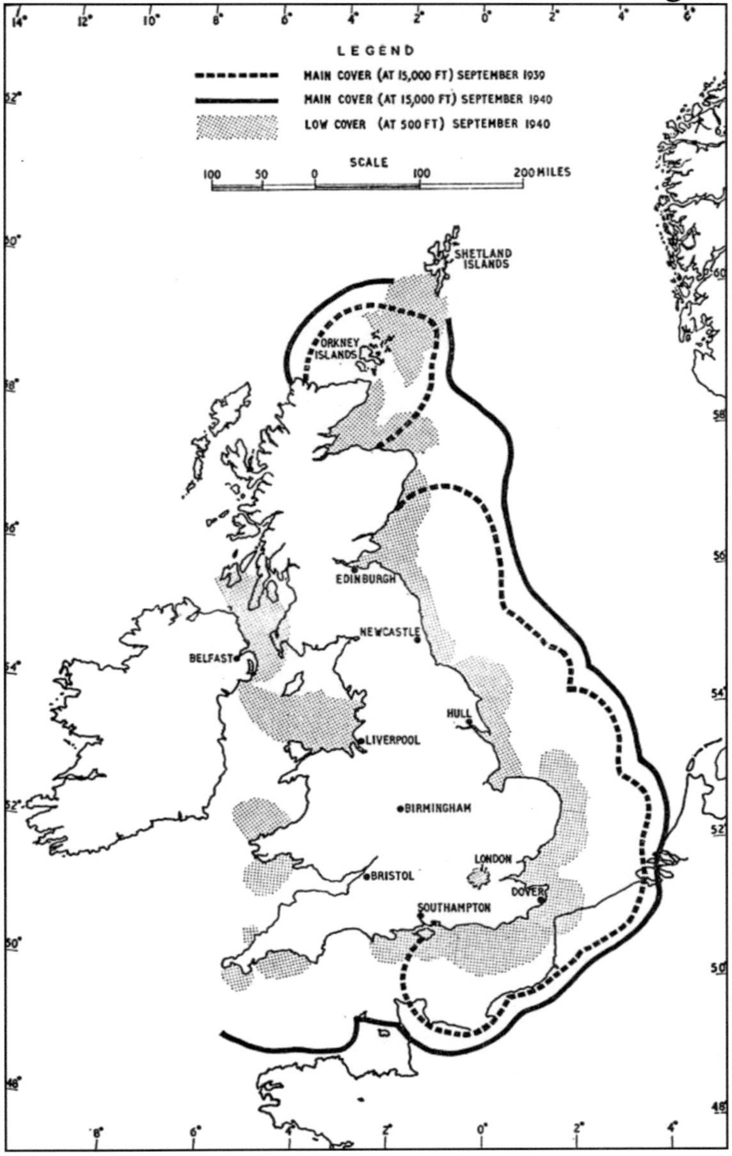

Source: Richards, D. (n.d.). Hyper war: The Royal Air Force 1939-1945, Vol. I: The Fight at Odds. Viewed September 13, 2018, Retrieved from http://www.ibiblio.org/hyperwar/UN/UK/UK-RAF-I/UK-RAF-I-6.html and http://enacademic.com/pictures/enwiki/67/Chain_home_coverage.jpg

Figure 14-12 Chain Home Radar Station

Source: Richards, D (2010). Hyper war: The Royal Air Force 1939-1945, Vol. I: The Fight at Odds. Retrieved from http://enacademic.com/dic.nsf/enwiki/219451/Chain_Home (codename)

"The principles of a primitive radar are formed. Figure 14-13 diagrams its functions. A burst of electromagnetic energy, oscillating at a predetermined frequency is generated and radiates into free space from an antenna. A clock is started. The electromagnetic energy propagates outward at the speed of light, reradiating (scattering) from objects it encounters along its path. Part of the scattered energy returns to the radar (is received) and can be detected there because it imitates the frequency and duration of the transmitted pulse." (Toomay, 1982) (Figure 14-14)

Figure 14-13 Simple Radar Block Diagram

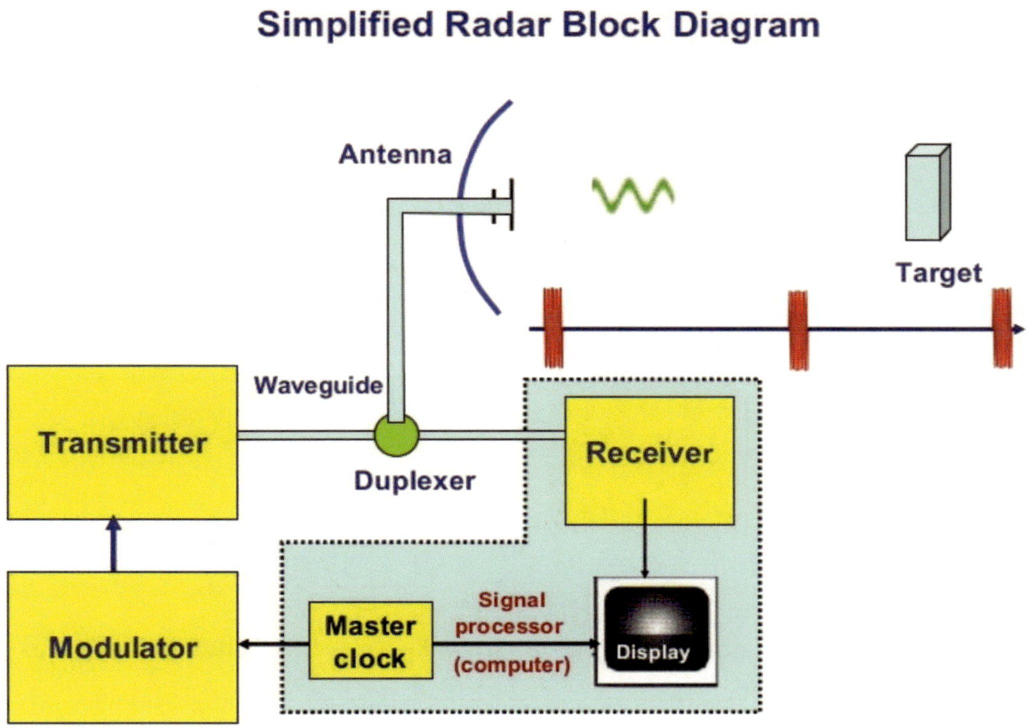

Source: Goel, R. (2014). Simplified Radar Block Diagram https://www.slideshare.net/remotesensor1/radar-transmitter-4-1

A full derivation of all the terms, the radar spherical geometry and derivations of subset equations are in all legacy and modern radar texts and papers.

Figure 14-14 Simple Surveillance Radar

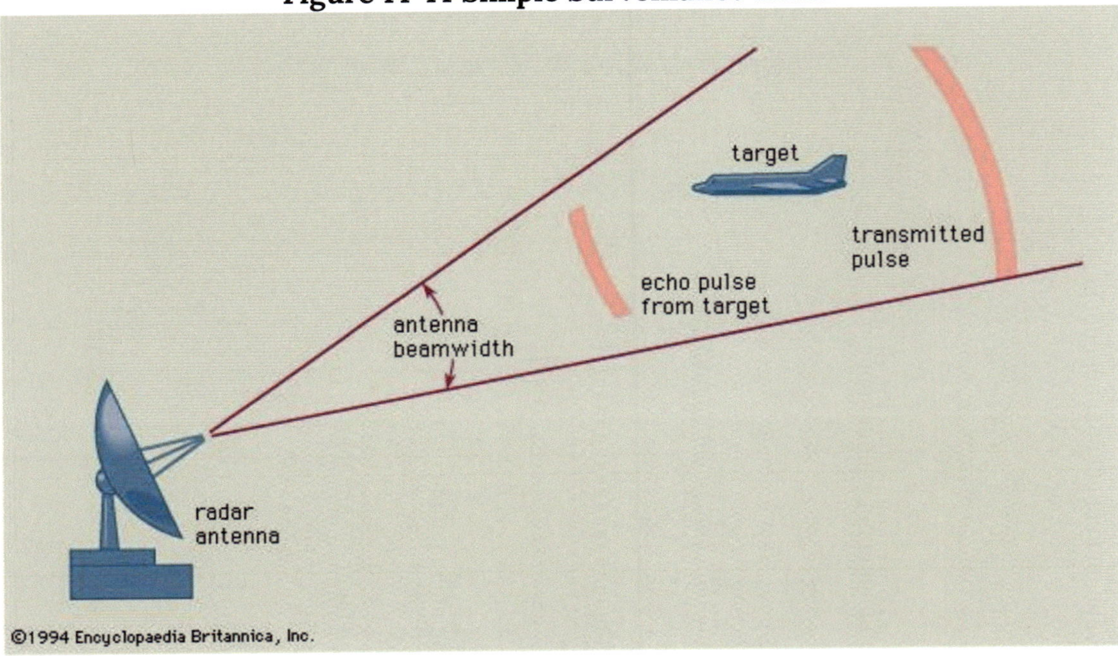

Source: Radar spherical Geometry to target. (1994). In *Encyclopedia Britannica*. https://www.britannica.com/technology/radar

The standard Radar Range Equation (RRE) is:

Eq. 14-4

$$S / N = (P\, G_T A_r \sigma) / [(4\pi)^2 R^4 K T_S L_S]$$

Where:

S / N = is one pulse received signal to noise ratio, dB
P = Isotropic source of an electromagnetic pulse of peak power, Mw
G_T = Gain of the transmit antenna, dB
A_r = Receive antenna effective area, m^2
σ = Radar Cross Sectional Area, m^2
R^4 = *Energy density received at detected target range*, R, *nm*
K = Boltzmann's constant (Noise component)
T_S = Measured noise temperature, Kelvin units above absolute zero
L_S = Losses existing in the system (lumped together), dB

Inherent in equation 14-14, is the fact that the range of the radar to a "detected object can be calculated by: $R = ct / 2$, where c is the speed of light (3×10^8 m/s) x time , in sec. also, λ = c

/ f, where λ is the wavelength in Hz, and frequency, f is the cycles/second for the sinusoidal oscillator." (Toomay, 1982)

The point of this diversion into Radar history was that the performance of both the jamming equation and the radar range equation are affected by a power of 4th exponent. This affects equipment design, cost, effectiveness of detection or capture.

LPI Communication Signals

The author gave the answer to the question never asked at the beginning of this chapter. The answer was LPI (*Low Probability of Intercept*) communications presented challenges to EW communication links. (Adamy D. -0., 2015) "**They are extremely dependent on interconnection with ground stations by command and data links**" And the question was how does the US protect its UAS/UAVs/drones from hostile take-over, considering the vulnerabilities of communication links from ground to UAS platform? Signals associated with LPI communications have unique modulations[13] designed to seriously perturb detection possibilities for the normal types of receivers. (Adamy D. -0., 2015) If designed correctly, a hostile receiver will not even detect the presence of the signal.

This is accomplished by spreading the frequency range over which the LPI signal is broadcast. (Adamy D. -0., 2015)[14] (Figure 14-15)

13. Signal Modulation is the process of varying one or more properties of a periodic waveform, called the carrier signal, with a modulating signal that typically contains information to be transmitted
14. These signals are call spread spectrum signals because of the frequency is spread over a larger range, so the signal PSD is reduced (Figure 14-15).

Figure 14-15 Spread Spectrum Signal

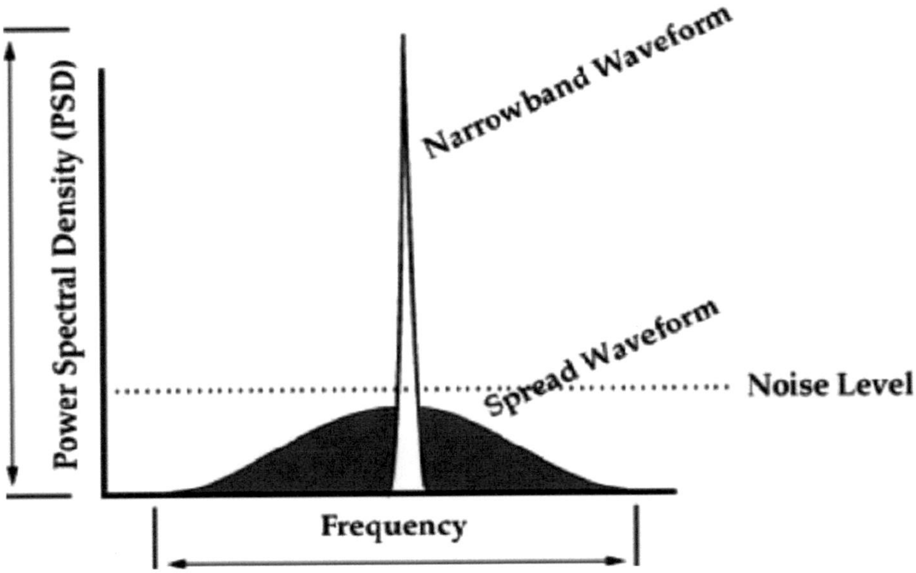

Source: Price, H. (1995). Digital Communications, column, NK6K, from QEX. http://www.qsl.net/n9zia/ss.qexss.html

Figure 14-15 shows the effect of a second modulation applied to the narrowband waveform to spread its frequency over a larger part of the spectrum and reduce the power spectral density to less than *random* noise. (Adamy D. -0., 2015)

Four types of spreading modulations are used: (Adamy D. -0., 2015)[15]

- *Frequency Hopping*: The transmitter periodically hops to a pseudo-randomly selected frequency. The range is much greater than the bandwidth of the signal carrying the information to be communicated. (Adamy D. -0., 2015)
- *Chirp*: The transmitter is rapidly tuned across a frequency range that is significantly wider than the information bandwidth. (Adamy D. -0., 2015)
- *Direct Sequence Spread Spectrum* (DSSS): The signal is digitized at a rate much higher than required to carry the information. This spreads the energy of the signal across a wide bandwidth. (Adamy D. -0., 2015)
- *Complex combinations*: Experimental models combining two or more spreading modulations made up of frequency hopping, chirp, and DSSS.

15. Each of the spreading modulations is a book in itself: Frequency hopping – About 8,080,000 results (0.40 seconds) ; Chirp -About 192,000 results (0.22 seconds) and DSSS – About 1,110,000 results (0.56 seconds). Therefore, discussion of LPI is limited.

LPI Restrictions

There are several restrictions on the effective use of LPI as a counter EW tactic for UAS.

LPI equipment and software may be cost-ineffective in smaller UAS systems.

- The spreading modulator in the receiver must be synchronized with the spreading modulator in the transmitter to reverse the spreading modulation so that the signal can be returned to the same bandwidth (information bandwidth) it had before the "secrecy operation";
- The synchronization requires that both the modulator and demodulator use the same pseudo-random function based on same digital code sequence;
- the codes transmitted and received must be in phase; and
- synchronization may require a delay before the transmission begins.

Discussion Questions /Assignment

Chapter 14 concentrated on UAS, EW, and LPI. However, there is a closely related science that intersects with EW and that is Cyber. There are distinct parallels and intersections between Cyber and EW. For instance, the sister of signal spreading techniques is encryption. See Figure 14-16 showing the intersection of Cyber, EW, and Spectrum Warfare designated as Cyber Electromagnetic Activities (CEA)[16] Figure 14-17 puts CEA in the perspective of total war.[17][18]

Student will research CEA and its parallels to EW (start with FM 3 – 38 Cyber Electromagnetic Activities in CANVAS or use Google to find the free PDF) How do these intersections support both friendly and hostile actions on UAS systems in all classes? Develop a PowerPoint presentation with your answers for class submission. Look for tools like cyber offensive weapons against key UAS systems and cyber defensive weapons/countermeasures that can be used to thwart the cyber weapons that you have found in Open Source literature (Non- CLASSIFIED). Try to develop a taxonomy around your findings.

16. FM 3-38 (2014)
17. Askin, O., Irmak, R, and Avseyer, M. (14 May 2015)
18. CEA AKA Cyber electronic warfare

Figure 14-16 Cyber Electromagnetic Activities

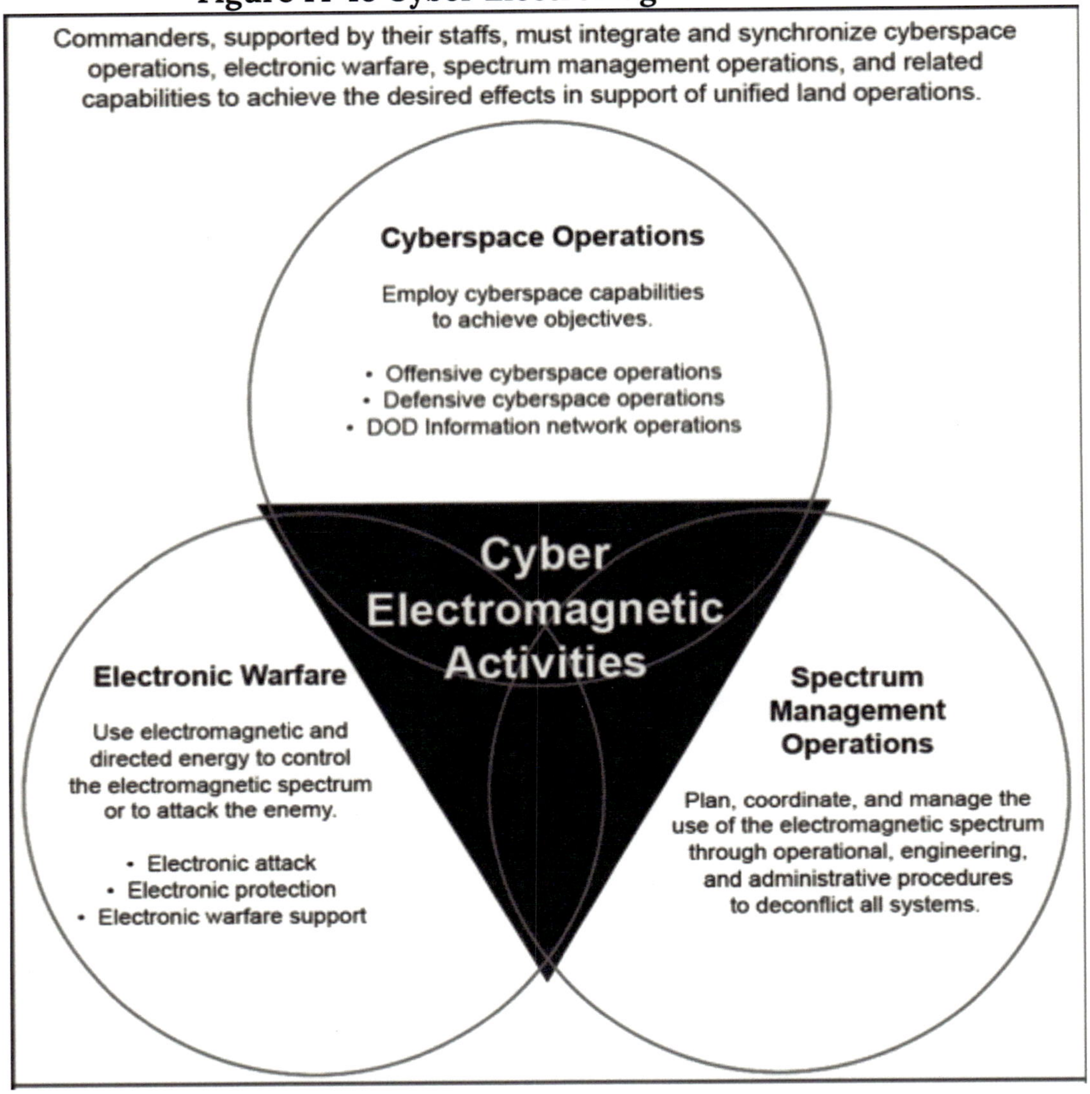

Source: US Army. (2014). FM 3-38 (2014) Cyber Electromagnetic Activities (US Army)

Figure 14-17 CEA / CEW in the view of Total War

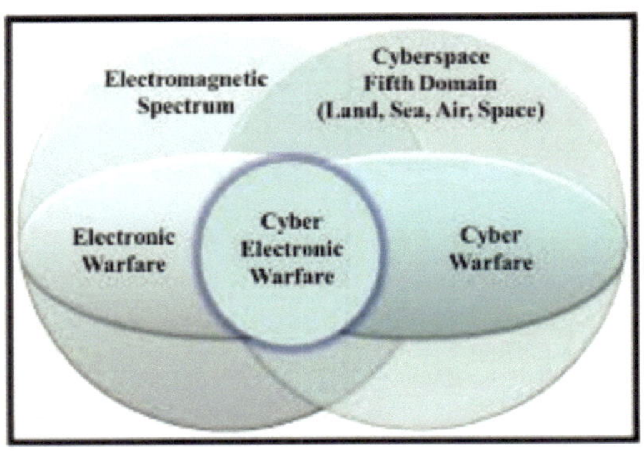

Source: Askin, O., Irmak, R, and Avseyer, M. (14 May 2015). Cyber warfare and electronic warfare integration in the operational environment of the future: cyber electronic warfare. Proceedings Vol 9458, Cyber Sensing 2015; 94580H (2015) SPIE Defense + Security, 2015, Baltimore, MD. https://doi.org/10.1117/12.2189351

Bibliography

Adamy, D. -0. (2015). EW 104 EW *against a New Generation of Threats*. Boston: Artech House.

Adamy, D. (2001). EW 101 A *First Course in Electronic Warfare*. Boston, MA: Artech House.

Adamy, D. (2009). EW 103 *Tactical Battlefield Communications Electronic Warfare*. Boston, MA: Artech House.

Adamy, D. L. (2004). EW 102 A *Second Course in Electronic Warfare*. Boston: Artech House.

Alford, L. (2000). Cyber Warfare: Protecting Military Systems. *Acquisition Review Quarterly*.

Army, U. (1992, November 23). US Army Field Manual FM 34-40-7. *Communications Jamming Handbook*.

Barker, W. (2003, August). SP 800-59 *Guidelines for Identifying an Information System as a National Security System*. Retrieved from NIST: https://csrc.nist.gov/publications/detail/sp/800-59/final

Burch, D. (2015). *RADAR for Mariners*. New York: McGraw-Hill.

C4ISystems. (2013). *basics-of-information-operations*. Retrieved from: http://c4isys.blogspot.com/2013/11/basics-of-information-operations-24.html

DoD. (2018). *Dictionary of Military Terms*. Retrieved from JCS.Mil: http://www.jcs.mil/doctrine/dod_dictionary/

DoD-01. (2018). JP 1-02. Retrieved from Department of Defense Dictionary of Military and Associated Terms: www.dtic.mil/doctrine/new_pubs/jp1_02.pdf

DoD-02. (2018). *Information Operations (IO) in the United States*. Retrieved from JP 3-13 : http://www.dtic.mil/doctrine/new_pubs/jp3_13.pdf

DoD-03. (2015). *Unmanned Systems Roadmap* 2013 *to* 2038. Retrieved from DTIC: http://www.dtic.mil/dtic/tr/fulltext/u2/a592015.pdf

Fitts, R. (1980). *The Strategy of Electromagnetic Conflict*. Los Altos, CA: Peninsula Publishing.

Kaye, T. a. (2001, September 30). ACHIEVING INFORMATION DOMINANCE:. Retrieved from DODCCRP-Space and Naval Warfare Systems Center San Diego: http://www.dodccrp.org/events/2002_CCRTS/Tracks/pdf/026.PDF

Merrick, K. (2016). Future Internet. 10.3390/fi8030034 *Review*, 8(3), p. 34.

Moir, I. a. (2006). *Military Avionics Systems*. New York: Wiley Aerospace Series.

MORS. (2018). *Military Operations Research Society* . Retrieved from http://www.mors.org/meetings/oa_definition.htm

NASA. (2018). *Unmanned Aircraft Systems* (UAS) *Integration in the National Airspace System* (NAS) *Project*. Retrieved from NASA: https://www.nasa.gov/feature/autonomous-systems

Nichols, R. K. (2008, September 05). Counterintelligence & Sensitive Compartmented Information Facility . (SCIF) *Needs – Talking Points*.

Pettit, R. (1982). ECM *and* ECCM *Techniques for Digital Communication Systems*. Belmont, CA: Lifetime Learning Publications .

Toomay, J. (1982). RADAR *for the Non – Specialist. London; Lifetime Learning Publications*. London: Lifetime Learning Publications.

Wiley, R. G. (1993). *Electronic Intelligence: The Analysis of Radar Signals, 2nd ed.* Norwood, MA: Artech House.

Readings

Adamy, D. (2001) EW 101 *A First Course in Electronic Warfare*, Boston: Artech House.

Adamy, D. (2009) EW 103 *Tactical Battlefield Communications Electronic Warfare*, Boston: Artech House.

Adamy, D. (2003) *Introduction to Electronic Warfare Modelling and Simulation*, Boston: Artech House.

Adkins, B.N. Major, USAF. (April 2001) *The Spectrum of Cyber Conflict from Hacking to Information Warfare: What is Law Enforcement's Role?* Air Command and Staff College, Air

University, [AU/ACSC/003/2001-04]. Maxwell Air Force Base, Alabama

Alford, L. D., Jr., USAF, Col. (2000) Cyber Warfare: Protecting Military Systems, *Acquisition*

Review Quarterly, Spring 2000, V.7, No. 2, P,105, http://www.dtic.mil/dtic/tr/fulltext/u2/A487951.pdf

Angelov, P. (2012). *Sense and avoid in UAS research and applications*. Hoboken, N.J.: Wiley.

Anonymous, JP 3-13 (Joint Publication) and pertains to Information Operations (IO) in the United States. http://www.dtic.mil/doctrine/new_pubs/jp3_13.pdf

Anonymous, (May 1995) *Introduction to Information Systems Security* (INFOSEC) *Guidebook*,

Askin, O., Irmak, R, and Avseyer, M. (14 May 2015) Cyber warfare and electronic warfare integration in the operational environment of the future: cyber electronic warfare. Proceedings Vol 9458, Cyber Sensing 2015; 94580H (2015) SPIE Defense + Security, 2015, Baltimore, MD. https://doi.org/10.1117/12.2189351

Austin, R. (2010) UAVS *Design, Development and Deployment*, New York: Wiley.

Bageshwar, B.L., and Euteneur, E.A. (2015) Multi-intruder aircraft, multi-sensor tracking system, at 2015 IEEE/AIAA 34th Digital Avionics Systems Conference (DASC)

Barnhart, R.K., Hottman, S.B, Marshall D.M., and Shappee, E. (2012) *Introduction to Unmanned Aircraft Systems*. New York: CRC Press.

Burch, D. (205) RADAR *for Mariners*. New York, McGraw-Hill.

Cagalj, M. (2014) FELK 19: *Security of Wireless Networks*: University of Spain PPTX Presentation.

Chain Home coverage 1939-1940 English Coast, http://enacademic.com/pictures/enwiki/67/Chain_home_coverage.jpg

Coats, D [DNI]. (February 13, 2018) *Statement for the Record; Worldwide Threat Assessment of the Intelligence Community*, released through Director of National Intelligence Office.

Count upon Security: *The Five Steps of the Intelligence Cycle*, https://countuponsecurity.com/2015/08/15/the-5-steps-of-the-intelligence-cycle/ DLA 07282018

Cuomo, S. Maj., (September 2017) *Guardian Angel: Transforming the MAGTF and naval services*, USMC Association Bulletin, Volume 101, Issue 9.

Department of Defense – *Unmanned Systems Roadmap* 2013 *to* 2038, http://www.dtic.mil/dtic/tr/fulltext/u2/a592015.pdf

Department of the Navy, Naval Information Systems Management Center, NAVSO P-5239-01,

Module 01, MAY 1995. Last accessed 9-3-08 from www.everyspec.com/USN/NAVY+(General)/download.php?spec=NAVSO_P-5239-01.002606. PDF

DoD Dictionary http://www.jcs.mil/doctrine/dod_dictionary/

Drone Kill https://www.computerweekly.com/blog/Public-Sector-IT/Drone-kill-communications-net-illustrated

Encyclopedia Britannica, (1994) Radar spherical Geometry to target https://www.britannica.com/technology/radar

FAA Booklet: Introduction to TCAS II Version 7.1, (2018) see: https://www.faa.gov/documentLibrary/media/Advisory_Circular/TCAS%20II%20V7.1%20Intro%20booklet.pdf

Fitts, R.E., (1980) *The Strategy of Electromagnetic Conflict*, Los Altos, CA: Peninsula Publishing.

FM 3-38 (2014) *Cyber Electromagnetic Activities* (US Army)

Gelbart, A., Redman, B.C, Light, R.S., Schwartzlow, C.A., and Griffis, A.J. (2002) Flash LiDAR based on multiple-slit streak tube imaging. LiDAR VII. *Laser Radar Technology and Applications* 4723, pp 9-18.

Glossary of NIST Terminology: https://nvlpubs.nist.gov/nistpubs/Legacy/IR/nistir7298r1.pdf

Headquarters Marine Corps, *Marine Corps Operating Concept: How an Expeditionary Force Fights and Wins in the 21st Century*, (Washington, DC: September 2016)

Information Assurance definition (additional) taken from and last accessed on 9-13-08 http://www.pcmag.com/encyclopedia_term/0,2542,t=information+assurance&i=44936,00.asp

Kissel, R. ed. (February 2011) NIST Glossary of Key Information Security Terms, NIST IR 7298 Revision 1, Ghttps://nvlpubs.nist.gov/nistpubs/Legacy/IR/nistir7298r1.pdf, DLA 07292018

LOS Overview: http://www.uastrain.com/guides/large-certifiable-uas-integration-in-the-national-airspace-system/civil-and-state-uas-types-and-missions-case-studies/

Marshall, D. M., Barnhart, R.K., Shappee, E., & Most, M. (2016) *Introduction to Unmanned Aircraft Systems*, 2nd Edition. New York: CRC Press.

Merrick, K, *Hardhienata*, M., *Shafi*, K. *and Hu*, J. (*July* 22,2016) “A Survey of Game Theoretic Approaches to Modelling Decision-Making in Information Warfare Scenarios”

Future Internet **2016**, 8(3), 34; doi:10.3390/fi8030034

Moir, I and Seabridge, A. (2006) *Military Avionics Systems*, New York: Wiley Aerospace Series.

NASA Unmanned Aircraft Systems (UAS) Integration in the National Airspace System (NAS)

National Institute for Cybersecurity Education (NCIE): http://csrc.nist.gov/nice/

National Institute for Cybersecurity Careers and Studies (NICCS): http://niccs.us-cert.gov/

Nichols, R. K. (Sept. 5, 2008) *Cyber Counterintelligence & Sensitive Compartmented Information Facility* (SCIF) *Needs – Talking Points*, Private Memo to R. Bruce McBride.

Nichols, R., McBride, R.B., Johnsen, J.J. (March 31, 2009) *Feasibility Proposal: M.S. Degree*

Program in Cybersecurity and Computer Forensics School of Business and Justice Studies, Economic Crime, Cybersecurity and Justice Studies, Utica, New York: Utica College Project, https://www.nasa.gov/feature/autonomous-systems DLA 07282018

Pettit, R.H. (1982) ECM *and* ECCM *Techniques for Digital Communication Systems*. Belmont, CA: Lifetime Learning Publications

Political Assassination Robots http://www.panafricanistinternational.org/?p=1390

Price, H. (1995) Digital Communications, column, NK6K, from QEX, http://www.qsl.net/n9zia/ss.qexss.html

UAS Traffic Management (UTM) https://dronelife.com/2018/02/12/utm-conversation-amit-ganjoo/

Reich, M., Carr, C., and Gunsch, G. (2002) *An Examination of Digital Forensic* Models, International Journal of Digital Evidence Fall 2002, 1, 3.

Simple Radar PPTX by Linkedin SlideShare (2018) https://www.slideshare.net/remotesensor1/radar-transmitter-4-1

Toomay, J.C. (1982) RADAR *for the Non – Specialist*. London; Lifetime Learning Publications

USMC Vision Guardian Angel: https://www.mca-marines.org/gazette/guardian-angel-uas

US Military Global Security AFP, 281009, *Political Assassination Robots* http://www.panafricanistinternational.org/?p=1390

Vatis, M. A. (1998). *Cybercrime, transnational crime, and intellectual property theft. Statement for the record before the Congressional Joint Economic Committee.* http://www.ilspi.com/vatis.htm DLA = 07282018.

Wikipedia IO Joint Operations: https://en.wikipedia.org/wiki/File:IO_Integration_into_Joint_Operations_-_Notional.jpg

Wiley, R. G. (1993) *Electronic Intelligence: The Analysis of Radar Signals*, 2nd ed. Norwood, MA: Artech House.

Private Memos / Meetings

Notes from meeting Prof. Nichols meeting with KSU president General Richard Myers, 26 February 2108 at 1330-1430 CST at KSUP Stevens Board Room.

Video Report

Video Report on anti-drone gun, Quote by Amy Hu. Data Expert Technology LTD, https://www.youtube.com/watch?v=o057LmNGsJA DLA 07312018

SECTION VI

UAS / UAV HOSTILE USE & COUNTERMEASURES

Chapter 15: Africa - World's First Busiest Drone Operational Proving Ground

Student Learning Objectives – Africa has become the drone investment -playground of many nations. The student will be introduced to activities of these geopolitical players (US, France, EU, Germany, Egypt and China) and the significance of their intentions. The history of drone investments / operations in Africa is directly a function of the growth of terrorist organizations and African economy.

Africa – Overview

Africa is a developing continent comprised of unstable states due to undeveloped economy, poor education, and unified government among the states. Africa's leaders want to see their country develop and become a world leader. They look to their long-term allies to solve their issues.

Radical Islam continues to spread and threaten the future of Africa. With the turbulence of state's government, insurgence groups have joined forces with terrorist organizations affiliated with radical Islam. Radical Islamic extremists are a global security threat. *Therefore, several countries fighting terror at home are also assisting Africa in the fight on terror*. A priority goal of many African leaders is to defeat terrorism. They feel this can be achieved by stopping terrorist organizations membership growth. Conflict on land is not the only issue facing Africa, maritime security is a huge factor in Africa's economic growth.

Other countries willingly assist Africa in protecting their waterways to reap the benefits of trade and profit. There are high stakes for China, European Union, and United States to ensure Africa's perimeter allows for safe passage of Commercial and Military vessels. Africa's land positioning and natural resources has made it an attractive global investment. This has given Africa the fastest growing economy in the world despite the lack of sustained infrastructure.

The ability to be flexible with technology testing and implementation has given Africa an advantage with the development of UAS. Now their infrastructure is being architected and constructed by technical companies and countries with an interest in a claiming Africa's resources as their own. *Africa is leading the* UAS *marketplace with the wiliness to formulate laws to suit the commercial* UAS *industry*.

Africa – The Facts

According to Theobald Barber, "there are over 3,000 protected areas in Africa, including 198

Marine Protected Areas, 50 Biosphere Reserves, and 129 UNESCO World Heritage Sites." (Staff-A, 2017)

"Africa has the second largest population in the world, at about one billion people." (Staff-A, 2017) Nigeria is ranked among the seven most populous countries, with 200 million citizens.[1] The United Nations has published Africa will have more than half of the world's population growth by 2050. "Africa is the world's poorest and most underdeveloped continent with a continental GDP that accounts for just 2.4% of global GDP. The national flag of Mozambique has the image of an AK-47 assault rifle embedded into it. It is one of only two national flags of UN member states to feature a firearm. The other is Guatemala." (Staff-A, 2017) Which could be an indication of their population's struggle for independence.[2] "About 41% of children in Africa aged between 5-10 years are actively involved in child labor. It is the hottest continent, with water scarcity impacting the lives of over 300 million Africans. (Staff-A, 2017)

Economics

Economy is a driver for the continent's decision to be accommodating to the UAS commercial community. When it comes to UAS development, Africa is the second most attractive investment destination in the world. See Figures 15-3 and 15-4.[3]

Africa is one of the most integrated regions in the world, ranking only behind Europe and Southeast Asia for economic integration. Africa is moving toward the negotiations for the establishment of a Continental Free Trade Area (CFTA) which will be the largest free-trade area in the world. See Figure 15-1.

The Spread of Radical Islam Across Africa

Africa – Al-Qaida and Islamic State

Starting in after the end of Qaddafi regime, Jihadists have spread from the Horn of Africa to Guinea-Bissau. Osama bin Laden in a statement in July 2006 clearly singled out Somalia as an important jihadist front of the future. Al-Qaida and Islamic State-linked groups in West Africa are now cooperating at the operational level and that there are connections with IS-affiliates in the Lake Chad region close to northeastern Nigeria and Maghreb-Sahel region. As a result of a three days of attacks in Mali (June 29 – July1, 2018), Islamic militants are in control of the area.

1. (Barber, 2018)
2. (The Guardian, 2015)
3. Figure 15-4 is an important figure as it relates to the investment return "payoffs" for UAS development by foreign and US governments.

They have closed schools and killed hundreds of civilians. This northeast location is a cornerstone to launch attacks against nearby countries. See Figure 15-2 Regional Conflicts Trends.

Figure 15-1 Africa: Economics

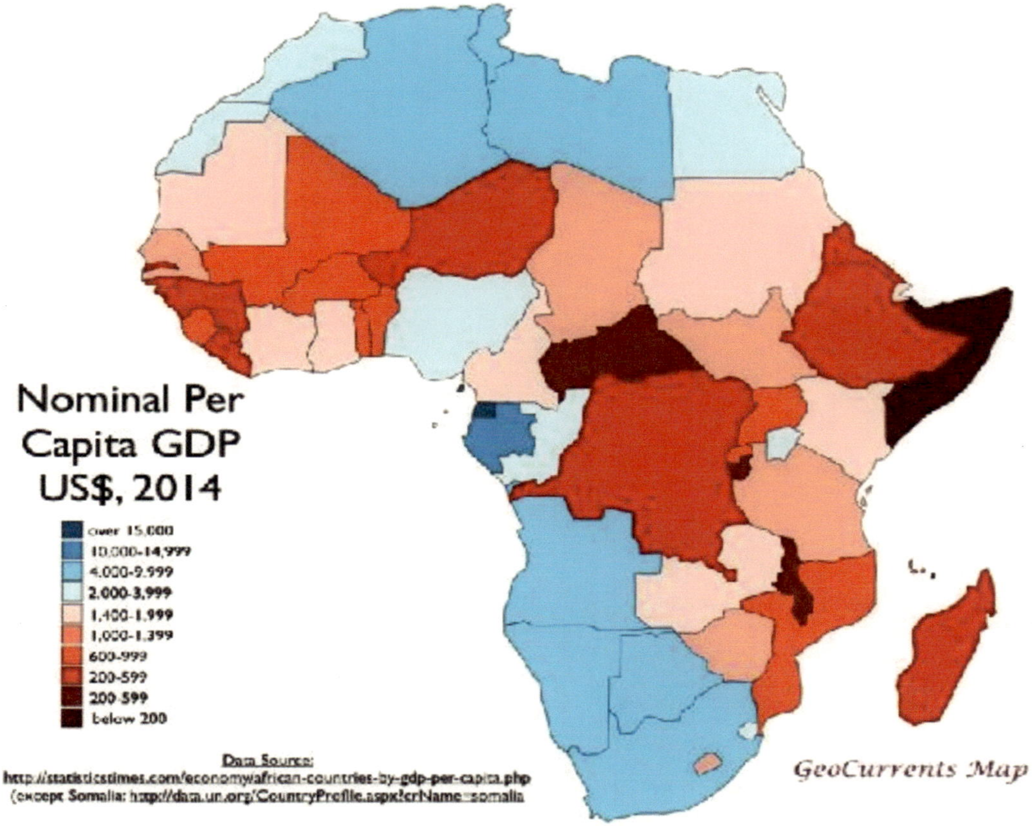

Source: Statisticstimes.com (Summer, 2018). List of African countries by GDP per Capita, http://statisticstimes.com/economy/african-countries-by-gdp-per-capita.php

The Spread of Radical Islam Across Africa

Africa – Al-Qaida and Islamic State

Salafi-Jihad Movement

Over the past 16 years the al Qaeda network, has become even stronger across Yemen, the Horn of Africa, Libya, and West Africa. Recently al Qaeda network recruited several groups across Middle East and North Africa during the Arab Spring's unrest. Their goal, the destruction of current Muslim societies using force and creation of what they regard as a true Islamic society.

By the joining together of smaller groups al Qaeda's survival even if the core group is defeated completely. Also, this type of formation makes it difficult for any country to defeat al Qaeda.

Figure 15- 2 Islamic Militant Groups in Africa[4]

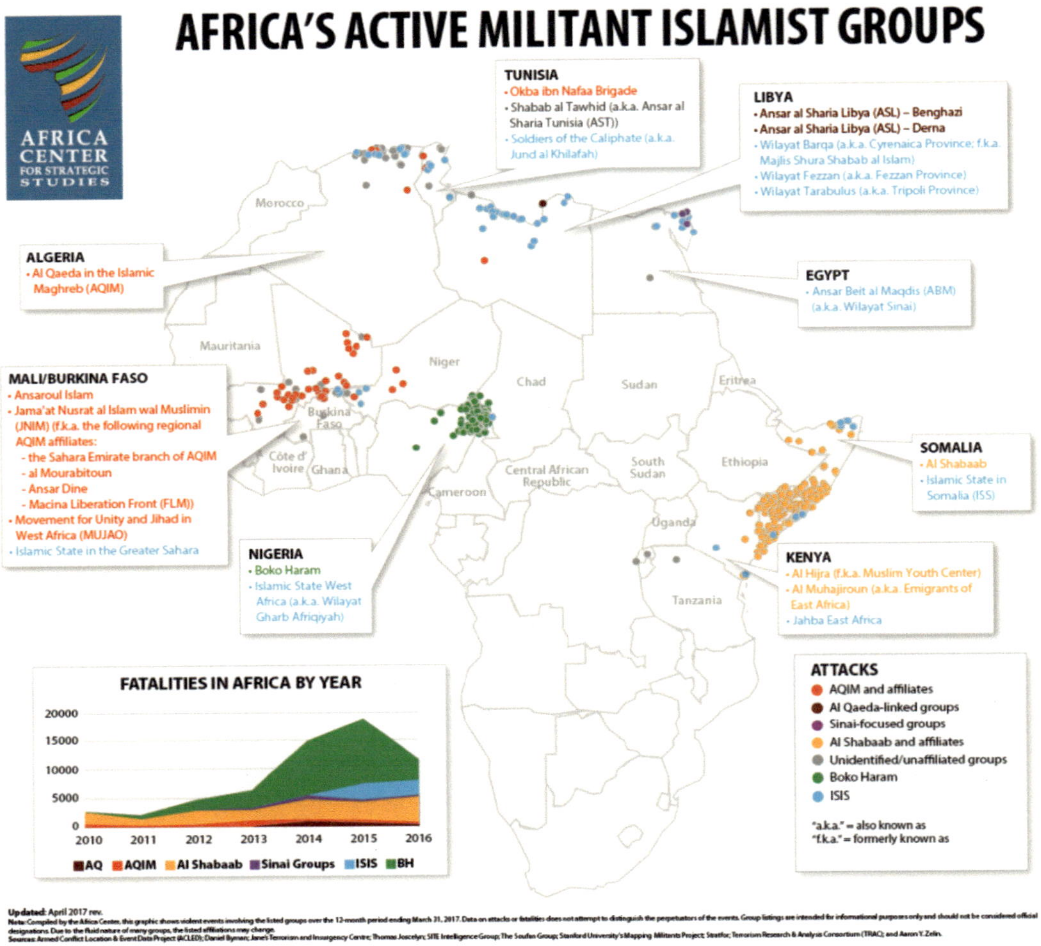

Source: Africa Center for Strategic Studies. (2017, April 26). Retrieved July 2018, from Map of Africa's Militant Islamist Groups http://africacenter.org/spotlight/map-africa-militant-islamic-groups-april-2017/

4. (Africa Center for Strategic Studies, 2017)

Figure 15-3 Africa -Population Distribution

Source: Barber, T. (2018, January 2018). Theobald Barber. Retrieved July 2018, from African Holidays http://www.theobaldbarber.com/africa-is-a-massive-continent-a-collection-of-55-countries/]

Figure 15-4 Africa: Primary Resources

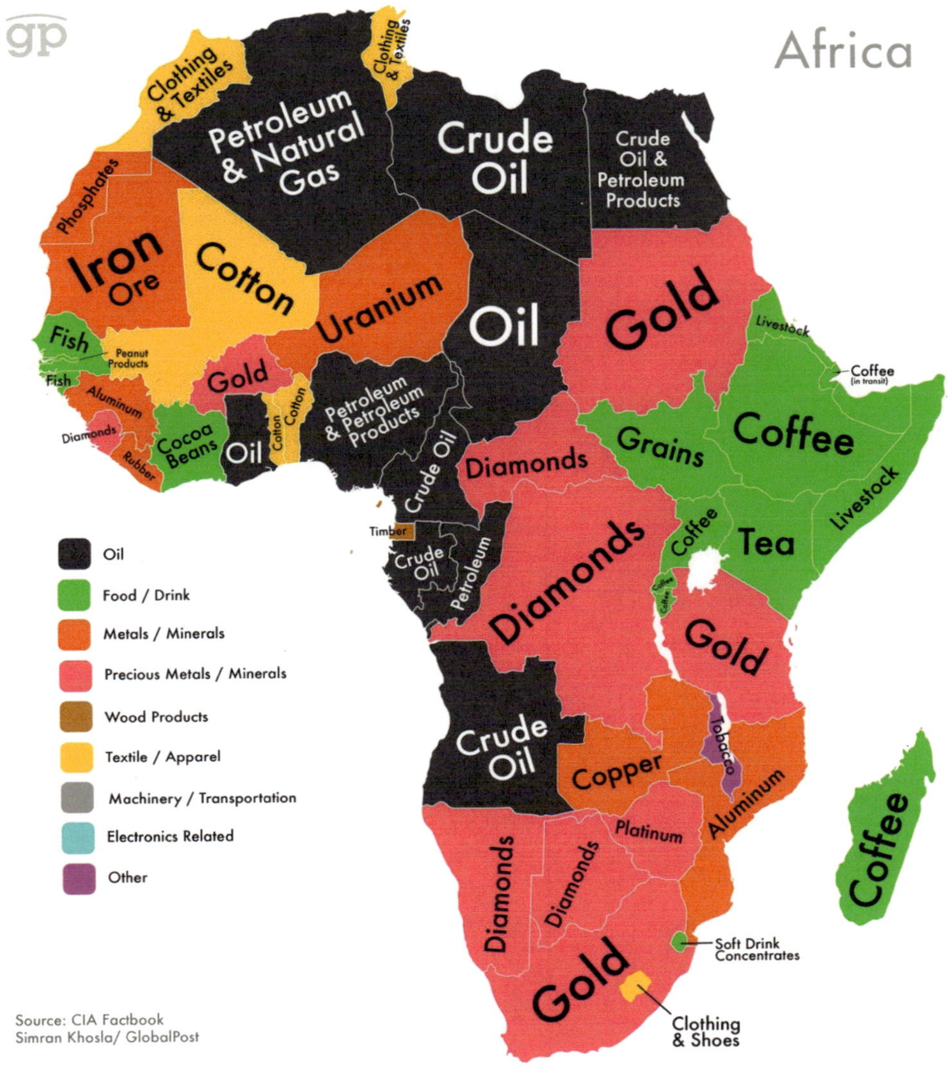

Source: Globalpost. (May 14, 2014). From *This map shows which export makes your country the most money* [Infographic on Africa the first product exported by every African state]. https://www.pri.org/stories/2014-05-14/map-shows-which-export-makes-your-country-most-money

Africa – *Katiba Macina Groups* (KM)

The KM Islam and Muslim groups operate mostly in Western Africa, in the center of Mali. Terrorist groups grow their ranks and recruit members through exploiting local conflicts to their own advantage. Growth continued because of absence of proper judicial systems, and to competition over natural resources.

Africa – Al-Qaida and Islamic State

The "Group for the Support of Islam and Muslims" (Jama'at Nusrat al-Islam wal-Muslimin) brings together four existing al Qaeda organizations under one banner. March 2, 2017 merged creating JNIM. This collective group is made up of members of Ansar Dine, Katiba Macina, al-Mourabitoun and al-Qaeda in the Islamic Maghreb (AQIM). Merger of al-Qaeda-linked groups in Mali might pose serious risks for regional security. JNIM operates in northern and central Mali including in the towns of Kidal, Timbuktu and Mopti. The merger of al-Qaeda-linked groups in Mali might pose serious risks for regional security. See Figure 15-5 – al-Qaeda in the Islamic Maghreb (AQIM).

Groups like al-Mourabitoun, with more experience, could share their expertise with new movements and field of explosives manufacturing. Combination of part of the groups increased attacks in 2016 by 150%. Currently, JNIM have six hostages a French, Australian, South African, Colombian, Swiss, and Romanian.

The group's largest attack, May 2017, JNIM disabled a communications tower outside of Gao, and used a suicide bomber to the Malian army base. The successful attack killed 7, wounded 16, and captured another 17 Malian soldiers. JNIM retreated from further terroristic action after the sight of French troops. The continue to wage attacks in northern Tunisia, using grenades, improvised explosive devices (IED), and firearms. See Figure 15-6.

Tuareg Rebellion [NMLA]

In 2012, Mali had experienced violence from ethnic Tuareg rebels who began a separatist insurgency and joined forces with Islamist militants to seize control. Similar attacks occurred in the last three months of 2017, in the central Mopti and Segou regions. This region has experienced more attacks then the five northern regions combined. In 2017, 22 Malian soldiers were killed. As a result, the Malian army imposed a ban on motorcycles and pickups in some areas to decrease opportunity for attacks. The attacks from the Militants use complex assaults with vehicles and landmines. The United Nations (U.N.) created Multidimensional Integrated Stabilization Mission in Mali (MINUSMA) after the National Movement for Liberation of Azawad (Tuareg Rebellion) [NMLA] began to invade Mali's Azawad region.

The U.N. mission, largest U.N. Peacekeeping mission with 13,000 members, protects the region from the combination of al-Qaeda in the Islamic Maghreb (AQIM) and Tuaregs. The U.N. troops continue to patrol and keep the peace along the Mali border.

Africa – Ansar Dine (AD)

"Mali based Tuareg, Al Murabitoon and Al Qaeda in the Islamic Maghreb's (AQIM) combination associate formed in December 2011. Its leader is part of JNIM. In 2013, Ansar Dine was recognized by the U.S. State Department. The group, AQIM, has since merged with the Movement for Unity and Jihad in West Africa (MUJAO) to fight against French and Malian forces." (Emad-Ceseden, 2013)

Figure 15 – 5 Salafi-Jihad Movement

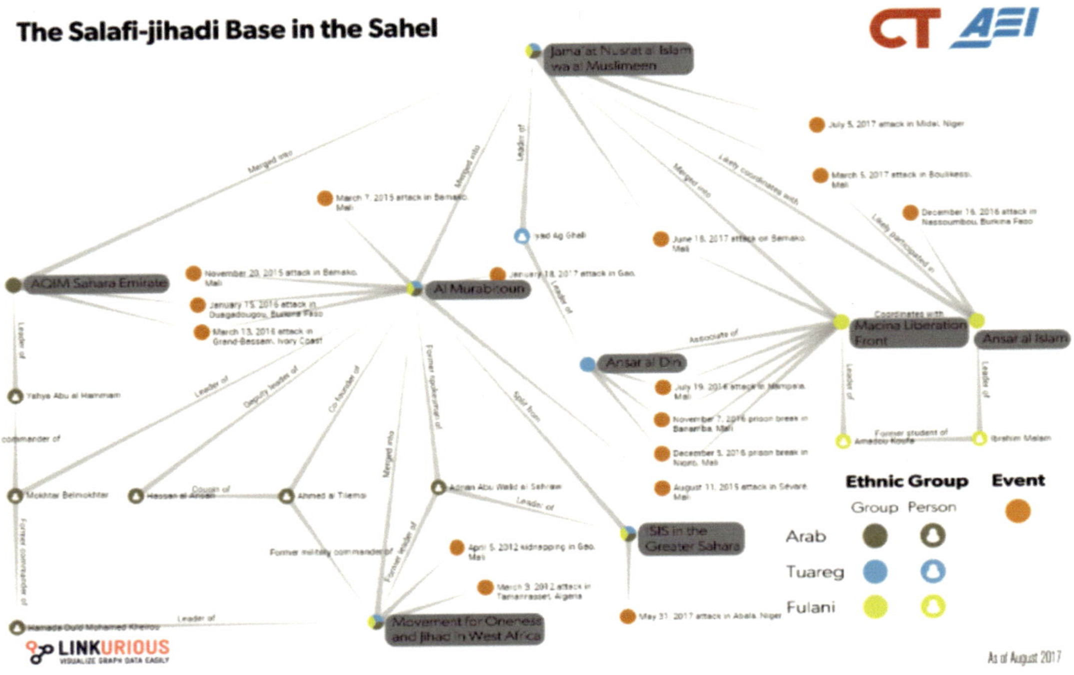

The Salafi-jihadi Base in the Sahel

Source: Critical Threats (CT). (summer, 2018). The Salafi-jihadi base in the Sahel – counter terrorism view. https://www.criticalthreats.org/analysis/The-Salafi-jihadi-Base-in-the-Sahel

Islamic State of Iraq and al Sham (ISIS)

Islamic State in the Greater Sahara Burkina Faso branch was recognized a new branch operating in the Sahel region of West Africa in 2016. A smaller section of the al Qaeda-associated al Murabitoon group gained ISIS leadership attention October 2016. The group executed several small-scale attacks in Burkina Faso and Niger. Including a Nigerien prison holding Boko Haram and al Qaeda in the Islamic Maghreb (AQIM) militants with a small team and at least one suicide bomber.

Figure 15-6 Terrorist Related Deaths in Africa

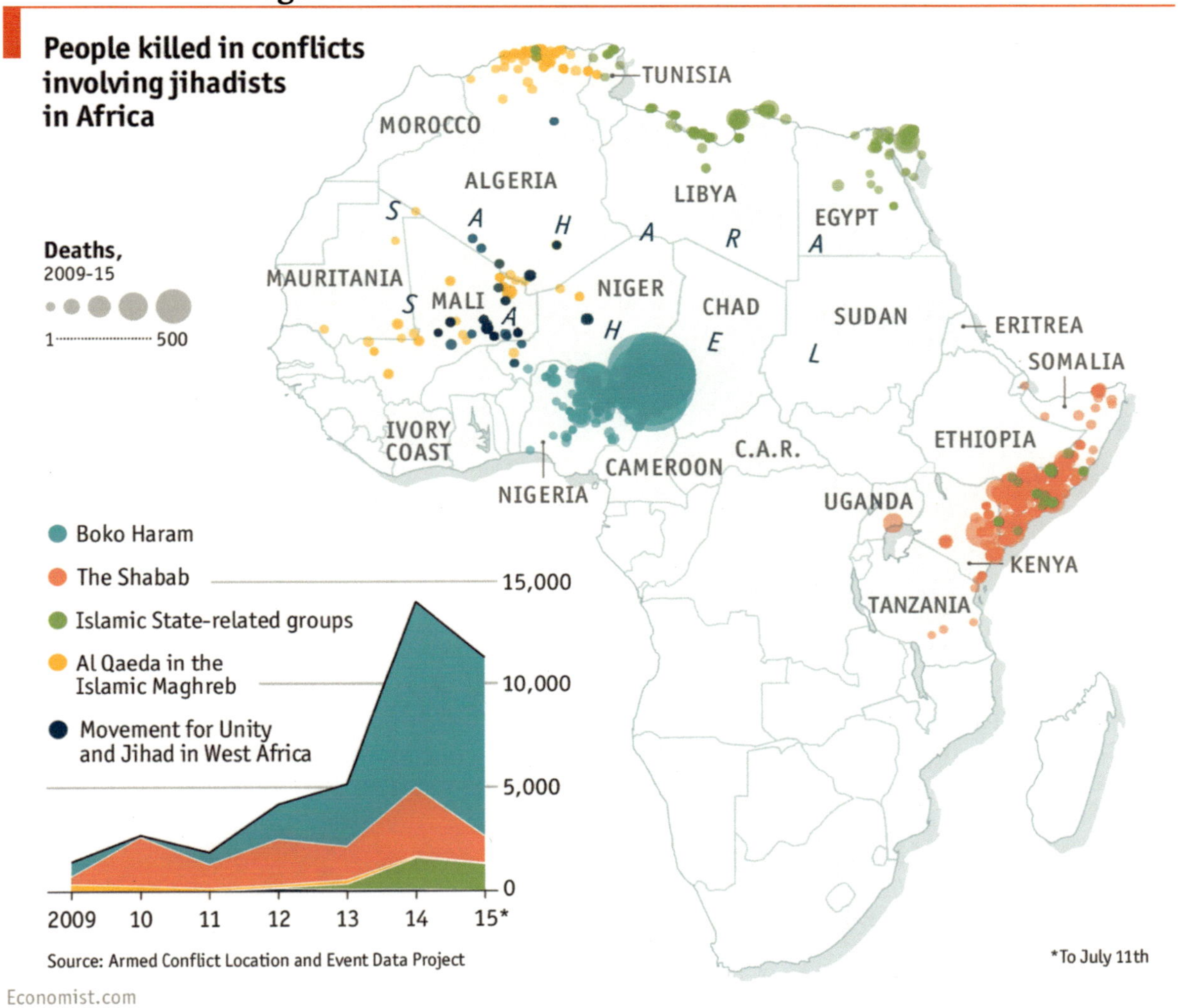

Source: Jihadists in Africa – a rising tide (July 20, 2015). In *The Economist*. https://www.economist.com/graphic-detail/2015/07/20/a-rising-tide

Figure 15-7 HOT SPOT – ARC OF INSTABILITY

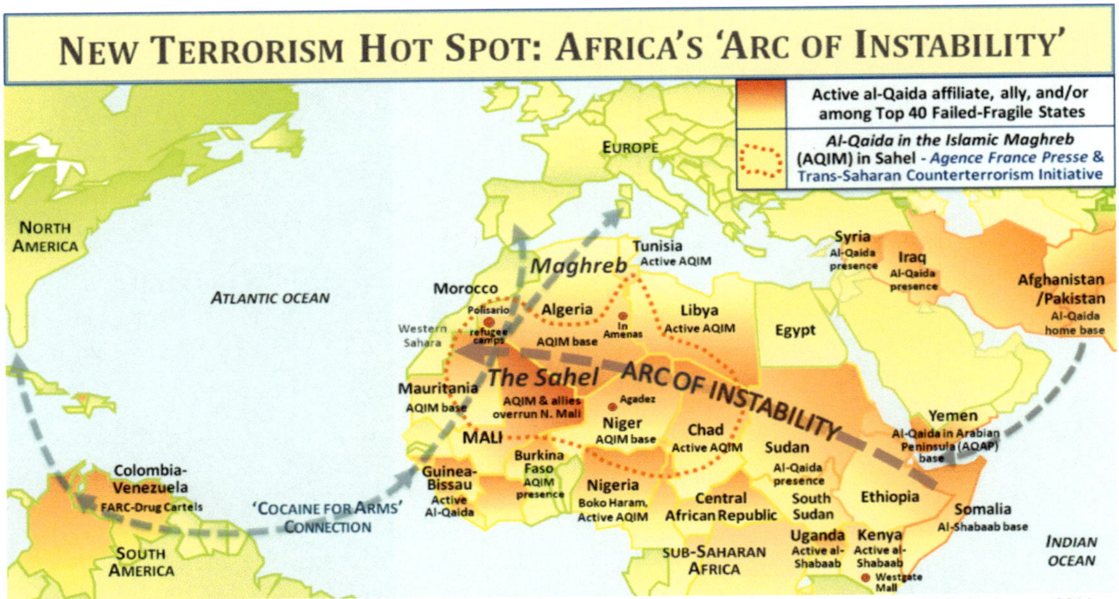

Terrorism in North Africa & the Sahel in 2013, by Yonah Alexander, Inter-University Center for Terrorism Studies (IUCTS). *January 2014*

Source: Terrorism: Two Reports Weigh In After President's State of the Union. (February 03, 2014). New terrorism hotspot, via *Morocco on the Move*. https://moroccoonthemove.com/2014/02/03/terrorism-two-reports-weigh-presidents-state-union-garth-neuffer/#sthash.uwiqLZys.dpbss

Africa – Counterterrorism Efforts

Why Fight Terror Groups in Africa?

Belgium has the highest number of foreign fighters in Syria of any Western European nation.

Because of France's presence in Africa, Al-Qaeda in the Islamic Maghreb (AQIM) has repeatedly issued calls for attacks against France. Since U.S. supports Israel, combined with the military involvement in Iraq and Afghanistan, *all terrorist groups challenge the US operations and facilities in Africa. See Figure 15-7.*

Joint European Union Counterterrorism

EU has deployed several Common Security and Defense Policy missions (CSDP) to provide some assistance in training armed forces, police and army, of these countries in addressing counterterrorism tactics and strategies. In April 2010, the EU launched a military training mission in Somalia (EUTM Somalia). By December 2016, the Council of European Union decided to prolong the mission in Somalia until 31 December 2018 with a budget of close to € 27 million. The commander of EUTM is from Italy. The mission has trained 5,000 Somalia soldiers (Somali National

Armed Forces (SNAF) to by end of 2016. The United Kingdom, in 2017, has committed military helicopters, surveillance aircraft, no troops considering Brexit. See Figure 15-28.

G5 Sahel – Five Africa States United

G5 Sahel (G5S) Joint Force, has membership of five states; Burkina Faso, Mali, Mauritania, Niger, and Chad. The battle with Al Qaeda is heavy in Africa's Sahel region. The G5S is comprised of up to 5,000 military and police personnel drawn from national battalions. Including the existing Liptako-Gourma task force (LGTF) established earlier this year by Burkina Faso, Mali, and Niger to secure their shared border region, with their headquarters located in central Mali.

In 2017, G5 Sahel (G5S) Joint Force was authorized by the African Union Peace and Security Council and the adoption of UN Security Council (UNSC) Resolution 2359. Mainly supported by France with 4,000 French troops deployed to the region. The U.S. pledged $60 million in support of the initiative. During an attack in June 2018, JNIM destroyed the entrance to G5S Headquarters, by a massive car bombing. See Figure 15-8.

Figure 15- 8 G5 Sahel – Five Africa States United

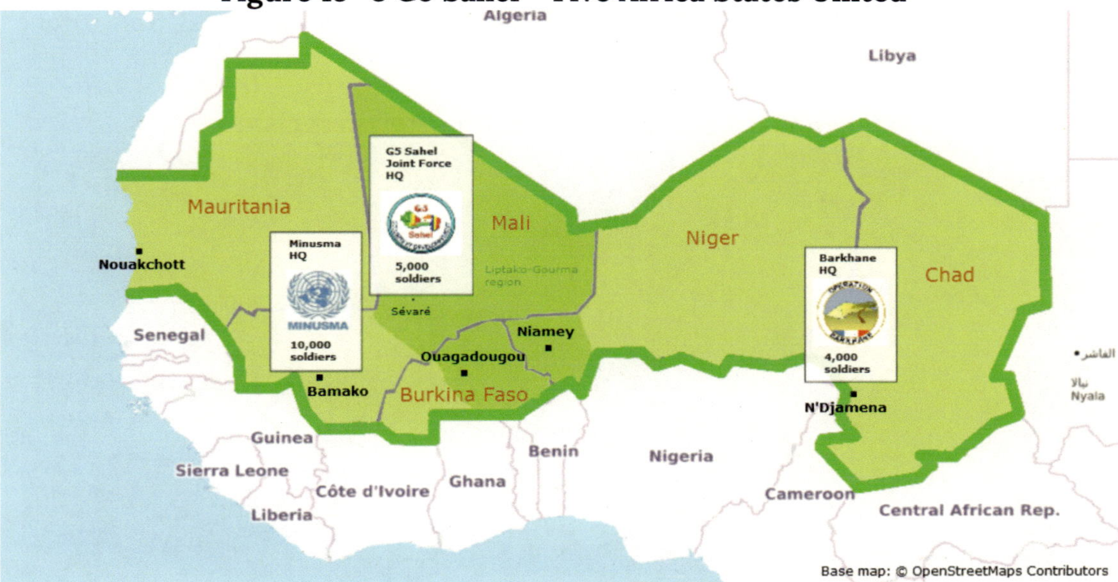

Source: Sahel Elite. (2017, November 16). Retrieved July 2018, from Sahel-Elite. https://httpsahel-elite.com/2017/11/16/mali-understanding-the-g5-sahel-joint-force-fighting-terror-building-regional-security/

United Nations Counterterrorism

UN has contracted Thales UK to operate 3 Hermes 900 drones out of Timbuktu to support UN peacekeeping missions in Mali. See Figure 15-9 In December 2015, Thales signed a three-year contract to support the UN. It was just renewed for another two years, through January 2020.

The primary mission of the Hermes drones are aerial surveillance for humanitarian convoys by UN to assist in protection from Al Qaeda in the Islamic Maghreb, Boko Haram and the Islamic State, See Figure 15-10.

Figure 15-9 An Elbit Hermes 900 drone at Timbuktu Airport, Mali

Source: Center for the Study of the Drone. (September 13, 2016). Retrieved from Drone Bases Updates, http://dronecenter.bard.edu/drone-bases-updates/

Figure 15-10 United Nations Counterterrorism

Source: Center for the Study of the Drone.(March 1, 2018). Drone Bases Updates. http://dronecenter.bard.edu/drone-bases-updates/

France – Operation Serval

Operation Serval began in 2012 by France. The operation's goal is to stop the jihadist incursion from Northern Mali into the Malian capital. Also, to stop all forms of passage, where jihadist groups between Libya and the Atlantic Ocean. French interest in Africa continues to grow with movement of French troops into Libya, Mali and the Central African Republic

France has acquired the MQ-9 Reaper to carryout Operation Serval. They wanted the MQ-9 for its ability to provide fire power and higher payload. However, the MQ-9 Reaper requires over a two-year timeframe for purchase through the U.S. weapon export policy and modifying them with European sensors. Because of U.S. stiff export drones, France's interest has dwindled. However, French special forces, both army and air force, possess mini-drones. The inventory includes the following, Israeli Elbit Skylark 1 and 1-LE, French Drac, Thales Spy Arrow, and U.S. AeroVironment Wasp. [See Figures 15-11, 15-12, 15-13 15-14.]

France – Operation Barkhane

Replacing Operation Serval in August 2014, France's new mission was to provide counterterrorism efforts in the Sahel region (G5S). France assisted the G5S region armed forces in fighting terrorist networks and prevented creation of terrorist safe-havens. France did this with

A mix of 20 mix of Gazelle, Puma and Cougar helicopters; 200 Armored vehicles, ten dedicated transport/reconnaissance aircraft, six fighter planes, and three Harfangs drones. In the G5S region, France established four permanent military bases. The headquarters of Operation Barkhane and air force in N'Djamena, a regional base in Gao, north Mali, a special-forces base in Burkina Faso's capital, an intelligence base in Niger's capital, an air base of Niamey. Niamey is an ideal location to hosts drones in charge of gathering intelligence across the entire Sahel-Saharan region.

Figure 15-11 France's MQ-9

Source: Larive, M. H. (2014, August 7). Welcome to France's New War on Terror in Africa: Operation Barkhane. Retrieved from http://nationalinterest.org/feature/welcome-frances-new-war-terror-africa-operation-barkhane-11029

Figure 15 -12 Thales Spy Arrow Drone

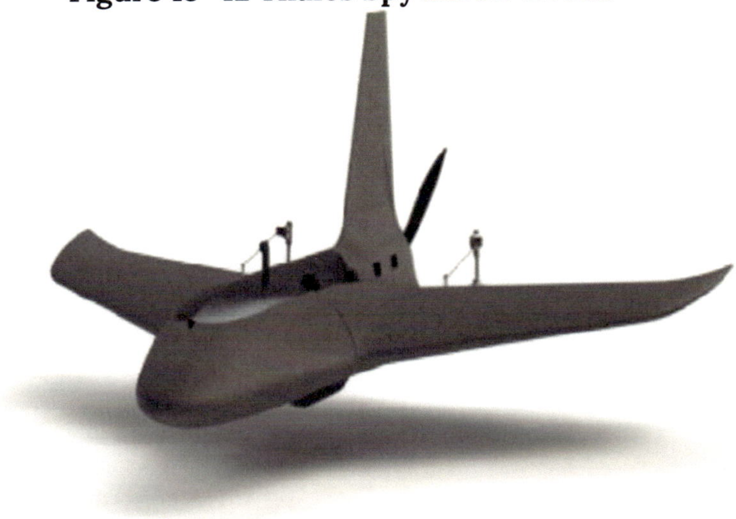

Source: Aviation Design. (2017). Thales Spy' Arrow Program. http://aviation-design-uav.fr/en/uav-thales-spy-arrow/

French EADS Harfangs

EADS Harfangs are modified Israeli IAI Herons, $25 million each for UAVs and ground stations. EADS/IAI's formal proposal to extend France's Heron-derived Harfang rent-a-drone service involves sensor upgrades, but no weapons. The experience in Mali and Libya are pushing France toward armed UAVs. It is possible to modify the Harfang UAVs to add RAFAEL's Spike-LR missiles, or MBDA's Viper Strike glide bombs. On January 8, 2018, a Harfang drone belonging to the 1/33 "Belfort" Drone Squadron landed for the final time on Cognac-Chateaubernard Air Base (BA 709), before its withdrawal from active service.

Figure 15 -13 Skylark 1 LE

Source: Skylark 1 LE Mini Unmanned Aerial vehicle (Mini-UAV). Nov 28, 2004. Retrieved from Defense Update, https://defense-update.com/20041128_skylark1-uav-2.html. (Image dated 2011.)

Figure 15-14 Harfangs or "Eagle"

Source: Drone Harfang. In *Actualités* (December 06, 2013). https://www.defense.gouv.fr/actualites/dossiers/le-bourget-2013/les-materiels-presentes/drone-harfang

France – West Africa

France has declared to focus on fighting Islamist militants as their primary foreign policy objective. France ordered six MQ-9 Reaper drones to replace EADS-made Harfang drones

The move to armed drones fits into a more aggressive policy. Estimated delivery by 2019

France working with Germany, Italy and Spain to develop a European drone field by 2025.

Trans-Sahara Counterterrorism Partnership (TSCTP)

The Trans-Sahara Counterterrorism Partnership was established in 2005. It is a U.S.-funded and implemented effort. "TSCTP partners include Algeria, Burkina Faso, Cameroon, Chad, Mali, Mauritania, Morocco, Niger, Nigeria, Senegal, and Tunisia." (Emad-Ceseden, 2013)

"TSCTP mission is to build the capacity and cooperation of military, law enforcement, and civilian actors across North and West Africa to counter-terrorism. With the TSCTP, the North and West African militaries will be enabled and enhanced their capacity to conduct counterterrorism operations. By integrating the ability of North and West African militaries and other supporting partners enables individual nations' border security to monitor, restrain, and interdict terrorist movements." (Emad-Ceseden, 2013) Also, by strengthening the rule of law, including access to justice, and law enforcement's ability to detect, disrupt, respond to, investigate, and prosecute terrorist activity. "The North and West African militaries, with TSCTP, can monitor and counter the financing of terrorism. In turn, reducing the limited sympathy and support among communities for violent extremism. TSCTP programs have worked to counter violent extremist radicalization and recruitment of youth, including educational and training courses in Algeria and Morocco, and extensive youth employment and outreach programs, community development, and media activities in Niger, Burkina Faso, and Chad." (Emad-Ceseden, 2013)

Partnership for Regional East Africa Counterterrorism (PREACT)

Funded by the U.S. government, PREACT, builds the capacity and cooperation of military, law enforcement, and civilian actors across East Africa for the mission of counterterrorism. PREACT has reduced the operational capacity of terrorist networks by developing a rule of law framework for countering terrorism in partner nations. In addition, PREACT has enhanced border security, countered the financing of terrorism, and reducing the appeal of radicalization and recruitment to violent extremism.

United States – West Africa

Supporting West Africa from terror actors, the U.S. has an Africa Command, located in the Nigerien air base in Niamey. The U.S. will use MQ-9 Reapers with precision-guided bombs and missiles, targeting Al Qaeda in the Islamic Maghreb, Boko Haram and the Islamic State. This is the second time that armed drones have been stationed and used in Africa by the U.S. See Figure 15-15.

Figure 15-15 United States – West Africa

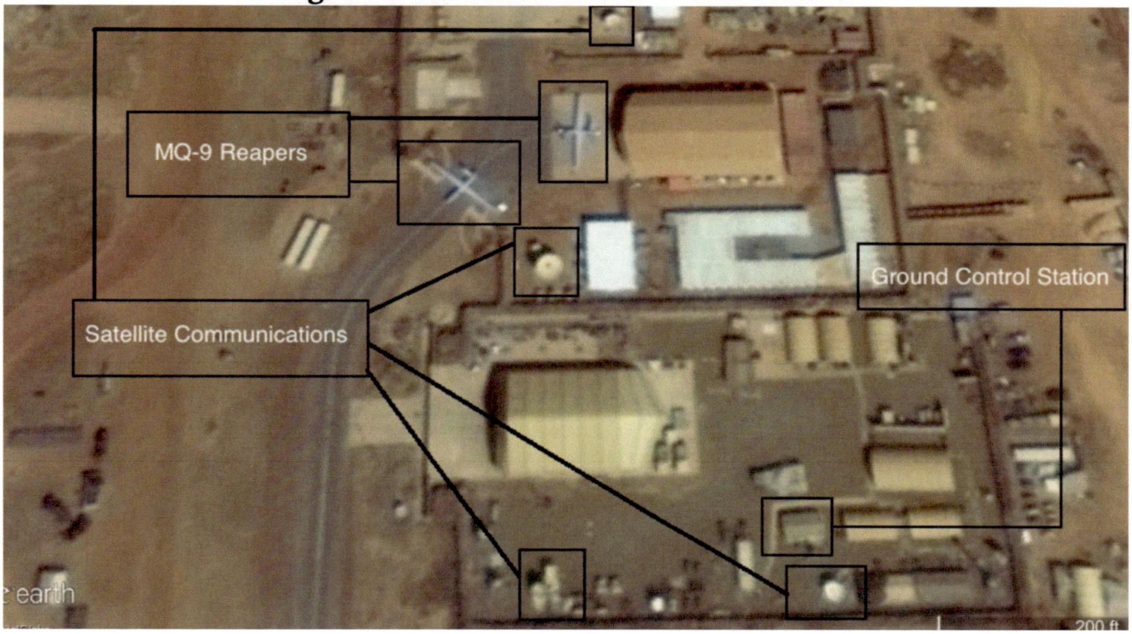

Source: Gettinger, D. (February 19, 2015). Features: How to Hunt for Drones. http://blogs.bard.edu/dronecenter/how-to-hunt-for-drones/ttp://blogs.bard.edu/dronecenter/how-to-hunt-for-drones/

Support from the U.S. has gone from in 2013, a 100 U.S. Troops assigned to Niger to 2018, 800 including special operations. U.S. Troops have been provided to support and facilitate intelligence collection and sharing with French forces, who are conducting CT operations in Mali. The U.S. provides guidance to local troops as they battle Boko Haram and al Qaeda. The U.S. is currently building a drone and airbase in the northern city of Agadez. In 2018, US Africa Command will launch its MQ-9 Reapers, nicknamed "hunter/killer" drones with advanced intelligence gathering capabilities. Niger was the only country in Africa granting the U.S. permission to build out a new U.S. airbase. The U.S. has maintained a presence at Guelmim Air Base in southern Morocco. Research suggested MQ-1 Predators have been operated here. See Figure 15-16.

Figure 15-16 Guelmim Air Base

Source: Vela, J.A. & Corbacho, J. (September, 2007). A new species of Lehua from Lower Ordovician of Dra Valley of Morocco. https://www.cia.gov/library/publications/the-world-factbook/geos/mo.html

United States – East Africa

On 25 June 2011, U.S. Predator drones attacked an al Shabaab training camp in **Somalia**, killing a senior a senior terror leader. Four Al-Shabaab fighters, including a Kenyan, were killed in a drone strike late February 2012. U.S. Drone strikes of Somalia Al-Shabaab fighters have increased: 14 in 2016 vs. 34 in 2017. As of January 2018, 24 more al-Shabaab fighters have been killed in 3 drone strikes.

The U.S. in Ethiopia has armed MQ-9 Reapers operating out of Arba Minch Airport. Since 2011, flights have been conducted with a primary mission to disrupt and eradicate terroristic operations in Ethiopia and Somalia. The U.S. Air Force has invested "millions" of dollars to upgrade the runway and airport in this area. See 15-17.

home to U.S. Special Operations forces and armed drones for operations in Somalia & Yemen

The Kenyan military has publicly denied it hosts U.S. drones or surveillance flights. Seychelles International Airport is home to an unknown number of MQ-9 Reapers. The drones carry out counter-terrorism missions in Somalia. There are has been two U.S. drone crashes at this area. See Figure 15-17.

United States – North Africa

U.S. wants to establish a drone base in North Africa, to fight against Libya based Islamic State (ISIS). U.S. drones used in Libya currently fly from NAS Sigonella on Sicily, Italy.

Libya based Islamic State (ISIS) targets attacks across Northern Africa.

However, the U.S. intelligence has a blind spot, thus the need for the additional drone base.

Figure 15-17 Seychelles International Airport

Source: Publicintelligence.net. (December 1, 2012). Located on the island of Mahé, Seychelles the Seychelles International Airport has hosted US drones since 2009. Since December 2011, two MQ-9's have crashed at this airport. https://publicintelligence.net/us-drones-in-africa/

In Kenya, the U.S. Navy is operating Camp Simba located near Manda Bay. The camp is current

ISIS in June 2015 executed an attack in Port Sousse 38 dead and the March 2015 Bardo Museum attack left 19 dead. U.S. Air Force MQ-1B Predators have been involved in reconnaissance and strike sorties in Operation Unified Protector. There are also some suggestions that a Predator was involved in the final attack against Muammar Gaddafi. The U.S. MQ-1Bs Predators along with MQ-9 Reapers returned to Libya in 2012, after the attack that killed the US Ambassador in Benghazi. The operation proved successful with the U.S. taking Libya from ISIS stronghold.

In May 2015, ISIS had 6,000 fighters in Libya. From August to December 2016, four-month air campaign over Sirte by three MQ-9 Reapers (flown from bases in Nevada, Tennessee and North Dakota) conducted 495 airstrikes.

United States – Central Africa

US Africa Command (AFRICOM) based in Cameroon has and inventory of MQ-1C drones.

AFRICOM claimed the base for location of being in the heart of the Central Africa counter-terrorism combat zone. See Figure 15-18.

Figure 15-18 AFRICOM Base in Cameroon

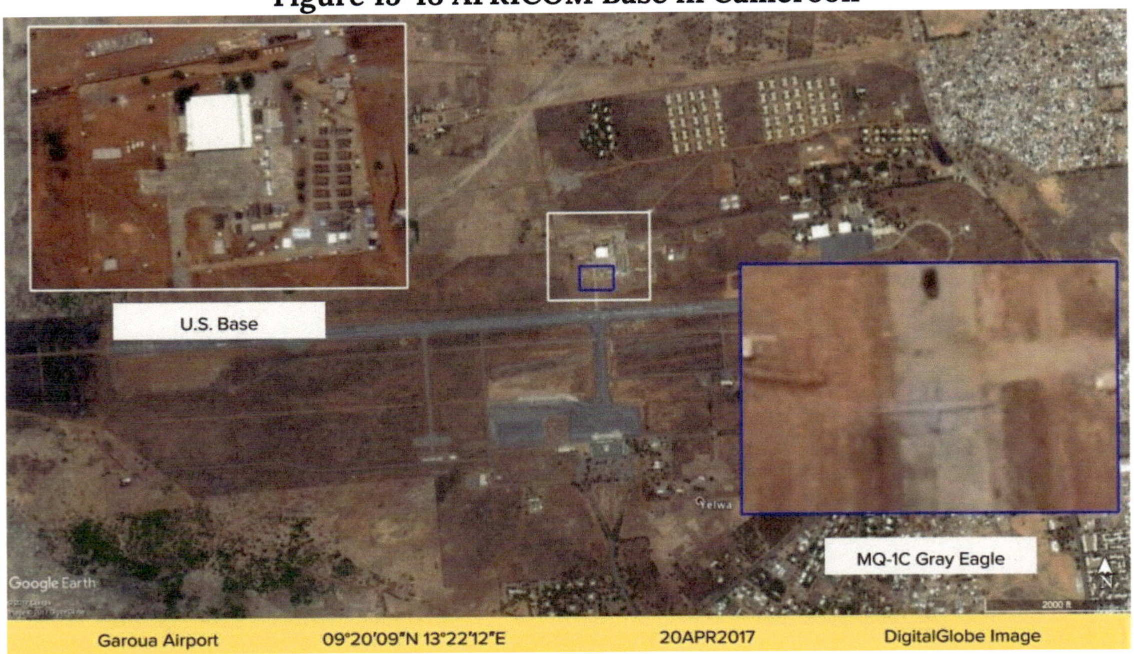

Source: Center for the study of Drone, S: (March 1, 2018), CT: Drone Bases Updates, URL: http://dronecenter.bard.edu/drone-bases-updates/

Cameroon

A satellite image of the U.S. drone base in Garoua, Cameroon. See Figure 15-19.

Figure 15-19 A satellite image of the U.S. drone base in Garoua, Cameroon

Source: Center for the study of Drone. (March 1, 2018). Drone Bases Updates. http://dronecenter.bard.edu/drone-bases-updates/]

Figure 15-20 China CH-5 Rainbow

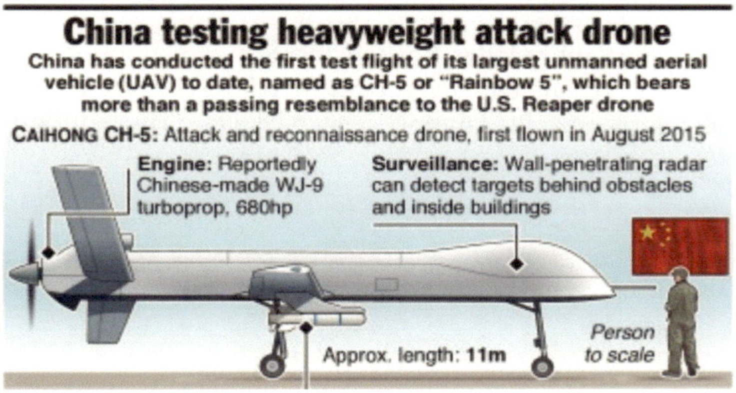

Source: China unveils Rainbow 5 mega killer #drone design – an annotated infographic. (September 9, 2015). In *Engineering & Technology Magazine*. Retrieved from https://engtechmag.wordpress.com/2015/09/09/china-unveils-rainbow-5-mega-killer-drone-design-an-annotated-infographic/

Figure 15-21 US MQ-9

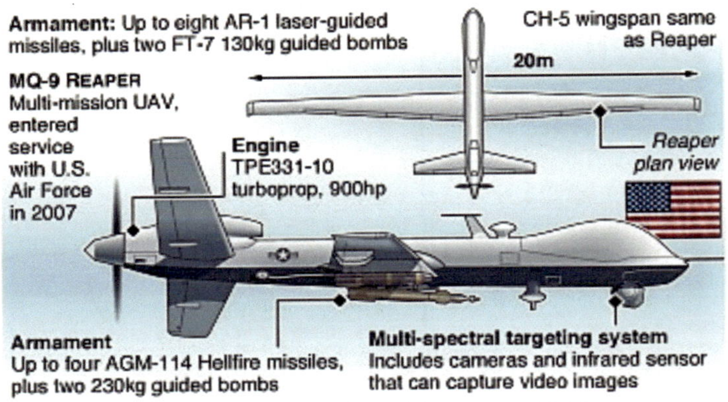

PERFORMANCE COMPARISON	*Caihong CH-5**	*MQ-9 Reaper*
Maximum takeoff weight	**3,000kg**	**4,760kg**
Payload	**900kg**	**1,700kg**
Maximum speed	**300km/h**	**480km/h**
Range	**2,000km**	**1,850km**
Endurance	**More than 30 hours**	**18-27 hours**

Sources: Chinese media, U.S. Air Force, Global Security *Estimated data © GRAPHIC NEWS

Source: China unveils Rainbow 5 mega killer #drone design – an annotated infographic. (September 9, 2015). In *Engineering & Technology Magazine*. Retrieved from https://engtechmag.wordpress.com/2015/09/09/china-unveils-rainbow-5-mega-killer-drone-design-an-annotated-infographic/

Figure 15-22 CAIG Wing Loong

Source: Aviation News. (June 18, 2017). Photo of Wing Loong Armed. https://www.ainonline.com/aviation-news/photos/aerospace/2017-06-18/paris-air-show-highlights

China Counterterrorism

Terrorism in Africa is not a high priority issue for China. However, the China-Africa Cooperative provides technical and financial assistance for the counterterrorism effort. China will not join missions associated with Western political agendas nor until terrorist attacks against Chinese entities in Africa or physical attacks in China by African terrorist groups. See Figure 15-20, 15-21 and 15-22. Compare the Chinese CH-5 to the MQ-9.

Israel Counterterrorism Efforts

"Africa and Israel have a long history of partnering with military operations. For example, Israeli anti-terrorist forces advised Kenyan units in connection with the 2013 attack on the Westgate Mall in the capital, Nairobi. Currently the countries are conduct operations together Al Qaeda and Palestinian militancy." (Emad-Ceseden, 2013)In 2016, Israeli counterterrorism and intelligence agencies set up a special command center in the Kenyan port city of Mombasa. Israel has successfully lead Kenyan operations against the Al Qaeda cells embedded in eastern and southern Kenya. Israel has trained the Elite guards that protect VIPs in Africa. See Figure 15-23.

East Africa union (EAU) is comprising of Israel and six East African states, Kenya, Ethiopia, Tanzania, Uganda, Rwanda and South Sudan; for to rid their countries of the jihadist Islamic State.

Kenyan army is receiving training from Israeli military instructors on tactics for battling Al Shabab, affiliates of Al Qaeda, inside Somalia and a supply of Israeli weapons tailored for anti-terror warfare. In return for military intelligence assistance for combating terror, the six East African governments are offering Israeli firms preferential treatment for developing their markets. August 2017, Israel offers to train drone operators for Ghana's Special Forces for tactical surveillance and counter-terrorism purposes. The Israeli officials said Israel was willing to finance the training of the drone operator program. Israel and Ghana cooperate on several defense matters. Ghana participates in the UNIFIL force in southern Lebanon, and has about 870 soldiers stationed there, stated a United Nations report in June. In addition, several Israeli defense companies operate in Ghana and provide services to Ghana military.

Figure 15-23 Israel -East Africa CT efforts

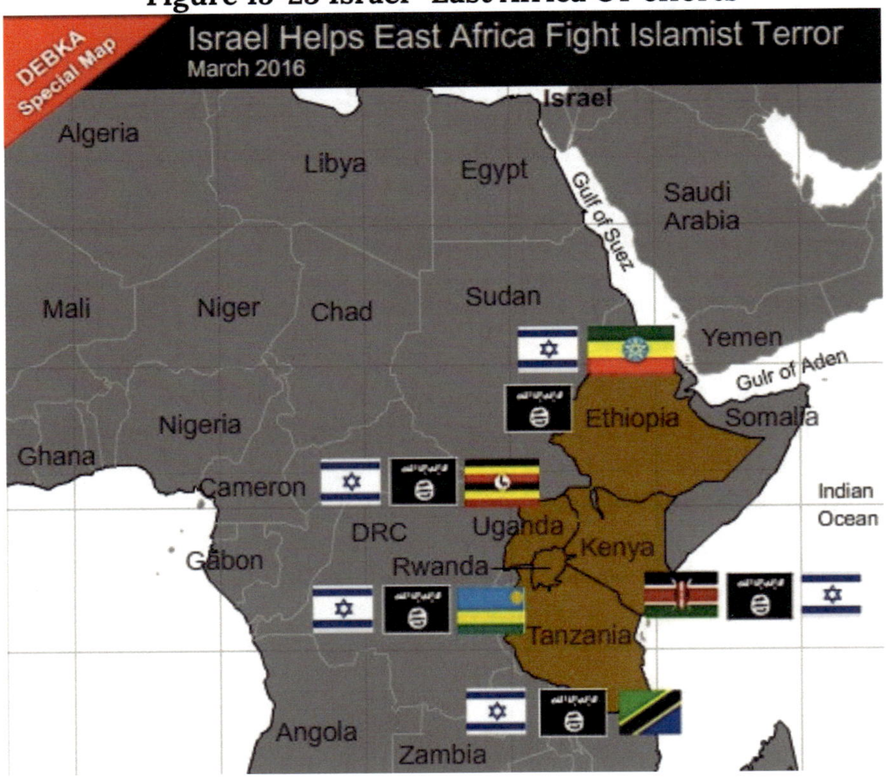

Source: East-African-union-form-counter-terror-task-force-Israel. (March 2016). In OSNET *Daily*, http://osnetdaily.com/2016/03/east-african-union-form-counter-terror-task-force-israel/

Figure 15-24 LUNA Drone

Source: By SSGT REYNALDO RAMON, USAF – Public Domain, https://commons.wikimedia.org/w/index.php?curid=1738164

Germany – West Africa

Germany plans to deploy Heron drones leased from IAI to Gao, Mali to join their reconnaissance and surveillance LUNA drones already there. After the August 13, 2017 terrorist attack in Burkina Faso, the country also accepted Germany's offer to train its soldiers in German military training camps. See Figure 15-24. Germany has committed to the African countries "defend their security and stability and fight against terror and organized crime" Germany Government allowed U.S. drones forces access to Ramstein AB, to fly Predator & Reaper drones, Global Hawk aircraft, under the condition that the Americans do nothing there that violates German law.

Pakistan – West Africa

Pakistan attraction to be involved in West Africa is simple, the trade of Nigeria liquefied natural gas and Pakistani farm tractors. West Africa welcomes the Pakistan military hardware to assist with the fight against Boko Haram, who represents ISIS in Nigeria. Boko Haram continually conducts attacks against Nigerian government forces. Pakistani assembled drones, like the Chinese CH-3, assist from the air.

Egypt – North Africa

The airbase, Bir Gifgafa, is 50 miles east of Suez Canal is home to Egypt's Wing Loong drones.

Drone deployment to Bir Gifgafa to combat Islamist insurgency on Sinai Peninsula in support of Northern Africa continues to grow with further construction of the airbase. See Figure 15-25.

Figure 15-25 Egypt – North Africa Satellite images from Nov 2016 show 4 Wing Loong drones

Source: Center for the study of Drone. (March 1, 2018). Drone Bases Updates http://dronecenter.bard.edu/drone-bases-updates/

Italy/France/United States – East Africa

Chabelley Airfield, Djibouti is the French built but U.S. military operated airfield, French allies use for training and as a divert location ten miles from the capital of the small African nation of Djibouti. U.S. Air Force's 870th Air Expeditionary Squadron run U.S. operations from airfield (with permission of French and Djiboutian government) Used for U.S. drone operations over Somalia and Yemen From September 2014 to February 2015 Italian Air Force operated MQ-1 Predators from Chabelley to support counter-piracy missions. See Figure 15-26.

Figure 15-26 Chabelley Airfield Djibouti

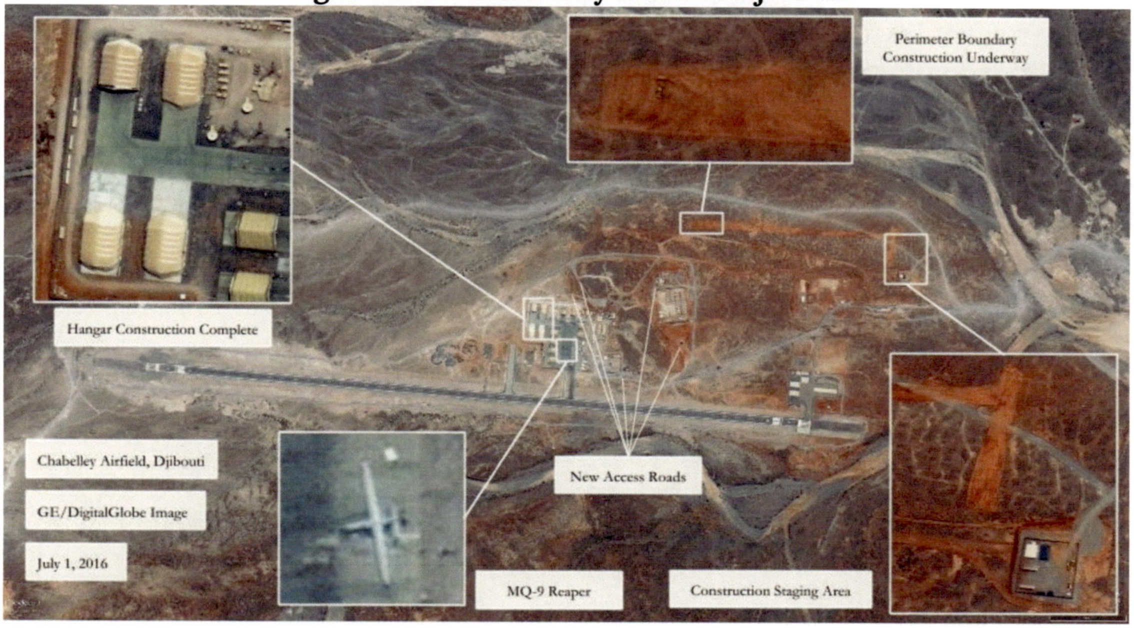

Source: Center for the study of Drone. (March 1, 2018). Drone Bases Updates http://dronecenter.bard.edu/drone-bases-updates/

Africa Maritime Piracy and Violence

Africa's Maritime Security

Maritime security has become a priority for most of Africa's regional communities since the increase in piracy. The importance of protecting their maritime territories needs to occur to protect regional economies. Africa realizes the "blue economy" will increase with the safeguarding of the area. The continent has a coastline of 18,950 miles, according to international law permits coastal states to claim as much as 200 nautical miles from their coastlines. The Southern African Development Community (South Africa, Mozambique and Tanzania) have coastlines on both the Indian and Atlantic oceans. The threats faced by the sub-region include illegal and irregular fishing and smuggling. As a result, the three states agreed to conduct joint patrols, military exercises and surveillance

China – Africa's Maritime

Port of Djibouti home to China's People's Liberation Navy (PLAN) new naval base in the Horn of Africa nation, which is home to approximately a thousand Chinese military personnel. PLAN escorts ships in the Gulf of Aden and the waters off the Somali coast and perform humanitarian rescues. The new PLAN naval base will provide a place for rest and rehabilitation for the Chinese troops taking part in U.N. peacekeeping mission. Will be used to ensure China's performance of missions, such as escorting, peace-keeping and humanitarian aid in Africa and west Asia. Images appear to show evidence of a squadron building, seven shelters for helicopters or UVA's along a 1,300-foot long apron, which is intended to accommodate aircraft for loading, refueling or maintenance. In Africa, Chinese business investment grew from $10 billion to more than $200 billion between 2000 and 2016, U.S. investment remained stagnant. See Figure 15-27.

Figure 15-27 China Navy Base – Horn of Africa nation

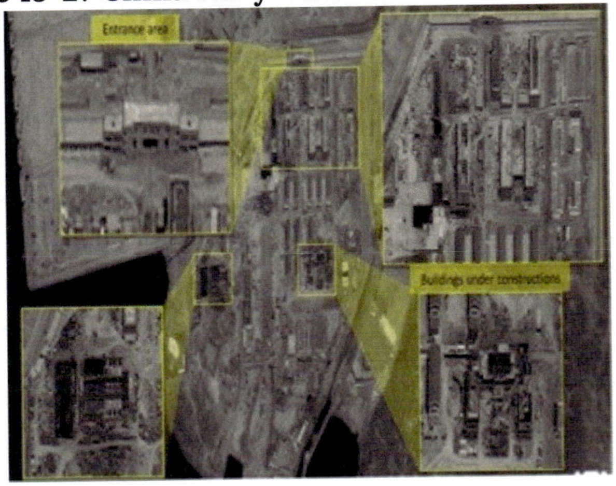

Source: Fox News. (July 12, 2017). Chinese personnel set sail for first African military base. http://www.foxnews.com/world/2017/07/12/satellite-images-show-progress-at-chinas-first-african-base.html

European Union Naval Force's (EUNAVFOR)

Operation Atalanta, was launched in 2008, as part of European Union Naval Force's (EUNAVFOR) anti-piracy naval mission. The operation runs from the sea off the Horn of Africa and in the Western Indian Ocean. It deters and disrupts piracy and armed robbery at sea and monitors fishing activities off the coast of Somalia. The operation supports other EU missions and international organizations working to strengthen maritime security and capacity in the region. EUNAVFOR deploys UAVs for intelligence, surveillance, target acquisition, and reconnaissance. On February 23, 2018, EUNAVFOR, responded to piracy attack reported by Motor Tanker Leopard Sun off coast of Somalia. Italian Airforce uses remote controlled Predator unmanned aerial system (UAS) in support of Operation Atalanta off coast of Somalia. To execute long-range surveillance and reconnaissance patrolling missions The Italian Air Force's Predator B UAS is a long-endurance, medium-high-altitude RPA designed for surveillance, military reconnaissance and targeting, and close air support missions over land or sea. Predator B has a single Honeywell turboprop engine, the aircraft can stay airborne for up to 27 hours at 50,000ft altitude.

Morocco's Commercial Activity in Africa

A Morocco-based startup Atlan Space has developed software to use drones for monitoring illegal maritime activity like illegal fishing or oil spills. Testing has been taking place in Uganda.

UAVs operated by this software can be launched and deployed into monitoring operations without having an aircraft operator.

Figure 15-28 Italian Air Force's Predator B UAS

Source: Air Force Technology. (September 10, 2014). Italian Air Force completes UAS sortie for EU NAVFOR's Operation Atalanta. https://www.airforce-technology.com/news/newsitalian-air-force-completes-uas-sortie-for-eu-navfors-operation-atalanta-4369489/

UNICEF *and Virginia Tech*

Virginia Tech Unmanned Systems Lab designed a drone, called EcoSoar, that could perform drug delivery flights. Virginia Tech worked with UNICEF to teach Malawian students to build EcoSoar. In June 2017, UNICEF and Government of Malawi conducted flight testing in the drone testing corridor in Kasung. The University hosted a two-day workshop for 13 Malawian students and faculty to teach the construction of the aircraft that is made of foam core (poster board) and 3D printed parts. See Figure 15-29.

Figure 15-29 Virginia Tech and Malawian Students

Source: UNICEF by Ong, A. (January 11, 2018). Malawi: Low-cost drone built by students delivers medicine over 19 km distance. https://blogs.unicef.org/innovation/malawi-low-cost-drone-built-students-delivers-medicine-19-km-distance/

Summary

Africa's rich natural resources attract other countries for trade in exchange for military protection and commercial growth of the continent. Terror groups continue to violently spread through strategic areas of the Africa to gain ideal land for attacking neighboring countries. The European Union, Russia, China, and the U.S. continue to peacekeeping support and military training of African troops in several regions. However, the future of U.K. support is not clear and U.S. Special Forces are considering reducing operations by half over the next three years. China is one of the largest aid donors to Africa across infrastructure, education, and military. Despite Africa's fight against radical Islam, their infrastructure and commercial UAS market continues to lead the globe.

Discussion Questions

1. Why is Africa ideal continent to test and grow innovation?
2. How can the spread of radical Islam hinder Africa's growth? Support from other countries?
3. Why does Africa have the potential to lead the UAS drone market?

Bibliography

Africa Center for Strategic Studies. (2017, April 26). *The Africa Center for Strategic Studies*. Retrieved July 2018, from Map of Africa's Militant Islamist Groups: http://africacenter.org/spotlight/map-africa-militant-islamic-groups-april-2017/

Barber, T. (2018, January 2018). *Theobald Barber*. Retrieved July 2018, from African Holidays: http://www.theobaldbarber.com/africa-is-a-massive-continent-a-collection-of-55-countries/

Ekwe, D. (2017, 20 June). *Steemit Exciting*. Retrieved July 2018, from Steemit: https://steemit.com/exciting/@ekwedavid/amazing-facts-of-africa

Sahel Elite. (2017, November 16). Retrieved July 2018, from Sahel Elite Mali: https://httpsahelelite.com/2017/11/16/mali-understanding-the-g5-sahel-joint-force-fighting-terror-building-regional-security/

Staff-A. (2017, December 16). *Volunteering Solutions Blog*. Retrieved from Africa Facts: https://www.volunteeringsolutions.com/blog/interesting-facts-about-africa/

The Guardian. (2015, October 20). *The Guardian*. Retrieved July 2018, from The Guardian Global Development: https://www.theguardian.com/global-development/2015/oct/20/two-thirds-of-worlds-illiterate-adults-are-women-report-finds

Readings

Africa Center for Strategic Studies. (2017, April 26). *The Africa Center for Strategic Studies*. Retrieved July 2018, from Map of Africa's Militant Islamist Groups: http://africacenter.org/spotlight/map-africa-militant-islamic-groups-april-2017/

Allani, R., Monan, D., Mueller, N., Puscas, I., & Watanabe, L. (2016). EU and Maghreb Countries: Counterterrorism Cooperation. Retrieved from Swiss Federal Institute of Technology Zurich website: https://www.files.ethz.ch/isn/130708/EU%20and%20Maghreb%20Countries.pdf

Anyadike, O. (2017, March 17). *UPDATED: A rough guide to foreign military bases in Africa. Retrieved February 17, 2018, from https://www.irinnews.org/feature/2017/02/15/updated-rough-guide-foreign-military-bases-africa

Aviation Week Network. (2018, February 5). China Commits to Becoming Global Power. Aviation Week. Retrieved from http://aviationweek.com/singapore-airshow-2018/china-commits-becoming-global-power

Barber, T. (2018, January 2018). *Theobald Barber*. Retrieved July 2018, from African Holidays: http://www.theobaldbarber.com/africa-is-a-massive-continent-a-collection-of-55-countries/

Beck, J. (2017, January 3). ISIL ramps up fight with weaponized drones. Retrieved from https://www.aljazeera.com/indepth/features/2016/12/isil-ramps-fight-weaponised-drones-161231130818470.html

Bekdil, B. E. (2017, March 7). Turkey gets additional drones to fight ISIS, Kurds. Retrieved from https://www.defensenews.com/smr/unmanned-unleashed/2017/03/07/turkey-gets-additional-drones-to-fight-isis-kurds/

Boko Haram, not ISIS, is world's deadliest, study finds. (2015, November 18). Retrieved from https://www.cbsnews.com/news/boko-haram-isis-worlds-deadliest-terrorist-group-study/

Bolduc, D. C., Puglisi, R. V., & Kaailau, R. (2017, May 29). The Gray Zone in Africa | Small Wars Journal. Retrieved from http://smallwarsjournal.com/jrnl/art/the-gray-zone-in-africa

Brimelow, B. (2017, November 16). Chinese drones may soon swarm the market – and that could be very bad for the US. Business Insider. Retrieved from http://www.businessinsider.com/chinese-drones-swarm-market-2017-11

Chase, M. S., Gunness, K. A., Morris, L. J., Berkowitz, S. K., & Purser, B. S. (2015). Emerging Trends in China's Development of Unmanned Systems (rr990). Retrieved from Rand National Defense Research Inst Santa Monica Ca Website: www.rand.org/t/rr990

Child, D. (2018, January 30). From Kigali to Khartoum: Africa's drone revolution. Retrieved from https://www.aljazeera.com/indepth/features/africa-drone-revolution-180123090528801.html

The China Africa Project Email Newsletter. (2018, January 21). Retrieved January 21, 2018, from http://www.chinaafricaproject.com/category/military/

China Power. (2017, December 8). Where is China targeting its development finance? | China Power Project. Retrieved from https://chinapower.csis.org/china-development-finance/

China Power. (2017, December 21). Is China contributing to the United Nations? mission? | China Power Project. Retrieved from https://chinapower.csis.org/china-un-mission/

Cooney, P. (2015, July 13). U.S. mulls drones in North Africa to monitor ISIS in Libya, Wall Street Journal reports. Retrieved from https://www.haaretz.com/u-s-mulls-drones-in-north-africa-1.5304009

Defense Industry Daily Staff. (2014, May 22). Apres Harfang: Frances Next High-End UAVs.

Retrieved from https://www.defenseindustrydaily.com/apres-harfang-frances-next-high-end-uav-06451/

Dorsey, J. M. (2018, January 13). Chinese interests at risk as Gulf crisis expands into the Horn of Africa. Retrieved from http://www.scmp.com/week-asia/geopolitics/article/2128064/gulf-crisis-expands-horn-africa-and-china-sits-eye-storm

Durden, T. (2015, May 10). China to Build Military Base in Africa Next to Critical Oil Transit Choke Point. Retrieved from https://www.zerohedge.com/news/2015-05-10/china-open-military-base-horn-africa-next-critical-oil-transit-choke-point

E&T Magazine. (2015, September 9). China unveils Rainbow 5 mega killer #drone design – an annotated infographic. Engineering & Technology. Retrieved from https://engtechmag.wordpress.com/2015/09/09/china-unveils-rainbow-5-mega-killer-drone-design-an-annotated-infographic/

Ekwe, D. (2017, 20 June). *Steemit Exciting*. Retrieved July 2018, from Steemit: https://steemit.com/exciting/@ekwedavid/amazing-facts-of-africa

Feickert, A. (2006). U.S. military operations in the global war on terrorism: Afghanistan, Africa, the Philippines, and Colombia (RL32758). Washington, D.C.: Congressional Information Service, Library of Congress.

Foley, T. J. (2018). Competition to Confrontation: Will Africa ignite conflict between the United States and China? Retrieved from http://go.galegroup.com/ps/i.do?p=ITOF&u=ksu&id=GALE%7CA524843354&v=2.1&it=r&sid=ITOF&asid=441530d8

Gady, The Diplomat, F. (2018, January 30). Is the UAE Secretly Buying Chinese Killer Drones? Retrieved from https://thediplomat.com/2018/01/is-the-uae-secretly-buying-chinese-killer-drones/

Gaffey, C. (2017, May 13). Kenya just opened a $4 billion Chinese-built railway, It's the largest infrastructure project in 50 years. Newsweek. Retrieved from http://www.newsweek.com/kenya-railway-china-madaraka-express-618357

Goh, B., & Doyle, G. (2018, February 9). U.S., Israeli drone makers keep wary eye on rising Chinese. Retrieved from https://uk.mobile.reuters.com/article/amp/idUKKBN1FS1E7

Goldstein, L. J. (2018, January 22). How to Avoid Making 'the Afghanistan Mistake' in Africa. Retrieved from https://www.yahoo.com/news/avoid-making-apos-afghanistan-mistake-002300753.html

Hillman, J. (2018, February 5). Opinion | The hazards of China's global ambitions. Retrieved from https://www.washingtonpost.com/news/theworldpost/wp/2018/02/05/obor-china-asia/?utm_term=.a5d6e77c1750

Huang, K. (2018, February 19). Chinese Rainbow 4 drones in use by foreign powers have 96pc strike rate in combat situations, paper says. South China Morning Post [International Edition]. Retrieved from http://www.scmp.com/news/china/diplomacy-defence/article/2133818/chinese-rainbow-4-drones-use-foreign-powers-have-96pc

Ipe, J., Cockayne, J., & Millar, A. (2010). Implementing the UN Global Counter-Terrorism Strategy in West Africa. Center on Global Counterterrorism Cooperation.

Jones, T. (2017). International Commercial Drone Regulation and Drone Delivery Services(RR1718z3). Retrieved from RAND Corporation website: https://www.rand.org/pubs/research_reports/RR1718z3.html

Karimi, F. (2017, October 18). US has hundreds of troops in Niger. Here's why. Retrieved from https://www.cnn.com/2017/10/18/politics/niger-american-troops-drones/index.html

Kharief, A. (2016, October 28). ANALYSIS: Just where are the US drone bases in North Africa? Retrieved from http://www.middleeasteye.net/news/drones-just-where-are-us-bases-northern-africa-2142422444

Kuo, L. (2018, January 8). Africa is changing China as much as China is changing Africa. Retrieved from https://qz.com/1168130/africa-is-changing-china-as-much-as-china-is-changing-africa/

Larive, M. H. (2014, August 7). Welcome to France's New War on Terror in Africa: Operation Barkhane. Retrieved from http://nationalinterest.org/feature/welcome-frances-new-war-terror-africa-operation-barkhane-11029

Lebur, C. (2018, January 24). Battle for land becomes Nigeria's biggest security challenge. Retrieved from https://www.yahoo.com/news/battle-land-becomes-nigerias-biggest-security-challenge-155539633.html

Lintner, B. (2017, November 23). China-India vie for a strategic slice of paradise. Asia Times. Retrieved from http://www.atimes.com/article/china-india-vie-strategic-slice-paradise/

Lubold, G., & Barnes, J. E. (2016, February 22). Italy Quietly Agrees to Armed U.S. Drone Missions Over Libya. The Wall Street Journal. Retrieved from https://www.wsj.com/articles/italy-quietly-agrees-to-armed-u-s-drone-missions-over-libya-1456163730

Maojo, V. (2015). AFRICA BUILD Report Summary (165776). Retrieved from University of Politics, Madrid website: https://cordis.europa.eu/result/rcn/165776_en.html

Michel, A. H., & Getting, D. (2018). Drone Year in Review: 2017. Center for the Study of the Drone at Bard College.

Military Factory Staff Writer. (2017, November 22). CASC CH-3 Rainbow Unmanned Combat Aerial Vehicle (UCAV) – China. Retrieved February 13, 2018, from https://www.militaryfactory.com/aircraft/detail.asp?aircraft_id=1378

Mitchell, C. (2017, August 3). Malawi's drone corridor challenges skepticism towards UAVs – African Business Magazine. Retrieved from http://africanbusinessmagazine.com/sectors/development/malawis-drone-corridor-challenges-scepticism-towards-uavs/

Monnier, O. (2018, February 5). Islamic State, al-Qaida support fuels attacks in West Africa. Retrieved from https://m.stamfordadvocate.com/business/article/Islamic-State-al-Qaida-support-fuels-attacks-in-12552019.php

Neethling, T. (2017, August 1). All about China's growing role in Africa. Retrieved from http://www.businessinsider.com/chinas-growing-role-in-africa-2017-8

O'Conner, T. (2018, January 19). China Military Training to Expand Across Asia and Around the World. Newsweek. Retrieved from http://www.newsweek.com/china-military-training-expand-across-asia-around-world-785045

Obi, P. (2018, February 5). Contemporary Issues in Africa – Vanguard News. Retrieved from https://www.vanguardngr.com/2018/02/contemporary-issues-africa/

Page, J. (2017, July 17). Unable to Buy U.S. Military Drones, Allies Place Orders with China. The Wall Street Journal. Retrieved from https://www.wsj.com/articles/unable-to-buy-u-s-military-drones-allies-place-orders-with-china-1500301716

Pairault, T. (2018, February 22). China in Africa: The Real Story. Retrieved from http://www.chinaafricarealstory.com

Paton, C. (2015, July 13). Isis in Libya: US wants drone base in North Africa to fight Islamic State. Retrieved from http://www.ibtimes.co.uk/isis-libya-us-wants-drone-base-north-africa-fight-islamic-state-1510596

Peter, M. (2015). Between doctrine and practice: The UN Peacekeeping dilemma. Global Governance, 21, 351-370. Retrieved from http://journals.rienner.com/doi/pdf/10.5555/1075-2846-21.3.351

Pike, J. (2017, February 3). Chang Hong (CH-4). Retrieved from https://www.globalsecurity.org/military/world/china/ch-4.htm

PR Newswire. (2017, September 15). Africa: Belgian Support for Innovative Agricultural Research. Retrieved from http://allafrica.com/stories/201709150763.html

Reel, M. (2016, March 23). Djibouti is Hot. Bloomberg Businessweek. Retrieved from https://www.bloomberg.com/features/2016-djibouti/

Sahel Elite. (2017, November 16). Retrieved July 2018, from Sahel Elite Mali: https://httpsahel-elite.com/2017/11/16/mali-understanding-the-g5-sahel-joint-force-fighting-terror-building-regional-security/

Sky News. (2018, February 5). US and China prepare for AI submarine warfare. Retrieved from https://news.sky.com/story/us-and-china-prepare-for-ai-submarine-warfare-11238090

Strategy Page. (2018). Warplanes: Instant Air Force for the Impoverished. Retrieved from https://www.strategypage.com/htmw/htairfo/20180219.aspx

The Guardian. (2015, October 20). *The Guardian*. Retrieved July 2018, from The Guardian Global Development: https://www.theguardian.com/global-development/2015/oct/20/two-thirds-of-worlds-illiterate-adults-are-women-report-finds

Thrall, Lloyd. (2015). China's expanding African relations: Implications for U.S. national security (rr905). Retrieved from RAND Corporation website: www.rand.org/t/rr905

Tsaigumi UAV: Nigerian Air Force develops new drone [Web log post]. (2018, February 6). Retrieved from https://www.africanmilitaryblog.com/2018/02/Tsaigumi-UAV.html?m=1

Turse, N. (2013, September 5). Tomgram: Nick Turse, AFRICOM's Gigantic "Small Footprint" | TomDispatch. Retrieved from http://www.tomdispatch.com/post/175743

Turse, N. (2015, October 15). Target Africa: America's expanding drone network. Retrieved from https://theintercept.com/drone-papers/target-africa/

Turse, N. (2015, November 17). The US Military's Best-Kept Secret. Retrieved from https://www.thenation.com/article/the-us-militarys-best-kept-secret/

Turse, N. (2016, July 11). In Africa, the U.S. Military Sees Enemies Everywhere. Retrieved from https://theintercept.com/2016/07/11/in-africa-u-s-military-sees-enemies-everywhere/

Turse, N. (2016, September 29). U.S. Military Is Building a $100 Million Drone Base in Africa. Retrieved from https://theintercept.com/2016/09/29/u-s-military-is-building-a-100-million-drone-base-in-africa/

Turse, N. (2018, January 29). Fitness Tracker Data Highlights Sprawling U.S. Military Footprint in

Africa. Retrieved from https://static.theintercept.com/amp/strava-heat-map-fitness-tracker-us-military-base.html

United Nations Peacekeeping. (2018). United Nations Peacekeeping. Retrieved February 16, 2018, from https://peacekeeping.un.org/en

U.S. Drone and Surveillance Flight Bases in Africa Map and Photos | Public Intelligence. (2013, February 23). Retrieved from https://publicintelligence.net/us-drones-in-africa/

U.S. Air Force MQ-9 Reaper Fact Sheet. (2018) http://www.af.mil/About-Us/Fact-Sheets/Display/Article/104470/mq-9-reaper/

Waddington, C. (2014, August 1). Understanding Operation Barkhane. Retrieved from https://www.africandefence.net/operation-barkhane-under-the-hood/

Wensink, W. (2017). The European Union's policies on counter-terrorism: Relevance, coherence and effectiveness: study (PE 583.124). Retrieved from European Parliament, Policy Department website: http://www.europarl.europa.eu/RegData/etudes/STUD/2017/583124/IPOL_STU(2017)583124_EN.pdf

Woolston, A. (2018, January 15). Overcoming the legal challenges to 'One Belt, One Road? Retrieved from https://www.globalrailwayreview.com/article/65484/legal-challenges-one-belt-one-road/

Worcester, M. (2016). Combating Terrorism in Africa. Retrieved from Institute for Strategic, Political, Security and Economic Consultancy, Berlin website: https://www.files.ethz.ch/isn/50103/Combating_Terrorism_Africa.pdf

Xinhua News Agency. (2017, March 25). Backgrounder: Major China-Africa infrastructure cooperation projects. Retrieved from http://go.galegroup.com/ps/i.do?p=ITOF&u=ksu&id=GALE%7CA487073405&v=2.1&it=r&sid=ITOF&asid=ea28e7ad

Zhang, A. (2017, December 28). Loaning Stability for Development: Chinese Aid and African Consequences | Harvard Political Review. Retrieved from http://harvardpolitics.com/world/loaning-stability-for-development-chinese-aid-and-african-consequences/

Zhao, C. (2018, February 5). China building artificial intelligence-powered nuclear submarine that could have 'its own thoughts,' report says. Newsweek. Retrieved from http://www.newsweek.com/china-building-artificial-intelligence-powered-nuclear-submarines-have-its-own-799351

Zhen, L. (2018, February 22). China boosts intellectual property protection? for its own tech at least. Retrieved from http://www.scmp.com/news/china/diplomacy-defence/article/2132135/drones-dredgers-stop-chinas-top-tech-falling-foreign

Zheng, S. (2017, October 17). China's Djibouti military base: logistics facility or geopolitical platform? Retrieved from http://www.scmp.com/news/china/diplomacy-defence/article/2113300/chinas-djibouti-military-base-logistics-facility-or-geopolitical-platform?/

Chapter 16: Chinese Drones in Spratly Islands, and Chinese Threats to USA forces in Pacific

Student Learning Objectives – The student will be re-introduced to the *problem* of countering hostile use of UAS, UV / Unmanned boats / UUV against U.S. national defense interests. This chapter focuses on the Spratlys; a tiny set of islands in the South China Seas. The Spratlys are the forefront of China's military expansion and control program (Corr, 2018). From this tiny island sanctuary drones and unmanned boats are the intelligence weapons of choice. Intrusions on US capital ships has already begun and could escalate to become the flash point for WW III.[1] Deployment of Chinese GPS spoofing cyber weapons via UAS against US Naval capital ships in the Spratly Area of Operations (AO) are author-theorized.

Location of the Spratly Islands and Their Strategic Importance

The Spratly Islands are a disputed group of islands, islets, cays, and more than one hundred reefs in the South China Sea. Named after British Whaling captain Richard Spratly in 1843, they represent only 490 acres spread over 164,000 square miles. The archipelago lies off the coasts the Philippines, Malaysia and China (WWF,2019)

1. Strictly authors speculation. Not supported by US official reports. Not the opinions of KSU or NPP press or co-authors. (See Discussion Question 3 in Chapter 3 Understanding Hostile Use and Cyber-Vulnerabilities of UAS: Components, Autonomy v Automation, Sensors, SAA, SCADA and Cyber Attack Taxonomy

Figure 16-1 Spratly Islands

Base 802086 (R01807) 9-95

Source: A Geographic Map of Spratlys, By Yuje – CIA, Public Domain, https://commons.wikimedia.org/w/index.php?curid=38157110 viewed September 12, 2018

Figure 16-2 Spratly Islands

Source: Google Earth. (2007). Spratly Islands. https://earth.google.com/web/

Although, there are some civilian settlements in the approximate 45 islands, all contain structures occupied by military forces from Malaysia, Taiwan (ROC), China (PRC), the Philippines, and Vietnam. Brunei has claimed an exclusive economic zone around the Louisa Reef (Spratly, 1955). Figures 16-1 and Figure 16-2 show the Spratly Islands. Officially they are in the South China Sea at 10 degrees N, 114 degrees E.

Target Drones

The Spratlys may be disputed in theory, but the undeniable winner in any real this AO would be China. China has made huge investments in defensive infrastructure, military, and unmanned aircraft, and boats to solidify its position in the Spratlys. China has one of the largest UAS intelligence operations in place in the Spratlys and regularly conducts drills (Staff, 6 Jul 2018). These drills simulate fending off an aerial attack. The drills, which involve three target drones making flyovers of a ship formation at varying heights and directions, are part of the on-going efforts to improve its real-life combat ability(Staff, 6 Jul 2018). The drones have been sent out several hundred times during more than thirty drills (Staff, 6 Jul 2018).

Shark Swarm and Wanshan Marine Test Field

China has tested an army of tiny drone ships that can "shark swarm" enemies during sea battles. It has a fleet of fifty-six unmanned craft sent out on maneuvers off the Wanshan Archipelago in the South China Sea (Barnes, 7 June 2018).[2] The Chinese firm Oceanalpha confirmed the drones were designed to overwhelm enemies in sea battles. A mothership controls the armed swarm (Barnes, 7 June 2018). Oceanalpha confirmed that the Wanshan Marine Test Field, was constructed sole purpose of conducting drone craft drills (Barnes, 7 June 2018).

Fast Drone Ship

In December of 2017, HiSIBI, a Chinese nautical firm, announced the development of the world's fastest drone ship, which can travel at 50 knots (58 mph).[3] The new speed drone is being tested in the Wanshan Marine Test field. The test field is still under construction and is believed to be the world's largest test field, covering over 297.9 square miles. Military observers have indicated that the test site for unmanned vessels was part of China's overall plans to develop autonomous systems for both civilian and military applications. The new test site dovetails with China's push to use technology to safeguard China's maritime interests (Staff writer, 6 July 2018).

Long-Range UUV

Tianjin University researchers completed a sea test of the Haiyan autonomous Unmanned Underwater Vehicle (UUV). It can endure for 30 days and has a 621.37miles range Lin, J & Singer, P.W., 4 June 2014). Just as the US Navy is conducting UUV research for facing off against China's growing Anti-Access Area Denial capabilities, the Chinese are building up these capabilities (Lin, J & Singer, P.W., 4 June 2014). UUVs cover a larger area, can operated more efficiency, use multiple sensors to monitor water temperature, conductivity, optical backscatter, and acoustics. In battle mode for detection of a stealthy submarine, using multiple sensor types increases the probability of finding the prey(Lin, J & Singer, P.W., 4 June 2014). Unlike fixed underwater sonar stations, UUVs can be rapidly deployed via ships or airdrops to new uncovered areas (such as

2. IBID Author note -this is more of a TEAM formation as discussed in chapter 3. Swarms do not have a team leader or Mother ship.
3. The author is Captain of /owns a recreational yacht, 36-foot CRYPTOWIZ, that can do supposedly 32-35 knots at peak performance top speed on dual Volvo-Penta GXI 315 Hp in-board engines. Above 23 knots is nuts for control (unless you have a death wish and / or married with wife and children on-board). Just imagine being in the rough South China Seas.

Taiwan Straits or South China Sea), where mobility complicates enemy efforts to disrupt and destroy them (Lin, J & Singer, P.W., 4 June 2014).

The Haiyan UUV is part of the deployed assets for an Underwater Great Wall, which would be a network of sensors on the seabed, coupled with long endurance UUVs to identify and destroy enemy submarines and mines. The sister fish-like Qianlong autonomous underwater vehicle (AUV) can dive to 14,800 feet indicates Chinese interests in deep-sea robotic ships (Katoch, 4 July 2018). These UUVs can also be used to attack targets anywhere in the Indian Ocean, in addition to collecting enemy submarine acoustics and oceanographic conditions for improving stealth and anti-stealth measures (Katoch, 4 July 2018).

Crisis Watch

The US and China are in a power struggle in the South China Sea centered around US countering Chinese military operations in the Spratlys. Defense Secretary General Mattis addressed some of the disputed issues at the Shangri-La Dialogue Asia security summit in Singapore 2 June 2018:

> U.S. Sec Defense Mattis outlined U.S. "Free and Open Indo-Pacific Strategy", consisting of expanded maritime security support for U.S. partners; helping regional navies become more interoperable with U.S. Navy; strengthening governance through defense engagements; and private sector-led development. Mattis said U.S. wants to work with regional multilateral institutions, particularly ASEAN; that new U.S. national security and defense strategies emphasize Indo-Pacific; said cooperation with China is "welcome wherever possible". Mattis criticized China's militarization of features in disputed Spratly archipelago. Also addressing Shangri-La Dialogue, China for first time publicly acknowledged that it was basing weapons and military personnel on disputed features it controls in Paracel and Spratly Islands, which it said are Chinese territory. Chinese military representative said Mattis's comments were "irresponsible" and that U.S. was the one militarizing, citing U.S. air and naval passages within twelve nautical miles of Chinese-controlled territory. U.S. 5 June flew two B-52 bombers over disputed Scarborough Shoal near Philippines; China sent ships and aircraft, said U.S. "stirring up trouble". Reuters 3 June reported U.S. considering stepping up its naval operations near disputed features. U.S. held annual Malabar naval exercise with India and Japan 7-16 June off coast of Guam and in Philippine Sea. Biennial U.S. Rim of the Pacific (RIMPAC) naval exercises began 27 June without China after U.S. late May rescinded China's invitation to participate. Citing satellite imagery dated 8 June, ImageSat International reported that China had redeployed surface-to-air missile systems to Woody (Yongxing) Island in Paracels. PLA navy 15 June carried out missile drills in South China Sea (SCS). UK and French defense ministers 3 June said they would send more naval ships through SCS to assert

right to freedom of navigation. Meeting with Sec Defense Mattis in Beijing 27 June, President Xi Jinping reasserted that China would not give up any of its territorial claims in SCS; also called for deepening military-to-military ties (Staff, June 2018, Crisis Watch).

In May 2018, the US disinvited China from the 27 nation International Naval Exercises (RIMPAC) in response to South China Sea aggression. The Pentagon claims evidence that the Chinese have deployed anti-ship missiles, surface-to-air missiles (SAM) systems, and electronic jammers to the Spratly Islands. The Chinese have landed bomber aircraft at Woody Island (Huang, 23 May 2018). This is along with the new drone systems and intelligence UAS assets discussed supra.

A Birds' Eye View

China has developed and deployed hi-tech drones for surveillance of population. These spy birds, code-named Dove, have completed more than 2000 test flights before deployment in real-life situations. Early versions of the bird robots had fixed wings and rotor blades. The Dove drones actual mimic the flapping action of birds, replicating about 90% of the real doves' movements and producing very little noise signature. Doves weigh only 0.441 pounds and have a wing-span of 19.685 inches. They can fly up to 24.855 mph for 30 minutes (Katoch, 4 July 2018). China is testing the Doves with facial recognition, stabilizing software, arming with explosives and increasing endurance for targeted Assassinations. The drones are being tested in Swarm formations (Katoch, 4 July 2018). Because of Chinese aggressiveness and its policy of ambiguity and deceit, the danger is clear and present (Katoch, 4 July 2018). See Figure 16-3 Chinese Dove Drone.

Red Drones over Disputed Seas

One of the best reports on how Chinese military uses unmanned drones as a means of power projection and surveillance in the contested South and East China Seas was written by (McCaslin, August 2017). China is currently undergoing a "drone" driven by heavy investment in the Chinese drone industry and by illegal acquisition of foreign drone technology (Katoch, 4 July 2018).

Figure 16-3 Chinese Dove Drone

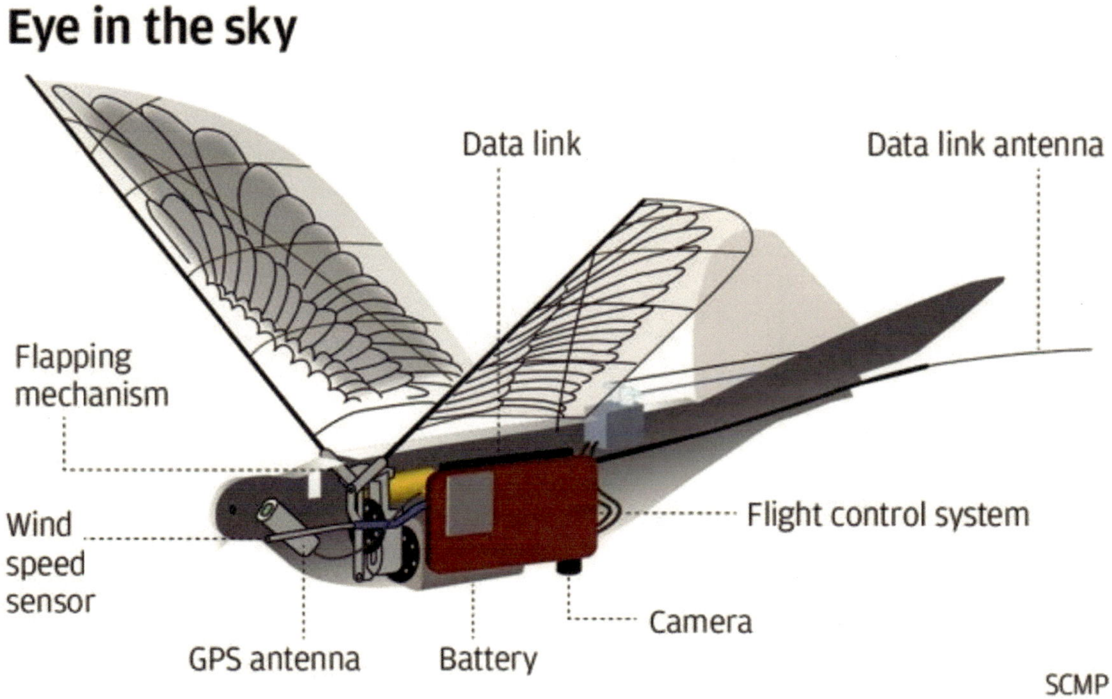

Source: Chua, M. (3 July 2018). In China, Dove Surveillance Drone Is Watching From The Sky Above. https://mikeshouts.com/chinas-dove-surveillance-drone/

US DOD predicts China will produce tens of thousands of drones by 2023 (DoD Report, 2015). Drone sightings and proper identification is important because of lack of international rules governing treatment of drones, including in areas where sovereignty is contested (Lehman, 29 August 2017).

The report documents four drones known to be used by PLAN: The S-100, ASN-209, BZK-005, and the GJ-1. All but the S-100 are Chinese-produced. The S-100 is made by Scheibel, in Austria (Lehman,29 August 2017). The drones discussed fill a variety of roles, from surveillance (S-100) to military / weaponized (GJ-1, aka Wing Loong I model) (Lehman, 29 August 2017).

One limiting factor facing Chinese power projection is the inability of their current inventory to runway launch from aboard the Chinese Navy's sole aircraft carrier. This limits the BZK-005 (primary mission surveillance) to be launched from land (McCaslin, 2017). The S-100 uses vertical take-off and landing (VTOL) system and does not have this problem. Additionally, drones can be launched from Chinese-controlled artificial islands in the contested areas [i.e. the Spratly Group.] (Lehman, 29 August 2017).

The author contends that the BZK-005 is suspected of being outfitted with cyber weapons to harass the US Naval forces in the Spratly AO causing chaos with the commercial and potentially US navy GPS systems.[4]

S-100 by Scheibel

Figure 16- 4 S-100 Drone Trajectories in Spratly Islands

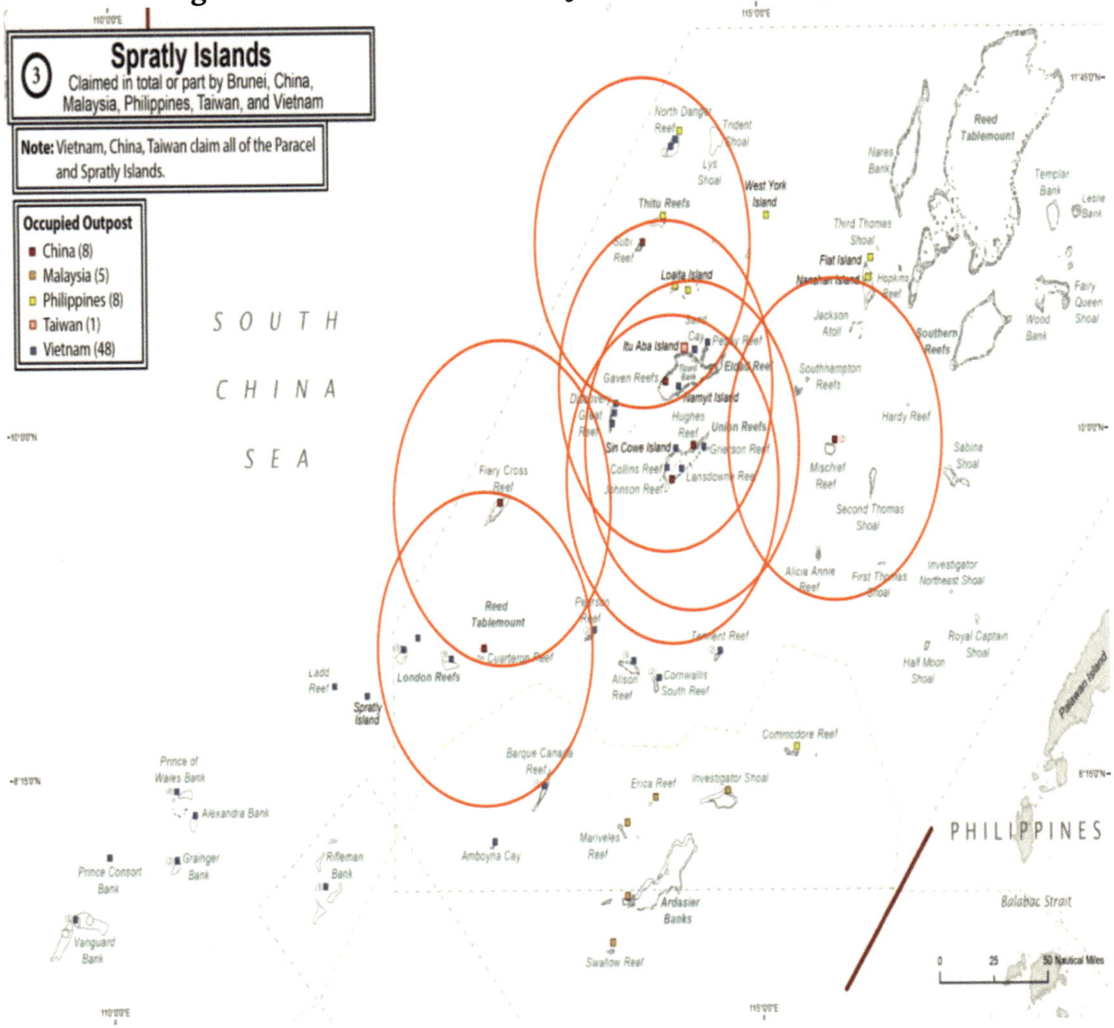

Source: McCaslin, I.B. (2017). Red Drones over Disputed Seas: A Field Guide to Chinese UAVs/ UCAVs Operating in the Disputed East and South China Seas. Released by Project 2049 Institute at: http://project2049.net/documents/Red%20Drones%20over%20disputed%20seas_PLA_project2049.pdf

4. This speculation is covered later in this Chapter.

Figure 16 -5 S-100 Chinese Drone

Source: Schiebel Camcopter S-100 at ILA 2010 (12 June 2010). By Matthias Kabel. CC BY-SA 3.0 https://commons.wikimedia.org/wiki/File:Schiebel_Camcopter_S-100_at_ILA_2010.jpg

The S-100 has an 18,000-foot ceiling, weighs 75 pounds armed with Thales Lightweight Multirole Missiles (LMM), with a range 60 to 125 miles and can be operational for 10 hours. They are generally launched from a PLAN Type 054 /054A frigate (McCaslin, 2017).

China uses the S-100 for intelligence, surveillance, and reconnaissance (ISR). They are equipped with Synthetic Aperture Radar (SAR), Maritime Radar, Signal Intelligence (SIGINT) and Communications Intelligence (COMINT) payloads. See Figures 16-4 and 16-5 for S-100 views and ranges.

ASN-209

The ASN -209 is a medium altitude, medium endurance (MAME) UAV. It has an operational ceiling of 16,404 ft, an operational range of 124.3 miles, and an endurance of 10 hours. The ASN-209 is deployed with a rocket booster on the back of a truck and lands via parachute (ASN-209, 2018) The ASN-209 is equipped with Tian Long -2 (TL-2) missile. The TL-2 has heat and fragmentation warheads. The ASN-209 is used for border surveillance, counter-terrorism, maintaining stability to target light armored vehicle, skiff or armed personnel (ASN-209, 2018) See Figure 16-6.

Figure 16-6 ASN-209 Chinese Drone

Source: Chinese ASN-209 Unmanned Aerial Vehicle (UAV). http://chinesemilitaryreview.blogspot.com/2011/10/chinese-asn-209-tactical-unmanned_20.html Viewed on September 12, 2018.

Figure 16 -7 BZK-005 Chinese Drone

Source: The Cyber Shafarat – Treadstone 71. (4 October 2017). Drone Wars! Threats, Vulnerabilities and Hostile Use. https://cybershafarat.com/2017/10/07/dronewars/

BZK -005

The BZK-005 is also a MAME drone specialized surveillance missions. It has an operational ceiling of 26,247 feet, with a maximum range 1491 miles and endurance of 40 hours. The range is limited by ground-based runways, i.e. Spratly Island group(McCaslin, 2017).

It is equipped with electro-optical, infrared, SAR, SIGINT and satellite communications systems, allowing real-time data transmission capability (McCaslin, 2017). See Figure 16-7 for BZK-005 view.

The BZK-005 range permits surveillance over the entire South China Seas if launched from Chinese – controlled islands (artificial and natural): Woody Island, Subi Reef, Mischief Reef, and Fiery Cross Reef (McCaslin, 2017).

GJ- 1 Chinese UCAV

The GJ-1 is also a MAME UAV converted to unmanned combat aerial vehicle (UCAV).

The GJ-1 can carry 441 pounds. It has SAR and electro-optical loadouts. It has been equipped with 8 different weapons systems, primarily, air-to-surface missiles and small diameter bombs (McCaslin, 2017). See Figures 16-8 and 16 -9 for GJ-1 identification views.

This model has an operational capability of 2,485 miles and can fly higher than any other military drone in the Spratly AO (McCaslin, 2017). Its endurance capability is classified.

Figure 16-8 GJ-1 Chinese UCAV Drone (Armed)

Source: Mil.huanqui.com. (February 6, 2018) Jane's: China sells the most advanced drones at the Singapore Air Show, http://mil.huanqiu.com/world/2018-02/11586378.html. See also https://www.youtube.com/watch?v=0QCNQfqkgDY

Figure 16-9 GJ-1 Chinese UCAV Drone (Armed)

Source: Mil.huanqui.com. (February 6, 2018) Jane's: China sells the most advanced drones at the Singapore Air Show, http://mil.huanqiu.com/world/2018-02/11586378.html. See also https://www.youtube.com/watch?v=0QCNQfqkgDY

Think of Chinese use of swarming drones on the seas, in the air, floating nuclear power plants, underwater mining, robot freighters and anti-submarine UUVs. In the author's view, they are leapfrogging US technology and antiquating defenses (Lehman, 29 August 2017).[5]

Interference with US Ships – Exploring the Cyberweapon deployed from UAS against US Capital Ships

It should be clear that the Chinese (PLAN) are heavily invested in military operations using unmanned aircraft and naval vessels in the Spratly Islands. This researcher has been tracking Chinese UAS and Intelligence assets /facilities / naval vessels since 2014. Figure 16-3 shows a glimpse of the deployment in the Spratly Area of Operations (AO). The black pin is the Spratly Islands group. Blue pins represent US Navy capital ships involved in either collisions or groundings in the AO. Red pins represent center of known Chinese UAS Intelligence elliptical paths. Green pins represent Chinese Intelligence facilities or seaborn assets. Figure 16-3 is not comprehensive. An exploded map view would show many more Chinese assets in the AO (Nichols & Carter, 4 May 2018).

Given the capabilities that Chinese (and US) UAS systems can deploy in almost any conditions and any location, it seems reasonable to this researcher, that the Chinese military might test their cyberweapons from their UAS in the Spratly AO coverage to harass US vessels and poten-

5. with additional author commentary

tially disrupt US Navy capital ships navigation systems. [This would be a natural priority for the Wanshan Marine Test Facility.] As a lesser alternative, the Chinese might take the 911 approach [i.e. turning planes into missiles loaded full of fuel and ramming them straight into fixed buildings] by disrupting (signal spoofing) the GPS /AIS unencrypted signals of huge commercial vessels and forcing them to act as Greek trireme vessels, colliding into the US Naval vessels in restricted maneuverable waters.[6]

Figure 16-10 Chinese UAS Chinese Intelligence Assets Deployment in Spratlys

Source: Nichols, R.K & Carter, C. (4 May 2018). RSCAD Presentation of Research to KSUP Faculty on Deployment of Chinese Cyber-weapons and GPS spoofing of Naval Vessels

6. Triremes were used in the Peloponnesian Wars to ram at about 4 knots at a 60-degree angle of attack. The greater the angle of attack the lesser the speed requirement for ramming. What is interesting is that the Athenians used a multi-trireme attack, an early predecessor to Swarm tactics. They also used grappling hooks to engage the enemy ships directly up close. This was the predecessor to piracy tactics in the 1500's – 1830's.

The Case for Cyber Weapon Spoofing of Legacy GPS Signals Affecting Us Navy and Commercial Vessels in Pacific

U.S Navy Vessel Collisions in the Pacific

In 2017 there was a chain of incidents/collisions involving four U.S. Navy warships and one U.S. Navy submarine.

On 17 June, the destroyer USS Fitzgerald collided with the ACX, a 30,000-ton container ship resulting in seven dead. Records show that the ACX turned sharply right at the time of collision. *The captain of the Philippine-flagged container ship accused the Navy destroyer of failing to heed warning signs before the crash.* Those warning signs came from the commercial vessels Automated Collision Systems (AIS) on the bridge. On 9 May, the guided-missile cruiser USS Lake Champlain collided with a South Korean fishing boat off the Korean Peninsula. There were no injuries (Department of the US Navy, Office of Chief of Naval Operations: 29 November 2017). On 31 January, the guided-missile cruiser USS Antietam ran aground dumping more than 1000 gallons of oil into Tokyo Bay. On 18 August, the ballistic-missile submarine USS Louisiana collided with the Navy Offshore Support Vessel in the Strait of Juan de Fuca. There were no injuries. "On 20 August, the guided-missile destroyer USS John S McCain collided with the 600-foot oil and chemical tanker Alnic MC at 0624 JST resulting in ten dead (Navy Office of Information, 11/1/2017)". (Weise E., 2017)

Navy Response

In all five incidents, the U.S. Navy blames their field leadership for not responding in an appropriate manner. This response means that the Skipper / XO / COB and at least 5 watch sailors on each Naval vessel (roughly 40 – 50 personnel including bridge staff plus 130 lookouts on the USS McCain because of ordered watch conditions) have been judged incompetent (Navy Information Office, 11/2/2017). Their careers are over, and some will face courts marshal and possible brig time. This response also implies that all five Navy vessels' radar, emergency positioning alert systems, AIS, sonar, and long-range collision avoidance equipment must have been functioning perfectly, without a catastrophic failure or interference of any kind. This conclusion assumes that none of the ships were in difficult maneuverable waters or serious traffic. The Navy blames funding, readiness and training. However, their response may not fully account for the commercial vessel accident data, actions required, or GPS positional data received (Olson, August 30, 2017).

The Navy Official Reaction regarding the possibility of Cyber-Weapon or Cyber-Attack

The Navy has downplayed the possibility of a Cyber Weapon or Cyber Attack. "Chief of Naval Operations (CNO) Admiral John Richardson said in a tweet on Monday 23 August, referring to the USS McCain and USS Fitzgerald collisions, "there was no indication of the possibility of cyber intrusion or sabotage was involved or that the Navy ships were hacked, but the review will consider all possibilities." (Weise E. , 2017) The Navy investigators after inspecting the physical damage to the USS McCain and USS Fitzgerald agree with the CNO's conclusions (Olson, August 30, 2017).

"Navy experts in the technology and researchers at University of Texas at Austin say there are certainly scenarios they can imagine in which GPS hacks could have been used to foil ships' navigations systems but emphasize there's no evidence such attacks took place in the case of the Navy collisions." (Weise E., 2017) "The technology to jam or misdirect navigational software is readily available, though the Navy uses a much more robust encrypted version of GPS that would be very difficult to disrupt." (Weise E., 2017)

The only way to spoof such a system is a *record and replay* attack, "where a recording is made of the encrypted location data being sent from GPS satellites to the naval ship. Replaying the recording at a slightly later time could fool a ship into thinking it is someplace else. This is a very sophisticated and difficult hack that requires multiple recordings of the navigation data stream from multiple angles, and then sending the recorded signal from two or more locations." (Weise E. , 2017) "To ensure that nearby ships do not also get the false data, it would have to be transmitted from close to the Navy ship being targeted, perhaps using multiple drones." (Weise E., 2017)

However, according to "Professor David Lust, former president of the Royal Institute for Navigation in the United Kingdom, "it takes two to Tango.... I" think you just have to attack the weakest of the pair, which is the commercial vessel." Commandeering the GPS of the cargo ship to get it to veer off course could cause collision, and it is a much easier hack." (Humphreys, 2009)

The Case for a Cyber Weapon

There appears to be valid evidence to support the theory that at least two of the U.S. Navy Warships, USS John McCain and the USS Fitzgerald AND/OR the commercial vessels involved were the on the wrong end of a Cyber-Weapon and were receiving incorrect GPS generated positional information. In agreement with Dr. Lust's conclusions, the Cyber Weapon may have been deployed by an adversary's UAS off a small nearby vessel. The author believes that the subject Cyber-Weapon is an advanced modular entity that can spoof the GPS signals received by all vessels in its range. J.S. Warner & R.G. Johnson established in 2013 that the cyber-security of many common automated navigational systems today lacks basic cyber-attack protection; vessels using incorrect data will make wrong decisions in terms of navigation and emergency responses, leading to potential collisions and deaths (Warner &Johnson, 2013).

Surfacing Questions

Spoofing is generation of false transmissions masquerading as P(Y) [the encrypted] Precise Signal that makes up the military vessel positioning basis, or unencrypted C/A [Civilian Acquisition] code from GPS satellites. In a virtual world tracking invalid data streams or non-integrity-based data is difficult, especially on three dimensional vessels moving in time. However, there may be more than one method to spoof a signal no matter how well it is encrypted. The cargo ships involved could have received unencrypted GPS ranging; a much less complex method than is required for military vessels.

Both ships do not need to be disabled or spoofed. All ships (military, commercial, recreational, specialized service) in international waters require detailed positional information. GPS systems accurately supply a 3-D position, velocity and time fix in all types of weather, 24 hours a day.

GPS satellite signals are ranging devices that deliver two signals made up of a civilian carrier, C/A code, NAV message, P-Code, and a military carrier. See Figure 16-11 (Balduzzi, et al 2014)

Figure 16-11 GPS Signals

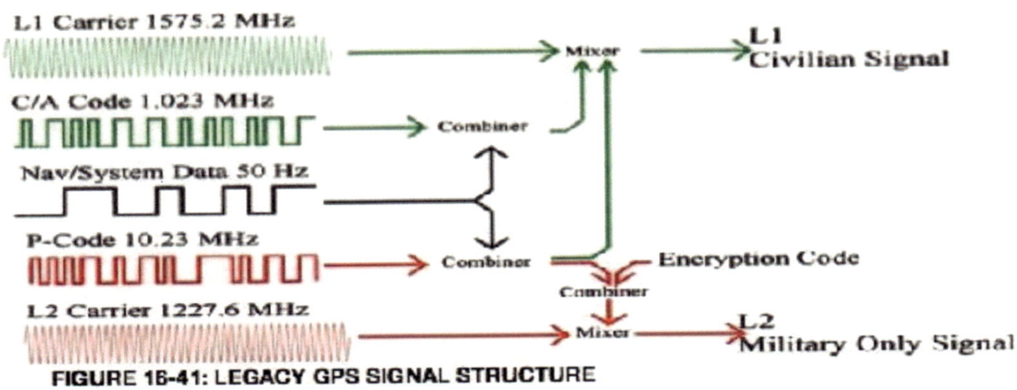

Source: Balduzzi, M. W. (2014). A Security Evaluation of AIS. Retrieved from Trend Micro. https://www.acsac.org/2014/program-final/oc_multifile/3/62.pdf

Delivered GPS signals also include a 50 Hz NAV message that is combined and ultimately mixed with the two codes to form the Civilian and Military Signals sent to the multiple radar and GPS receivers on both the Navy and Civilian vessels. Subframe 4 of the NAV message has a flag that tells the receiver that the P code is encrypted into the P(Y) code, thereby protecting the military signal. The Civilian signal has no such flag. Spoofing the civilian signal would be as simple as switching off the flag to make both the civilian and military components of the L1/L2 GPS signals unencrypted. If an adversary controls the signal the vessels are receiving, then the false position calculated by their receivers will be wrong regardless of encryption algorithms, military security enhancements or communication protocols used. Another spoofing method would overpower the real signal with a false one, then lock on and maintain access.

Because of cost, most systems on commercial vessels have legacy GPS systems. In the author's view, even if the GPS signals of the military vessels were not hacked the unencrypted C/A – L1 Civilian signal may have been. *It is also provable that this spoof is technically feasible* on the legacy systems. Experiments by Warner and Johnston out of Los Alamos, and surveys by

Schmidt, et al out of Queensland University clearly support the GPS/GNSS Cyberattack threat vector. (Warner, 2013) In 2013, Humphreys and his students successfully spoofed an $80MM Yacht's GPS system. (Humphreys, 2009)

What the physical damage indicates for the USS McCain and USS Fitzgerald is that both naval vessels appear to have collided on the starboard side. This leads to the theory that the Civilian vessels involved in crossing or approaching the US Naval vessels were relying on faulty information for their position. Further, the cyber weapon may have been delivered by small UAS from a nearby fishing or recreational vessel. It would be a perfect delivery vehicle: stealth, quiet, low radar signature, requiring only 1- 25 watts signal spoofing power. Since the true GPS signal strength reaching the surface of the Earth is about -160dBw (1x 10**-16 Watts), a 1-Watt GPS jamming spoof signal can over-ride C/A code acquisition for more than 620 miles (Line of Sight (LOS) to horizon.) (Warner, 2013)

Closest Point of Approach (CPA) Spoofing

"Collision avoidance is one of the primary objectives of using long range Automated Identification Systems (AIS), especially in open sea where port authority monitoring does not occur. CPA works by computing the minimal distance between two ships, at least one of which is in motion. CPA can be configured to trigger an alert (e.g., visually on the captain's console or acoustically via a siren) when a possible collision is detected so the ship can change course or speed or both." (Warner, 2013)

Figure 16-12 CPA Algorithm

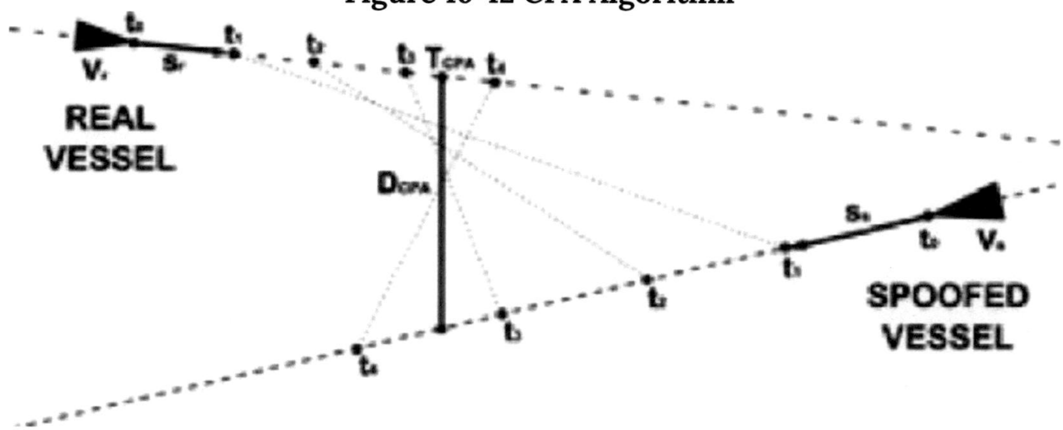

Source: Balduzzi, M. W. (2014). A Security Evaluation of AIS. Retrieved from Trend Micro https://www.acsac.org/2014/program-final/oc_multifile/3/62.pdf

"The CPA algorithm shown in Figure 16-12 allows ship captains to compute time and distance remaining before they collide with another ship, if the vessels are traveling at fixed speeds and courses. A CPA alarm is triggered if one of the two parameters is lower than the transponder's configured thresholds. **TCPA** refers to the amount of time left before reaching the CPA point,

DCPA refers to the distance between the vessels before they reach the CPA point, ***w(ti)*** refers to the distance between the vessels at a certain time **(*ti*), and *Sr* and Ss** are the vessels' vectors." (Balduzzi, 2014) "**CPA *spoofing involves faking a possible collision with a target ship***. This will trigger a CPA alert, which could lead the target off course to hit another vessel." (Balduzzi, 2014)

The question arises, what if the civilian vessel was given a **false position at [X0, Y0, T0; X1, Y1, T2] distance (range)** directly below the true point of collision (offset DCPA,) that the commercial vessel would receive [Z0), Z1 assumed negligible]? According to Warner, this range difference could be 2000 ft. or approximately 1/3 of an Nm! That is an enormous potential navigation error considering that normal legacy GPS signal is off only 9-15 feet at 95% RMS. See Figure 16-13.

Figure 16-13 CPA Algorithm Details

$$\begin{cases} T_{CPA} = \dfrac{-w(t_i)(S_r - S_s)}{|S_r - S_s|^2} \\ D_{CPA} = |w(t_i) + T_{CPA}(S_r - S_s)| \end{cases}$$

Source: Balduzzi, M. W. (2014). A Security Evaluation of AIS. Retrieved from Trend Micro. https://www.acsac.org/2014/program-final/oc_multifile/3/62.pdf

The starboard side collisions suggest that one of the vessels may have turned port or that the commercial ship tried to avoid a fake collision target received by turning starboard at the wrong time. The USS Fitzgerald report confirms this observation. These are huge vessels. Turning, stopping or reversing course on a dime are not possible. Decisions must be made well in advance of potential collision alerts. This is also why delivery of a cyber-weapon by UAS is so attractive. It would be a small bird *in the glasses* while attention was directed to the huge targets closing in on each other. In the chaos, the adversary wins.

How could be the GPS chaos to US Vessels be achieved?

The author believes, that for the spoofing GPS signal theory [targeting a commercial vessel by cyber weapon to give it a false position and potentially cause collision to itself or another vessel], to be possible. It would require an enemy Unmanned Aircraft System (UAS) to be launched from either a sea-based vessel or land-based intelligence station in the Spratly Islands. The methodology contemplated consists of three cyber-attack activities:

1) Breaking the existing AIS GPS commercial vessel receiver signal locks,

2) Locking the AIS GPS tracking device onto the GPS Simulator counterfeit signal,

3) Maintaining access by continued broadcasting of the fake GPS signal.

The problem is interesting because there are two three-dimensional maritime targets moving in time based on inaccurate or false ranging (GPS position) signals. The clocks used in GPS satellite systems are extremely accurate and present synchronization difficulties with the target naval / commercial vessel receivers. If it is possible to simulate and spoof the GPS signals to the commercial vessel using AIS collision avoidance systems (Cyber-weapon CONOP), then it is also possible that the US Navy may not have given proper attention to the non – personnel issues in their accident investigations.

Further, the possible delivery of such a Cyber-Weapon by close range UAS means that adversaries may have increased their knowledge management and understanding of U.S. Navy defensive systems. Using asymmetric warfare tactics and attacking the commercial traffic, which deploys legacy and cheaper GPS receivers, forces dependence on faulty information. Unfortunately, it is an effective tactic that bypasses much of the military modernization of GPS signals and satellites. This same possibility could affect military and commercial aircraft also, especially at airports where traffic speeds are reduced, and aircraft are closer to each other.

Discussion Questions

Consider the Signal Spoofing theory supra as the only Discussion Question for Chapter 16. Would the methodology work?

Bibliography

Balduzzi, M. W. (2014). A *Security Evaluation of* AIS. Retrieved from Trend Micro: https://www.acsac.org/2014/program-final/oc_multifile/3/62.pdf

Humphreys, T. e. (2009, January 1). *Assessing the Spoofing Threat: Development of a Portable Civilian GPS Spoofer.* Retrieved from Cornell University: https://gps.mae.cornell.edu/humphreys_etal_iongnss2008.pdf, Cornell University

Spratly. (1955). *Spratly Islands Fact Page.* Retrieved from Library of Congress: http://hdl.loc.gov/loc.gmd/g9237s.ct002223

Warner, J. &. (2013). *A Simple Demonstration That the Global Positioning System (GPS) is Vulnerable to Spoofing.* Retrieved from Journal of Security Administration: https://pdfs.semanticscholar.org/8ddb/89f56dd3e2ae265047822bc47cfb06815d9a.pdf, LAUR-03-6163

Weise, E. (2017, August 23). *could-hackers-behind-u-s-navy-collisions.* Retrieved from USATODAY: https://www.ruidosonews.com/story/tech/news/2017/08/23/could-hackers-behind-u-s-navy-collisions/594107001/

Readings

Adamy, D. (2001) EW 101 A *First Course in Electronic Warfare*, Boston: Artech House.

Adamy, D. (2004) EW 102 A *Second Course in Electronic Warfare*, Boston: Artech House.

Adamy, D. (2009) EW 103 *Tactical Battlefield Communications Electronic Warfare*, Boston: Artech House.

Adamy, D. (2015) EW 104 EW *against a New Generation of Threats*, Boston: Artech House.

Anonymous, (2017) GPS/SBAS *Signal Generator*, GSS4100, Spirent Communications Data Sheet. *Satellite* AIS, Exact Earth, Ltd.

Anonymous, (8/22/2017) *Nationwide Automatic Identification System*, www.navgen.uscg.gov

Anonymous, (8/22/2017) *Long Range Identification and Tracking* (LRIT) *Overview*, www.navgen.uscg.gov

Anonymous, (8/22/2017) *How* AIS Works, www.navgen.uscg.gov

Anonymous, (2015) *Satellite* AIS, Exact Earth, Ltd.

Anonymous, (6/21/2015) *Cyber Threats against the Aviation Industry*, in SCADA on April8, 2014, INFOSEC Institute.

Anonymous, (2012) A *Guide for Testers of GPS Devices and Systems*, spectracom, Test & Measurement technical Note, TN15-101A – What You Want to know about GPS.

Anonymous, (5/14/2012) *what is a GPS Simulator?* spectracom, Test & Measurement White Paper, WP08-101A.

Anonymous, (1/10/2014) GPS Signal Plan, Navipedia, http://www.navipedia.net/index.php/GPS_Signal_Plan

Anonymous, (4/2017) Counter-Unmanned Aircraft System Techniques, HQ, Department of the Army, https://fas.org/irp/doddir/army/atp3-01-81.pdf

Barker, B.C Capt., et.al. (2006) *Overview of the GPS M-Code Signal*, MITRE Report.

Barnes, T (7 June 2018) China Tests army of tiny drone ships that can 'shark swarm' enemies during sea battles. Independent. Retrieved 07072018 from https://www.independent.co.uk/news/world/asia/china-drone-ships-unmanned-test-video-military-south-sea-shark-swarm-a8387626.html.

Bay-Yen, J. (2000) Chapter 5: GPS C/A Code Signal Structure, Fundamentals of Global Positioning System Receivers: A Software Approach, New York: John Wiley.

Buesne, G & DeSanto, D. (2017) *GNSS Receivers and the Cyber-Threat: Lessons from the Information Security Community*, Spirent Communications, Baltimore, MD

Buesne, G & Holbrow, M. (6/29/2017) *GNSS Threats, Attacks and Simulations*, Spirent: PNT Advisory Board, Baltimore, MD

Chachak, E. (retrieved 9/1/2017) U.S. Naval Mishaps – Human Error or Cyber Malfeasance?

Corr, a. (2018) Great Powers, Grand Strategies: The New Game in the South China Sea. Naval Institute Press Annapolis.

Crosby, J. (12/16/2017) *here's What USNS Bowditch Does*, Inverse Innovation, https://www.inverse.com/article/25346-usns-bowditch-underwater-drone-stolen-china

CyberDB. https://www.cyberdb.co/u-s-naval-mishaps-human-error-or-cyber-malfeasance/

Department of the US Navy, Office of Chief of Naval Operations: (29 November 2017) Report on the USS Lake Champlain Collision https://www.documentcloud.org/documents/4316708-171129-USS-Lake-Champlain-Collision-Report.html

DoD Report: https://www.defense.gov/portals/1/documents/pubs/2015_china_military_Power_report.pdf

Easton, R.D. & Frazier, E.F. (2013) *GPS Declassified: From Smart Bombs to Smartphones*, University of Nebraska Press.

Editor, (8/31/2017) GPS Block IIIA, https://en.wikipedia.org/wiki/GPS_Block_IIIA

FCC Wireless Telecommunications Bureau, Marine VHF Radio Channels, per 47 CFR 80.371© and 80.373(f)

Fessenden, F. & Watkins, D. (6/18/2017) *the Path of the Container Ship that Struck a U.S. Destroyer*, NYT. https://www.nytimes.com/interactive/2017/06/18/world/asia/path-ship-hit-uss-fitzgerald.html?mcubz=3

Haider, Z. & Khalid, S. (8/2016) *Survey on Effective GPS Spoofing Countermeasures*, 6th International Conference on Innovative Computing Technology (INTECH 2016), https://www.researchgate.net/publication/313543601_Survey_on_effective_GPS_spoofing_countermeasures

Heath, T. (5/7/2015) How to Hack a Military Drone Parts I & II, Technology-Hackers, www.cybersecurityintelligence.com/blog/

Hodge, H. (8/23/2017) *why are Navy Ships colliding in the Pacific? Experts Weigh In*, Military.com

Huang, P. (23 May 2018) US Disinvites China from International Naval Exercise in Response to South China Sea Aggression. The Epoch Times. Retrieved from: https://www.theepochtimes.com/us-disinvites-china-from-international-naval-exercise-in-response-to-south-china-sea-aggression_2535152.html

Unmanned Aerial Vehicles and Other Systems to Civil GPS Spoofing, Submitted to the Subcommittee on Oversight., Investigations, and Management of the House Committee on Homeland Security.

Kao, Lee, Chang, and Ko. (2007) A Fuzzy Logic Method for Collision Avoidance in Vessel Traffic Service, *Journal of Navigation*, 60,17-31.

Katoch, P.C, Gen (4 July 2018) New Chinese Drones – formidable challenge. SPSMAI. Retrieved from: https://spsmai.com/experts-speak/?id=556&q=new-chinese-drones-formidable-challenge

LaGrone, S. (8/21/2017) *Chain of Events Involving U.S Navy Warships in the Western Pacific Raise Readiness, Training Questions*, USNI News

LaGrone, S. (1/31/2017) *Cruiser USS Antietam Runs Aground in Tokyo Bay*, Spills Oil, USNI News.

Lehman, C.F. (29 August 2017) Report: China Increasing Drone Operations in Disputed Seas, Freebeacon. Retrieved from http://freebeacon.com/author/charles-lehman

Lin, J & Singer, P.W. (4 June 2014) Not a Shark, But a Robot: Chinese University Tests Long-Range Unmanned Sub. Popular Science. Retrieved from: https://www.popsci.com/blog-net-

work/easter-arsenal/
not-shark-robot-chinese-university-tests-long-range-unmanned-mini-sub#page-3

McCaslin, I.B. (2017) Red Drones over Disputed Seas: A Field Guide to Chinese UAVs/ UCAVs Operating in the Disputed East and South China Seas. Released by Project 2049 Institute at http://project2049.net/documents/Red%20Drones%20over%20disputed%20seas_PLA_project2049.pdf

Navy Information Office (11/1/2017) Navy Releases Collision Report for USS Fitzgerald and USS John S McCain Collisions Story Number: NNS171101-07Release Date: 11/1/2017 9:01:00 AM

Navy Information Office (11/2/2017) Navy Releases Results of the Comprehensive Review of Surface Force Incidents Story Number: NNS171102-06Release Date: 11/2/2017 12:22:00 PM

News Correspondent, (8/22/2017) USS *McCain crash is* 4th *Navy Accident in Pacific this Year*, The Washington Post, AP.

News Correspondent, (8/31/2017) *DDG 51 Arleigh Burke Class Destroyer*, Military.com

News Correspondent, (8/21/2017) *CNO Orders Operational Pause, Review After Latest Ship Collision*, Military.com

News Correspondent, (8/21/2017) 10 *Sailors Missing*, 5 *injured after Destroyer Collides with Tanker*, Military.com

News Correspondent, (8/22/2017) *Remains of Navy Sailors found on USS John S McCain*, Military.com

News Correspondent, (8/17/2017) *Navy Fires Commander, XO from USS Fitzgerald for Fatal Collision*, Military.com

News Correspondent, (7/21/2017) *Investigation Faults Navy in Fitzgerald Collision Report*, Military.com

News Correspondent, (6/20/2017) *Stories of Fitzgerald Sailors Killed in Destroyer – Container Ship Crash*, Military.com

News Correspondent, (6/16/2017) US Navy Destroyer Collides with Japanese Merchant Ship, Military.com

News Correspondent, (5/09/2017) US *Navy Ship Collides with South Korean Fishing Boat*, Military.com

News Correspondent, (1/31/2017) *Oil Spill in Tokyo Bay After Navy Cruiser Runs Aground*, Military.com

Nichols, R.K & Carter, C. (4 May 2018) RSCAD Presentation of Research to KSUP Faculty on Deployment of Chinese Cyber-weapons and GPS spoofing of Naval Vessels

Nichols, R.K (8/31/2017) *Stand By for a whole slew of military short articles on the Navy Collisions (my students only)*, Private memo to COT799 & CMST 455.

Nichols, R.K. & Lekkas, P.L. (2002) *Wireless Security: Threats, Models, Solutions*, New York, McGraw Hill.

Olson, W. (August 30, 2017) *Adm* No *Evidence of Hacking in McCain Fitzgerald Collisions* pdf. Stars and Stripes.

Ranganathan, A, et.al, SPREE A *Spoofing Resistant* GPS *Receiver*, Department of Computer Science, ETH Zurich, Switzerland, Zurich Information Security and Privacy Center.

Richardson, J. Adm., (8/31/2017) No *Evidence of Hacking in McCain and Fitzgerald Collisions*, Military.com

Schallhorn, K., (9/1/2017) US Military crashes, collisions in the Pacific, FoxNews. http://www.foxnews.com/us/2017/08/28/us-military-crashes-collisions-in-pacific.html

Schmidt, D. et.al., (5/2016) A *Survey and Analysis of the* GNSS *Spoofing Threat and Countermeasures*, ACM Computing Surveys, Vol 48, No 4, Article 64

Shouts, M. (3 July 2018) Science blog. https://mikeshouts.com/chinas-dove-surveillance-drone/

Sickle, J.V. (8/25/2017) GEOG 862 GPS *and* GNSS *for Geospatial Professionals*, Lessons 1-10 complete, Penn State University, College of Earth and Mineral Sciences https://www.e-education.psu.edu/geog862/node/1407 [Superb Course on the subject]

Staff (6 Jul 2018) Chinese navy deploys drones in South China Seas missile drills. Diplomacy and Defense article. https://www.scmp.com/news/china/diplomacy-defence/article/2150957/chinese-navy-deploys-drones-south-china-sea-missile/

Staff writer. (6 July 2018), China starts work on world's biggest test site for drone ships near South China Sea. Today. Retrieved from: https://www.todayonline.com/world/china-starts-work-worlds-biggest-test-site-drone-ships-gateway-south-china-sea.

Staff (June 2018) Crisis Watch. Retrieved from map overlay, https://www.crisisgroup.org/crisiswatch

Sterling, J. 8/21/2017) *A Spate of US Navy warship accidents in Asia since January*, CNNNEWS. http://www.cnn.com/2017/08/21/politics/navy-ships-accidents/index.html

YouTube Gongji GJ-1 UAV, https://www.youtube.com/watch?v=0QCNQfqkgDY

Volpe, J.A, (8/29/2001) *Vulnerability Assessment of the Transportation Infrastructure Relying on the Global Positioning System, Final Report*, Office of Assistant Secretary for Transportation Policy, U.S. Department of Transportation, John A Volpe Transportation Systems Center.

Warner, J.S. & Johnson, R.G. (2003) *GPS Spoofing Countermeasures, Journal of Security Administration*, LAUR-03-2384, Los Alamos, NM: Los Alamos National Laboratory

Weise, E. (8/23/2017) *Could Hackers Be Behind the U.S. Navy Collisions?* USATODAY.

WWF. (2019, July 25). *South China Sea, between the Philippines, Borneo, Vietnam, and China.* Retrieved from worldwildlife.org/ecoregions/: https://www.worldwildlife.org/ecoregions/im0148

S-100. Images, https://en.wikipedia.org/wiki/Schiebel_Camcopter_S-100, Retrieved 08082018.

ASN-209, https://en.wikipedia.org/wiki/Aisheng_ASN-209, retrieved 08082018

BZK-005, https://en.wikipedia.org/wiki/Harbin_BZK-005, retrieved 08082018

Patents

Berry, R. & Cook, C. (2016) *Detection of wireless data jamming and spoofing*, US 9466881 B1

SECTION VII

TECHNOLOGY UPDATES

Chapter 17: High - Altitude Platforms (HAPS) – A Promise not Reached

Student Learning Objectives

The student will be introduced to High-Altitude Platforms (HAPS) for wireless communications. This overview will include the key design considerations, architecture, channel considerations, link budget, opportunities and challenges (not met) for this promising technology. Not covered are the detailed channel modelling calculations or the complex antenna designs required. The latter two subjects can be found in Chapters 3 and 4 of (Alejandro Aragon-Zavala, 2008)

Introduction

For three decades wireless communications designers have researched the inclusion of unmanned aircraft systems into their network architectures "to provide cost- effective wireless connectivity for devices without infrastructure coverage. Compared to terrestrial communications or satellites, low altitude UAS are generally faster to deploy, more flexibly reconfigured and have better communication channels due to presence of short-range LOS links. However, the use of highly mobile and energy constrained UAVs for wireless communications introduces new challenges." (Yong Zeng, 2016) Table 17-1 shows HAPS capabilities compared to terrestrial and satellite systems for telecommunications. (Jesus Gonzalo, 2018)

Missions

"HAPS are capable of providing services that complement, compete or replace those currently offered by airplanes, satellites and terrestrial networks." (Jesus Gonzalo, 2018) Three key areas on the list are telecommunications, Earth observation, and GNSS.

Telecommunications

HAPS platforms hold a promise for improvement of existing communications systems both in capacity and coverage. (Jesus Gonzalo, 2018) This same promise has been made for three decades. Technology has moved from 2nd, 3rd and 4th Generation wireless systems to 5th Generation systems. 5G software (aka 5G NR for New Radio) refers to software under the 3GPP industry association standard or the ITU IMT-2020 international requirements. (Romano, 2017) 2G, 3G and 4G technologies are associated with their respective advances: GSM, UMTS, LTE, LTE . (See Abbreviations list for definitions) (Romano, 2017) Advanced telecommunications services that can be offered from HAPS platforms are in various stages of development. Table 17-2 gives an overview of HAPS enhanced services that are envisioned by developers.

Table 17-1: HAPS Capabilities Compared to Terrestrial and Satellite Systems for Telecommunications

Issue	High Altitude Platform
Deployment	Faster deployment than space-based platforms. Less "build -out" than terrestrial networks. Very fast response to emergency situations
Upgrading	Access to platform/ payload after deployment enables service upgradeability like terrestrial. Enhanced flexibility and adaptability
Link Budget	Shorter distances to HAPS makes the Link budget favorable compared to satellite links. Smaller antenna coverage area permits high focus on areas of interest getting capacity higher density (x100) than GEO Satellites
Signal Processing	HAPS are quasi-stationary. This significantly reduces the Doppler shift due to platform motion
Ground Terminals	Smaller / simpler terrestrial terminals than satellite exhibit data rates
Antenna Pointing & Directivity	Mobile LTE services and TETRA are based on omnidirectional links
Latency	Very low, equivalent to terrestrial networks. Round-trip time ~ 0.26 ms versus ~30ms for LEO and ~250 ms for GEO
Geographic Coverage	Hundreds of miles per platform (~125 miles radius) between terrestrial (few miles) and space GEO (up to 33% of the Earth surface)" (Jesus Gonzalo, 2018)

Source: (Jesus Gonzalo, 2018)

Table 17-2: HAPS Platform Advanced Telecommunications Services in various stages of engineering and development (Jesus Gonzalo, 2018) (D, 2010)

Dream	Service
Direct-To-Home (DTH)	DTH broadband: useful in unserved areas with no infrastructure or poor connectivity. Mimics a satellite or terrestrial tower
Trunking	Large number of users under a HAPS footprint can connect and share a single satellite connection. Good balance between coverage and signal degradation
Backhauling	HAPS provides very high capacity backhaul links between network nodes (cell towers) and backbone. Costly optical fiber or terrestrial microwave links are avoided
High Throughput	HAPS service to Offload congested GEO spot beams
Tactical	Communication usually in UHF, HAPS services are scalable, agile, reliable, affordable, defendable, rapidly deployable and requires minimum in theater ground infrastructure. (T.C. Dozer, 2008)
Mobile Broadband	Normally provided by terrestrial wireless networks. If none available existing satellite (Iridium, Inmarsat, etc.) can provide. HAPS provides a higher capacity equivalent due to favorable link budgets.
5G	HAPS infrastructure supports 5G services." (Jesus Gonzalo, 2018)[1]

Source: (Jesus Gonzalo, 2018)

The deployment of services shown in Table 17-2 "have been hindered due to limitations derived from the telecom bands assigned to HAPS by the World Radio Conference (WRC) when providing such services according to Resolution 122 from WRC-07 and Resolutions 145 and 150 from WRC -12. Resolution 809 from WRC-15 in 2019 discusses the appropriate regulatory actions for HAPS within existing fixed-service allocations." (Jesus Gonzalo, 2018)

Table 17-3 shows a more detailed view of the basic characteristics of terrestrial, satellite and HAPS systems. (Alejandro Aragon-Zavala, 2008) Several of these characteristics are covered later in this chapter.

1. In 2019, there are only five companies in the world offering 5G radio hardware and 5G systems for carriers: Huawei, ZTE, Nokia, Samsung, and Ericsson. In April 2019, the Global Mobile Suppliers Association had identified 224 operators in 88 countries that are actively investing in 5G. (Romano, 2017)

Table 17-3: Basic Characteristics of Terrestrial, Satellite and HAPS Systems (Alejandro Aragon-Zavala, 2008)

ISSUE	TERRESTRIAL	SATELLITE	HAPS
Propagation Delay	Not an issue	Large delay causes noticeable impairment in voice communications for	Low, since altitude for HAPS is much lower for satellites
Health and safety	Low power handsets are used	GEO and MEO. High power handsets needed to overcome large path losses	Similar to terrestrial except for large coverage areas
Technology risk	Mature technology	New technology for LEO & MEO. GEO behind terrestrial in cost, volume & performance	Terrestrial wireless technology supported by spot beams. Research on smarter antenna in progress
Deployment timing	Development staged -substantial build-out to provide coverage	Entire system needs to be built to operate	BIG advantage: needs only one platform and one ground station to initiate operations
System growth	Easy upgradable. Cell splitting to increase capacity	Capacity is increased by adding new satellites. Hardware upgrades are possible if replacing satellites	Spot beam resizing and adding more platforms used to increase capacity; hardware upgrades easier than satellites
Complexity	User terminals are mobile. Operations well understood	Mobile satellites in LEO and MEO are complex.	Modern mobility platforms; operations not complex. Platforms need refueling.
RF Channel quality	Good signal quality through proper antenna placement	GEO distance limits spectrum. Ricean fading.	Free space like channel at distances ~ terrestrial
Indoor coverage	Might be achieved. Research in progress on outdoor-to-indoor penetration	Not available due to large path loss at satellite communication frequencies	Coverage via repeaters but not outdoor-to-indoor penetration
Breadth of geographical coverage	A few miles per base station	Large regions in GEO. Global for MEO and LEO	100's of miles per platform

Cost	Varies. Much lower than satellite systems	>$200MM for GEO; ~$2 Billion for LEO	~$50MM but less than terrestrial network
Cell diameter	0.06214 -> 0.6214 miles	31 miles for LEO; 310 miles for GEO	0.6214 to 6.214 miles

Source: (Alejandro Aragon-Zavala, 2008). Distances converted to miles by author.

Earth Observation

HAPS can be used as an "effective platform for Earth Observation (EO) payloads. It provides useful capabilities for many services and complementing satellite and conventional aircraft (manned and unmanned)." (Jesus Gonzalo, 2018) "Space sensors can map large areas worldwide. They offer a relatively course resolution. They suffer operational constraints due to weather (clouds), and fixed-timing acquisitions." (Jesus Gonzalo, 2018) Regular aircraft are more flexible but also costlier. When continuous monitoring by MALE aircraft requires deployment of multiple platforms. (Jesus Gonzalo, 2018) EARSC has identified a variety of EO services that fit well with a HAPS solution. (EARSC, 2015) HAPS has the advantage when required coverage area is local or regional. "HAPS platforms can fly for very long periods, up to several months and its capacity for EO is comparable to conventional aircraft." (Jesus Gonzalo, 2018)

GNSS

"HAPS platforms provide functionality for navigation systems:

- Additional ranging sources to assist and improve position
- Network node to provide data from an external source
- Reference stations for network RTK (Real Time Kinematic) and PPP (Precise Point Positioning) types of services
- Additional sensor platform to perform radio occultation and / or GNSS reflectometry measurements."(Jesus Gonzalo, 2018)

UAV-Aided Wireless Communications

Figure 17-1 to 3 from (Yong Zeng, 2016) illustrates three typical use cases of UAV-aided wireless communications, discussed below.

UAV-aided ubiquitous coverage

"Figure 17-1 shows two example scenarios where HAPS plays a significant role in recovered wireless communications. HAPS UAVs can be deployed where there is a need to assist the existing infrastructure and provide seamless wireless coverage. Two examples are rapid service recovery after partial or complete infrastructure damage due to natural disasters and base station offloading in extremely crowded areas (ex. NFL stadium support)." (Yong Zeng, 2016) (Osseiran, Dec 2014)

Figure 17-1: & 17-2: UAV-aided ubiquitous coverage with overloaded / malfunctioning base station and UAV-aided relaying

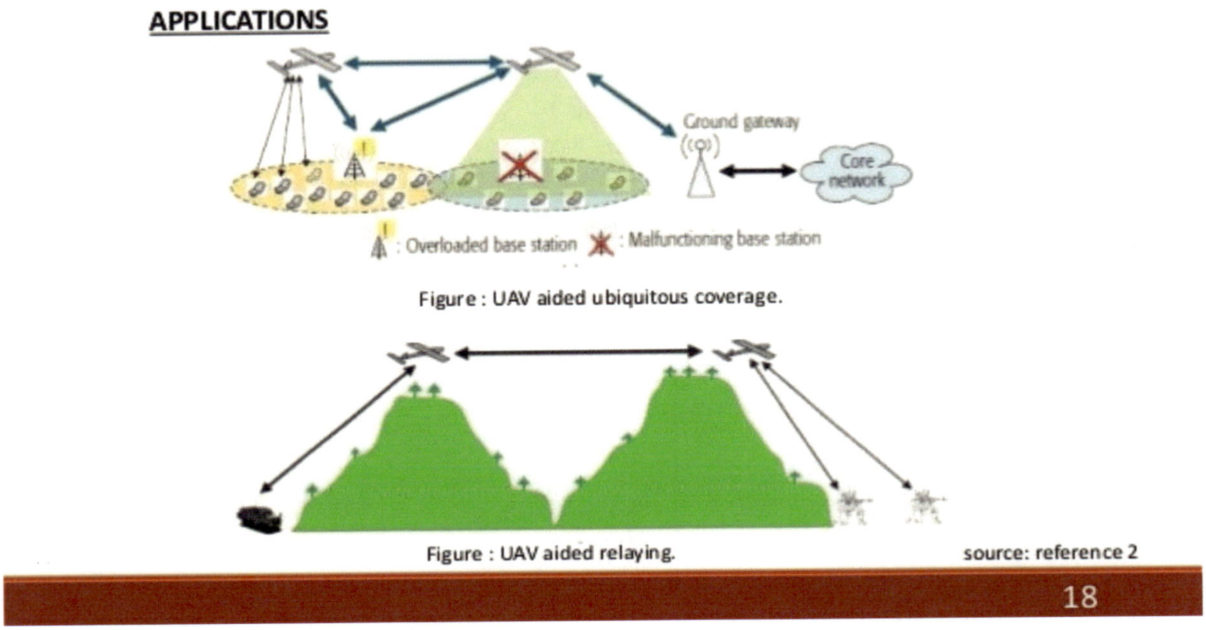

Source: (Yong Zeng, 2016) and (Y. Zeng, May 2016.)

UAV - aided relaying

"Figure 17-2 shows the case where UAVs are deployed to provide wireless connectivity between two or more distant users or user groups without reliable direct communication links. From a military example, links between frontline and the command center for emergency purposes would qualify for HAPS treatment." (Y. Zeng, May 2016.)

UAV - aided information dissemination and data collection

"HAPS UAVs may be dispatched disseminate or collect delay tolerant information to / from a large number of distributed wireless devices." (Y. Zeng, May 2016.) In Kansas, the agriculture industry uses HAPS for precision agriculture applications and water distribution.

Figure 17-3: "UAV – aided information dissemination and data collection

Source: (Y. Zeng, May 2016.)

Challenges

HAPS for wireless communications has faced some stiff design challenges. Along with the normal communications links as in terrestrial systems, additional control and non-payload (CNPC) links with more stringent latency and security requirements are needed for HAPS UAV systems." (Yong Zeng, 2016) "These are needed for safety-critical functions, real-time control, and collision avoidance. The result is that effective resource management and security mechanisms must be designed for HAPS communications UAVs." (Y. Zeng, May 2016.)

"Mobility is also a design challenge. HAPS requires highly mobile environment. This translates to highly dynamic network topologies. The latter are sparsely and intermittently connected. (Brown, Dec 2008) Effective multi-UAV coordination or UAV swarm operations need to be designed for reliable network connectivity." (N. Goddemeir, June 2015) The author notes that in this textbook, Swarm operations have a very negative side too. One chapter has been devoted to C-UAS swarm operations.

"Another interesting challenge stems from the size, weight and power (SWAP) constraints of HAPS UAVS. These limit their communication, computation and endurance capabilities. To tackle SWAP constraints, energy -aware UAV deployment and operation mechanism are needed under SCADA control to efficiently use the energy and replenishment thereof." (Yong Zeng, 2016).

"Cell interference is not negligible. Mobility and lack of fixed backhaul links under centralized control means that neighboring cells with UAV – enabled aerial base stations is more challenging than that of terrestrial systems. Interference management techniques and software requires expensive and complex customization for UAV -aided cellular coverage." (Yong Zeng, 2016)

Simple HAPS UAV Network Architecture

It is time to explore the HAPS UAV basic networking architecture, channel characteristics, design considerations, performance enhancers to exploit mobility. We then take on the Link budget and work an example. There are hundreds of diagrams for HAPS network structure. The author has chosen a figure from (Yong Zeng, 2016) because it the easiest to build upon. "Figure 17-1 shows the generic networking architecture of wireless communications with HAPS UAVS. [2]Three basic types of links exist: the CNPC link, backhaul links and the data link." (Yong Zeng, 2016)

Control and Non-Payload Communications Link (CNPC)

"CNPC links are essential to ensure the safe operation of all UAV systems. "CNPC links are highly reliable, low – latency and provide secure two-way communications." (Yong Zeng, 2016)

Figure 17-4: Basic HAPS networking architecture of UAV-aided wireless communications"

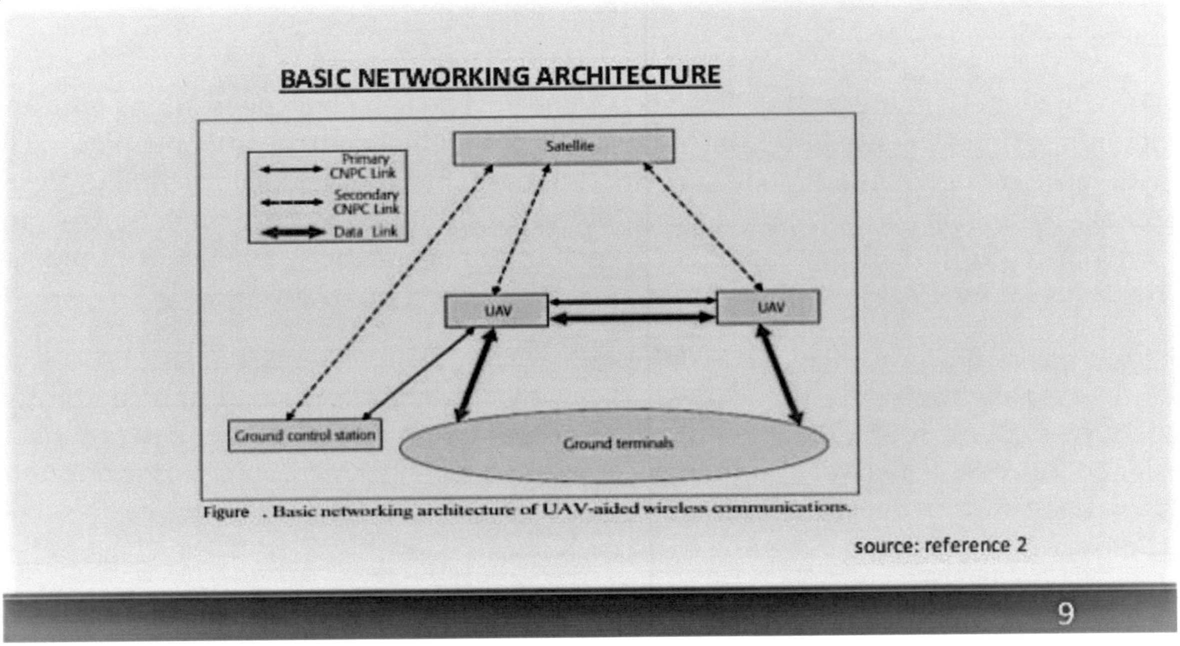

Source: (Yong Zeng, 2016)

"CNPC links usually have low data rate requirements. They exchange safety critical information among HAPS UAVS and ground control stations (GCS), such as dedicated mobile terminals mounted on ground vehicles." (Y. Zeng, May 2016.)

"CNPC information flow can be categorized into four types:[3]

2. Only in this chapter is the designation "UAV" used interchangeably with "UAS."
3. A fifth type is ATC (Air Traffic Control) links within a controlled airspace or near an airport. (Yong Zeng, 2016)

- Command and Control from GPS to UAVs
- Aircraft status report from UAVs to ground
- Sense – and – avoid (SAA) information among UAVs
- Necessary in case of emergency human intervention requirements

CNPC links operate in protected spectrum

Currently there are two bands that have been allocated:

- L-band (960-977 MHz)
- C-band (5030 -5091 MHz)

These bands apply to the primary GCS to UAVs (primary CNPC links) because of low delay factor. Secondary CNPC links via satellite act as backup to enhance reliability and robustness. (Yong Zeng, 2016) A key requirement for CNPC links is enhanced security. Efforts must be made to avoid the *ghost control scenario* (think James Bond film Goldeneye) where UAVs are controlled by a hostile force or unauthorized hacker. CNPC have enhanced authentication protocols and physical security." (Y. Zeng, May 2016.)

Backhaul Links

Backhaul links are intermediate links that connect the core network or internet backbone to the small peripheral subnet systems at the edge of the system. "They are essentially a subset of the data links category that process dozens of gigabits per second in the UAV- gateway wireless backhaul." (Yong Zeng, 2016)

Data Links

"Data links support mission-related communications for the ground terminals: These include: terrestrial base stations (BSs), mobile terminals, gateway nodes, wireless sensors," (Y. Zeng, May 2016.) "and SCADA communications." (Randall K. Nichols, 2018) Refer to Figure 17-1 for HAPS UAV – aided ubiquitous coverage. "The data links maintained by the UAVs support the following communication modes:

- Direct mobile- UAV communications for BS offloading or during BS malfunction
- UAV-BS and UAV-gateway wireless backhaul
- UAV – UAV wireless backhaul (Rappaport, 2014)

Data links support a wide range of bit rate / sec capacity. They have a higher tolerance for latency and security. They can reuse spectrum allocations in a given band and / or have dedicated spectrum for enhanced performance." (Yong Zeng, 2016)

Channel Characteristics, Propagation and Channel Modelling

This section will give only a "flavor" of the subject of channel characteristics, propagation and modeling. For a detailed mathematical look at this interesting subject, consult chapter 3 in (Alejandro Aragon-Zavala, 2008). Subjects / factors covered include: Free space loss, Multipath, Rayleigh criterion, Rain attenuation, Gaseous adsorption, scintillation, general case of channel modelling, geometric characterization, platform altitude effects, two-ray model, Rice factor, Rician PDF (probability density function), statistical characterization, Lognormal PDF, LOS conditions, UHF channels, wideband models, Markov chains, Chapman-Kolmogorov equation, Semi-Markovian processes, switched broadband channel model, Politecnico di Torino (Polito) Multipath channel model, SHF and clear sky models, shadowing, time series, Gaussian model, fading mitigation techniques, uplink and downlink power control, on-board beam shaping, adaptive methods for coding and modulation, digital transmission rate reduction, site, platform, frequency and time diversity; and open, closed and hybrid loop models. (Alejandro Aragon-Zavala, 2008) Chapter 3 is a treasure house of information.

HAPS UAS / UAV systems work because of effective antenna design and placement. Like channel modeling, this subject is outside the scope of this chapter 17. In that same reference (Alejandro Aragon-Zavala, 2008) is an excellent chapter 4 on antennas for HAPS. (Alejandro Aragon-Zavala, 2008)

If chapter four is not enough to fill your palette, the author recommends the superior textbook by McNamara entitled: *Introduction to Antenna Placement & Installation*. (Macnamara, 2010)

"Both CNPC and data links in HAPS UAV-aided communications consists of two types of unique channels, UAV – ground and UAV – UAV." (Alejandro Aragon-Zavala, 2008)

UAV-Ground Channel

Research on HAPS UAV – Ground channels is still ongoing after three decades. "This compares unfavorably to similar channel research for piloted aircrafts for aeronautical applications." (Sun, June 2015) "Whereas piloted aircraft systems ground stations are usually in open areas with tall antenna towers, the HAPS UAV -ground links operate in a more complex environment. UAV-ground links do not always have LOS links available." (Alejandro Aragon-Zavala, 2008) They may be blocked by terrain, buildings, or the airframe itself. The latter introduces a severe signals – link delay during aircraft maneuvering (a process called *airframe shadowing*). "Shadowing can be significant during mission-critical operations." (Matolak, April 2015)

For low-altitude UAVs, the UAV-ground channels may constitute a number of multi-path components due to reflection, scattering, and diffraction by mountains, ground surface, and foliage. (Yong Zeng, 2016) Multi-path propagation is when radio signals reach an antenna by two or more paths causing interference, phase shift of the primary signal, or destructive fading. This effect can be modelled by the Rayleigh or Rician Distributions depending on conditions. (Alejandro Aragon-Zavala, 2008)

"For HAPS UAVS operating over the desert or sea, the two-ray model is used because of dominance of the LOS and surface reflection components." (Alejandro Aragon-Zavala, 2008) [See Chapter 3 of (Alejandro Aragon-Zavala, 2008)].[4]

HAPS UAV – UAV Channel

"The UAV – UAV channel is dominated by the LOS component. Multi-path fading due to surface reflection is limited compared to the UAV – Ground channel experience. UAV – UAV channels experience higher Doppler frequencies than their UAV – ground counterparts due to large relative velocity between UAVs. These characteristics have a direct influence on the spectrum allocation for UAV – UAV links." (Yong Zeng, 2016) There are two competing theories. "The dominance of LOS links suggests that the emerging mmWave (5G) communications should be employed to achieve high-capacity UAV – UAV wireless backhaul. (Yong Zeng, 2016) Or the high relative velocity between UAVs coupled with higher frequency in the mmWave band could lead to excessive Doppler shift." (Yong Zeng, 2016) [5]

4. "The stochastic Rician fading model consists of a deterministic LOS component with certain statistical distribution properties. Depending on the environment surrounding the ground terminals and frequency, the UAV-ground channels exhibit widely varying Rician factors – i.e. Power ratio between LOS and the scattered components. Typical values are 15 dB for L-band and 28 dB for C band in hilly terrain. " (Yong Zeng, 2016) "Rician factors and typical values may be found in Chapter 3 reference" (Alejandro Aragon-Zavala, 2008).
5. The author has always been fascinated by the Doppler effect. Henderson gives a decent description of the shift (effect). (Henderson, T., 2017). "The Doppler effect (or the Doppler shift) is the change in frequency or wavelength of a wave in relation to an observer who is moving relative to the wave source. It is named after the Austrian physicist Christian Doppler, who described the phenomenon in 1842. A common example of Doppler shift is the change of pitch heard when a vehicle sounding a horn approaches and recedes from an observer. Compared to the emitted frequency, the received frequency is higher during the approach, identical at the instant of passing by, and lower during the recession. The reason for the Doppler effect is that when the source of the waves is moving towards the observer, each successive wave crest is emitted from a position closer to the observer than the crest of the previous wave. " (Henderson, T., 2017). "Therefore, each wave takes slightly less time to reach the observer than the previous wave. Hence, the time between the arrival of successive wave crests at the observer is reduced, causing an increase in the frequency. While they are traveling, the distance between successive wave fronts is reduced, so the waves bunch together. Conversely, if the source of waves is moving away from the observer, each wave is emitted from a position farther from the observer than the previous wave, so the arrival time between successive waves is

From the Designers Shoes

What would a designer of a communications system based on HAPS need to know? "Certainly, the designer would need to know the components, subsystems and interfaces involved and the sensors and instruments which could be part of the HAPS mission. In addition, the spectrum regulations and link budgets for licenses bands (clear sky and rain) would be fully worked using typical systems parameters for HAPS." (Alejandro Aragon-Zavala, 2008) Maybe we can take a helicopter view and get a handle on the design problem.[6] Start with the stratosphere segment where we will play.

Stratosphere Segment

"The stratosphere segment includes platform requirements to establish communications with ground facilities and other HAPS. This segment has two main elements, the *payload* and the *bus*.

The payload is the mission." (Alejandro Aragon-Zavala, 2008) It is the raison de entre into telecommunications and infrastructure. "The bus provides resources. Its functions are:

- Payload must be pointed in the correct direction, coverage to right users and interference restricted,
- Payload must be operable,
- Data from payload must be communicated to the GCS and back,
- Payload must operate reliably over a design time-period. An energy source must be provided to enable performance by payload functions."(Alejandro Aragon-Zavala, 2008)

Figure 17-5 shows the subsystems that form the stratospheric segment for HAPS.

Platforms

"Environmental conditions present in the stratosphere affect HAPS (whether they are manned / unmanned aircraft or LTA airships). These conditions are atmospheric pressure, air density, specific lift, and wind speed." (Alejandro Aragon-Zavala, 2008)

increased, reducing the frequency. The distance between successive wave fronts is then increased, so the waves spread out." (Henderson, T., 2017).

6. The author has reviewed / researched a wide variety of books, papers and conference proceedings to build this chapter. The author has chosen as a primary reference (Alejandro Aragon-Zavala, 2008) because of its clarity in terms of student preparation and best "bang for the buck". Two other newer references are worth mentioning only if your wallet is unlimited: Ibrahim's *Optimization Methods for User Admissions and Radio Resource Allocation for Multicasting over High Altitude Platforms* (Ibrahim, 2019) and Grace's Broadband Communications via High Altitude Platforms (Mohorcic, 2010)

Figure 17-5: Subsystems of HAPS Stratosphere Segment

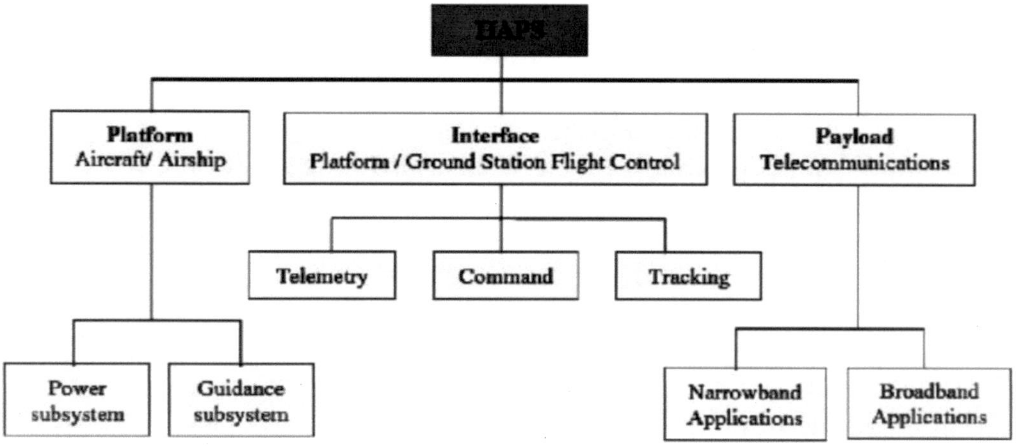

Source: (Alejandro Aragon-Zavala, 2008), page 142, renamed Figure 5.1

Many services require geostationary HAPS. "The platform must be kept in a control box with dimensions of 800 x 400 x 1000 m (0.4971 x 0.2485 x 0.6214 miles). The size of the platform determines the maximum payload weight at an altitude up to 20 km (12.427 miles). Both types of HAPS can be powered by different power systems, fuel, electric motors or solar energy. (Alejandro Aragon-Zavala, 2008) The sun is the preferred energy source. The drawback for sun is efficient storage of energy for long term use. The planned mission duration can be from 6 months to 5 years. Two types of platforms can be used in the stratosphere: *aerostatic* (LTA) and *aerodynamic* (UAV)." (Alejandro Aragon-Zavala, 2008)

Aerostatic Platforms (LTA)

"Aerostatic platforms carry heavier payloads ~1000 kg (2204.6 lbs). They have precise station-keeping for nominal environmental conditions and plenty of area for power generation. Medium- strong winds are a problem. Ground stations and infrastructure is expensive and complex. High-strain hull material is needed – a definite disadvantage." (Alejandro Aragon-Zavala, 2008)

Aerodynamic Platforms (UAVs)

"UAVs are less expensive than LTAs with easy station-keeping in turbulent conditions. They require less ground infrastructure. Unfortunately, they have cannot carry much weight. Payloads are limited to less than 100 kg (~220 lbs). Power is also restricted (up to 1200W). Continuous movement of the platform is required, which translates to more power consumption." (Alejandro Aragon-Zavala, 2008)

Platform Choice – Key Designer Issues

“Key issues to be considered by the designer when choosing the platform:

- Cost – deployment, acquisition and operations
- Environmental compatibility – emissions, re-usable energy, non – pollutants
- Power – fuel powered or solar-powered
- Service offering – affects payload capacity”(Alejandro Aragon-Zavala, 2008)
- “Technological similarity to space-based systems – autonomous operations, payload types, accommodation, reliability, payload operations, communications budget.”(Alejandro Aragon-Zavala, 2008) [7]

Telecommunications Payload

“The telecommunications payload consists of phased array antennas (transmit / receive) for gateway links with the ground and or terrestrial subscriber stations with a bank of processors that handle receivers, multiplexing, switching and transmitting functions. The payload can use different multiple access techniques.” (Alejandro Aragon-Zavala, 2008) Table 17-4 dives deeper into the what the designer must be aware of in terms of HAPS communication payload constraints / requirements / elements / subsystems. The list is extensive, and our hypothetical designer is still not in deep water.

7. “The author recommends UAVs for this issue alone. He disagrees with the aerostatic platforms recommended by Stratos project. (ESA-ESTEC Contract 162372/02/NL/US, September 2005)” (Alejandro Aragon-Zavala, 2008)

Table 17-4: HAPS design communication payload constraints / requirements / elements / subsystems.
(Alejandro Aragon-Zavala, 2008)

ISSUE	Constraint / Requirement / Element
"System Constraints	Considerations different from satellite or terrestrial systems (refer to Table 17 -1)
Customer requirements	Mission lifetime, connectivity with other operating systems, coverage area served, control station site location, capacity, services offered, availability, (Alejandro Aragon-Zavala, 2008)
Technical	Maximum available transmit power, receiver sensitivity, interference, environment, available components and spares, trade-off between cost and technology performance and noise performance. (Alejandro Aragon-Zavala, 2008) See for definitions / derivations: (Adamy, EW 101 A First Course in Electronic Warfare, 2001); (Adamy, EW 102 A Second Course in Electronic Warfare, 2004) (Adamy, EW 103 Tactical Battlefield Communications Electronic Warfare, 2009)
International Regulations	HAPS systems must control interference between different systems (satellite, terrestrial) and ensure compatibility between national systems connected end-to-end. (Alejandro Aragon-Zavala, 2008)
Antenna subsystem	Affects total mass and stability. Consider critical. (Alejandro Aragon-Zavala, 2008). See Chapter 4 in (Alejandro Aragon-Zavala, 2008)
HAPS transponder	Along with antenna subsystem constitutes the communications payload, includes low and high passband filters, beamforming, low noise amplifiers (LNA), multiplex method, FDM (frequency division multiplexing for UL), power amplifiers / demultiplexer for DL. (Alejandro Aragon-Zavala, 2008) See chapter 4 in (Alejandro Aragon-Zavala, 2008)
LNA	Low-noise temperature and sufficient gain keeps contributions small from prior stages in transponder. (Alejandro Aragon-Zavala, 2008)
Frequency converters	Frequency change devices also known as mixers, to differentiate traffic for UL /DL. (Alejandro Aragon-Zavala, 2008)
IF processor	Provides HAPS transponder gain. (Alejandro Aragon-Zavala, 2008)
Filters	Limit spurious adjacent signals and noise. (Alejandro Aragon-Zavala, 2008)
Transmitters	Amplifying signals to level of DL transmissions. (Alejandro Aragon-Zavala, 2008)
Payload system performance	Performance parameters on individual equipment. (Alejandro Aragon-Zavala, 2008)

Key payload electrical parameters	1) "Antenna coverage area, gain, dimensions, electrical efficiency and feeder losses 2) Figure of merit G/T – ratio of the receive antenna gain to system noise temperature ~ affects link performance 3) EIRP – Effective isotopic radiated power ~ determines the power capability of the HAPs 4) Power per backhaul carrier in user link." (Alejandro Aragon-Zavala, 2008) See Chapter 3 in (Alejandro Aragon-Zavala, 2008)
Other important parameters	1) "Isolation between channels ~to reduce potential adjacent channel interference issues 2) Spurious outputs ~ to reduce interference levels to contiguous wireless systems 3) Amplifier linearity 4) Group delay variation – time delay experienced by the modulating waveform passing through equipment. This causes signal dispersion and degrades performance." (Alejandro Aragon-Zavala, 2008)

Source: (Alejandro Aragon-Zavala, 2008) compressed text information from pp. 144-146

Telemetry, Tracking and Command (TT & C)

"The interface between the ground station flight control and HAPS platform is performed by TT and C subsystems. This subsystem is responsible for two-way flow of information from HAPS to the GCS. It also links HAPS to the Flight Control Station (FCS) via LOS conditions and under spectrum control as regulated by the ITU-R. Table 17-5 shows the functions of the TT & C Subsystem." (Alejandro Aragon-Zavala, 2008)

Table 17-5: Functions of TT & C Subsystem

TT & C Function	Comment
Telemetry and Data Acquisition	Telemeter of data is conveyed by telemetry signals continuously from HAPS and received by GCS. 3 Classes: Housekeeping, attitude and payload
Housekeeping data	Check on health and operating status of platform on-board equipment
Attitude data	Sensor output determines attitude and control
Payload data	Information received and processed as part of the communications payload: temperature, power, voltages, currents, telemetry monitoring
Command functions	Operational control of the UAS by commands from GCS. Acceptance of command relayed back to GCS Some operations are very time-dependent / some automatic.in emergency, commands may be sent in the blind – without confirmation that the uplink lock has been achieved. Classified as low-level commands; high-level commands; proportional commands
Low-level on-off commands	Logic to reset / set logic flip-flops
High – level on-off commands	Operates RF waveguide or latching relay
Proportional commands	Used to reprogram memory locations on the on-board computer or setting attitude control
Tracking data	Used to determine the precise vehicle position and velocity relative to the FCS." (Alejandro Aragon-Zavala, 2008)

Source: (Alejandro Aragon-Zavala, 2008) compressed text information from pp. 146-148

Avionics

Avionics systems are needed for guidance, attitude and stabilization control of the HAPS. "This is accomplished by sensors that measure the orientation with respect to a reference and its angular departure from this reference measurement. HAPS attitude is specified by means of three Euler angles, known as yaw, pitch, and roll. These angles measure rotations around the z, y, and x axis, respectively." (Alejandro Aragon-Zavala, 2008) "Attitude measurements and 'fixes' are transmitted to the payload via the on-board sensors." (Alejandro Aragon-Zavala, 2008)

Electrical Power Subsystem

"Power for the stratospheric vehicles is the most fundamental requirement for HAPS. Power system failure means failure of the HAPS mission. Three main components make up this system: Primary power system (solar array or fuel cells or photo-hydrogen energy systems); secondary power systems (batteries); and power management, distribution and control. Power management operates with all power systems whose characteristics and needs change in time. The electrical bus provides /regulates a variety of voltages to meet the needs of the equipment." (Alejandro Aragon-Zavala, 2008)

Ground Segment

"The HAPS ground segment (GS) is composed of terrestrial subsystems required by the platform. Functions of the GS include (Alejandro Aragon-Zavala, 2008):

- Tracking to determine the position of a HAPS
- Telemetry operations to acquire and record HAPS data and status
- Commanding operations to interrogate and control the various functions of HAPS
- Data-processing operations and engineering reporting
- Communications links to other GS, gateways and processing centers.

The main hardware components of a HAPS GS are: antenna, transceiver, LNA, HPA, data-recorders, computers and peripherals and control consoles." (Alejandro Aragon-Zavala, 2008) Also, GS require software, simulation capabilities, emergency waypoint data, and people.

Spectrum Allocation for HAPS

"Stringent conditions of non-interference and protection are imposed between the HAPS system and other systems using same or adjacent frequency bands, fixed service (FS), and fixed satellite service (FSS) via a GEO satellite orbit." (Alejandro Aragon-Zavala, 2008) ITU-R recommendations for HAPS at 2GHz and other services is presented in (Alejandro Aragon-Zavala, 2008) Table 5.2 p. 158. Table 17-6 gives the Frequency spectrum available for HAPS prior to May 2019 ITU meeting.

Frequency, GHz	Area / Country	Service	Sharing service (primary allocation)	Reference in RR
47.9 48.2	Global	FS (uplink & downlink)	FS, FSS, MS	5.552A
47.2 -47.5 31.0 -31.3	Region 2	FS (uplink)	FS, MS	5.543A
27.5 – 28.35 2.160 – 2.170	Regions 1 & 3	FS (downlink); IMT -2000 (base station)	FS, FSS, MS FS, MS	5.537A
2.110 -2.160	Global		FS, MS, space research	5.388A
2.010 -2.025	Regions 1 & 3		FS, MS	
1.885 – 1.980	Global		FS, MS	
"Region 1: Europe, Africa, Russia & Middle East; Region 2: North & South America; Region 3: Asia & Pacific			FS: Fixed service; FSS: Fixed satellite service; MS: Mobile service" (Alejandro Aragon-Zavala, 2008)	

Table 17-6: Frequency spectrum available for HAPS prior to May 2019 ITU meeting (Alejandro Aragon-Zavala, 2008)

Source: Table 5.1 from (Alejandro Aragon-Zavala, 2008)

HAPS Link Budget

We now come to most interesting part of this chapter, calculating the Link Budget for an example of HAPs. "Link budget analysis is the real first step in designing a HAPS. It determines the essential system parameters, size of antennas, power amplifier characteristics, link availability, fade margins and effects of rain and weather conditions. In order to perform an accurate link budget analysis, knowledge of several factors affecting the link must be ascertained: power amplifier gain, noise factors, transmit antenna gain, slant path angles, atmospheric losses, receive antenna and amplifier gains and noise performance (noise factor), cable losses, climatic factors (essential at frequencies above 10 GHz.)" (Alejandro Aragon-Zavala, 2008) Link budget analysis is clearly presented in Adamy's Electronic Warfare texts. (Adamy, EW 101 A First Course in Electronic Warfare, 2001), (Adamy, EW 102 A Second Course in Electronic Warfare, 2004) and (Adamy, EW 103 Tactical Battlefield Communications Electronic Warfare, 2009) "The goal of a HAPS link budget is to guarantee a successful implementation of haps communication link. This section will focus on just the uplink analysis. In reality, the transmission of radio waves is between two GS with access to a HAPS transponder, one transmitting and one receiving, via HAPS. So, the link consists of two sections, the uplink from the terrestrial GS transmitter to HAPS and the receiving terrestrial GS. HAPOS performance is affected by rain attenuation losses (and signal degradation) – especially between 27/32 and 47/48 GHz. For 1.8 /2.1 GHz bands the multipath and shadowing effects are the significant impairments." (Alejandro Aragon-Zavala, 2008) (Adamy, EW 102 A Second Course in Electronic Warfare, 2004) "HAPS QOS (Quality of Service) is influenced by antenna performance, service availability, receiver noise performance, and available power. All these must be accounted for if the stratospheric segment is to provide an efficient payload for transmission and reception in HAPS communications systems." (Alejandro Aragon-Zavala, 2008)

One-Way Link Budget Analysis

"The HAPS link gains and losses used for performance evaluation may be expressed as a one-way link block diagram, as shown in Figure 17-3." (Alejandro Aragon-Zavala, 2008) (Xiaoyang Liu, 2016) In the following analysis, "*lower-case* variables represent parameters which are factors, e.g. gain as a factor, g; and *upper-case* variables are parameters in decibels, e.g. gain expressed in dB, G. See note about refresher in decibel mathematics." (Alejandro Aragon-Zavala, 2008)[8]

8. "Decibel mathematics is used to compare values that differ in many orders of magnitude. Any number N can be expressed in decibels (dB) is a logarithm base 10. To multiply linear numbers, we add their logarithms base 10. To divide linear numbers, subtract their logarithms base 10. To raise a number to the nth power, we multiply its logarithm base 10 by N. To take the nth root of linear number, we divide its logarithm by N. It is desirable to handle all link budget calculations as early as possible in decibels. To convert the linear number N to decibels:N(dB) = 10 log 10 (N). to convert dB to a linear number we use the relation N = 10 N (dB/10). "Adamy (Adamy, EW 101 A First Course in Electronic Warfare, 2001) Nichols (Randall K. Nichols, 2018) gives examples. "dB equations normally take the form of one of these equations: 1) A(dBm) /_ B(dB) = C(dBm); 2) A(dBm) -B(dBm) = C(dB) and 3) A(dB) = B(dB) /_ N log where N is a number not in dB. Remember that any value in decibels is a RATIO converted to a logarithm. The denominator is usually a known reference base number or value." (Adamy, EW 101 A First Course in Electronic Warfare, 2001)

Figure 17-6: Schematic of a HAPS Link Budget [Corrected]

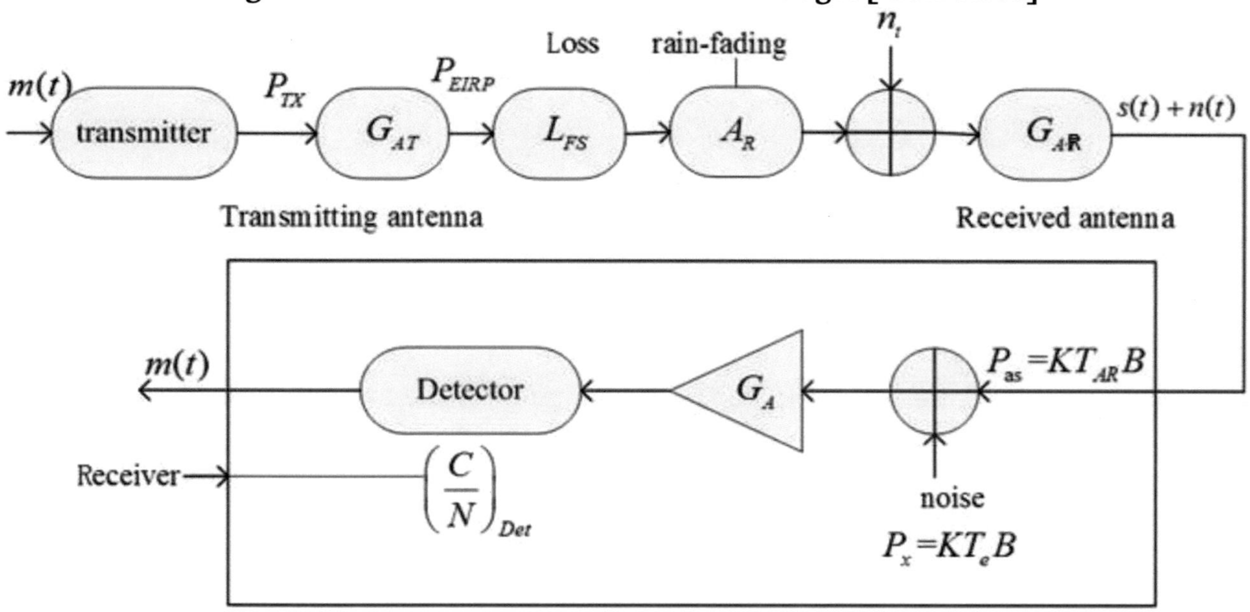

Source: (Alejandro Aragon-Zavala, 2008) [9]

"The performance of a HAPS system is directly related to how high the signal to noise ratio (S/N) at the receiver can be achieved. The received signal power, CRX, is expressed:

Eq. 17-1

$$C_{RX} = PEIRP \times gAR / LFS$$

Where:

PEIRP = transmitter effective isotropic radiated power, watts,

gAR = receiving antenna gain as a factor,

LFS = free- space loss, as a factor" (Alejandro Aragon-Zavala, 2008)

"The available noise power N, in watts at the input of the ideal amplifier is defined as:

9. GAR = Gain of receiver. Nomenclature corrected by author. Original paper showed GAT which is the transmit antenna. This would represent the two antenna gains as equal which is unlikely and not a robust solution

Eq. 17-2

$$N = k \times Ts \times B$$

Where:

N = available noise power, watts
K = Boltzmann's constant (1.38 x 10 -23 J/K)
B = IF equivalent bandwidth, Hz
Ts = Receiving system noise temperature, Kelvin" (Alejandro Aragon-Zavala, 2008)

And, the receiving noise temperature "Ts is a function of the antenna noise temperature TAR and effective input noise temperature Te, all in degrees Kelvin." (Alejandro Aragon-Zavala, 2008)

Eq. 17-3

$$Ts = T_{AR} + Te$$

Rearranging to get the power to thermal noise:

Eq. 17-4

$$C / N = P_{EIRP} \times g_{AR}^{10}$$

In (Alejandro Aragon-Zavala, 2008), receiver gain term, g_{AR} is represented at the receiver as g_{RX} without difference of value.

Eq. 17-4 can be expressed in a more convenient format, decibels:

Eq. 17-5

$$\text{“}(C/N)_{dB} = P_{EIRP} - Ls - A_R\,(G/Ts)_{dB} \quad - 10 \log (kB)$$

Where: $(G/Ts)_{dB}$ represents the figure of merit of the receiver, in dB, L_{FS} is the free-space loss, in dB, and P_{EIRP} is the EIRP in dBW." (Alejandro Aragon-Zavala, 2008)

"A measure of system performance of a digital transmission system is the bit error rate (BER). BER is a function of the energy per bit over thermal noise power spectral density ratio (Eb / No). The latter may be calculated or measured.

10. $/ k \times Ts \times B \times L_{FS}$

The energy per bit is given by:

Eq. 17-6

$$Eb = C \times Tb$$

Where: C = average power, watts. Tb is the time required to send one bit, secs. It now possible to calculate the carrier-to-noise ratio (C/ N):

Eq. 17-7

$$(C/N) = Eb /Tb / N_0 \times B = Eb \times R / N_0 \times B$$

Where: R = 1 /Tb is the bit rate (b/s)

Playing with Eq. 17-4 we get: (Alejandro Aragon-Zavala, 2008)

Eq. 17-8

$$Eb / N_0 = P_{EIRP} \times g_{AR} / k \times Ts \times R \times L_{FS}$$

The carrier-to-noise spectral density ratio (C/N0) is useful parameter that applies to any one-way RF link. C/N_0 in decibels is expressed:

Eq. 17-9

$$(C/ N_0)_{dB} = P_{EIRP} - Ls - A_R (G /Ts)_{dB} - 10 \log (k \times R)$$

Converting to Eb / N_0 is:

Eq. 17-10

$$(C/ N_0)_{dB} = (Eb / N_0)_{dB}\, 10 \log (R)$$

This conversion requires knowledge of the symbol/ bit rate and C/N. " (Alejandro Aragon-Zavala, 2008) Our primary reference (Alejandro Aragon-Zavala, 2008) gives further equations which define MAR, link margin of error and further manipulate the (Eb / N0) dB. These are left for the student to cogitate. (Alejandro Aragon-Zavala, 2008) on p. 162.

We have reached the key point in our link budget analysis. We can define the *uplink* (UL) decibel equation from the GTS transmitter to HAPS receiver and the decibel *downlink* (DL) equation from HAPS transmitter to GTS receiver. These become the working equations.

Uplink equation:[11]

Eq. 17 – 11

$$(C/N_0)_{dB,UL} = P_{EIRP,ES} - L_{FS,UL} - A_R\,(G/T_s)_{dB,HAPS,fom,/k} - k_{dB} - R_{dB,UL}$$

Downlink equation:[12]

Eq. 17 – 12

$$(C/N_0)_{dB,DL} = P_{EIRP,HAPS} - L_{FS,DL} - A_R\,(G/T_s)_{dB,ES,fom,/k} - k_{dB} - R_{dB,DL}$$

Table 17–7 shows an example "HAPS link budget analysis for Ka -band for clear sky." (Alejandro Aragon-Zavala, 2008) The primary reference works out many examples at for different bands, weather, elevation angles and other factors. The student is left to explore at leisure.

11. Nomenclature: ES = Earth Station, FS = Free space loss in UL; AR = Rain attenuation, fom = figure of merit, R = data rate, dB for HAPS uplink. Rain attenuation valid for frequency above 10 GHz.
12. $P_{EIRP,ES}$ is the EIRP for HAPS, FS = Free space loss in DL, fom = figure of merit in DL,

Table 17–7 shows an example "HAPS link budget analysis for Ka -band for clear sky."
(Alejandro Aragon-Zavala, 2008)

Parameter	Units	Uplink	Downlink	Uplink	Downlink
Elevation Angle	deg	20		90	
Frequency	GHz	31.28	28	31.28	28
Bandwidth	MHz	20	20	20	20
Transmit antenna					
Power output	dBW	-16.3	-14.5	-16.3	-15.2
Gain	dBi	35	29.5	35	16.5
EIRP	dBW	18.7	15	18.2	18.2
Distance	Km	58.5	58.5	20	20
Free space loss	dB	157.7	156.7	148.4	147.4
Rain attenuation	dB	0	0	0	0
Availability	%	100	100	100	100
Atmospheric gas attenuation	dB	0.4	0.4	0	0
Receive antenna gain	dBi	29.5	35	16.5	35
Received power	dBW	-110.9	-108.1	-114.2	-112.2
Noise temperature	K	700	500	700	500
Interference power density	dBW/MHz	-150.2	-151.6	-150.2	-151.6
Receiver losses	dB	2.5	2.5	2.5	2.5
Available C/N0	dB/MHz	86.3	90.6	83	13.3
Data rate	Mb/s	13.3	13.3	13.3	13.3
Required Eb/N0, BER =10 -6	dB	10.5	10.5	10.5	10.5
Coding gain	dB	5	5	5	5
Obtained Eb. / N0	dB	5.5	5.5	5.5	5.5

Obtained C/N0	dB	76.7	76.7	76.7	76.7
Link Margin	**dB**	**9.6**	**13.9**	**6.3**	**9.8**
Where:					
Uplink bit rate for elevation angles (UBR)	Mb/s				
At 20 degrees	Availability %	UBR		At 90 degrees at availability%	UBR
	99.40	20			20
	99.50	12.9			16.2
	99.60	7.4			12
	99.70	3.5			8.1

Source: (Alejandro Aragon-Zavala, 2008) pp. 176 -177 [13]

Discussion Questions

HAPS represents so much promise for telecommunications and emergency networking. Researchers have been designing and calculating for three decades. Lots of projects. Lots of money invested. A fascinating technology. But conversion into real projects and utility for the user seems to be still in limbo. Any ideas?

1. What stratospheric factors most effect HAPS performance?
2. Refer to Table 17 -7. Change the analysis from clear sky to hard rain. What parameters would be affected? Assume the winds have shifted from calm to gale. What parameters would be affected? Shift the frequency to the SHF -band (47 /49 GHz). What would you

13. The (Alejandro Aragon-Zavala, 2008) reference was published in 2008 where the world was enjoying LTE 3G and Intel processors of 4th generation cores. Data rates have risen since then. The world is LTE 5G deployment and on-board computers use 8th gen Intel cores with special ASICs and FPGA hardware. Bit rates obviously much more. However, the author is a believer in not reinventing the wheel. (Alejandro Aragon-Zavala, 2008) represents the best product of about 112 papers and texts written in 2016-2019 and reviewed by the author for student benefit and understanding. The basic principles of Link budget analysis for satellites, weapon systems, and HAPs have not changed that much over the year. Nomenclature has been upgraded but the link balance analysis stays the same. A similar argument can be made for chemical engineering with their material and heat balances or in statics where the forces on a unit are conserved.

expect the Link budget analysis to look like?

Bibliography

Adamy, D. (2001). EW 101 A *First Course in Electronic Warfare.* Boston: Artech House.

Adamy, D. (2004). EW 102 A *Second Course in Electronic Warfare.* Boston: Artech House.

Adamy, D. (2009). EW 103 *Tactical Battlefield Communications Electronic Warfare.* Boston: Artech House.

Alejandro Aragon-Zavala, J. L.-R.-P. (2008). *High-Altitude Platforms for Wireless Communications.* Chichester, West Sussex, UK: John Wiley & Sons.

Brown, E. F. (Dec 2008). Airborne Communication Networks for Small Unmanned Aircraft Systems. *Proc. IEEE, vol* 96, *no* 12, pp. 2008-17.

D, G. a. (2010). *Broadband Communications via High Altitude Platforms.* New York City, NY: John Wiley & Sons.

EARSC. (2015). A Taxonomy for the EO Services Market: enhancing perception and performance of the EO service industry. EARSC *Issue* 2.

ESA-ESTEC Contract 162372/02/NL/US. (September 2005). STRATOS: *Stratospheric Platforms a definition study for ESA Platform, Final Report,* 1-34. ESA-ESTEC .

Giordano, N. (2009). *College Physics: Reasoning and Relationships.* New York City, NY: Cengage Learning. pp. 421–424.

Goddemeir, K. D. (June 2015). Role-based Connectivity Management with Realist Air to Ground Channels for Future Applications. IEEE *Vehic. Tech. Mag.* Vol 10, *no* 2, pp. 79-85.

Henderson, T. (2017). The Doppler Effect – Lesson 3, Waves. *Physics tutorial. The Physics Classroom.* Retrieved from Henderson, T. (2017). "The Doppler Effect – Lesson 3, Waves". Physics tutorial. The Physics Classroom. Retrieved September 4, 2017.: Henderson, T. (2017). "The Doppler Effect – Lesson 3, Waves". Physics tutorial. The Physics Classroom. Retrieved September 4, 2017.

Ibrahim, A. (2019). *Optimization Methods for User Admissions and Radio Resource Allocation for Multicasting over High Altitude Platforms* . Memorial University of Newfoundland , Canada: River Publications.

Jesus Gonzalo, D. L. (2018, March 15). On the Capabilities and Limitations of High Altitude Pseudo-Satellites. *Progress In Aerospace Sciences*, 37-56. doi:https://doi.org/10.1016/j.paerosci.2018.03.006

Macnamara, T. M. (2010). *Introduction to Antenna Placement & Installation.* New York City, NY : John Wiley & Sons.

Matolak, R. S. (April 2015). Initial Results for Airframe Shadowing in L-band and C-band Air -Ground Channels. *Proc. Integrated Commun,, Navigation, and Surveillance Conf*, (pp. pp. 1-8).

Mohorcic, D. G. (2010). *Broadband Communications via High Altitude Platforms.* New York City, NY: John Wiley & Sons.

Osseiran, A. (Dec 2014). Scenarios for 5G Mobile and Wireless communications: the vision of the METIS Project. *IEEE Communications Magazine*, Vol 52, *no* 5, pp. 26-35.

Possel, M. (2017). Waves, motion and frequency: the Doppler effect. *Einstein Online*, Vol. 5. *Max Planck Institute for Gravitational Physics, Potsdam, Germany.*

Randall K. Nichols, J. J. (2018). *Unmanned Aircraft Systems (UAS) in the Cyber Domain: Protecting USA's Advanced Air Assets.* Manhattan, KS: New Prairie Press.

Rappaport, T. (2014). *Millimeter Wave Wireless Communications.* New York City, NY: Prentice Hall.

Romano, G. (2017, October 5). *Preparing the Ground for IMT-2000.* Retrieved from 3gpp.org/news-events/: https://www.3gpp.org/news-events/3gpp-news/1901-imt2020_news

Staff. (2008). FINAL ACTS WRC-07. *World Radiocommunication Conference.* Geneva: ITU.

Staff. (2012). FINAL ACTS WRC-12. *World Radiocommunication Conference.* Geneva: ITU.

Staff. (2019). FINAL ACTS WRC-15. *World Radiocommunication Conference.* Geneva.

Sun, W. M. (June 2015). Unmanned Aircraft Systems:Air-Ground Channel Characterization for future applications. *IEEE Vehic. Tech Mag.* Vol 10, No 2 , pp. 79-85.

T.C. Dozer, D. A. (2008). High Altitude Platforms for VHDR in-theater communications. *IET Seminar on Military Satellite Communications Systems.*

Xiaoyang Liu, C. L. (2016). High Altitude Platform Station Network and Channel Modeling Performance Analysis. *Mathematics and Computer Science.* Vol. 1, No 1, pp. 10-16. doi:Xiaoyang Liu, Chao Liu, Wanping Liu, Xiaoping Zeng. High Altitude Platform Station Network and Channel Modeling Performance Analys10.11648/j.mcs.20160101.13

Zeng, R. Z. (May 2016.). Wireless communications with unmanned aerial vehicles: opportunities and challenges. IEEE *Communications Magazine.vol.* 54, *no.*5, pp. 36-42.

Yong Zeng, R. Z. (2016). Wireless Communications with Unmanned Aerial Vehicles: Opportunities and Challenges. IEEE *Communications Magazine*, 36-42.

Chapter 18: C-UAS and Large Scale Threats

Chapter 18 will introduce developing trends for countering illegal / rogue drone use within the United States using counter-UAS (C-UAS) technology and systems.

Student Learning Objectives. Upon completion of this chapter, students should be able to:

- Further understand what Counter-UAS (C-UAS) technology is
- What civil restrictions are currently in place and what agencies can use C-UAS
- How the government is attempting to ease restrictions for C-UAS use
- How Congress through HR302 plays a role in easing restrictions for C-UAS
- How and why government agencies use C-UAS
- What types of threats currently exist?
- The growing potential for countering large UAS

Countering Emerging Unmanned Air System Threats

The FAA has forecasted that the unmanned air systems market will see exponential growth over the next 20 years. Commercial 'hobbyist' drones are expected to double from 1.1 million to 2.4 million systems. The larger sized commercial UAS fleet will grow from 111,000 in 2017 to roughly 452,000 by 2022 with the number of registered pilots climbing from 74,000 to over 300,000 by 2022 (Miller, 2018). With this accelerated growth, (Figure 18-1) there have been -numerous incidents around the world involving safety and privacy issues. These incidents have ranged from intruding on personal property to the complete shutdown of major international airports. (Perez-Pena, 2018) These incidents stemmed from UAS being in either the wrong place at the wrong time or the systems were intentionally violating airspace for reasons unknown. The purpose of this study is to examine the current and future methods that are and will be used to safely identify, counter and intercept potentially hazardous drones and to review national policy regarding counter UAS (C-UAS) practices and procedures.

Figure 18-1: UAS Market Growth 2018 -2036

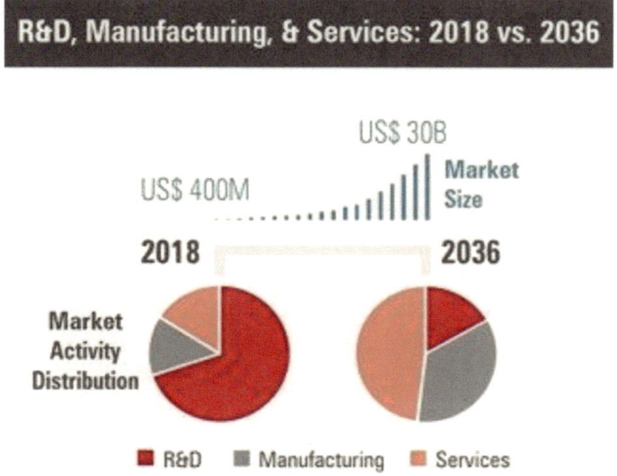

Source: (AIA & Avascent Report, 2018)

The demand for ever larger systems continues to push the UAS industry to meet this need. (see Figure 18-2) Increased payload capacity also increases the destructive potential for UAS systems when misused for malicious purposes. As these systems continue to grow in capability, so too will C-UAS to match.

Figure 18-2: UAS MTOW Market Analysis 2018 – 2036

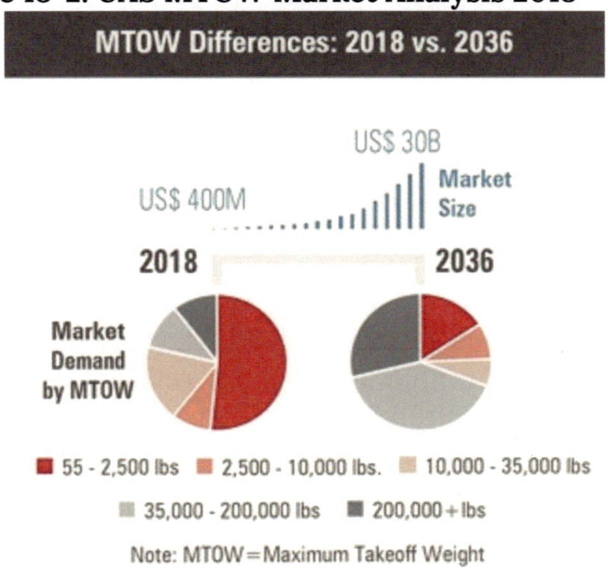

Source: (AIA & Avascent Report, 2018)

As of October 2018, the United States government implemented its first policy that directly affects drone usage. Known as the FAA Reauthorization act of 2018, HR 302 directs the FAA to begin drafting the way ahead for a UAS usage framework. Subtitle B – Unmanned Aircraft Systems, contains all directed actions the FAA will have to address to include C-UAS operations. (115th-congress, 2018) Most of the directed actions involve developing a practical way forward for C-UAS practices within the US and its territories. As of this moment, very few if any national policies are in place that regulate or even guide what can and cannot constitute C-UAS technology. Nor is there any governance on where and when C-UAS tech can be used outside of common laws that are already in use. It remains vague as to how a drone and its use is governed as personal property.

C-UAS is more than just using a system to knock a drone from the sky. The users must take into consideration the implications of what could happen if a UAS falls from the sky and damages property or injures people. If a drone is hovering over a large crowd for example, steps must be taken to allow the drone to be successfully removed without harming bystanders. This was the case recently when a drone was spotted dropping leaflets over a pro football game in Nov 2018. The operator was eventually identified and arrested but nothing could be done about the UAS while it was in flight. (Laris, 2018) This incident could have been far worse if the UAS operators' intentions were more sinister in nature.

The US population has a vested interest in the future of UAS and C-UAS operations. Civil and -military agencies, academic researchers as well as the everyday citizen already have access to UAS technology. These systems are well on their way to becoming part of our daily lives and there is currently no limit to what they can achieve. Knowing when, where, and having the ability to stop UAS misuse will be paramount.

Introduction

Counter Unmanned Air Systems is an array of linked technology designed to detect, identify / analyze a threat and if necessary, intercept unmanned air systems / drones. Rogue and hostile drone use is becoming an exponentially increasing concern for the Department of Homeland Security (DHS) and the Department of Defense (DoD). Illicit drones have also stirred the general public into seeking technological means of preventing UAS from violating privacy or exploiting intellectual property. The recent passing of HR 302, the FAA Reauthorization Act of 2018, has directed the FAA to conduct multiple inter agency reviews in order to determine the current status of C-UAS affairs in the United States. It also directs the FAA to work with the DOD to glean current best practices and procedures for acquiring and operating C-UAS systems for future use with the US and its territories. The intent for this research is to illustrate the current lack of legislation and regulation regarding the use of C-UAS within the US as well as review current and future technology trends in use. This research also aims to review C-UAS technology under development with the Department of Homeland Security (DHS) and the Department of Defense (DOD).

Current Civil Restrictions / Policy, Directed Reviews from HR 302

"There are several legal impediments to utilizing counter-UAS technology. The Communications Act of 1934 prohibits the use of unlicensed radio equipment such as jammers or other devices that interfere with communication, such as the UAS command link." (Embry Riddle Aeronautical University, 2018) "The general public is further prohibited to manufacture, import, market, sell or operate jamming equipment in the U.S. under 47 CFR 2.803." (Embry Riddle Aeronautical University, 2018) "Finally, 18 USC section 32 imposes imprisonment or fines upon those that damage, disable, or destroy civil aircraft which can be interpreted to include drones." (Embry Riddle Aeronautical University, 2018) "Operators may also be subject to liability associated with tort claims arising from the potential collateral damage, injury, or adverse effects of counter UAS activities." (Embry Riddle Aeronautical University, 2018) Such liability issues may include interference caused by jamming equipment or damage / injury caused by the forced disabling of the offending unmanned aircraft. (Wallace, 2018)

Steps to Easing Restrictions

In 2016, Congress passed the National Defense Authorization Act of 2017. In Sec. 1697,

"Congress codified new authority for military leaders to mitigate UAS threats. The statute gave relatively broad powers for the armed forces to disrupt control, intercept, seize, disable, damage, and destroy offending aircraft." (Embry Riddle Aeronautical University, 2018) The security risk posed by unmanned aircraft has not gone unnoticed by commercial entities either. Stadiums and other open-air public gatherings are recognizing the need for counter-UAS activities. On November 28, 2017, Tracy Mapes was arrested after flying a small UAS over NFL game at both the Levi Stadium and Oakland Coliseum two days earlier. (Gomez, 2017) The unmanned aircraft allegedly dropped leaflets over the stands at Levi Stadium. After reviewing surveillance footage of the initial incident, law enforcement personnel anticipated the alleged perpetrator would try the same activity at the nearby Oakland Coliseum. Santa Clara Police Lt. Dan Moreno highlighted the risk of UAS operations over the crowded areas stating, "A drone can lose control and injure someone in the crowd or drop material that may be harmful. We are evaluating our security practices with state and federal authorities to make sure this doesn't happen again." (Gomez, 2017) This and other incidents have led to the need for more robust yet public friendly C-UAS technology.

HR 302: FAA Reauthorization Act of 2018

For this theme the author spent a great deal of time reviewing two sections within HR 302

passed by congress in October of 2018:

- Section 364, U.S. C-UAS Review of Interagency Coordination Processes
- Section 365, Cooperation Related to Certain C-UAS Technology (115th-congress, 2018)

From these sections, the author was able to infer what the government was trying to accomplish. Congress directed the review of how the FAA and other agencies were using C-UAS within the United States and its territories. The reviews directed specific considerations to address, the most significant of these are listed below:

- Safety in the national airspace
- Protecting individuals and property on the ground
- Coordination procedures and protocols with the FAA during the operation of C-UAS systems
- Adequate training for personnel operating C-UAS systems
- Best practices for the consistent operation of C-UAS systems to the maximum extent practicable (115th-congress, 2018)

The last significant piece of information from HR 302 was that Congress directed the FAA to specifically conduct a review of any additional authorities needed by the FAA to effectively oversee the management of C-UAS systems within the US and its territories. (115th-congress, 2018)Section 365 deals directly with the deployment of C-UAS systems in the national airspace. Congress also directed the Secretary of Transportation to consult with the Secretary of Defense to streamline deployment of C-UAS systems by drawing upon the expertise and experience of the DOD in acquiring and operating C-UAS systems consistent with the safe and efficient operation of the national airspace system. (115th-congress, 2018) Once the directed reviews are complete and made public, C-UAS procedures will become standardized and regulated by the FAA. Industry will then be able to evolve, and narrow down existing trends moving forward and develop C-UAS technology that complies with these new standards. This moves industry one step closer to opening the door for broader general public use of C-UAS.

C-UAS and the Department of Homeland Security

To increase safety, lower costs and increase efficiency more and more private companies are looking to add drones to their workforce. These drones have the potential for use in dangerous aerial inspection jobs as well as local deliveries normally performed by humans. The largest hurdle for these companies is how to incorporate the large volume of drones necessary for these jobs safely into the national airspace without collisions or infringement on privacy. (Sullivan-Nightingale, 2015) This increased interest ultimately means more drones will soon be flying around localities and neighborhoods. This increase in drone activity will continue to pose a problem for privacy, security and general public safety. Presently, UAS and drones are an inexpensive way to gain real time situational awareness over any venue to include large gatherings and sporting events. Hovering drones pose a real danger to the crowd below should they malfunction or loose link with their pilot. (Warwick, 2016) Until recently, most C-UAS systems used kinetic means to physically remove a drone from the airspace. Researchers are now able to identify, track and if needed commandeer the drone and land it safely. Current technology can also be used to triangulate the offending pilot's location to pass on if notifying the authorities is necessary. (Warwick, 2016) With the eventual increase in drone activity, local government authorities will need access to C-UAS in order to maintain the security and safety of its citizenry.

C-UAS and the Department of Defense

The militarization of drone use began during the opening years of the Global War on Terror (GWOT) in the early 2000's. Before this time, the average citizen had no concept of what UAS capabilities were at the time. Large systems such as the MQ-1 Predator and RQ-9 Reaper have since become the face of GWOT. Their ability to conduct long endurance reconnaissance and precision strike operations vs large scale bombing campaigns has made drone proliferation a priority in most of the militaries of the world moving forward. China has declared that it will have produced over 20,000 drones by the year 2020 in order to protect its ever growing "One Road," initiative as it seeks to expand is economic influence in the region. Other countries that would be considered potentially adversarial to the United States have also dramatically increased drone acquisitions within the past decade. Simplicity and low cost of UAS technologies make their proliferation difficult to control. (See Figure 18-3 and 18-4)

Figure 18-3: Unmanned Systems Funding by Service *Source*: (Krasnov, 2017)

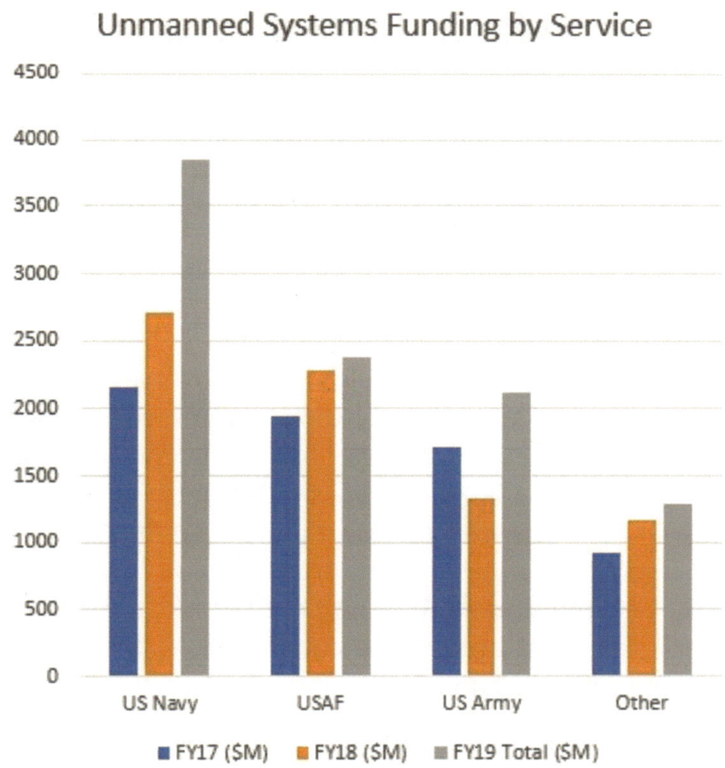

Figure 18-4: UAS Market Size by Destructive Mitigation Type

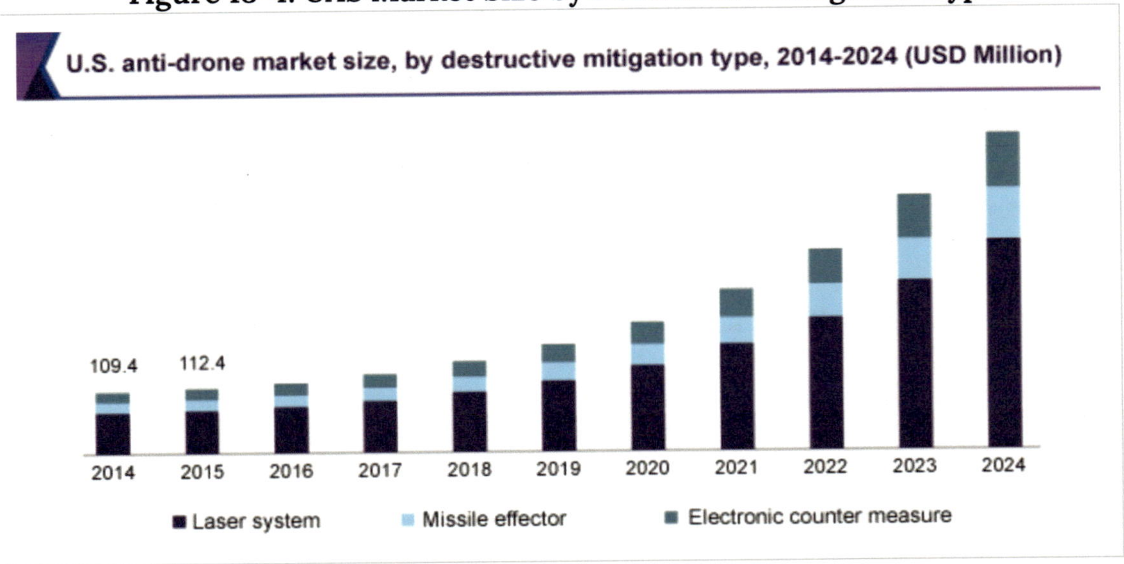

Source: (Grandview, 2018)

As more and more drones prepare to enter the future battlespace, the US Department of Defense (DoD) has begun to re-look its short- and long-range air defense capabilities. Vulnerabilities for each service have recently been identified based on the services given specific mission and current drone technology available to enemy combatant to observe and counter their operations. The DoD has specifically focused on brigade and below sized levels for the Army and Marines, individual vessels for the Navy and large airfield infrastructure for the Air Force. Each service has made C-UAS technology and system acquisitions a top priority in recent years and will remain a priority for the near future.

SWARMS

The threat(s) UAS pose continue to evolve with the most recent development of drone "SWARMS." Drone swarm is used in two separate ways, cooperative and coordinated swarming. For a cooperative swarm, drones perform complex functions en masse such as land surveying, mapping, complex entertainment displays, manned / unmanned teaming, single attack performed by several drones against one target. Coordinated SWARMS, drones perform separate tasks aimed to achieve one goal. Examples include transmission line repair and complex attacks against multiple targets at the same time. Swarming has been carried out by low slow and small drones with varied success, the most notable was an attack against a Russian air base in Syria in January 2018, where several home-built drones carrying crude munitions penetrated Russian defenses and successfully damaged aircraft and equipment on the base. This attack proves drone SWARMS are difficult to track and engage effectively using current systems. The US DoD has taken great strides to create and expand C-UAS doctrine and update existing systems.

The most difficult challenge faced by the DoD is developing and implementing C-UAS across a Joint Force. Each service has its own methods, technologies and systems but the hard part is getting them all to talk and coordinate with each other. Some of these challenges include being able to identify current enemy UAS, integrating short range defense (SHORAD) and operating with the absence of robust joint doctrine that allows commanders to wield all forms of C-UAS within their respective battle space.

AI and Machine Learning

Advances in C-UAS tech has begun to see emerging technology, like artificial intelligence (AI), as a means of supporting non-lethal counter-UAS applications. Citadel Defense Company recently launched its newest counter-UAS specific system, Titan. The system is designed for anti-drone scenarios such as drug trafficking, espionage, cyber-attacks and attacks on airports. Titan provides the user real-time information, identifying and classifying a single approaching unmanned aerial vehicle or a larger swarm. The system selectively applies precise countermeasures to induce the UAV(s) to land or return to its home base. Citadel Defense uses machine learning, artificial intelligence and software defined hardware technology to rapidly address new threats and protect people and assets. (Rees, 2019)

C-UAS and the General Public

Reviewing currently available C-UAS technology yielded three major types that best suite usage by the general public in a civil domain. These systems have the lowest chances of posing any potential harm bystanders:

- Local radar systems
- Integrated microphone / RF signature detection systems
- Acoustical systems for passive detection / offensive destruction

Local Radars have seen success as recently as this year's super bowl in Atlanta, GA. The FAA declared a restricted flying zone for one mile around Mercedes Benz Stadium. Local radar was used to track and if needed, direct integrated systems to remove the drone. The radars lead to the successful interception and safe landing of multiple drones without incident. (O'kane, 2019)A commonly occurring theme is that most vendors still do not know what they need as far as C-UAS systems and the demand for security experts with this knowledge continues to grow. (Warwick, 2016) The need for C-UAS systems extends to non-standard facilities such as correctional institutions. In recent years, prisons have seen an increase in illicit drone activity in the form of contraband drops to inmates. By integrating RF detection systems / jammers linked with passive microphones, correctional facilities have been able to identify when and where drones conduct these drops. (Dedrone, 2018)

The final emerging C-UAS tech trend is acoustical systems. Small teams in academia are currently working to see how ultra-high frequency noise that cannot be heard by the human ear, can be used to safely disrupt drones while in flight. There has been success in disrupting the microscopic gyroscopes (MEMS) within the drone motors causing the rotor heads to spin at differing speeds resulting in the drone becoming unstable and crashing. (Son, 2015) In its current form this technology is best used as an area denial system. Acoustical technology will most likely be used to defend an area from a multiple drone's swarm and could possibly be coupled to part of an existing network of C-UAS systems to form a layered defense for a private facility.

Emerging Threat of Large Civil UAS

The next evolution for the UAS industry will to begin to dramatically increase payload capacities. This would enable more economical uses for delivery services to begin delivering large goods right to consumers or deliver goods across long distances reducing costs of aerial transport. As this trend seeks to take hold of the UAS market, C-UAS researchers will have to begin to visualize the threats scenarios these systems will pose and how they can be safely countered. (See Figures 18-5 to 18-7)

Figure 18-5: UAS Market Growth Predictions by Civil Sector

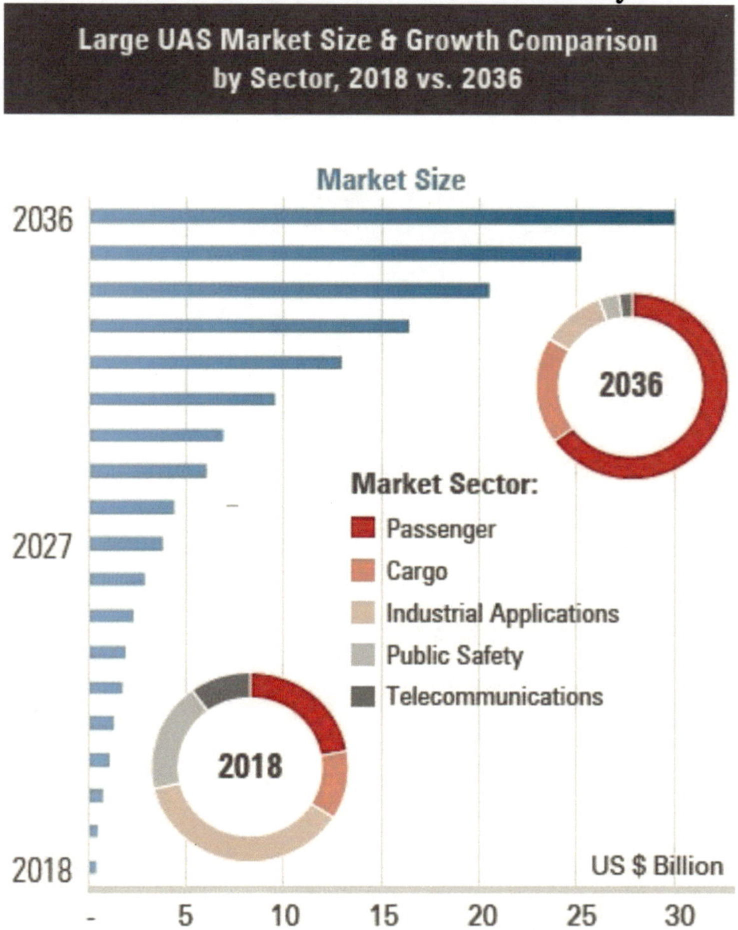

Source: (Unmanned Airspace.info, 2019)

Figure 18-6: Think Bigger: Large UAS and the Next Major Shift in Aviation

Source: (News, 2019)

Results

Current Restrictions / Policy, Directed Reviews from HR 302

Once the final reports from HR 302 are released, the FAA in coordination with the DOT and DoD will begin to standardize current best practices. This will establish rules and regulations that industry can then use to guide its research and development as it clamors to meet market need. Drone use will continue to grow exponentially as shown by market research as applications for UAS' continues to be developed. It is the opinion of the researcher that once the FAA has established a firm precedent for civil C-UAS operations, the C-UAS industry will see very similar exponential growth to match the UAS industry.

Figure 18-7: Defining Large UAS

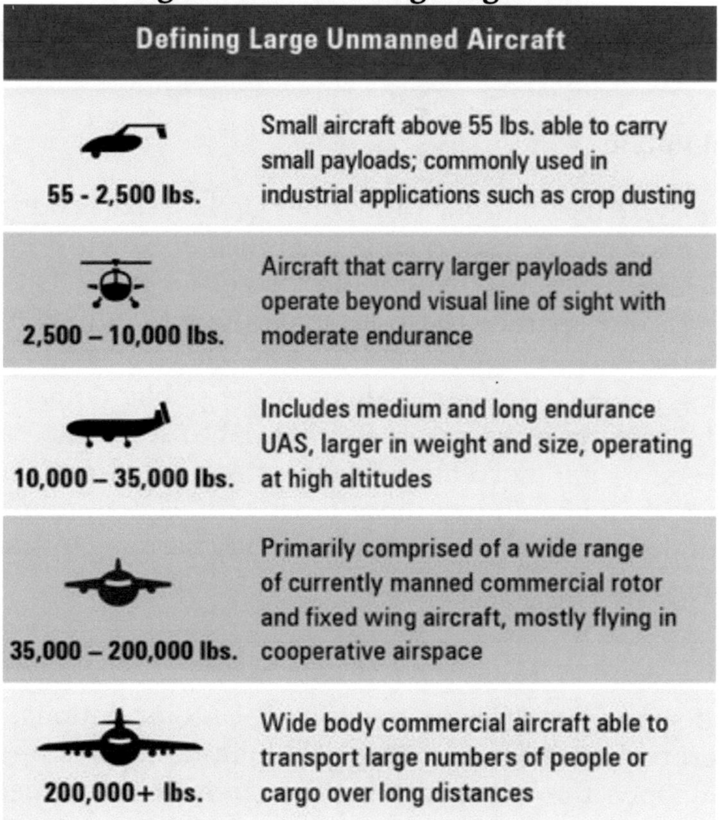

Source: (AIA & Avascent Report, 2018)

C-UAS and the Department of Homeland Security

Civil authorities will soon have no choice and will have to address / budget for local government use of C-UAS in order to protect local infrastructure and public gatherings. The threat(s) posed to open space venues and facilities will require affordable and adaptable systems that can be used to protect the public at large. Fortunately, private industry has already been developing integrated systems that are designed for this purpose(s). As drones become increasingly sophisticated, potential adversaries and drug cartels can also potentially use these systems to conduct cross border incursions.

C-UAS and the Department of Defense

Addressing current gaps for the DoD and finding ways forward for joint doctrine integration will be crucial for the Defense Department to stay ahead of growing threats posed by peer / near peer state actors and rogue organizations / individuals. The DoD was the first organization to see and begin to adapt to the growing threat of drone use. There guidance and expertise will be invaluable for the FAA and DHS to expand C-UAS from. With the largest budget for C-UAS technology development, the DoD will maintain its current position as the industry leader for C-UAS in the US.

C-UAS and the General Public

As the commercial and private use of drones and other UAS systems continues to grow, systems to limit their use will inevitably see use to limit the areas in which they can fly. With several forms of non-kinetic C-UAS technology emerging, the author predicts that the general public will soon be able to own and operate these systems once Congress has formalized further authorizations allowing the FAA to regulate C-UAS use. General safety for the public is paramount when considering using C-UAS technology in the public domain. The three emerging technologies discussed, I feel, have the greatest potential to serve the public safely and effectively. Large open-air vendors and correctional institutions are two of the many organizations along with private citizens who could use these types of C-UAS systems. They will be able to maintain their privacy and security while not inadvertently harming bystanders or creating collateral damage in the general area around them.

Conclusion(s)

The literature collected shows the growing need for C-UAS for national defense, civil defense and private property protection market(s). The only limiting factors appear to be the lack of government regulation. Once the FAA has identified and recommended to Congress C-UAS standard operating procedures and best practices, stemming from HR 302, Congress will then create legislation that will help guide the C-UAS industry in developing new and better technology for civil use. Outside of the DOD, private industry has recognized the growing need for C-UAS and has already developed technologies that allow for the safe removal of drones from private airspace. Once the regulations are in place, the C-UAS industry is ready to grow exponentially.

The rapid pace of UAS development will soon introduce larger UAS to the public airspace. This will further drive the need to develop systems that can counter / deter there use for nefarious reasons. Large UAS will be able to inflict more severe damage to varying degrees of structures and venues if they are repurposed or digitally hi-jacked. Currently the only way to combat large UAS is using kinetic means which is only operated by the DoD and is in very sparse location across the country. As this sector grows, so too will the need for larger and more integrated systems across the US.

Counter-UAS will continue to grow exponentially to match the capabilities of the UAS industry.

This growth will be bolstered once Congress and the FAA formalize procedures and operating parameters for these systems.

Bibliography

115th-congress. (2018). *115th Congress. (2018). HR 302 FAA Reauthorization Act. Sec 364 and 365.* Retrieved from www.congress.gov/bill/115th-congress/house-bill/302/text : https://www.congress.gov/bill/115th-congress/house-bill/302/text

49 U.S. Code §40103, 49 U.S. Code §40103 Sovereignty and use of airspace (U.S. Code July 5, 1994).

Abramson, E. (2016). *Ethical Dilemmas in the Age of AI.* Retrieved from Abramson, E. – know-mail.me/blog: https://www.knowmail.me/blog/ethical-dilemmas-age-ai/

Adamy, D. -0. (2015). EW 104 EW *against a New Generation of Threats.* Boston: Artech House.

Adamy, D. (2001). EW 101 A *First Course in Electronic Warfare.* Boston, MA: Artech House.

Adamy, D. (2001). EW 101 A *First Course in Electronic Warfare.* Boston: Artech House.

Adamy, D. (2004). EW 102 A *Second Course in Electronic Warfare.* Boston: Artech House.

Adamy, D. (2009). EW 103 *Tactical Battlefield Communications Electronic Warfare.* Boston: Artech House.

Adamy, D.-9. (1998, Jan). Lesson 4: the basic link for all EW functions. (electronic warfare)(EW Reference & Source Guide). *Journal of Electronic Defense, Jan 1998 Issue.*

Administrator. (2015, June 15). *Standing Wave and Fundamental Frequency.* Retrieved from Electronics Hub: https://www.electronicshub.org/?s=fundamental+frequency

Administrator. (2019, May 17). *Harmonic Frequencies.* Retrieved from electronicshub.org: https://www.electronicshub.org/harmonic-frequencies/

AIA & Avascent Report. (2018, April 23). *Think Bigger: Large Unmanned Systems and the Next Major Shift in Aviation.* Retrieved from www.avascent.com: https://www.avascent.com/2018/02/think-bigger-large-unmanned-systems-and-the-next-major-shift-in-aviation/

Alejandro Aragon-Zavala, J. L.-R.-P. (2008). *High-Altitude Platforms for Wireless Communications.* Chichester, West Sussex, UK: John Wiley & Sons.

Alford, L. (2000). Cyber Warfare: Protecting Military Systems. *Acquisition Review Quarterly.*

Anon. (2019). *Saudi Arabia grants citizenship to robot Sophia.* Retrieved from dw: Saudi Arabia grants cihttps://www.dw.com/en/saudi-arabia-grants-citizenship-to-robot-sophia/a-41150856

Asimov, I. (1950). "*Runaround*". I, Robot (*The Isaac Asimov Collection ed.*). New York City: Doubleday.

Atherton, K. D. (2019). Can the Pentagon sell Silicon Valley on AI as ethical war? . C4ISRNET.

Austin, R. (2010). "*Design for Stealth*", *Unmanned Aircraft Systems UAVS Design Development and Deployment.* New York: John Wiley and Sons.

Brown, E. F. (Dec 2008). Airborne Communication Networks for Small Unmanned Aircraft Systems. *Proc. IEEE, vol* 96, *no* 12, pp. 2008-17.

Burch, D. (2015). RADAR *for Mariners.* New York: McGraw-Hill.

Cameron, J. &. (Director). (1991). *Terminator* 2: *Judgement Day* [Motion Picture].

Chapman, A. (2019, May 31). *GPS Spoofing.* Retrieved from Tufts University – Tech Notes 2017: https://sites.tufts.edu/eeseniordesignhandbook/files/2017/05/Red_Chapman.pdf

Cornell University Legal Information Institute. (2019, June 5). *But-for test.* Retrieved from law.cornell.edu: https://www.law.cornell.edu/wex/but-for_test

Cornell University Legal Information Institute. (2019, June 5). *Intervening Cause.* Retrieved from law.cornell.edu: https://www.law.cornell.edu/wex/intervening_cause

Cornell University Legal Information Institute. (2019, June 5). *Personal Jurisdiction.* Retrieved from law.cornell.edu: https://www.law.cornell.edu/wex/personal_jurisdiction

D, G. a. (2010). *Broadband Communications via High Altitude Platforms.* New York City, NY: John Wiley & Sons.

Daniel-Cornel TĂNĂ, S. (2018). The Impact of the Development of Maritime Autonomous Systems on the Ethics of Naval Conflicts. *Annals: Series on Military Sciences*(2), 118-130.

Dedrone. (2018). *Correctional Facilities.* Retrieved from www.dedrone.com/solutions: https://www.dedrone.com/solutions/correctional-facilities

Deloitte Center for Government Insights analysis. (2018, June 18). *The future of regulation. Principles for regulating emerging technologies.* Retrieved from Deloitte Insights:

https://www2.deloitte.com/insights/us/en/industry/public-sector/future-of-regulation/regulating-emerging-technology.html

Dewey, D. (2017, July 14). *Drone crashes into LDS temple in Utah; raises questions of airspace rules.* Retrieved from eastidahonews.com: https://www.eastidahonews.com/2017/07/drone-crashes-lds-temple-utah-raises-questions-airspace-rules/

Diversity of citizenship; amount in controversy; costs, 28 U.S. Code §1332 (United States Congress June 25, 1948).

DJI. (2019, June 5). *DJI Enterprise.* Retrieved from Enterprise DJI.com: https://enterprise.dji.com/civil-protection

DLSR Pros. (2019, June 3). *Best Drones (UAVs) for Firefighting in 2019.* Retrieved from dlsrpros.com: https://www.dslrpros.com/dslrpros-blog/best-drones-firefighting-2019/

DoD. (2018). *Dictionary of Military Terms.* Retrieved from JCS.Mil: http://www.jcs.mil/doctrine/dod_dictionary/

DoD. (2018). *Joint Publication (JP) 3-01 Countering Air and Missile Threats.* Washington, DC: DoD.

DoD-02. (2018). *Information Operations (IO) in the United States.* Retrieved from JP 3-13 : http://www.dtic.mil/doctrine/new_pubs/jp3_13.pdf

DoD-03. (2015). *Unmanned Systems Roadmap 2013 to 2038.* Retrieved from DTIC: http://www.dtic.mil/dtic/tr/fulltext/u2/a592015.pdf

Drones, Q. S. (2017, July 11). *quadcopters-have-hit-the-sound-barrier/.* Retrieved from quadstardrones.com: https://quadstardrones.com/2017/07/11/quadcopters-have-hit-the-sound-barrier/

EARSC. (2015). A Taxonomy for the EO Services Market: enhancing perception and performance of the EO service industry. *EARSC Issue 2.*

Embry Riddle Aeronautical University. (2018, June 16). *ERAU Common Documents.* Retrieved from ERAU: www.commons.erau.edu

Entokey, a. G. (2019, May 16). *entokey.com/acoustics-and-sound-measurement/.* Retrieved from entokey.com/acoustics-and-sound-measurement/: https://entokey.com/acoustics-and-sound-measurement/

ESA-ESTEC Contract 162372/02/NL/US. (September 2005). *STRATOS: Stratospheric Platforms a definition study for ESA Platform, Final Report,* 1-34. ESA-ESTEC .

European Union. (2019, May 2019). *About the regulation and data protection.* Retrieved from ec.europa.eu: https://ec.europa.eu/commission/priorities/justice-and-fundamental-rights/data-protection/2018-reform-eu-data-protection-rules_en

Federal Question, 28 U.S. Code §1331 (United States Congress June 25, 1948).

FEMA. (2013). *Lessons Learned from the Boston Marathon Bombings: Preparing for and Responding to the Attack.* Retrieved from www.fema.gov: http://www.fema.gov/media-library-data/20130726-1923-25045-1176/lessons_learned_from_t

Filippo Santoni de, S. &. (2018). Meaningful Human Control over Autonomous Systems: A Philosophical Account. *Frontiers in Robotics and AI.* doi:10.3389/frobt.2018.00015

Fleetwood, J. (2017). Public Health, Ethics, and Autonomous Vehicles. *American Journal of Public Health*, 107(4), 632-537.

Fortuna, C. (2017, 12 02). *Autonomous Driving Levels 0–5 + Implications.* Retrieved from cleantechnica.com: https://cleantechnica.com/2017/12/02/autonomous-driving-levels-0-5-implications/

Gelfand. (2004). "*Physical Concepts*", *Hearing an Introduction to Psychological and Physiological Acoustics, 4th ed.* New York City.

Gelfand, S. A. (2009). *Essentials of Audiology, 3rd Edition.* Stuttgart, DE: Thieme.

Giordano, N. (2009). *College Physics: Reasoning and Relationships.* New York City, NY: Cengage Learning. pp. 421–424.

Gomez, M. &. (2017). *Man suspected of flying drone over 49ers, Raiders games arrested.* Retrieved from securityinfowatch.com/news/12383982: www.securityinfowatch.com/news/12383982/man-suspected-of-flying-drone-over-49ers-raiders-games-arrested

Grandview. (2018). *US Anti-Drone Market Size, by destructive Mitigation Size.* Retrieved from Grandviewresearch.com: www.Grandviewresearch.com

Guardbaum, S. (1994). The Nature of Preemption. *Cornell Law Review*, 767, 771.

Harris Aerial. (2019, June 5). *Carrier HX8 Sprayer Drone.* Retrieved from harrisaerial.com: https://www.harrisaerial.com/carrier-hx8-sprayer/

Heinman, C. (2019). *Hearing Loss Tests Patrient D v-105.* Carlisle, PA: Brown Optical Hearing Aid Service.

Henderson, T. (2017). The Doppler Effect – Lesson 3, Waves. *Physics tutorial. The Physics Class-*

room. Retrieved from Henderson, Tom (2017). "The Doppler Effect – Lesson 3, Waves". Physics tutorial. The Physics Classroom. Retrieved September 4, 2017.: Henderson, Tom (2017). "The Doppler Effect – Lesson 3, Waves". Physics tutorial. The Physics Classroom. Retrieved September 4, 2017.

Hern, A. (2017, 1 12). *Give robots 'personhood' status, EU committee argues*. Retrieved from The Guardian: www.theguardian.com/technology/2017/jan/12/give-robots-personhood-status-eu-committee-argues

Hubbard, R. K. (1998). *Boater's Bowditch*. Camden, MA: International Marine.

Ibrahim, A. (2019). *Optimization Methods for User Admissions and Radio Resource Allocation for Multicasting over High Altitude Platforms* . Memorial University of Newfoundland , Canada: River Publications.

IEEE . (2017). The IEEE Global Initiative on Ethics of Autonomous and Intelligent Systems Announces New Standards Projects. *Telecom Standards*, 27(11), pp. 4-5. .

Jesus Gonzalo, D. L. (2018, March 15). On the Capabilities and Limitations of High Altitude Pseudo-Satellities. *Progress In Aerospace Sciences*, 37-56. doi:https://doi.org/10.1016/j.paerosci.2018.03.006

Johnson, O. &. (2012). *Ethics: Selections from Classic and Contemporary Writers*. Boston, MA: Cengage Learning.

Jones, T. (2017). *International Commercial Drone Regulation and Drone Delivery Services*. Santa Monica: The Rand Corporation.

Kanowitz, S. (2019, 05 15). *Toward the deployment of ethical* AI. Retrieved from Government Computer News. : Kanowitz, S. (2019). Toward the dephttps://gcn.com/articles/2019/05/15/ethical-ai-idc.aspx?s=gcntech_200519

Kirk, J. (2015, August 5). *sounds-can-knock-drones-sky*. Retrieved from www.computerworld.com.au/article/581231: https://www.computerworld.com.au/article/581231/sounds-can-knock-drones-sky/

Knight, W. (2018). Nine charts that really bring home just how fast AI is growing. *MIT Technology Review* .

Krasnov, V. (2017, November 08). *Arm-takes wing*. Retrieved from https://blog.cloudflare.com/: https://blog.cloudflare.com/

Laris, M. (2018, May 10). *Stadium and Team Owners See Drones as Major Threat*. Retrieved from washingtonpost.com/local/trafficandcommuting;nationalinclude: https://www.washingtonpost.com/local/trafficandcommuting;nationalinclude;/stadium-and-team-owners-see-drones-as-major-league-threat/2018/05/10/83e0b954-50ad-11e8-84a0-458a1aa9ac0a_story.html?noredirect=on&utm_term=.e6aebf20ac9a

Legal Information Institute – Cornell University. (2019, May 31). *Strict Liability* . Retrieved from Legal Information Institute: https://www.law.cornell.edu/wex/strict_liability

LRAD. (2019, May 189). *LRAD 450XL Datasheet*. Retrieved from LRADX: http://www.lradx.com/wp-context/uploads/2015/05/LRAD_datasheet_450XL.pdf

MacGregor, D. S. (2018). *Colorado Causes of Action: Elements, Defenses, Remedies, and Forms.* Denver: Bradford Publishing Co. .

Macnamara, T. M. (2010). *Introduction to Antenna Placement & Installation.* New York City, NY : John Wiley & Sons.

Mahon, J. (2012). *Classical Natural Law Theory* St. Thomas Aquinas (1227-1274) – *the "Angelic Doctor" Lecture.* Retrieved from Mahon, J. (2012). Classical Natural Law Theory St. Thomas Aquinas (1227-1274) – the Philosophy of Law. : Mahon, J. (2012). Classical Natural Law Theory St. Thomas Aquinas (1227-http://home.wlu.edu/~mahonj/PhilLawLecture1NatLaw.htm

Marbury v. Madison, 5 U.S. 137 (United States Supreme Court February 23, 1803).

Matolak, R. S. (April 2015). Initial Results for Airframe Shadowing in L-band and C-band Air -Ground Channels. *Proc. Integrated Commun,, Navigation, and Surveillance Conf*, (pp. pp. 1-8).

McCullogh v. Maryland, 17 U.S. 316 (United States Supreme Court March 6, 1819).

Merriam-Webster. (2019, May 17). Merriam-Webster Online Dictionary.

Merriam-Webster, Inc. (2019). *Definition of Ethics.* online: Merriam-Webster, Inc. Retrieved from Definition of Ethics. (2019a). Online: Merriam-Webster, Incorporated.: Definition of Ethics. (2019a). Online: Merriam-Webster, Incorporated.

Middleton, C. (2018). *SAP launches ethical A.I. guidelines, expert advisory panel.* Retrieved from internetofbusiness.com: Middleton, C. (2018). SAP launches ethical A.I. guidelines, expert advisory panel. Retrieved from https://internetofbusiness.com/sap-publishes-ethical-guidelines-for-a-i-forms-expert-advisory-panel/

Miller, P. C. (2018, March 27). *FAA Forecasts Phenomenal Growth for* UAS. Retrieved from www.uasmagazine.com/articles/1833/: http://www.uasmagazine.com/articles/1833/faa-forecasts-phenomenal-growth-for-uas-industry

Misselhorn, C. (2018). Artificial Morality. Concepts, Issues and Challenges. *Society*, 55(2), 161-169.

Mohorcic, D. G. (2010). *Broadband Communications via High Altitude Platforms.* New York City, NY: John Wiley & Sons.

Monahan, K. (2004). *The Radar Book: Effective Navigation and Collision Avoidance.* Anacortes, WA: Fineedge Publications.

Muspratt, A. (2018, November 22). *New global drone standards proposed.* Retrieved from Defence iQ: https://www.defenceiq.com/defence-technology/news/new-global-drone-standards-proposed

Goddemeir, K. D. (June 2015). Role-based Connectivity Management with Realist Air to Ground Channels for Future Applications. *IEEE Vehic. Tech. Mag.* Vol 10, *no* 2, pp. 79-85.

National Conference of State Legislatures. (2018, September 10). *Current Unmanned Aircraft State Law Landscape.* Retrieved from NCSL.org: http://www.ncsl.org/research/transportation/current-unmanned-aircraft-state-law-landscape.aspx

NBC Today Show. (2018, May 9). *How peeping drones could be spying on you without you knowing it.* Retrieved from Today.com: https://www.today.com/video/how-peeping-drones-could-be-spying-on-you-without-you-knowing-it-1229001795967

Newman, L. H. (2017, August 7). *THE ARMY GROUNDS ITS DJI DRONES OVER SECURITY CONCERNS.* Retrieved from WIRED: https://www.wired.com/story/army-dji-drone-ban/

News, U. T. (2019, April 21). *Explosive Growth in Large UAS Operations, Predicts New Report If Regulators Think Big.* Retrieved from Unmannedairspace.info: https://www.unmannedairspace.info/uncategorized/explosive-growth-large-uas-operations-predicts-new-repo

Nichols, R. K. (1996). *Classical Cryptography Course Volume I / II.* Laguna Hills, CA: Aegean Park Press.

Nichols, R. K. (1996). *Classical Cryptography Course, Volume I.* Laguna Hills, CA: Aegean Park Press.

Nichols, R. K. (2018). *Unmanned Aircraft Systems (UAS) In the Cyber Domain: Protecting USA's Advanced Air Assets. 1st Ed.* Manhattan, KS: New Prairie Press.

Nichols, R. K. (2019, March 14). Hardening US Unmanned Systems Against Enemy Counter Measures. *7th Annual Unmanned Systems Summit*. Alexandria, VA, USA: PPTX presentation , self.

Nichols, R. K., Mumm, H. C., Lonstein, W. D., Ryan, J. J., & and Carter, C. (2018). *Unmanned Air-*

craft Systems (UAS) in the Cyber Domain: Protecting USA's Advanced Air Assets. Manhattan, KS: New Prairie Press (NPP) eBooks. 21.

NIST. (September 2012). *Guide for Conducting*. Washington, DC: GPO.

North Carolina Department of Transportation. (2019, May 30). *Law & Regulations*. Retrieved from NCDOT.GOV: https://www.ncdot.gov/divisions/aviation/uas/Pages/laws-regulations.aspx

O'kane, S. (2019, March 18). *Drones are Already Being Confiscated Near the Super bowl*. Retrieved from The Verge.

Osseiran, A. (Dec 2014). Scenarios for 5G Mobile and Wireless communications: the vision of the METIS Project. *IEEE Communications Magazine*, Vol 52, no 5, pp. 26-35.

O'Sullivan, J. L. (1845). The Great Nation of Futurity. *United States Magazine and Democratic Review* Vol 6 Issue 23, pp. 426-430.

Perez-Pena, R. (2018, December 27). *Gatwick Airport Drone: Lots of Guessing, but Not Many Answers*. Retrieved from NY Times: https://www.nytimes.com/2018/12/27/world/europe/gatwick-airport-drone.html

Pierson. (2019, May 16). *tuning-fork-waves-sound*. Retrieved from airfreshener.club – Pierson Education: https://airfreshener.club/quotes/tuning-fork-waves-sound.html

Porter, J. D. (2019, June 8). *jdporterlaw.com/intellectual-property-law/*. Retrieved from jdporterlaw.com: http://www.jdporterlaw.com/intellectual-property-law/

Possel, M. (2017). Waves, motion and frequency: the Doppler effect. *Einstein Online*, Vol. 5. *Max Planck Institute for Gravitational Physics, Potsdam, Germany*.

Pricewaterhousecoopers, LLP. (2018). *Skies without limits – Drones- taking the UK's economy to new heights*. London: Pricewaterhousecoopers, LLP.

PROTECTION OF CERTAIN FACILITIES AND ASSETS FROM UNMANNED AIRCRAFT, H. R. 302 (United States Congress January 3, 2018).

Proyas, A. (Director). (2004). *I, Robot. In. Hollywood*, CA. [Motion Picture].

Ramzy, A. &. (2008). *Tainted-Baby-Milk Scandal in China*. Retrieved from content.time.com/time/world/article/: http://content.time.com/time/world/article/0,8599,1841535,00.html

Randall K. Nichols and Lekkas, P. C. (2002). *Wireless Security: Threats, Models, Solutions*. New York City, NY: McGraw Hill.

Randall K. Nichols, D. (2018). Chapter 20 Acoustic CM & IFF Libraries V SWARMS Rev 1 05142019. In R. K. Nichols, H. C. Mumm, W. D. Lonstein, & J. S. Hood, *Unmanned Aircraft Systems (UAS) in the Cyber Domain: Protecting USA's Advanced Air Assets, 2nd ed.* Manhattan, KS: NPP Press.

Randall K. Nichols, D. (2019 for publication). Unmanned Aircraft Systems in the Cyber Domain: Protecting USA's Advanced Air Assets, 2nd ed. In H. M. Randall K. Nichols, *Chapter 18 Audiology, Acoustic Countermeasures against Swarms and Building IFF Libraries* (p. 2nd ed.). Manhattan, KS: For Publication, NPP.

Randall K. Nichols, J. J. (2018). *Unmanned Aircraft Systems (UAS) in the Cyber Domain: Protecting USA's Advanced Air Assets.* Manhattan, KS: New Prairie Press.

Rappaport, T. (2014). *Millimeter Wave Wireless Communications.* New York City, NY: Prentice Hall.

Rees, M. (2019, April 9). *New Counter-UAS System Utilizes AI and Machine Learning.* Retrieved from Unmanned Systems News.

Ricker, D. (2017, July 1). *Navigating drone laws has become a growing and lucrative legal niche.* Retrieved from ABA Journal: http://www.abajournal.com/magazine/article/drone_law_attorneys

Rupprecht, J. (2017). *7 big problems with counter-drone technology (drone jammers, anti-drone guns, .* Retrieved from jrupprechtlaw.com: https://jrupprechtlaw.com/drone-jammer-gun-defender-legal-problems

Said Emre Alper, Y. T. (December 2008). Compact Angular Rate Sensor System Using a Fully Decoupled Silicon-on-Glass MEMS Gyroscope. JOURNAL OF MICROELECTROMECHANICAL SYSTEMS, VOL. 17, NO. 6.

Sanchez, M. (2019, June 4). *No Drones.* Retrieved from Unspalsh.com: https://unsplash.com/photos/oMqswmrie4Y

Schroeder, A. (2018, February 1). *Localizing Humanitarian Drones: Robotics & Disaster Response from the Maldives to Malawi.* Retrieved from medium.com: https://medium.com/radiant-earth-insights/localizing-humanitarian-drones-robotics-disaster-response-from-the-maldives-to-malawi-a1f362432cb1

Signia. (2019, May 16). *Signia Hearing Aids.* Retrieved from Signia Hearing Aids – Hear across America: www.signiausa.com

Singer v. City of Newton, 284 F. Supp. 3d 125 (U.S. District Court Massachusetts September 21, 2017).

Son. (2015). *Rocking Drones with Intentional Sound Noise.* Retrieved from USINEX Symposium. 24, 881.: https://www.usenix.org/systems/files/confrence/usenixsecurity15/

Sood A.K. & Enbody, R. (2014, December 19). *https://www.georgetownjournalofinternau-s-military-defense-systems-the-anatomy-of-cyber-espionage-by-chinese-hackers.* Retrieved from georgetownjournalofinternationalaffairs.org/online-edition: https://www.georgetownjournalofinternationalaffairs.org/online-edition/u-s-military-defense-systems-the-anatomy-of-cyber-espionage-by-chinese-hackers

Sovereignty and use of airspace, 49 U.S. Code §40103 (United States Congress July 5, 1994).

Staff. (2008). FINAL ACTS WRC-07. *World Radiocommunication Conference.* Geneva: ITU.

Staff. (2012). FINAL ACTS WRC-12. *World Radiocommunication Conference.* Geneva: ITU.

Staff. (2016, April 17). *Equal Loudness Contours.* Retrieved from Gutenberg Organization: http://central.gutenberg.org/article/WHEBN0001046687/Equal-loudness%20contour

Staff. (2019). FINAL ACTS WRC-15. *World Radiocommunication Conference.* Geneva.

Stone, Z. (2007, 11 7). *Stone, Z. (2017). Everything You Need To Know About Sophia, The World's First Robot Citizen. Retrieved from https://www.forbes.everything-you-need-to-know-about-sophia-the-worlds-first-robot-citizen.* Retrieved from Forbes: https://www.forbes.com/sites/zarastone/2017/11/07/
everything-you-need-to-know-about-sophia-the-worlds-first-robot-citizen/#1667784246fa

Studios, D. D. (2017). Boaters Ref. USA.

sUAS News. (2018, March 2). *RAS Consulting & Investigations hire Jeff Parisse to offer sophisticated UAS security and surveillance services.* Retrieved from suasnews.com: https://www.suasnews.com/2018/03/
ras-consulting-investigations-hire-jeff-parisse-offer-sophisticated-uas-security-surveillance-services/

Sullivan-Nightingale, D. (2015). *Unmanned Aerial Systems: Risks & Opportunities in the Workplace. Professional Safety,* 6(3), 34-42.

Sun, W. M. (June 2015). Unmanned Aircraft Systems:Air-Ground Channel Characterization for future applications. IEEE *Vehic. Tech Mag.* Vol 10, No 2 , pp. 79-85.

T.C. Dozer, D. A. (2008). High Altitude Platforms for VHDR in-theater communications. IET *Seminar on Military Satellite Communications Systems.*

The Shepard News Team. (2018, September 12). *Liteye Receives Follow-on Contract for C-AUDS*

– DB – *Digital Battlespace.* Retrieved from Aerospace, Defense and Security News and Analysis – Shephard Media, The Shepard Press, Ltd: www.shephardmedia.com/news/digidigital-battle-space/liteye-receives-follow-contract-c-auds

Toomay, J. (1982). *RADAR for the Non – Specialist. London; Lifetime Learning Publications.* London: Lifetime Learning Publications.

TRS, S. (2018, July 10). *Tontechnik-Rechner-Sengpielaudio.* Retrieved from Tontechnik-Rechner-Sengpielaudio Calculator: www.sengspielaudio.com/calculator-wavelength.htm

UAV Coach. (2019, May 30). *Drone Laws in South Carolina* (2019). Retrieved from UAVcoach.com: https://uavcoach.com/drone-laws-south-carolina/

United States Constitution Article VI, Sec.2 (United States of America September 17, 1787).

Uni-wuppertal. (2019, May 15). *Inverse Square Law, General.* Retrieved from hydrogen.physik.uni-wuppertal.de/hyperphysics/: http://hydrogen.physik.uni-wuppertal.de/hyperphysics/hyperphysics/hbase/forces/isq.html

Unmanned Airspace.info. (2019, June 17). *US Market Growth Predictions for Civil Sector.* Retrieved from UnmannedAirspace.info: https://www.unmannedairspace.info/counter-uas-industry-directory/

Urban, T. (2018). *Teach Your Robots Well: Will Self-Taught Robots Be the End of Us?* Retrieved from www.worldsciencefestival.com: Urban, T. (2018). Teach Your Robots Well: Will Self-Taught Robots Be the End of Us? Retrieved from https://www.worldsciencefestival.com/programs/teach-robots-well-will-self-taught-robots-end-us/

Usenix.org. (2019, 6 9). *MEMS, Drones, & Sound Sourcing.* Retrieved from Usenix.org: www.usenix.org

Wallace, R. e. (2018, April 9). *Exploring Counter-UAS Operations. A Case Study of the 2017 Dominican Republic Festival Presidente.* Retrieved from researchgate.net/publication/325209812: https://www.researchgate.net/publication/325209812_Exploring_Commercial_CounterUAS_Operations_A_Case_Study_of_the_2017_Dominican_Republic_Festival_Presidente

Warwick, G. (2016). *Rapid Defense. Aviation Week & Space Technology,* 178(9), 31.

WebFinance, Inc. (2019). *Definition of Ethics.* (2019*b*). online: Online: WebFinance, Inc.

Weise, E. (2017, August 23). *could-hackers-behind-u-s-navy-collisions.* Retrieved from USATODAY: https://www.ruidosonews.com/story/tech/news/2017/08/23/could-hackers-behind-u-s-navy-collisions/594107001/

Wong, C. (2017). *Top Canadian researcher says AI robots deserve human rights*. Retrieved from Wong, C. (2017). Top Canadian researcher says AI robots deserve human rights. Retrieveit-business.ca: Wong, C. (2017). Top Canadian researcher says AI robots deserve human rhttps://www.itbusiness.ca/news/top-canadian-researcher-says-ai-robots-deserve-human-rights/95730

Wordpress. (2012, 08 29). *The True Sign of Intelligence*. Retrieved from deepthinkings.wordpress.com: http://deepthinkings.wordpress.com/2012/08/29/the-true-sign-of-intelligence/

Wright, T. (2017, August 11). *You've Been Warned: Keep Your Drones Away From Military Bases*. Retrieved from Air & Space, Smithsonian: https://www.airspacemag.com/daily-planet/keep-your-drones-away-military-base-180964451/

Wyvern, T. (2018). *National Critical Intelligence Estimate: Counter Unmanned Aircraft Systems (C-UAS) in the* US. Salina, KS: KSUP.

Xiaoyang Liu, C. L. (2016). High Altitude Platform Station Network and Channel Modeling Performance Analysis. *Mathematics and Computer Science*. Vol. 1, No 1, pp. 10-16. doi:Xiaoyang Liu, Chao Liu, Wanping Liu, Xiaoping Zeng. High Altitude Platform Station Network and Channel Modeling Performance Analys10.11648/j.mcs.20160101.13

Zeng, R. Z. (May 2016.). Wireless communications with unmanned aerial vehicles: opportunities and challenges. *IEEE Communications Magazine.vol.* 54, *no.*5, pp. 36-42.

Yong Zeng, R. Z. (2016). Wireless Communications with Unmanned Aerial Vehicles: Opportunities and Challenges. *IEEE Communications Magazine*, 36-42.

Yunmonk Son, H. S. (2015, August 12-14). Rocking Drones with Intentional Sound Noise on Gyroscopic Sensors. *Proc. 24th Usenix Security Symposium*. Washington, DC: USENIX. Retrieved from https://www.usenix.org/conference/usenixsecurity15/technical-sessions/presentation/son

Zetter, K. (2015). So, The NSA Has An Actual SKYNET Program . WIRED *Magazine(Online)*. . Retrieved from Zetter, K. (2015). So, The NSA Has An Actual SKYNET Program WIRED Magazine(Online).

Further Readings

Knowles, J. (2017). Technology Survey: A Sampling of Counter-UAS Systems. *Journal of Electronic* Defense, 40(9), 37.

Lee, C. (2018). Pentagon Exploring Counter-UAS Software. *National Defense Journal*, 102(775), 11.

Mason, et al. (2009). Assimilating Unmanned Aircraft Systems. *Air and Space Power Journal*, 23(2), 5-10, 126-127.

Nichols, Randall et. Al. (2018) *Unmanned Aircraft Systems (UAS) in the Cyber Domain: Protecting USA's Advanced Air Assets*. Kansas State Polytechnic Universities Libraries.

Palmer, T. S.; Geis, J. P. (2017). Defeating Small Civilian Unmanned Aerial Systems to Maintain Air Superiority. *Air & Space Power Journal*, 31 (2).

Park, et al. (2018). Aerial Release of Rhinoncomimus Latipes (Coleoptera: Curculionidae) to Control Persicaria Perfoliata (Polygonaceae) Using an Unmanned Aerial System. *Pest Management Science*, 74 (1), 141-148.

Patterson, D. R. (2017). Defeating the Threat of Small Unmanned Aerial Systems. *Air and Space Power Journal*, 31(1), 15-25.

Urban, J. (2018). What is the Eye in the Sky Actually Looking at and Who is Controlling It? An International Comparative Analysis on How to Fill Cybersecurity and Privacy Gaps to Strengthen Existing U.S. Drone Laws. *Federal Communications Law Journal*, 70(1), 0-6.

Warwick, G. (2016). Measured Response. *Aviation Week & Space Technology*, 178(5), 19.

Chapter 19: Audiology, Acoustic Countermeasures against Swarms and Building IFF Libraries

Student Learning Objectives

The student will explore the use of acoustic countermeasures against hostile UAS (especially Swarms) and their dual use as IFF vectors for UAS characterization. The student will take brief sojourn into the science of audiology to understand why acoustic countermeasures actually work.

Problem

The Risk of success of Terrorist Attacks on US Air Defense Systems (ADS) via sUAS / UAS is higher and improving because of commercial capabilities and accessibility. Advanced small drones capable of carrying sophisticated imaging equipment, significant (potentially lethal) payloads and performing extensive Intelligence, Surveillance, and Reconnaissance (ISR) missions are readily available to civilian market. **They pose a significant threat to civilian and military UAS operations and safety in the NAS. The highest threats to ADS are presented by hostile UAS Swarms.**

Problem Solution

The author's research suggests that UAS Swarms can be both identified (IFF) and destabilized / mitigated /eliminated / countered in the air by applying harsh acoustic countermeasures at resonance frequencies. UAS (in any formation – especially Swarms) present detectable acoustic signatures that can be collected in an IFF sound libraries and like fingerprints or DNA they are unique to the make, model and origin manufacturer. Once identified as hostile, UAS (Swarm units) may be destabilized by harsh – explosive amplitude acoustic countermeasures to the MEMS or rotor base of the UAS's causing destabilization of the UAS and grounding. Emergency and waypoint recovery functions do not work under this approach.

Review of key points from Chapter 8 Designing UAS Systems for Stealth

Detection Signatures

Recall from Chapter 8 Stealth, that UAS / UAVs are detected by their ***signatures***: noise (acoustic), optical (visible), infrared (thermal) and radar (radio). "These acoustic or electromagnetic emissions occur at the following wavelengths: (Austin, 2010)

A) Noise (acoustic) [16 m-2 cm, or 20 – 16000 Hz]

B) Optical (visible) [0.4 – 0.7 um]

C) Infrared (thermal) [0.75 um – 1 mm]

D) RADAR (radio) [3 mm – 3 cm]" (Austin, 2010)

In the discussion on stealth, it was presented that "If the designer is to reduce the vehicle detectability to an acceptable risk level, it is necessary to reduce the received emissions or reflection of the above wavelengths (expressed as frequencies) below the threshold *signature* value. A good portion of the UAS signatures are a function of the operating height of air vehicle." (Austin, 2010) The concept of frequency as a fifth realm was elucidated in terms of targets, battlespace, and wavelengths. One of the parameters, range was a serious limitation on performance. Range has a significant impact on radio transmission. Depending on the environment, the strength of a received signal, T, is a function of the square or fourth power of a distance, d, from the transmitter. (Adamy D. -0., 2015) The EMS was presented with emphasis on sound frequencies, many out of human hearing range. The author's experiments were using DJI Phantom 4 at 400 ft. This is not a tactical distance for a countermeasure. However, the LRAD made by LRAD Corporation is effective to a mile. It was described in the C-UAS chapter.

Chapter 8 presented that for the UAS designer the upper end of noise – Stealth acceptability 17,150 Hz. **The Stealth range is 20 Hz – 17,150 Hz.**

Essentials of Audiology

The question is why would hitting a UAS going at 100 mph or more be susceptible to a loud noise hitting the MEMS under the rotors or the rotors themselves? Why would this same noise or variation thereof be capable of characterization of the UAS's of a hostile or friendly power? It is not something we can just take for granted without understanding the essentials of audiology underlying the process.

For the Birds

The author recently purchased two "Cadillac: hearing aids made by Signia. (Signia, 2019) They came with an Iphone application that permits all kinds of variation in frequency, tone, loudness, dampening factors, etc. and can be adjusted for universal, TV, reverberating rooms, chaos, party atmosphere, stadium, quiet, music hall and many other customizations. Sitting outside on my deck, I was able to match the frequency of chirping birds and graph their frequency as heard through my hearing aids. This is a different learning process from when the author wrote about speech cryptography in my Classical Cryptography Course, Volume I. [CCCI] (Nichols, Classical Cryptography Course, Volume I, 1996) In CCCI, he was trying to characterize and encrypt / decrypt speech patterns and make a library of those vowel -consonant patterns. Here the author is concerned with creating resonant frequencies to disrupt the delicate unprotected circuitry of a moving UAS vehicle in the air.

The author was able to jimmy-rig a transmitter at the frequencies below 16 MHz and send frequency tones about 50 feet at the birds in my backyard. My purpose was to emulate Identification Friend or Foe (IFF) conditions to the various birds feeding at my porch and trees nearby. In a non-scientific attempt to see what the sent signals (at bird frequencies matched to the birds' chirping) would cause any, none, confusion, chaos, friendly, or hostile reactions. The rig was not capable of sending exact signals at resonance frequencies. (So, the birds didn't explode or lose flying ability.) The object was to understand the audiology principles involved and to test conditions within the hearing range. Stop laughing for a moment. In the space of just two days, the author was able to characterize the following bird species: White-breasted Nuthatch, House Finch, American Robin, Blue Jay, Northern Mockingbird, House Wren, Tufted Titmouse, Baltimore Oriole, Morning Dove, Black-capped Chickadee, Northern Cardinal, White-throated Sparrow, a Common Flicker Woodpecker and a most irritating Red-Bellied Woodpecker (RBW) (who if exploded, would not cause the author tears). The author taped these sounds and tested the taped version through the computer on three of the above species with similar effects. Some of the larger birds reacted harshly / threatened. The smaller birds acted is if they were in love and wouldn't shut-up even when the frequency stimulus was stopped. The RBW was unimpressed and kept up his chaotic bug – finding noises. Interestingly, only one bird was hit with my frequency simulator, while he was flying between trees – the House Finch. For a few brief moments the bird seemed to wobble and drop in a decreasing angular pancake fashion, and then recovering in midair. It is possible that the frequency transmitted was close to his resonating chirp – frequency. Then again, he might just have been yawning.

Audiology Fundamentals

The science of sound is called *acoustics*, which a branch of physics. Table 19-1 displays the principal physical quantities in MKS, cgs, and English units. Table 19-1 can be found in most engineering, physics or medical textbooks. (Entokey, 2019) It is the starting point of a trip uphill to resonance frequencies.

Table 19-1: Principal Physical Properties (Entokey, 2019) and (Gelfand S. A., 2009)

Quantity	Formula	MKS (SI)Units	Cgs Units	Comments	English Units
Mass (M)	M	kilogram (kg)	gram (g)	1kg = 103 g 1kg = 2.2046 lbs	pounds (lbs)
Time (t)	t	seconds, (s)	s		s
Area (A)	A	m2	cm2	1 m2 = 104 cm2	ft2
Displacement (d)	d	meter (m)	centimeter (cm)	1m = 102 cm	ft
Velocity (v)	v = d/t	m/s	cm/s	1 m/s = 102 cm/s	ft/s
Acceleration (a)	A = v/t	m/s2	cm/s2	1 m/s2 = 102 cm/s2	ft/s2
Force (F)	F = MA = Mv/t Mv = Momentum	kg x m/s2 newton (N)	g x cm2 dyne	1N = 105 dynes	1 lbf = 1 lb x 32.174049 ft-lbs /s2 = 9.80665 m/s2
Pressure (p)	p = F/ A	N /m2 Pascal (Pa)	dynes /cm2 microbar (µbar)	20 µPa = 2 x 10-5 N/m2 reference value	Psi = lbf /in2 1 N/m2 = 0.000145 psi
Work (W)	W =Fd	N x m Joule	dyne x cm erg	1 j = 107 erg/s Energy -capability to do Work. Potential energy for a body at rest and kinetic energy for a body in motion.	BTU [British Thermal Unit] 1 BTU = 1055.056 joules
Power (P)	P = W/t = Fd/t =Fv	Joules/s watt (w)	erg/s watt (w)	1 w = 1 J/s = 107 erg/s	1 watt = 3.412 BTU/hr

Intensity (I)	I = P/A I = P / 4?r2 Based on sphere radius	w/m2	w/cm2	10-12 *w/m2* *reference value*

Source: (Entokey, 2019) and (Studios, 2017)

Figure 19-1: Inverse Square Law, Sound Intensity

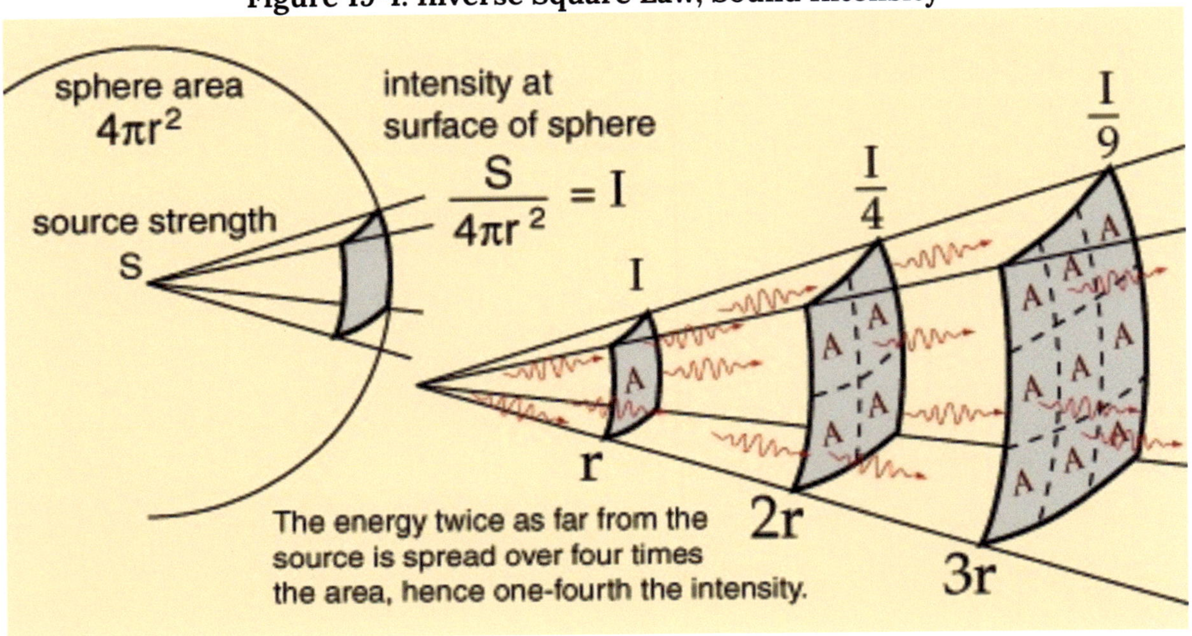

Source: (Uni-wuppertal, 2019) [Author revision for background color]

Intensity and Inverse Square Law

"Sound radiates outward in every direction from its source. This constitutes a sphere that gets larger and larger with increasing distance from the source." (Entokey, 2019) Figure 19-1 shows the relationship between Intensity and the Inverse Square Law. (Uni-wuppertal, 2019) Intensity (I) (power divided by area) decreases with distance from the original source because of finite amount of power is spread over increasing surface area. (Entokey, 2019) Proportionately less power falls on the same unit of area with increasing distance from the source. (Gelfand, 2004)

"Four important relationships to note are that power is equal to pressure squared, P = p2, pressure is equal to the square root of power, p = √ P, intensity is proportional to pressured squared, I ≈ p2 and pressure is proportional to intensity, p ≈ √I. This makes it easy to convert between sound intensity and sound pressure." (Entokey, 2019) These relations yield a few more to relate sound pressure, sound intensity and distance r. Given to pressures p1 and p2 at distance r1 and r2, they are proportional: p2 / p1 = r1 / r2; and factoring in intensities at I1 and I2, gives I2 / I1 = (r1 /r2)2.

Finally, r2 / r1 = p2 / p1 = √I1 / I2. (TRS, 2018)

Decibels (Adamy D. , 2001) (Gelfand S. A., 2009)

Sound magnitudes, intensities, and pressures vary over an enormous range. We use decibels (dB) to express sound values. Decibels takes advantages of ratios and logarithms. Ratios are used so that physical magnitudes can be stated in relation to a reference value that has meaning to us. The reference point chosen is the softest sound that can be heard by normal people. The reference value has an intensity of 10-12 w/m2 (10-16 w/cm2). In terms of sound pressure, the reference value is: 2 x 10-5 N/m2 or 20 µPa (2 x 10-4 dynes/cm2). An interesting Geek bar bet is what is the logarithm of all 2:1 ratios, 8:4, 20, 20:10, 100:50, etc.? Even though the distance between absolute numbers gets wider, 1,4,10, 50…, the logarithms of the 2:1 ratios are the same at 0.3. Another interesting factoid about ratios is the units generally cancel out.

The general decibel formula in terms of power level (PL) is as follows (Gelfand, 2004):

Eq. 19-1

$$\mathbf{PL = 10 \log P / Po}$$

Where P = power of the sound measured, and Po is the reference power to be compared.

The general decibel formula in terms of power level (IL) is as follows (Gelfand, 2004):

Eq. 19-2

$$\mathbf{IL = 10 \log I / Io}$$

Where I = intensity of the sound measured, and Io is the reference intensity to be compared. Io is given as 10^{-12} w/m^2 .

The general decibel formula for sound pressure level (SPL) is obtained by replacing all of the intensity values with the corresponding values of pressure squared because ($I \approx p^2$).

Eq. 19-3

$$\mathbf{SPL = 10 \log p^2 / p_o^2}$$

Where p is the measured sound pressure (in N/m^2) and p_o is the reference sound pressure of 2 x 10^{-5} N/m^2 . A more convenient form of this equation recognizes that log x^2 = 2 log x. (Gelfand, 2004)

Eq. 19-4

$$\mathbf{SPL = 20 \log p / p_o.}$$

Equation 19-4 is the common formula for SPL. A couple of observations a positive decibel value means that the sound pressure level is greater than the reference. The decibel value of the reference is 0 because reference value / reference value = 1 and 10 log 1 = 0. This does not mean no sound, it just means the sound measured is equal to the reference point. A negative value of decibels means that the sound magnitude is lower than the reference. (Gelfand S. A., 2009)

Figure 19-2 shows common decibel and Intensity levels within the hearing range. This does not consider environment, frequency differences or noise (discussed presently). It does show the ease of which decibels may be used to rank the sound intensity levels which vary greatly in magnitude. [1]Hearing aids are effective from about 6 – 90 decibels. Above 90 dB, they can dampen but not eliminate the very loud sounds unless there is complete loss of hearing.

1. 3 dB is an interesting cutoff datum. Because decibels are logarithmic and the log (base 10) of 3 =).477 ≈ 50% power. So, 3 decibel cutoff is where the power drops by approximately half. 3 dB implies ½ of the power. An increase of 3dB doubles the sound intensity but a 10-dB increase is required before sound is perceived to be twice as loud. A small increase in decibels represents a large increase in intensity. For example, 10 dB is 10x more intense than 1 dB, while 20 dB is 100x more intense than 1 dB. (Adamy D. , 2001)

Figure 19-2 shows common decibel and Intensity levels within the hearing range

Approximate sound levels and intensities within human hearing range

Source of sound	Intensity level (dB)	Intensity ($W\ m^{-2}$)	Perception
jet plane at 30 m	140	100	extreme pain
threshold of pain	125	3	pain
pneumatic drill	110	10^{-1}	very loud
siren at 30 m	100	10^{-2}	
loud car horn	90	10^{-3}	loud
door slamming	80	10^{-4}	
busy street traffic	70	10^{-5}	noisy
normal conversation	60	10^{-6}	moderate
quiet radio	40	10^{-8}	quiet
quiet room	20	10^{-10}	very quiet
rustle of leaves	10	10^{-11}	
threshold of hearing	0	10^{-12}	

Source: (Carter, 2012)

The Nature of Sound

"*Sound is defined as a form of vibration that propagates through the air in the form of a wave. Vibration is the to-and-fro motion (aka oscillation) of an object.* Some examples are playground swing, tuning fork prong, air molecules and UAS rotor blades [circular motion]. The vibration is called *sound* when it is transferred from air molecule to air molecule. This transfer may be simple like a tuning fork or a very complex pattern like the din in a school cafeteria. Naturally occurring sounds are very complex." (Entokey, 2019) UAS sounds are not natural and supported by machinery, hardware and software. Three weaknesses of the UAS are the MEMS, gimbal assembly and rotors. Although stealth mechanisms may be employed to reduce noise emissions, the former parts are exposed. They do produce discernable signatures.

Figure 19-3: Tuning for Oscillations

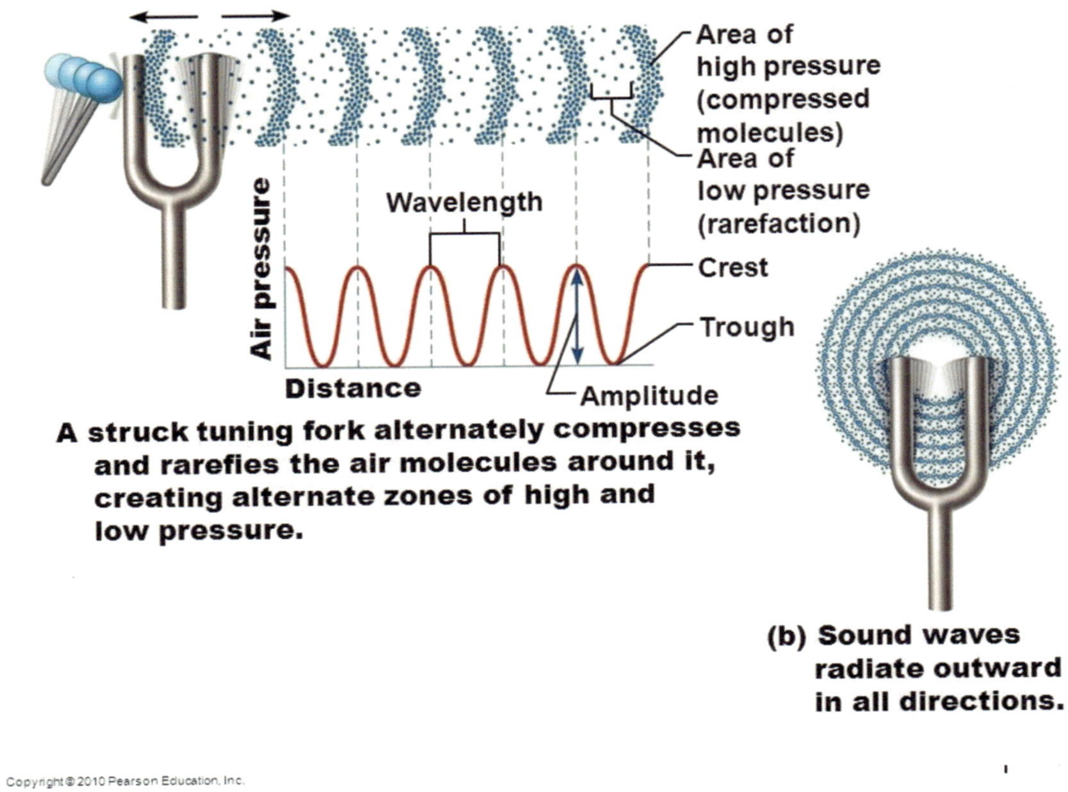

Source: (Pierson, 2019)

A tuning fork illustrates the oscillations of sound. After being struck, the tuning fork vibrates with a simple pattern that repeats itself over time. (Entokey, 2019) Figure 19-3 shows that the tuning fork when struck exerts a force on the air molecules which alternatively exerts a high pressure (compression) and a low pressure (rarefaction) zones. The zones exhibit wave amplitude and wavelength as a function of air pressure and distance. The sound wave is distributed in 360 degrees through the air.

Figure 19-4: Tuning fork oscillations over time

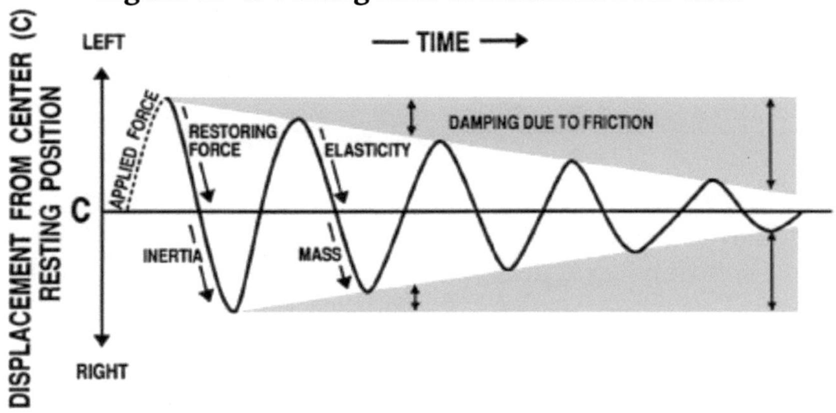

Source: (Entokey, 2019)

Figure 19-4 diagrams tuning fork oscillations over time. Sounds that are associated with simple harmonic motion are called pure-tones. Vertical displacement amount of the tuning fork prong displacement around its resting position. Distance from left to right represents progression of time. One complete round-trip or replication of an oscillating motion is called a cycle. The number of *cycles* occurring in one second is the *frequency.* The duration of one cycle is called its *period.* This form of motion occurs when a force is applied to an object having properties of elasticity and inertia. Simple harmonic motion (SHM) shows the same course of oscillations as in Figure 19-4 because they repeat themselves at the same rate until friction causes dampening of the waveform. (Entokey, 2019) and (Gelfand S. A., 2009)

Other Parameters of Sound waves

Probably the most useful SHM waveform is the sinusoidal wave or sine wave.[2]

The number of times a waveform repeats itself in one second is known as the frequency or cycles per second (CPS). (Gelfand S. A., 2009)Two useful relationships are: f = 1/ t or t = 1/f; where f is the frequency in cps and t is the period in seconds. *Amplitude* denote the magnitude of the wave. The *peak- to – peak* amplitude is the total vertical distance between negative and positive peaks. The peak amplitude is the distance from the baseline to one peak. The magnitude of sound at any instant is the *instantaneous amplitude.* Wavelength (λ) is the distance traveled between one peak and the next. (Gelfand, 2004)

Wavelength formula is: λ = c / f, where c is the speed of sound in air (344 m/s. f is the frequency of sound in Hz. Similarly, frequency is inversely proportional to wavelength or f = c / λ. (Gelfand

2. It is left to the reader to obtain any standard trigonometry text to see all the parameters of the well-known sine wave.

S. A., 2009) Another interesting sound parameter is *Pitch*. Pitch is the quality of sound and especially a musical tone governed by the rate of vibrations producing it. It is the degree of highness or lowness of sound. (Merriam-Webster, 2019)

Complex waves

When two or more pure-tone waves are combined, the result is *a complex wave*. (Gelfand, 2004) They may contain any number of frequencies. Complex periodic waves have waveforms that repeat themselves. If they don't they are *aperiodic*. Combining waves may reinforce themselves or cancel themselves whether they are in phase or out. The lowest frequency component of a complex periodic wave (like a combination of sign waves) is called its *fundamental frequency*. (Gelfand, 2004)

Harmonics are whole number or integral multiples of the fundamental frequency. Waveforms show how amplitude changes with time. (Gelfand, 2004) Fourier's Theorem shows that complex sound waves can be mathematically dissected into its pure tones.

Of more interest to UAS designers are aperiodic sounds which are made up of components that are not harmonically related and do not repeat themselves over time. The extreme cases of aperiodic sounds are transients and random noise. A *transient is an abrupt sound* that is very brief in duration. *Random noise* has a completely random waveform, so it contains all possible frequencies in the same average amplitude over the long run. *Random noise* is also called white noise like white light because all possible frequencies are represented.

Patient D v-105 (Heinman, 2019)

Let's look at some of the discussed sound parameters in a real patient. Data was collected by audiologist in an anechoic[3] sound chamber.[4] Refer to Figures 19-5 and 19-6. Note that the standard ANSI 1969 runs from 250 Hz to 10,000 Hz. Average hearing range is from 250 Hz – 8,000 Hz. Patient D v-105 has lost hearing in the upper ranges. Patient hearing declines precipitously after 2,000 Hz in both ears. The patient requires a louder (more dB) threshold stimulus as frequencies go above 1500 Hz in the right ear and above 1000 Hz in the left. Patients most comfortable level

3. Anechoic means free from echo. An anechoic chamber provides a non-reflective, non-echoing, echo-free room designed to completely absorb reflections of either sound or electromagnetic waves. They are also isolated from waves entering from their surroundings. Sound interactions / countermeasures for UAS testing are ideally done in an anechoic chamber. NASA Langley has a fully equipped room in Va. Patient D v-105 was only able to hear sounds entering his / her ears by headset connected to an audiometer. The audiometer is a machine that evaluates hearing acuity. The patient responds to test frequencies and sounds by pushing a red feedback button.
4. HIPPA Privacy Act releases signed.

is 60 dB in both ears. This corresponds to 400 Hz in right ear and 3,500 Hz in the left. Patient experience pain / discomfort at 95 dB. From Figure 19-2, 60 dB is moderate intensity normal conversation. 95 dB is a siren on a police car. Patient can pick up speech sounds at 15 /20 dB or comparable to a quiet room. Patient D v-105 experiences difficulty in differentiating consonants above 2,000 Hz or k, t, f, s, and th. (Heinman, 2019)

Because of pitch, Patient D -v105 would be able to hear and differentiate many bird calls and chirping below 4,000 Hz and would be able to hear the bird noises above 4,000 Hz with hearing aid assistance. UAS noise levels range from 250 Hz to 20,000 Hz so Patient would not be able to hear the UAS directly but could IFF with the appropriate LRAD.[5]

5. LRAD is a Long-Range Acoustic Device (weapon). It was discussed in detail in Chapter 8. "The 450XL model (LRAD, 2019) can produce 140 dB at 1m which is equivalent to 108.5 dB at 37.58 m." (Yunmonk Son, 2015)

Figure 19-5: Patient D v-105 Hearing loss

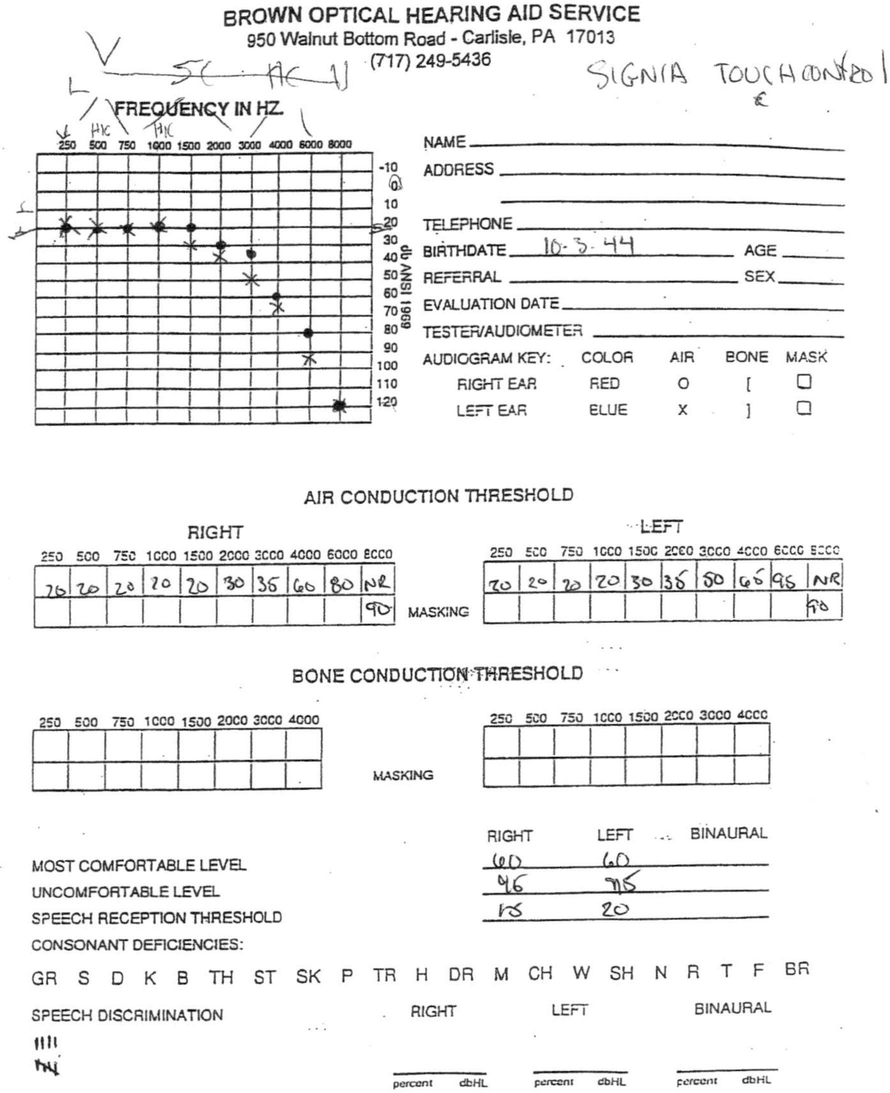

BROWN OPTICAL HEARING AID SERVICE
950 Walnut Bottom Road - Carlisle, PA 17013
(717) 249-5436

SIGNIA TOUCHCONTROL

FREQUENCY IN HZ.

250 500 750 1000 1500 2000 3000 4000 6000 8000

dB ANSI 1969: -10 0 10 20 30 40 50 60 70 80 90 100 110 120

NAME
ADDRESS
TELEPHONE
BIRTHDATE 10-3-44 AGE
REFERRAL SEX
EVALUATION DATE
TESTER/AUDIOMETER

AUDIOGRAM KEY:	COLOR	AIR	BONE	MASK
RIGHT EAR	RED	O	[	☐
LEFT EAR	BLUE	X	]	☐

AIR CONDUCTION THRESHOLD

RIGHT

	250	500	750	1000	1500	2000	3000	4000	6000	8000
	20	20	20	20	20	30	35	60	80	NR
MASKING										90

LEFT

	250	500	750	1000	1500	2000	3000	4000	6000	8000
	20	20	20	20	30	35	50	65	95	NR
MASKING										90

BONE CONDUCTION THRESHOLD

	250	500	750	1000	1500	2000	3000	4000
RIGHT								
MASKING								

	250	500	750	1000	1500	2000	3000	4000
LEFT								
MASKING								

	RIGHT	LEFT	BINAURAL
MOST COMFORTABLE LEVEL	60	60	
UNCOMFORTABLE LEVEL	96	95	
SPEECH RECEPTION THRESHOLD	15	20	

CONSONANT DEFICIENCIES:

GR S D K B TH ST SK P TR H DR M CH W SH N R T F BR

SPEECH DISCRIMINATION	RIGHT	LEFT	BINAURAL
	percent dbHL	percent dbHL	percent dbHL

Source: (Heinman, 2019)

Figure 19-6: Patient D v-105 Pitch v Loudness V Consonant Loss

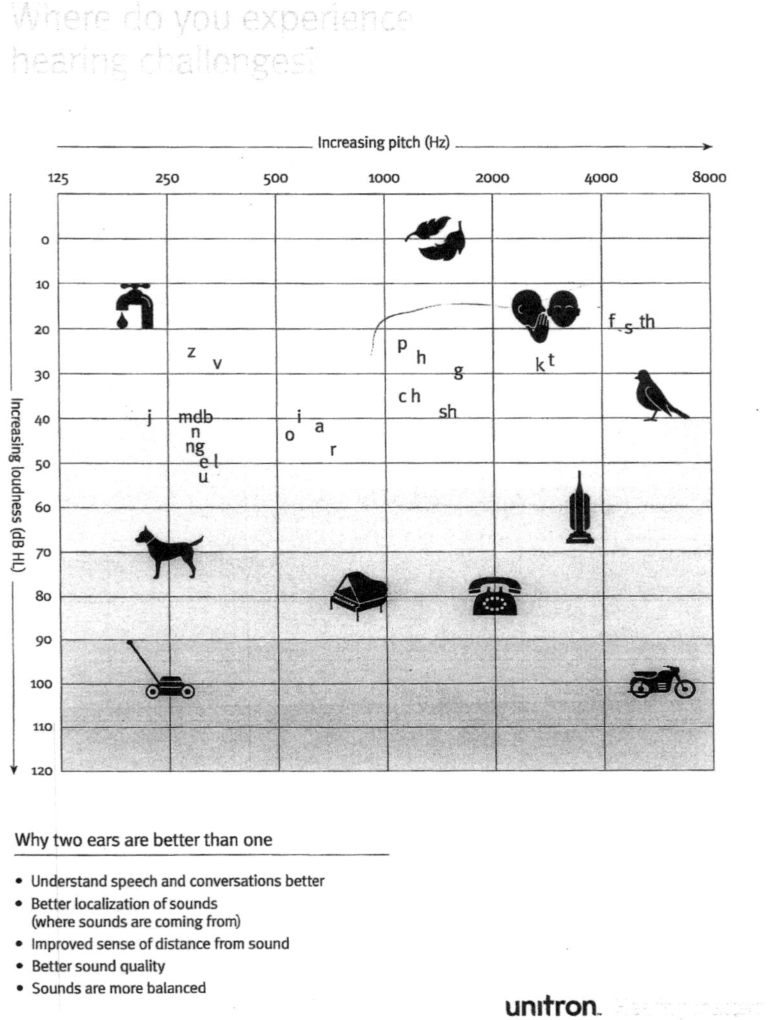

Source: (Heinman, 2019)

Standing Waves and Resonance

We have arrived at the crux of the acoustic CM discussion, the natural or resonating frequency.

"The frequency(ies) at which a body or medium vibrates most readily is called its natural or resonant frequency(ies)." (Gelfand S. A., 2009) Differences in resonance frequency ranges enable different devices to act as filters by transmitting energy more readily for certain high, low, or bandpass frequencies. UAS with multiple rotors circulate the rotors to gain lift and / or hold steady / or descend in altitude. Four, six, eight – rotor UAS maintain control via internal SCADA systems and send critical information through a MEMS component located at the bottom of rotors. Rotor frequencies are coordinated, monitored, and position -controlled through the MEMS and in-board computers. Even though the rotor(s) motions are spinning in circular fashion, the sound wave generation is not curvilinear, or aperiodic but transferred up through the Y axis and back again to its base as it ascends in altitude. There is a tendency to maintain equilibrium in terms of position of the UAS.

The author contends that the UAS rotor systems act like vibrating strings and resonance frequency information can be approximated by this approach. An example of a vibrating spring is when you "pluck" a guitar. The waves initiated move outward toward the two tied ends of the string. The waves are then reflected back, and they propagate in the opposite directions. The result is a set of waves that are moving toward each other, resulting in a perturbation sustained by continuing reflections from the two ends. The superimposed waves interact and propagate and appears as a pattern that is standing still. Peaks (maximum displacement) and no displacement (baseline crossings occur at fixed points along the string. Places along the spring where zero displacement in the standing wave pattern are called nodes. (Gelfand, 2004) Locations where the maximum displacement occurs are called antinodes.

Figure 19-7: Standing wave

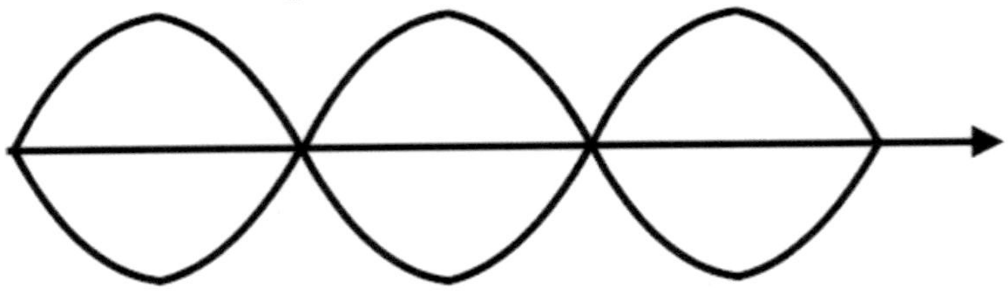

standing wave pattern for a string

Source: (Administrator, 2015)

"The *fundamental frequency* is defined as the lowest frequency of a periodic waveform. It is generally denoted as 'f'. The lowest resonating frequency of a vibrating object is called as fundamental frequency." (Administrator, 2015)

"*Harmonic is a frequency, which is an integer multiple of the fundamental frequency*. The forced resonance vibrations of an object are caused to produce standing waves. At the natural frequency it forms a standing wave pattern. These patterns are created at specific frequencies, they are called Harmonic Frequencies or Harmonics." (Administrator, 2015)

"The sound produced by a waveform at its harmonic frequency is very clear, and at other frequencies we get noise, and cannot hear the clear sound of waves. Harmonics may occur in any shaped wave forms, but mostly they occur in sine waves only. Non – sinusoidal waveforms, like triangular and sawtooth waveforms are constructed by adding together the harmonic frequencies. The word harmonic is generally used to describe the distortions caused by different undesirable frequencies called noise, of a sine wave." (Administrator, 2015)

"Node and antinodes occur in a wave form. So, the waves have harmonic frequency in them. The fundamental frequency is the smallest frequency in a harmonic. Hence there is only a single anti-node occurs between them. This Antinode is middle of the two nodes. So, from this we can say that the guitar string produces longest wavelength and the lowest frequency." (Administrator, 2015)

"The lowest frequency produced by any instrument is called the *fundamental frequency*. This is also known as first harmonic of the wave. In words of fundamental frequency, harmonics are the integer multiples of the fundamental frequency." (Ex: f,2f,3f,4f, etc.... are harmonics.) (Administrator, 2015)

"Because of multiple integers of fundamental frequency, we will have n number of harmonics like 1st harmonic, 2nd harmonic,3rd harmonic, and so forth." (Administrator, 2015) "The fundamental frequency is also called as *First harmonic*. In the first harmonic, we have two nodes and one anti -node. he numbers of antinodes are equal to the integer multiples of specific harmonics. i.e., for 1st harmonic we have 1 antinode, for 2nd harmonic we have 2 antinodes etc." (Administrator, 2015)

The formula for the string's resonant frequency Fo is:

Eq 19-5

$$Fo = 1 / 2L \times \sqrt{T} / M$$

Where Fo is resonance frequency in Hz, T is Tension, M is Mass, $L = \lambda / 2$ and $f = c / \lambda$ and c = speed of sound. L = length of the string. (Gelfand, 2004) The strings lowest resonant frequency is $f = c / 2L$ but Eq 19-5 considers that the speed of sound is different for a vibrating string than it is for air.

UAS / Acoustic Countermeasures FAQ (Nichols, Hardening US Unmanned Systems Against Enemy Counter Measures, 2019)

The reader has questions, so here are the most likely questions and answers.

In terms of UAS Countermeasures, why are Acoustics so important? (Nichols, Hardening US Unmanned Systems Against Enemy Counter Measures, 2019)

They are important because:

- Offensive systems use ultrasonic frequency resonance

- Cannot be heard by humans when used to intercept a drone
- Passive systems are difficult, if not impossible, to detect
- Able to identify and track drone based on acoustic and/or visual signature
- Acoustic detection systems are limited in range ~ 350 ft to 500 ft due to environmental variables BUT commercial companies like LRAD, Corporation have developed long range acoustic devices which can detect a UAS up to a mile away at altitude.
- Can be a cost-effective way to defend a small area –especially against Swarm Attacks

Acoustic Signature Reductions (Nichols, Hardening US Unmanned Systems Against Enemy Counter Measures, 2019)

Can the UAS signatures be reduced?

- Noise may be the first warning of its presence
- However, it may not immediately be directionally/locatable for detection
- UAS noise emanates predominantly from vortices, tips of wings, rotors, or propellers
- Lowering wing span or blade span enhances acoustical stealth
- Conventional propulsion systems are a concern because of the noise of combustion
- Electric motors develop virtually no noise
- Reducing mass and aerodynamic drag of the UAS reduces noise generation

What are the Acoustic Detection Issues?

Detection relies on uniqueness of the UAS and hearing capabilities at low frequencies:

- Detects drones by recognizing the unique sounds produced by their motors
- Rely on a library of sounds produced by known drones, which are then matched to sounds detected in the operating environment, however
- The human ear is a problem for the designer
- It is most sensitive to frequencies around 3500 Hz and can hear sound down to a practical threshold of 10 dB
- For a given sound pressure level, attenuation of sound with distance in air and insulating material varies as the square of the sound frequency
- Low frequency sound presents a greater problem for UAS stealth design
- The greater noise problem is posed by smaller UAS using piston engines
- Sound comes from their internal combustion and exhaust systems
- Sound emission can be reduced with sound-absorptive materials, silencers and mufflers and by directing the intake and exhaust manifolds upward
- Level of sound detected depends on the level of background noise for sound contrast
- Limited range to 500 feet (experimental and research – not commercial or military)
- Noisy backgrounds (airports, city downtown) limit detection & interdiction

- Drone tuning (changing the stock propellers) limits detection / Interdiction

Is Acoustic Quieting possible?

"Yes, under certain conditions:

- Disguise sounds from sensors to eliminate its noise and passive echoes
- "Acoustic superiority" used by the Navy to mask detection of U.S. submarines
- Acoustic technology is "passive," meaning it is engineered to receive pings and "listen" without sending out a signal which might reveal their location to an enemy
- Increased use of lower frequency active sonar and non-acoustic methods of detecting."(Nichols, Hardening US Unmanned Systems Against Enemy Counter Measures, 2019)

What is an Acoustical attack on the UAS Gyroscope?

There are two possibilities: compromising the sound source or drone on drone attack:

Compromising the Sound Source

- UAM with speakers (consider police and military operations or search-and-rescue operations)(Usenix.org, 2019)
- Counter the source of the sound from the speaker with different frequency sound
- Jamming attack aims to generate ultrasonic noises and cause continuing vibration of the membrane on the sensor, which make the measurements impossible
- Level of noise causes performance degradation

Drone on Drone Attack

- Taking a picture of a moving object from UAM
- An adversary drone equipped with a speaker could steer itself toward a victim drone and generate a sound with the resonant frequency of the victim's gyroscope to drag it down(Usenix.org, 2019)

How has the Long-Range Acoustic Device (LRAD) used as a sonic weapon? (LRAD, 2019)

It has been used primarily for denying GPS navigation:

GPS Denied Navigation

- GPS navigation relies on measuring the distance or delay, to several known transmitters in order to triangulate the mobile receiver's position

- GPS-denied environment presents navigation challenges for UAV and UAM
- These areas include urban canyons, forest canopy, etc.
- Strike Resonance frequency – which disrupts balance (vehicles tilt, orientation & rotation)

MEMS

What is a MEMS and how does it relate to the UAS gyroscope?

As shown in Figure 19-7 MEMS Gyroscope,

- MEMS (Microelectromechanical-Systems) gyroscopes are located in the rotor systems of most drones
- Visualization of a MEMS gyroscope is a single proof mass suspended above a substrate
- The proof mass is free to oscillate in two perpendicular directions, the drive and sense (Said Emre Alper, December 2008)

Figure 19-8: MEMS Gyroscope

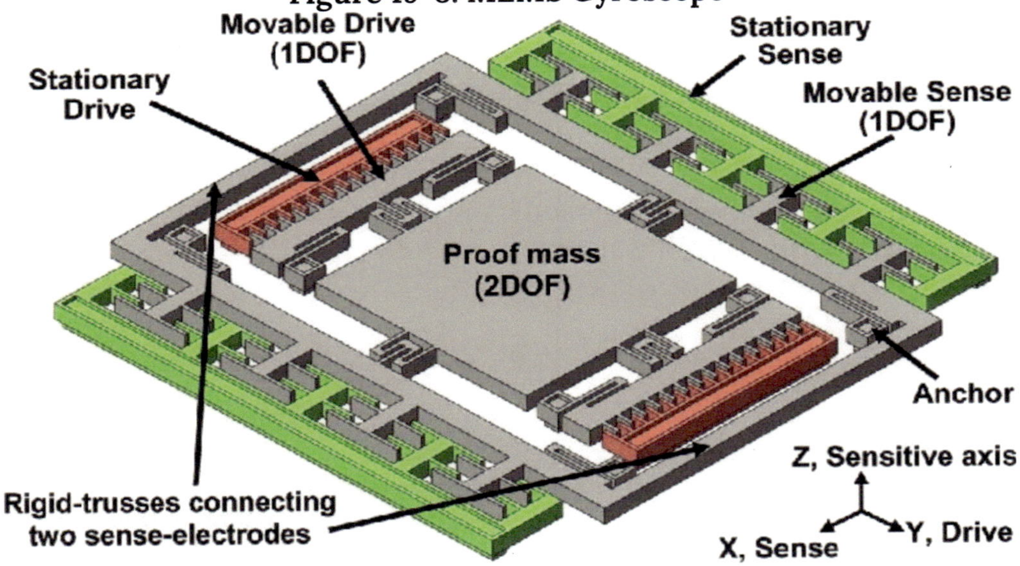

Source: (Said Emre Alper, December 2008)

A very interesting presentation on MEMS is available at (Said Emre Alper, December 2008).

Resonance Effects on MEMS

Achieving resonance frequencies can have a significant effect for countering hostile UAS:

- *MEMS Gyroscope can be degraded using harsh acoustic noise*
- *MEMS Gyroscope has a resonant frequency that is related to the physical characteristics of*

its structure (Usenix.org, 2019)

- MEMS gyroscopes have resonant frequencies much higher than can be heard (audible and ultrasonic ranges)
- *Unexpected resonance output caused from an attack will cause the rotor system to malfunction*
- *Rotors will spin at differing speeds causing the drone to become unstable and crash*

What is Resonance Tuning?

- "In the operation of MEMS gyroscopes, the bending changes the capacitance between the sensing mass and the sensing electrode, and this capacitance change is sensed as the output of the gyroscope
- By using an additional feedback capacitor connected to the sensing electrode, the resonant frequency and the magnitude of the resonance effect can be tuned
- Resonance can be induced by a malicious attacker, if resonant frequencies exist in gyroscopes"(Nichols, Hardening US Unmanned Systems Against Enemy Counter Measures, 2019)

What is the "so what" for Acoustics? Here are the author's thoughts:

- "Passive detection is much cheaper and cost effective to operate vs a complex radar system for a single installation (limited by detection range ~350ft)
- MEMS gyroscopes contained in rotor systems are very susceptible to malfunction when struck with rough noise that resonates inside the MEMS
- Offensive acoustic systems are currently mounted, could become man portable
- Offensive systems are not detected by National ELINT assets
- Not looking for acoustic energy signatures, enemy can remain hidden from detection when using acoustics."(Nichols, Hardening US Unmanned Systems Against Enemy Counter Measures, 2019)

Are there Countermeasures for Acoustic attack on Gyroscope?

"Yes, some but not totally effective:

- Physical Isolation – provide physical isolation from the sound noise
- Surrounding the gyroscope with foam would also be a simple and inexpensive countermeasure
- Differential Comparator
- Using an additional gyroscope with a special structure that responds only to the resonant frequency, the application systems can cancel out the resonant output from the main gyroscope

- Detect and cancel out analog sensor input spoofing against CIEDs."(Nichols, Hardening US Unmanned Systems Against Enemy Counter Measures, 2019)

South Korean experiment

A paper by Yunmonk son, et. Al. From the Korean advanced institute of science and technology (KAIST), in the authors judgement, is the seminal paper on taken down drones using sound noise on gyroscope sensors! (Yunmonk Son, 2015) It is required reading for my students.

(Yunmonk Son, 2015) describes the relationship between *Sound Pressure Level* (SPL) and Sound Amplitude and derives the attack distance, d as (in dB) :

Eq. 19-6

$$\mathbf{SPL = SPL_{ref}\ 20\ log\ (A\ /\ A_{ref})}$$

Where SPL = sound pressure level, SPLref is the reference, A and Aref are the amplitudes of the source and reference point. Using the real-world experiments (Yunmonk Son, 2015) found that:

Eq. 19-7

$$\mathbf{SPL = SPLref\ -\ 20\ log\ (d\ /\ dref)}$$

Where d, dref are the attack scenario distances.

KAIST under (Yunmonk Son, 2015) primary conclusions are:

1. "Many sensing and actuation systems trust their measurements and actuate according to them. Unfortunately, this can lead to security vulnerabilities that cause critically unintended actuations.
2. The sound channel can be used as a side channel for MEMS gyroscopes from a security point of view.
3. 15 kinds of MEMS gyroscopes were tested, and seven of them were found vulnerable to disruption using intentional noise.
4. The output of the vulnerable MEMS gyroscopes was found using a consumer-grade speaker to fluctuate up to dozens of times as a result of the sound noise.
5. Authors found that an attacker with only 30% of the amplitude of the maximum sound noise could achieve the same result (disruption) at the same distance.
6. At 140 decibels, it would be possible to affect a vulnerable drone up from around 40 meters away,
7. Some drone gyroscopes have resonant frequencies in both the audible and ultrasonic frequency ranges, making them vulnerable to interference from intentional sound noise.
8. Authors found that accelerometers integrated with MEMS gyroscopes were also affected

by high-power sound noise at certain frequencies."(Yunmonk Son, 2015)[6]

NOISE

Loud and abrupt sound as a countermeasure also brings the problem of exposure and loss. Chapter 17 of (Gelfand S. A., 2009) discusses the effects of noise and hearing conservation. Chapter 20 of (Gelfand S. A., 2009) discusses occupational standards. Safety is an important topic but outside the scope of this writing.

UAS Collaboration – SWARM

Time to wrap-up the chapter. Recall that the authors previously defined in Chapter 3, a UAS SWARM as a uniform mass of undifferentiated individual's w/o Chief at automation level 4 or 5. SWARMs exhibit the following advantages:

- Efficient based on numbers, emergent large group behaviors, and reactions
- Not controllable or automated, decentralized intelligence
- Think shoal of fish w/ evolving local rules; highly resistant form
- Not changing based on survivability of members, no hierarchy

SWARM Countermeasures include disruption, i.e. changing the Strategic Global View of SWARM (its only real vulnerability), complete Defender collaboration with multiple kinetic and non-kinetic countermeasures, and use of Acoustic Countermeasures for identification as friend or foe (IFF) based on a library of manufacture detection signatures and complete , abrupt rotor disablement by attacking the SWARM units with resonant, loud (100-140 dB) sound noise aimed directly at the MEMS gyroscopes or close by on the unit. [Think of glass breaking at resonance frequency or a submarine under depth charge attack. The former breaks by super- excited molecules in the glass and literally shakes apart. The latter is destroyed by violent shaking of the submarine so that its parts break and flooding ensues. It is not necessary to hit the submarine directly because explosions in water, hence sound waves and explosive forces, carry very far and effectively to the target.]

6. Author note: although not specified in (Yunmonk Son, 2015), according to chapter author research and experimentation, the frequencies turn out to be the resonance frequencies. So agrees Dr. Kim at KAIST. "You would think that the gyroscopes used in unmanned aircraft systems (UAS) would have been designed to have resonant frequencies above the audible spectrum – i.e., above 20 kHz – but Kim and his team found that some have not." (Yunmonk Son, 2015) In the case of a gyroscope, "you can get it to spit out very strange outputs, as researcher Yongdae Kim, a professor in the electrical engineering department of the Korea Advanced Institute of Science and Technology (KAIST), told ComputerWorld" (Kirk, 2015) An example of resonance frequency and breaking glass can be found on youtu.be at https://youtu.be/BE827gwnnk4

Discussion Questions

1) This chapter explores the use of acoustic countermeasures against UAS. The authors content that every manufactured UAS has unique sound detection signatures. Further these can be libraried and used in a search algorithm to IFF the UAS group or SWARM. At the DoD 7th Annual Summit, (Nichols, Hardening US Unmanned Systems Against Enemy Counter Measures, 2019) the author found that several contractors are actually doing this and building databases. BUT they refuse to share their data because it is proprietary. Assuming this situation cannot be changed, suggest two ways to get around this problem not involving legal actions. What type of research project would you propose to meet an 85% detection criteria that would suffice as an initial IFF database for evaluation?

2) Along with attacking the MEMS gyroscopes to disable the UAS rotor, propose an experiment to use acoustic countermeasures on the UAS internals, such as SCADA, payload, navigation, internal clocks, internal computer, battery, etc. Perhaps loud noise can disrupt additional UAS features?

3) This chapter has discussed sound in the in the extended hearing ranges from 10 Hz to 20,000 Hz. Many UAS are designed for higher frequencies, i.e. ultrasonic and hypersonic. Propose an experiment to test sound disruption effects at the higher frequencies. (Drones, 2017) Quad Star Drones has some interesting "takes" on hypersonic flight and Mach 0.8 speeds.

Bibliography

Adamy, D. -0. (2015). EW 104 EW *against a New Generation of Threats*. Boston: Artech House.

Adamy, D. (2001). EW 101 A *First Course in Electronic Warfare*. Boston, MA: Artech House.

Adamy, D.-9. (1998, Jan). Lesson 4: the basic link for all EW functions. (electronic warfare)(EW Reference & Source Guide). *Journal of Electronic Defense, Jan 1998 Issue.*

Administrator. (2015, June 15). *Standing Wave and Fundamental Frequency*. Retrieved from Electronics Hub: https://www.electronicshub.org/?s=fundamental frequency

Austin, R. (2010). *"Design for Stealth", Unmanned Aircraft Systems UAVS Design Development and Deployment*. New York: John Wiley and Sons.

Burch, D. (2015). RADAR *for Mariners*. New York: McGraw-Hill.

Carter, A. (2012, May 24) EEWeb. Retrieved from https://www.eeweb.com/profile/andrew-carter/articles/the-sound-intensity on 16 July 2019.

Drones, Q. S. (2017, July 11). *quadcopters-have-hit-the-sound-barrier/*. Retrieved from quad-stardrones.com: https://quadstardrones.com/2017/07/11/quadcopters-have-hit-the-sound-barrier/

Entokey, a. G. (2019, May 16). *entokey.com/acoustics-and-sound-measurement/*. Retrieved from entokey.com/acoustics-and-sound-measurement/: https://entokey.com/acoustics-and-sound-measurement/

Gelfand. (2004). *"Physical Concepts", Hearing an Introduction to Psychological and Physiological Acoustics, 4th ed.* New York City.

Gelfand, S. A. (2009). *Essentials of Audiology, 3rd Edition.* Stuttgart, DE: Thieme.

Heinman, C. (2019). *Hearing Loss Tests Patrient D v-105.* Carlisle, PA: Brown Optical Hearing Aid Service.

Hubbard, R. K. (1998). *Boater's Bowditch.* Camden, MA: International Marine.

Kirk, J. (2015, August 5). *sounds-can-knock-drones-sky.* Retrieved from www.computer-world.com.au/article/581231: https://www.computerworld.com.au/article/581231/sounds-can-knock-drones-sky/

LRAD. (2019, May 189). *LRAD 450XL Datasheet.* Retrieved from LRADX: http://www.lradx.com/wp-context/uploads/2015/05/LRAD_datasheet_450XL.pdf

Merriam-Webster. (2019, May 17). Merriam-Webster Online Dictionary.

Monahan, K. (2004). *The Radar Book: Effective Navigation and Collision Avoidance.* Anacortes, WA: Fine Edge Publications.

Nichols, R. K. (1996). *Classical Cryptography Course, Volume I.* Laguna Hills, CA: Aegean Park Press.

Nichols, R. K. (2019, March 14). Hardening US Unmanned Systems Against Enemy Counter Measures. *7th Annual Unmanned Systems Summit*. Alexandria, VA, USA: PPTX presentation , self.

Pierson. (2019, May 16). *tuning-fork-waves-sound.* Retrieved from airfreshener.club – Pierson Education: https://airfreshener.club/quotes/tuning-fork-waves-sound.html

Said Emre Alper, Y. T. (December 2008). Compact Angular Rate Sensor System Using a Fully

Decoupled Silicon-on-Glass MEMS Gyroscope. JOURNAL OF MICROELECTROMECHANICAL SYSTEMS, VOL. 17, NO. 6.

Signia. (2019, May 16). *Signia Hearing Aids*. Retrieved from Signia Hearing Aids – Hear across America: www.signiausa.com

Staff. (2016, April 17). *Equal Loudness Contours*. Retrieved from Gutenberg Organization: http://central.gutenberg.org/article/WHEBN0001046687/Equal-loudness contour

Studios, D. D. (2017). Boaters Ref. USA.

Toomay, J. (1982). RADAR *for the Non – Specialist. London; Lifetime Learning Publications.* London: Lifetime Learning Publications.

TRS, S. (2018, July 10). *Tontechnik-Rechner-Sengpielaudio*. Retrieved from Tontechnik-Rechner-Sengpielaudio Calculator: www.sengspielaudio.com/calculator-wavelength.htm

Uni-wuppertal. (2019, May 15). *Inverse Square Law, General*. Retrieved from hydrogen.physik.uni-wuppertal.de/hyperphysics/: http://hydrogen.physik.uni-wuppertal.de/hyperphysics/hyperphysics/hbase/forces/isq.html

Usenix.org. (2019, 6 9). *MEMS, Drones, & Sound Sourcing*. Retrieved from Usenix.org: www.usenix.org

Weise, E. (2017, August 23). *could-hackers-behind-u-s-navy-collisions*. Retrieved from USATODAY: https://www.ruidosonews.com/story/tech/news/2017/08/23/could-hackers-behind-u-s-navy-collisions/594107001/

Yunmonk Son, H. S. (2015, August 12-14). Rocking Drones with Intentional Sound Noise on Gyroscopic Sensors. *Proc. 24th Usenix Security Symposium*. Washington, DC: USENIX. Retrieved from https://www.usenix.org/conference/usenixsecurity15/technical-sessions/presentation/son

Readings

Adamy, D. (2001) EW 101 A *First Course in Electronic Warfare*, Boston: Artech House.

Adamy, D. (2004) EW 102 A *Second Course in Electronic Warfare*, Boston: Artech House.

Adamy, D. (2009) EW 103 *Tactical Battlefield Communications Electronic Warfare*, Boston: Artech House.

Adamy, D. (2015) EW 104 EW *against a New Generation of Threats*, Boston: Artech House.

Adamy, D. (2003) Introduction to Electronic Warfare Modelling and Simulation, Boston: Artech House.

Austin, R. (2010) UAVS *Design, Development and Deployment*, New York: Wiley.

Hubbard, R.K (1998) Boater's Bowditch, Camden, MA: International Marine.

Burch, D. (2005) RADAR *for Mariners*. New York, McGraw-Hill.

Monahan, K (2004) *The RADAR Book: Effective Navigation and Collision Avoidance.* Anacortes, WA: Fine edge Publications.

Toomay, J.C. (1982) RADAR *for the Non – Specialist*. London; Lifetime Learning Publications

Chapter 20: Legal and Regulatory – Where It Was, Where It Is and What's Ahead?

Student Learning Objectives:

This chapter presents students with an updated set of considerations relating to the history, methods and objectives of regulating Unmanned Aerial Systems (UAS). Since the first edition of this textbook, the impact of legal and regulatory effect upon the industry has predictably increased both domestically and globally. In this chapter students will examine the regulatory history of UAS as well as those laws and rules presently under consideration and those which may be enacted in the future. Students will also examine the impact which new laws and regulations will have upon new technologies and delicate balance between too little regulation and too much. Finally, students be challenged to decide whether or not they would choose to enact regulations based upon events which may occur in the future and the considerations underlying those decisions.

Introduction

Since the first edition of *UNMANNED AIRCRAFT SYSTEMS (UAS) IN THE CYBER DOMAIN: PROTECTING USA'S ADVANCED AIR ASSETS* there have been numerous laws, regulations and enactments relating to the operation, identification and responsibilities of UAS in airspace globally. In order for today's student to understand the complexity and consequences of regulating UAS it is necessary to examine how we have arrived at this point and what were the underlying causative events that led to their enactment. Armed with this information students will be better equipped to navigate the complexities of the interaction between law and technology and the delicate balance between public safety, national defense and the inevitable development of new and more complex automated systems.

As a parallel consideration students should be cognizant that they will be entering an industry where UAS operation, regulatory, insurance, legal and other related industries is expected to grow exponentially in decade. According to a 2018 report by the consultancy firm PricewaterhouseCoopers, by 2030 it is estimated that the Drone/UAS industry will provide a 42 Billion Pound (53 Billion US Dollars) increase in the Gross domestic Product of the United Kingdom alone.[See Figure 20-1.]

Figure 20-1: UK Predicted Uplift in GDP by 2030

Source: (Pricewaterhousecoopers, LLP, 2018)

Where vast amounts of money flow it is a certainty that disputes will occur, risks need to be insured and legal responsibility will need to be established and adjudicated. Lawyers, Judges, Politicians, Manufacturers, Regulators, are but a few of the many areas of potential growth which will provide vast opportunity for those who choose a career in the UAS legal field. (Ricker, 2017) And for those who choose to follow a career path in other areas, not related to law and regulation, it will still be important for them to have a solid foundation in understanding the interface between UAS law and regulation with the rest of the overall UAS industry.

Current Regulatory Overview

As a preliminary matter the focus of this chapter will be focused upon the legal and regulatory structure of the UAS industry in the United States. This is not done out of any preference or bias but rather recognition that the number of separate regulations continues to grow rapidly with laws being enacted on local, state, federal and international levels. Since it would be impossible to study and digest UAS laws in each nation on earth this chapter will focus upon the broad conceptual considerations of regulatory frameworks as opposed to detailed examination of each and every separate regulation. While the FAA has exclusive authority over the use of airspace in the United States, as of September 2018, 44 states have adopted some sort of UAS regulation. (Sovereignty and use of airspace, 1994) Simultaneously the Federal Aviation Administration and other parts of the federal government continue to promulgate rules and regulations which collectively make the legal landscape quite difficult to navigate and even harder to predict. (National Conference of State Legislatures, 2018) Just as was discussed in the first edition of this text UAS are inherently mobile technology so that they are being used both within, across and beyond local, state and international borders not to mention their use over international waters and unaffiliated or neutral territories. The following figures demonstrate the overlapping web like structure of laws which currently confront those in the UAS industry. [See Figure 20-2]

Figure 20-2: Domestic UAS Regulatory Matrix

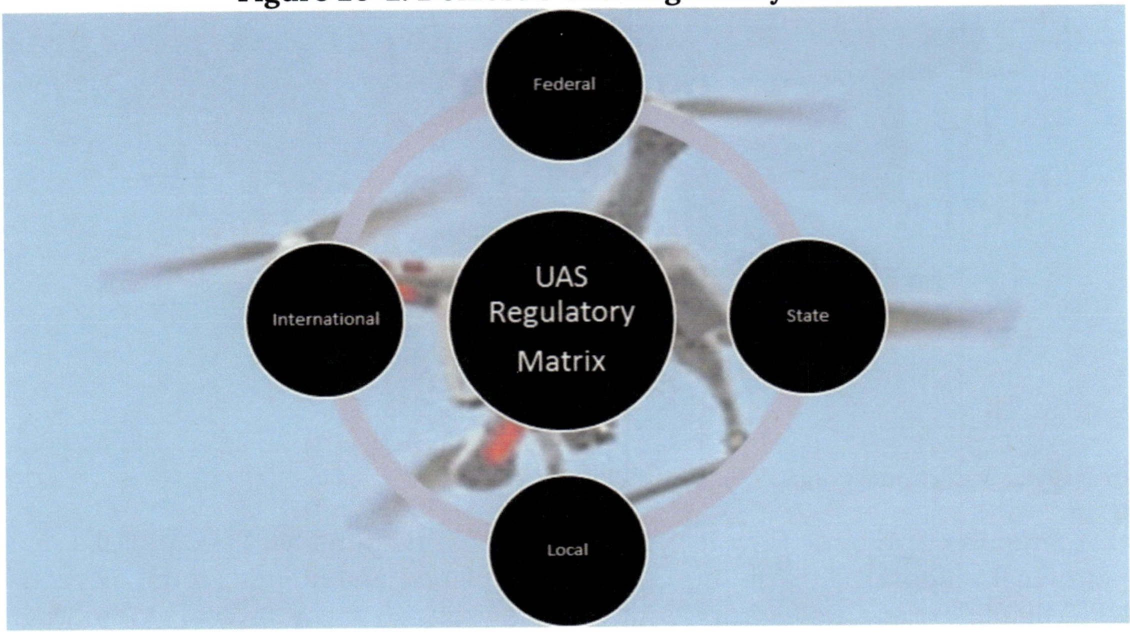

Source: (Sovereignty and use of airspace, 1994)

From a global perspective, as of January, 2018, according to medium.com of 233 countries being tracked just over one half have enacted some form of UAS regulation on a nationwide basis while over 100 nations having yet to have imposed any. (Schroeder, 2018) Most nations also have, states, provinces, prefects or other subdivisions not to mention agencies, authorities and even NGO's which all can promulgate more regulation. The reality is same globally as it is within the United States, an overlapping web of laws and regulations with a high likelihood that some portions of each may conflict with provisions of others. While the problem is currently one that is more burdensome and confusing than stifling innovation or a direct concern for public safety, with the increased of autonomous transportation the risk of more serious consequences grows with each new enactment.

Figure 20-3: Radiant Earth 2018 Drone Regulation Statistics

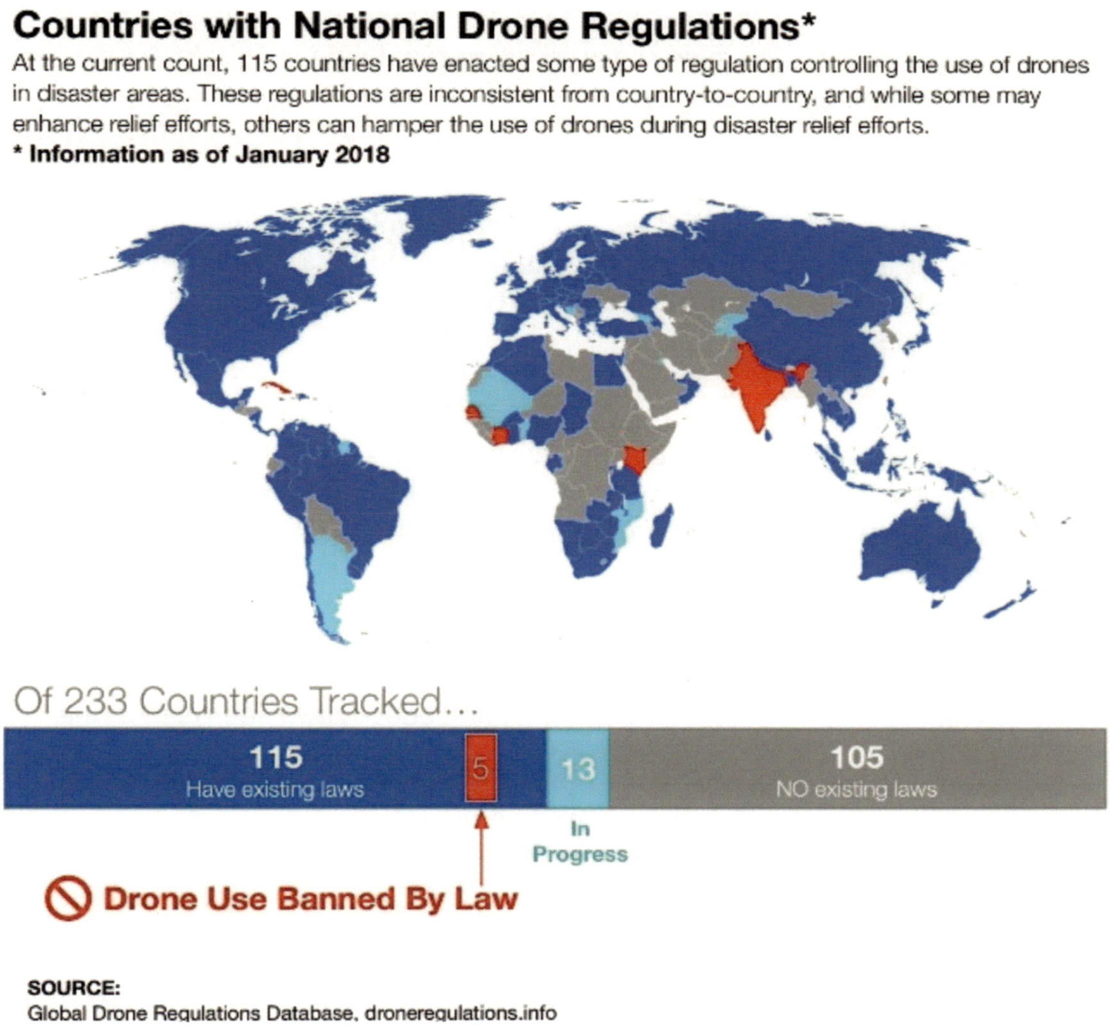

Source: (Database, 2018)

Some may ask "why not have a global set of rules for all nations regarding the UAS industry?" In reality as likely as conflict of regulations is, it is even more likely to expect that each nation's particular self-interest, autonomy, culture, mores and even form of government make such omnibus regulation a remote prospect at best. Just consider how regulations of UAS in China or North Korea might differ from those in the United States, United Kingdom and Australia. For example, a recreational UAV pilot might feel free to operate a drone over Times Square, Piccadilly Circus or the Sydney Opera House with some degree of registration and identification[1].

1. Since popular public locations can be considered "soft targets" for terrorists or criminals it would certainly

Contrast those locations with Tiananmen Square in China or Kim Il-sung Square in Pyongyang, North Korea. How would the CPRC or DPRK security forces react to such a flyover by a hobbyist? Hence the need to be extremely sensitive when regulating UAS to local customs, politics, mores and governing style.

Future Regulatory Framework

In order to achieve the most homogenous global UAS regulatory framework sense to focusing upon certain standardized critical areas which are generally considered "must have" in terms of UAS/UAV legislation seems to be most practical. Those general categories of regulation would consist of the following:

1. **UAS Pilot Licensing**
2. **Aircraft Registration with Oversight Agency (ies)**
3. **Areas of Restriction, Limited Use of Outright Prohibition**
4. **Liability, Operational "Rules of the road" and penalties for violation**. (Jones, 2017), (Muspratt, 2018)

In the United States the regulatory hierarchy is established on a pyramid model where the generalized power to regulate flows from the Congress who then empowers, by legislation the authority to certain government agencies to enact and adopt rules and regulations.

Next, based upon consultation with governmental experts as well as private sector or NGO input, which is vital to ensure that the imposition of regulations does not unduly burden the industry, public or stifle the safe development of the nascent technology. Figure 20-4 depicts just a portion of the hierarchy involved in the legislative and regulatory process in the United States.[2]

A similar regulatory hierarchy exists in each of the fifty states as well as territories, commonwealths and protectorates. While a number of jurisdictions have enacted various legal framework for UAS, still many more are actively considering and debating the imposition of operational, licensure, geographic, insurance, liability and criminal laws and regulations. Not only are such regulations often viewed as a public safety function, it also appears that with each registration, license and permit states can help offset budgetary deficits and create new job opportunities for state workers to administer these type of ministerial requirements. Just look

make sense to limit such overflight to known or approved aircraft with a heavily vetted pedigree of reliability, trustworthiness and piloting skill

2. Students should note that this diagram is more of an example than a complete list of agencies, departments and sub bureaucratic creations which all impact the UAS industry on a federal level.

to the study in the first edition of this text of how the automobile led to an entire massive industry in the early 1900's to where trillions of dollars are currently in play. (Nichols, 2018)

Figure 20-4: Portion of the hierarchy involved in the legislative and regulatory process in the United States

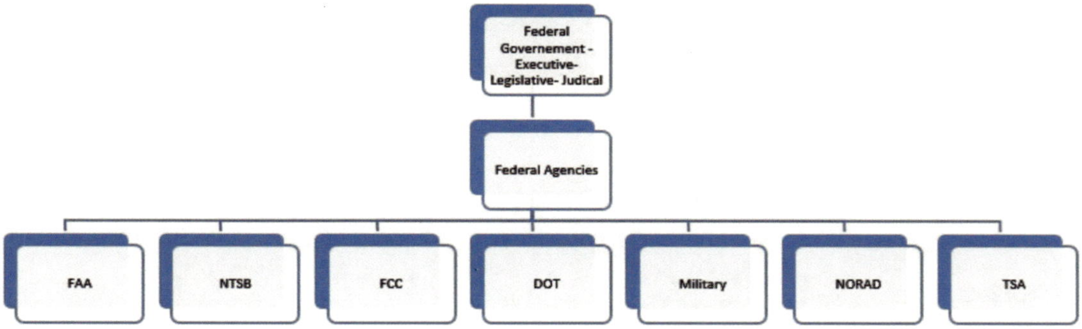

Source: (Lonstein, 2019)

Table 20-1 below is an example of the breadth and effect of many state drone laws.

Table 20-1: North Carolina Regulatory Framework

Knowledge Test	To fly in North Carolina, an operator must pass the N.C. Department of Transportation's' UAS Operator's Knowledge Test to ensure their safety and the safety of those around them (N.C. G.S. 63-95).
Permit	With the passing of the UAS Operator's Knowledge Test as a prerequisite, commercial operators may request a North Carolina UAS permit from the NCDOT Division of Aviation (N.C. G.S. 63-96).
Surveillance	It is illegal to use an unmanned aircraft system to take or distribute images of a person or their home without their consent (N.C. G.S. 15A-300.1 and N.C. G.S. 14-401.25).
Weapons	Attaching a weapon to a UAS is a Class E felony (N.C. G.S. 14-401.24).
Hunting & Fishing	Operators may not use a UAS to disrupt wildlife resources or the lawful taking of wildlife. It is also against the law in North Carolina to use a UAS in the process of taking wildlife resources (N.C. G.S. 113.295).
Interference with Manned Flights	Operators may not damage, disrupt the operation of or otherwise interfere with manned flights (N.C. G.S. 14-280.3).
Launch & Recovery Sites	It is illegal to launch or recover a UAS from either private or state property without the consent of the property's owner (N.C. G.S. 15A-300.2). Local and federal property have their own laws and regulations governing the launch and recovery of UAS.
Prisons	It is illegal to fly a UAS over any prison - state or federal - in North Carolina (N.C. G.S. 15A-300.3).

Source: North Carolina Department of Transportation

The North Carolina Regulatory Framework touches many areas of life and law. From hunting and fishing to privacy rights to licensing of pilots. It also contains provisions that are criminal in nature, others civil law and still others ministerial such as the licensure and registration requirement. Students should be aware that common or "case" law will provide interpretation of laws of which UAS professionals must keep abreast.

Below the state level comes local and regional authorities and governments which all can exercise various regulatory powers within their particular jurisdiction.

Finally input from industry associations, trade groups, manufacturers, NGO's and other private organizations which can all impact in in many ways regulate the UAS/UAV industry in the United States. The interaction between the UAS industry, commercial and private operators and the current and future regulatory structure should not be underestimated. Last is the overarching impact which the military and national defense establishment has upon the UAS industry. Students must be aware that military and national defense jurisdiction is largely border to border and beyond and in many instances trumps civilian authorities.

Figure 20-5: Principles for Future Regulation

Source: www.airspacemag.com

What is clear that a certain amount of regulation will always be needed however too much of a good thing can be worse than nothing at all. Regulators must be deliberative before, cautious during and responsive after regulation if we are to minimize negative and maximize the positives in UAS and AI regulation. Figure 20-5 is a good example of an agile regulatory framework and roadmap as it aims to find a broad consensus while recognizing the likelihood of unintended consequences. If we are to achieve maximum regulatory efficacy, while minimizing the risk of unintended consequence it will be essential to move quickly when such issues arise. This is particularly important in such a highly technological area where new developments will always outpace the law.

When students encounter the challenges of regulation and jurisprudence of technology, they may be wise to remember the writings of John L. O'Sullivan 1n 1847 who wrote:

Government should have as little as possible to do with the general business and interests of the people. If it once undertakes these functions as its rightful province of action, it is impossible to say to it 'thus far shalt thou go, and no farther.' It will be impossible to confine it to the public interests of the commonwealth. It will be perpetually tampering with private interests and sending forth seeds of corruption which will result in the demoralization of the society. (O'Sullivan, 1845)[3]

Conflict of Laws

With so many laws and regulations being promulgated internationally, domestically and locally it is inevitable that some of them will come into conflict with others. For example, North Carolina requires a knowledge test to pilot a done while South Carolina currently has no regulations. (UAV Coach, 2019) When neighboring jurisdictions have different regulations or as in the case of the Carolina's one state has regulations and the other does not, the prospect of an operator of a UAV encountering a situation of conflict of laws grows. Adding to the likelihood of conflict of laws is the national authority of the FAA over airspace. (49 U.S. Code §40103, 1994) What are the principles of law that can help resolve these disputes?

Federal supremacy is a constitutionally established doctrine that establishes the supremacy of Federal Law over conflicting state or local laws. It was a clause specifically included in the original text of the United States Constitution in 1776.

"This Constitution, and the Laws of the United States which shall be made in Pursuance thereof; and all Treaties made, or which shall be made, under the Authority of the United States, shall be the supreme Law of the Land; and the Judges in every State shall be bound thereby, any Thing in the Constitution or Laws of any State to the Contrary notwithstanding." (United States Constitution Article VI, Sec.2, 1787) (Marbury v. Madison, 1803)

From the Supremacy Clause flows the doctrine of Federal Preemption. In 1819 the United States Supreme Court held in McCullough v. Maryland (McCullogh v. Maryland, 1819) that federal legislation or regulation shall be deemed superior to and controlling over any state or local law. Sim-

3. Others have been reputed to have stated "The best government is that which governs least" including Thomas Jefferson, John Locke and Henry David Thoreau however it appears that O'Sullivan was the first to have penned the generalized concept.

ply put, "Preemption is a doctrine of American constitutional law under which states and local governments are deprived of their power to act in a given area, whether or not the state or local law, rule or action is in direct conflict with federal law." (Guardbaum, 1994)

In 2017 the Supremacy Clause and Federal Preemption came directly into play in the UAS industry when a medical doctor/ inventor Michael Singer intended drones to deliver medical supplies and services in the greater Newton Massachusetts area. At or about the same time the City of Newton enacted regulations regarding UAV operation in the jurisdiction. The following provisions were challenged by Dr. Singer in federal court.

1. Registration with the city of Newton in addition to the previous requirements established by FAA Regulations under Part 107;
2. Prohibition of the operation of pilotless aircraft under 400 feet in altitude;
3. Operation over private property: and
4. Prohibition of Beyond Line of Sight (BLOS) operation. (Singer v. City of Newton, 2017)

In holding the cities regulation invalid under the Supremacy Clause and doctrine of Federal Preemption the court held in part:

"The Ordinance limits the methods of piloting a drone beyond that which the FAA has already designated, while also reaching into navigable space. See Newton Ordinances § 20–64(c (1) (b). Intervening in the FAA's careful regulation of aircraft safety cannot stand; thus subsection (c) (1) (b) is preempted." (Singer v. City of Newton, 2017)

Regulatory legal conflict is not the only concern for UAS operators. Liability and insurance claims resulting from drone activity is likely to become a very significant area of litigation and court decisional law. Figure 20-5 is a photo of a drone which crashed into the Church of Latter-Day Saints in Draper, Utah in July of 2017.

Although the size of the drone was thankfully incapable of causing much damage to the property, the case is illustrative of the very likely prospect of larger UAVs crashing or coming into contact with structures, people or other flying and ground based vehicles. Courts will be hearing many cases and will have to blend the common law concepts such as assault, trespass and property privacy rights with statutes and regulations which may add liability to UAV pilots and/ or owners. In fact, many states, cities and town may attempt to enact laws which make drone owners "strictly liable" for any harm caused by UAV activity. Strict liability exists when a defendant is liable for committing an action, regardless of what his/her intent or mental state was when committing the action.

Figure 20-6: UAV Crash into LDS Church

Source: eastidahonews.com

Strict liability legislation is often reserved for harm caused by certain inherently dangerous activity which as a matter of public policy, has been designated as needed heightened liability rules to protect the public. One of the most likely areas of strict liability imposition upon UAS operators is in mid-air collisions where the risks become greater and the potential of damage or death is also greatly increased.

Students should be aware that conflicts of laws and regulation will most certainly increase with the expected exponential growth in commercial and hobbyist use of UAV's. While they may not directly be involved in the litigation or legislation of UAS rules, they should be keenly aware that there will be many instances where UAS activity may be subject to many laws some of which have conflicting provisions.

A significant portion of Chapter 2 of the First Edition of this text was devoted to the subjects of balancing the need for regulation with the possibility that over-regulation can inhibit the development of this nascent technology and the many benefits that may come from it. While the principles discussed in the prior edition are important for students to understand and assess, the passage of time can help us focus the areas of law where over or conflicting regulations can have real world consequences and inhibit innovation. The Singer decision is most likely just the beginning of an increased friction between the need to regulate and the desire to innovate. (Singer v. City of Newton, 2017)

One area which seems to be especially concerning is the prospect of laws being enacted in other parts of the world which establish liability for those globally. One of the most recent examples of such broad reaching legislation is the European Union's enactment of the General Data Protection Regulation (GDPR).

The three main objectives of GDPR are:

1. This Regulation lays down rules relating to the protection of natural persons with regard to the processing of personal data and rules relating to the free movement of personal data.
2. This Regulation protects fundamental rights and freedoms of natural persons and in particular their right to the protection of personal data.
3. The free movement of personal data within the Union shall be neither restricted nor prohibited for reasons connected with the protection of natural persons with regard to the processing of personal data. (European Union, 2019)

While on the surface on may ask, how does a data protection law enacted by the European Union have any impact on the UAS industry in the United States or other non-EU nations? As it is said, the devil is in the details.

GDPR defines personal data as:

"Personal data is any information that relates to an identified or identifiable living individual. Different pieces of information, which collected together can lead to the identification of a particular person, also constitute personal data." (European Union, 2019)

The definition of personal data is quite broad and when combined with the fact that any online form of communication may impact jurisdictional boundaries of the European Union, the potential of GDPR consequence from an online connected activity grows. For example, imagine you are a recreational drone pilot who likes to fly over beautiful homes on the South Florida

Coastline. You stream the flight video using Live4 or other streaming technology and fly by many beautiful homes along the fight.[4]

Along the flight, homes, mailboxes, house numbers, license plates and street sighs are all broadcast live globally. Additionally, there are a significant European citizens who own property in South Florida. Once recorded by the onboard camera and data processed for streaming in theory the UAV owner and/or pilot has in fact committed a GDPR violation.

4. https://www.today.com/video/how-peeping-drones-could-be-spying-on-you-without-you-knowing-it-1229001795967

Figure 20-7: Today Show Residential Drone Privacy

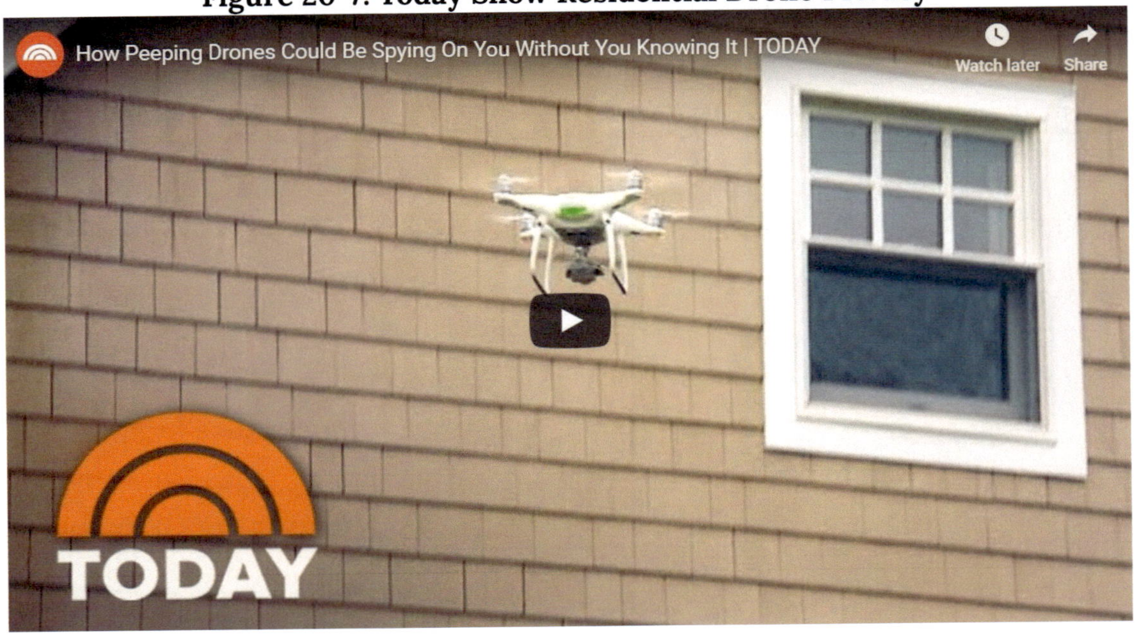

Source: NBC Today Show. (2018, May 9)

In terms of penalties they are significant and can easily cause havoc with commercial drone package delivery system in its infant stages. Although the possibilities of conflicts and confusion increase with each new law or decision, student should try to at least maintain a modicum of familiarity with recent laws, decisions and trends. Though by no means complete, some websites students may wish to visit include https://uavcoach.com/drone-laws/, http://www.ncsl.org/research/transportation/2017-unmanned-aircraft-systems-uas-state-legislation-update.aspx and https://www.americanbar.org/groups/air_space/.

Putting It Together – Where Law Meets Reality

As a general matter from the author's perspective much of the legal and regulatory process is about prescribing behavior and assessing blame when the laws are violated. In order to adequately operate within any quadrant of the greater UAS industry it is vital that students understand that each and every decision they make can have a legal consequence. To that end when designing aircraft, creating products, developing processes or even drafting regulations, consideration must be given to its legal consequences and the secondary effect upon the greater UAS industry.

Scenario 1

Randy, a drone hobbyist in the Rocky Mountains created local business of selling real-time and recorded video of crimes scenes, fires, and other serious disasters. Over time local television and print media would turn to Randy to acquire drone footage of breaking news in order to enhance their coverage. After a while Randy upgraded his aircraft, cameras and skills, He even formed a corporation called Dragon Vision to being to commercialize and become a true UAV video and photo vendor. Television, realtors, newspapers, even the local police were among many customers who began to regularly pay for his services.

During a particularly dry season the mountains near Randy's Rocky Mountain home a series of deadly wildfires, caused by a careless hiker with a cigarette butt which was not extinguished before he flicked it into the pine straw. Within moments, fueled by gusty winds the cigarette lit the pine straw and within moments thousands of acres of pristine forest were ablaze, including portions of the Rocky Mountain National Park. Local fire-fighters immediately summoned state and federal resources and simultaneously media bean to flock to the region. Frank was called by media and even the local fire authorities to help them by using his UAV fleet to capture images, provide streaming video and even identify "hot-spots" where new fires were igniting by using an infrared thermal imaging scanners on one of this aircraft.

Figure 20-8: DJI Matrice 210 V2 With Zenmuse XT2 Thermal & Zenmuse Z30 Visual

Source: Enterprise DJI.com

Cameras

Randy, ever the Good Samaritan immediately went into action without any agreements with any of the consumers of his content, just the usual handshake. As the fire spread Randy streamed live and also took still photos for the various fire departments and news outlets. As the fire continued to be spread by stronger winds, he found that the drone was increasingly difficult to control because of smoke, heat, uplift wind gusts from the fire and rotor wash from firefighting helicopters.

The result? Firefighting aircraft were unable to drop payloads until the confusion and uncertainty caused by Randy's drone could be resolved. Unfortunately, the delay let to the fire expanding and either partially or totally was responsible for the loss of thousands of acres of forest and hundreds of homes. Thankfully no humans were seriously injured or killed, and all pets and livestock had been relocated.

What are some of the laws may be implicated in the case of Randy's UAV misadventure in the Rocky Mountains? Who might be subject to these claims other than Randy?

Implicated Federal Laws

- **49 U.S. Code §40103. Sovereignty and use of airspace**
- **FAA Regulation Part 107 Requiring Registration of UAV and Certification of a Commercial Pilot of a UAV, Daylight Only, Under 100 MPH, below 400 feet and in line of sight.**
- **43 CFR 9212.1(f), it is illegal to resist or interfere with the efforts of firefighter(s) to extinguish a fire.**
- **UAS operation prohibited within all National Parks 36 CFR 2.17 (a) (3)**

State Statutes

Title 18 – Criminal Code

- **18-8-104. Obstructing a peace officer, firefighter, emergency medical services provider, rescue specialist, or volunteer.**

Colorado Misdemeanor & Felony Criminal Trespass – Third Degree Criminal Trespass (18-4-504)

Proposed Colorado Law:

- **SECTION 1. In Colorado Revised Statutes, add 18-7-802 as follows:**
- **18-7-802. Criminal invasion of privacy by the use of a device – penalty.**

Common Law Implications

- **"Invasion of Privacy**

Unreasonable intrusion upon the seclusion of another

(1) Another person has intentionally intruded, physically or otherwise;

(2) Upon the seclusion or solitude of the plaintiff;

(3) The intrusion would be offensive to a reasonable person; and

(4) The invasion was a cause of plaintiff's damages. " (MacGregor, 2018)

General Negligence

(1) "The defendant owed a legal duty to the plaintiff;

(2) The defendant breached that duty; and

(3) The breach resulted in injury to the plaintiff." (MacGregor, 2018)

Nuisance

(1) "The defendant unreasonably and substantially interfered with;

(2) The plaintiff's use and enjoyment of his property." (MacGregor, 2018)

Products Liability -Misrepresentation

(1)" The defendant sold the product while engaged in the business of selling the product for resale, use, or consumption;

(2) The defendant misrepresented a fact concerning the character or quality of the product that would be material to purchasers of the product or members of the public at large;

(3) The plaintiff or a third party purchased the product and reasonably relied on the misrepresentation; and

(4) The plaintiff suffered damages as a result of his or a third party's reasonable reliance on the misrepresentation. (Porter, 2019)

Products Liability –Misrepresentation

(1) The defendant owed a legal duty of care to the plaintiff;

(2) The defendant breached that duty; and

(3) The defendant's breach resulted in injuries to the plaintiff." (MacGregor, 2018)

Products Liability – Strict Liability

(1) "The product is in a defective condition unreasonably dangerous to the user or

(2) consumer or to the consumer's property;

(3) The seller is in the business of selling such a product;

(4) The design defect caused the plaintiff's injury; and

(5) The plaintiff incurred damages as a result." (MacGregor, 2018)

Trespass

(1)" Intentional physical intrusion on the plaintiff's property;

(2) Without proper permission; and

(3) Legal entitlement to possession of property by the party claiming trespass." (MacGregor, 2018)

Figure 20-9: No Drone Operation Rocky Mountains

Source: (MacGregor, 2018)

Possibly Liable Parties:

First and foremost, the hiker is would most certainly be deemed to have some degree of criminal or civil liability for being the cause of the fire itself. Had he not flicked a lit cigarette into the dry pine straw reason dictates the fire would never have occurred. This is known as the "But-for test" which asks, "but for the existence of the hiker's action of flicking a burning cigarette onto pine straw in a dry nation park, would the forest fire have occurred?" If the answer is yes, then factor X is an actual cause of result Y. Other jurisdictions use the similar concept of "proximate cause" which asks whether the defendant's actions are closely enough related to the result to make the defendant responsible. (Cornell University Legal Information Institute, 2019) Under either test the hiker would likely be found to be proximate or actual cause of the fire.

Randy is the next logical target for possible criminal and civil exposure, not only could he be fined or criminally charged by the FAA if his UAV was not registered, he could also face charges from the National Park Service for operating in a restricted area. He may also be charged with a violation of the federal law prohibiting UAV interference with firefighters. Similarly, the Colorado laws against obstructing a firefighter, criminal trespass and criminal invasion of privacy may also cause criminal liability on a state level. His potential liability stems from the legal concept of intervening or contributory cause. In this case whatever Randy did occurred after and as a result of the Hiker causing the fire. His liability will be predicated upon the concept of intervening or contributory cause. Intervening cause is defined as an event that occurs after a party's improper or dangerous action and before the damage that could otherwise have been caused by the dangerous act, thereby breaking the chain of causation between the original act and the harm to the injured person. The result is that the person who started the chain of events may no longer be considered fully or partially responsible for damages to the injured person since the original action is no longer the proximate cause. (Cornell University Legal Information Institute, 2019)

What other parties have possible liability to those property owners and the National Park Service itself for damages sustained or exacerbated during the fire? Was there a defect or improper design that led to the inability of Randy to navigate and control the UAV which caused the delay in dispersing fire retardant in some way caused by, DJI, Zenmuse or their component suppliers?

The prospect of liability on behalf of DJI, Zenmuse or their suppliers will likely flow from the concept of product liability and breach of warranty. This type of liability blends elements of tort or negligence law with principles of contractual law where a product was not designed, manufactured or perform in the manner promised in Figure 20-9.

From a civil liability perspective Randy may be the target of both federal and state litigation. Students must also be aware that there are concurrent and special jurisdictional rules for claims arising on federal property or between citizens of different states. Specifically, at least two different types of subject matter may result in Randy being named as a defendant in a Federal Court action.

Jurisdiction: Subject Matter

The first example of subject matter jurisdiction in federal courts (what this dispute is about – its "subject matter") is known as Federal Question Jurisdiction. The Federal Question jurisdiction reads: The district courts shall have original jurisdiction of all civil actions arising under the Constitution, laws, or treaties of the United States. (Federal Question, 1948)

Figure 20-10: DJI Public Safety Applications

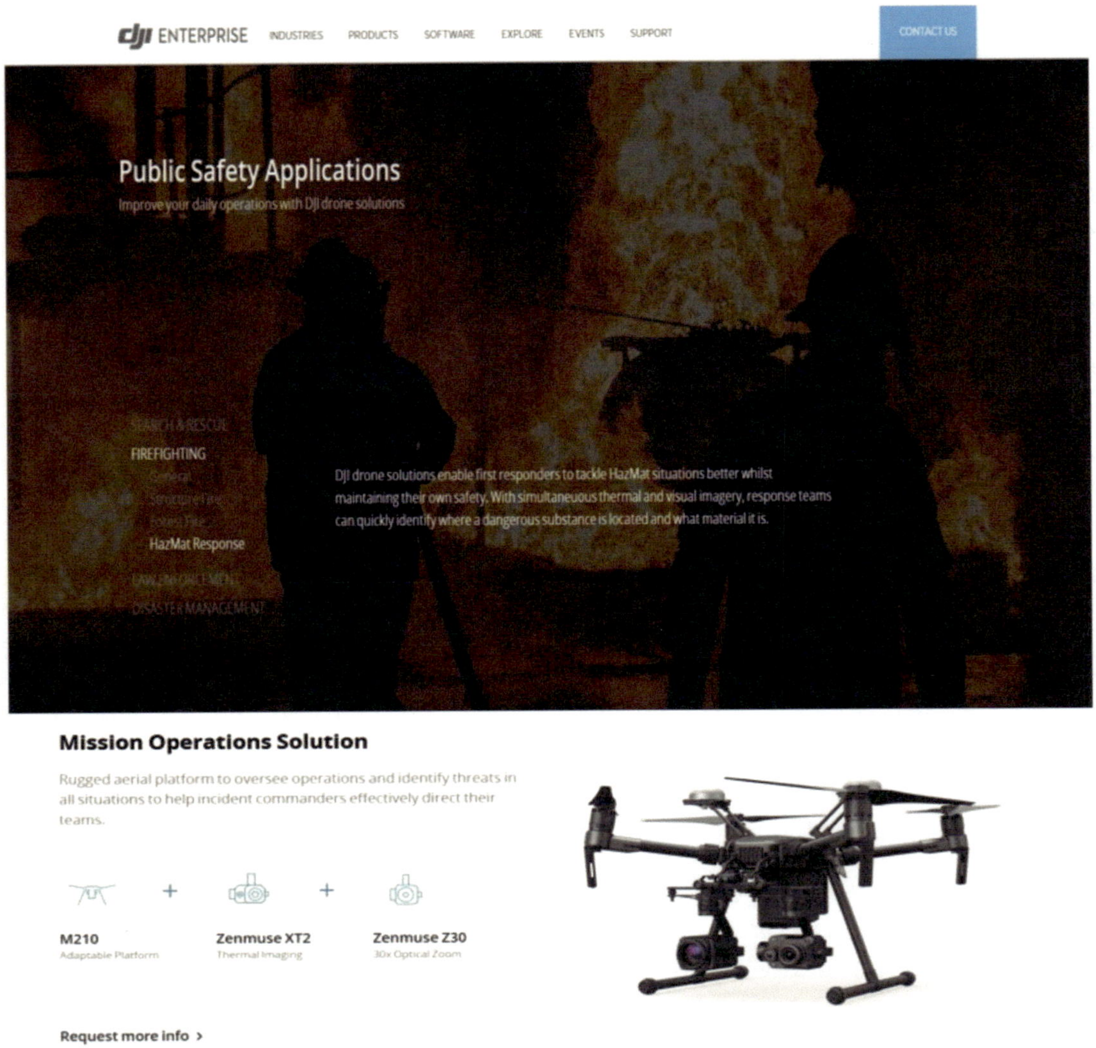

Source: DJI Enterprise. Retrieved from dlsrpros.com

<u>A second type is called diversity jurisdiction</u>. The Diversity Jurisdiction statute reads in applicable part:

(a) The district courts shall have original jurisdiction of all civil actions where the matter in controversy exceeds the sum or value of $75,000, exclusive of interest and costs, and is between–

(1) Citizens of different States;

(2) citizens of a State and citizens or subjects of a foreign state, except that the district courts shall not have original jurisdiction under this subsection of an action between citizens of a State

and citizens or subjects of a foreign state who are lawfully admitted for permanent residence in the United States and are domiciled in the same State;

(3) Citizens of different States and in which citizens or subjects of a foreign state are additional parties; and

(4) A foreign state, defined in section 1603(a) of this title, as plaintiff and citizens of a State or of different States. (Diversity of citizenship; amount in controversy; costs, 1948)

When applying these principles to Randy's drone operation in Colorado and particularly, the consequences caused by it, the Federal and State Courts would have Subject Matter jurisdiction to adjudicate claims as delineated above.

Personal Jurisdiction

Broadly speaking "Personal jurisdiction refers to the power that a court has to make a decision regarding the party being sued in a case. Before a court can exercise power over a party, the U.S. Constitution requires that the party has certain minimum contacts with the forum in which the court sits. International Shoe v Washington, 326 US 310 (1945). So, if the plaintiff sues a defendant, that defendant can object to the suit by arguing that the court does not have personal jurisdiction over the defendant." (Cornell University Legal Information Institute, 2019)

Practically speaking Personal Jurisdiction asks the following question, is it fair to expect that one's actions have an effect or consequence in a location which under the notions of fairness and reason would cause an actor to perceive the likelihood of being subject to the jurisdiction of a court in that location? When applied to Randy's drone activities during the fire it is clear that a reasonable person who was engaging in drone flights during a wildfire in the Rocky Mountain National Park might expect that civil or criminal proceedings resulting from his actions could be instituted in Colorado State or Federal District Courts.[5]

Although it is unrealistic for students to become deeply immersed in UAS jurisprudence, it is essential that students remain cognizant of the interplay between the law and autonomous systems domestically and internationally.

Scenario 2

It is January 2021 and the Inauguration ceremonies are about to begin throughout the greater Washington, DC region. As a result of the 2016 Presidential Election there is still a significant amount of political unrest related to the claim that other nations interfered in some fashion

5. There may be many other appropriate jurisdictions for litigation involving Randy's actions in the national park. Injuries to visitors from nearby states or even nations may well have a right to bring claims in other jurisdictions outside Colorado. For the purpose of brevity, we have limited our inquiry to one state.

with the electoral process. In an abundance of caution the various military, federal, state and local stakeholders have devised a layered and agile defense strategy. As part of the analysis many experts have expressed concern that the use of UAS in crowded, public "soft targets" has been repeatedly mentioned in terror network chatter intercepted in the months immediately after the November 2020 election. After Inauguration Day concludes without incident, a significant portion of the security infrastructure begins to stand down and for the most part initial debriefing calls the plan a success.

The following weekend is the National Football League NFC Championship Game between the Washington Redskins and San Francisco 49ers at Fedex Field just east of the District of Columbia. During pre-game festivities a concert starring Beyoncé and Jay-Z. In addition to the 82,000-fan standing room only crowd are 500 local high school dancers who are on the field as part of the performance. Once the concert concludes and the performers are slowly filing off the field player introductions with pyrotechnics begins. The air is now hazy and filled with smoke.

Figure 20-11: Carrier HX8 Sprayer Drone Over FedEx Field

Source: www.aviewfromyseat.com; www.harrisaerial.com/

She noticed something abnormal in the sky over the light stanchions on the northwestern corner of the stadium. Immediately upon spotting the sprayer drone her mind races and she immediately text the photo in Figure 20-11 to her contacts at the Department of Homeland Security.

You are the Agent-In-Charge of this jurisdiction and are called on to immediately assess what, if any actions Counter Unmanned Aviation Systems (CUAS) can technologically, legally and ethically be taken against the drone and its operators.

Legal Considerations

According to the Preventing Emerging Threats Act of 2018, the following authority was granted by congress as follows:

EC. 1602. PROTECTION OF CERTAIN FACILITIES AND ASSETS FROM UNMANNED AIRCRAFT.

(a) In General.—Subtitle A of title II of the Homeland Security Act of 2002 (6 U.S.C. 121 et seq.) is amended by adding at the end the following:

"SEC. 210G. PROTECTION OF CERTAIN FACILITIES AND ASSETS FROM UNMANNED AIRCRAFT.

"(a) Authority.—Notwithstanding section 46502 of title 49, United States Code, or sections 32, 1030, 1367 and chapters 119 and 206 of title 18, United States Code, the Secretary and the Attorney General may, for their respective Departments, take, and may authorize personnel with assigned duties that include the security or protection of people, facilities, or assets, to take such actions as are described in subsection (b)(1) that are necessary to mitigate a credible threat (as defined by the Secretary or the Attorney General, in consultation with the Secretary of Transportation) that an unmanned aircraft system or unmanned aircraft poses to the safety or security of a covered facility or asset.

"(b) Actions Described.—

"(1) IN GENERAL. —The actions authorized in subsection (a) are the following:

"(A) During the operation of the unmanned aircraft system, detect, identify, monitor, and track the unmanned aircraft system or unmanned aircraft, without prior consent, including by means of intercept or other access of a wire communication, an oral communication, or an electronic communication used to control the unmanned aircraft system or unmanned aircraft.

"(B) Warn the operator of the unmanned aircraft system or unmanned aircraft, including by passive or active, and direct or indirect physical, electronic, radio, and electromagnetic means.

"(C) Disrupt control of the unmanned aircraft system or unmanned aircraft, without prior consent, including by disabling the unmanned aircraft system or unmanned aircraft by intercepting, interfering, or causing interference with wire, oral, electronic, or radio communications used to control the unmanned aircraft system or unmanned aircraft.

"(D) Seize or exercise control of the unmanned aircraft system or unmanned aircraft.

"(E) Seize or otherwise confiscate the unmanned aircraft system or unmanned aircraft.

"(F) Use reasonable force, if necessary, to disable, damage, or destroy the unmanned aircraft system or unmanned aircraft. (PROTECTION OF CERTAIN FACILITIES AND ASSETS FROM UNMANNED AIRCRAFT, 2018)

Upon review of this law it seems clear to you that there is adequate authority to engage in some sort of CUAS activity, however numerous legal and ethical considerations still must be addressed. Often referred to as Counter Unmanned Aviation Systems (CUAS) technology, this security service works to protect spaces such as public stadiums and arenas, amusement parks, casinos or airports, where the presence of drones could be illegal, dangerous or even deadly. (sUAS News, 2018)

Decisions & Dilemmas for Student Consideration

1. Who is the owner of the aircraft?
   If you decide to act against a UAV prior to identifying its owner or function you will be taking a significant risk if it is an authorized but not properly disclosed vehicle.Identifying ownership is critical however delay can be a catastrophic mistake, endangering thousands if not millions from nuclear, chemical or biological attack. Even if the UAV is harmless using some type CUAS which is obvious to the public thereby risking a mass stampede of over 82,000 spectators.
2. What is its purpose?
   Will destroying it because a failure of a critical function related to the game that was mistakenly not disclosed to authorities?If there is a legitimate purpose perhaps assisting in the broadcast by spraying an anti-fogging agent on the overhead television camera suspended on wires above the field.
3. What is the payload?
   If a harmful agent as suggested above will a CUAS actually exacerbate the damage and injury caused by dispersing the agent in the payload before its nature is known. ?
4. Should the public be advised?
5. What will be the consequence of delay in acting?

Conclusions

Although it is impossible to cover all the legal considerations resulting from the growth of autonomous systems, hopefully students will take away a better understanding of the interface of law and technology. The contents of this chapter as well as chapter 2 of the First Edition should serve as a sampling of the techno-legal considerations which will confront all of us as we move forward in a world with increasing automation. Although there are no hard and fast rules to guide students when considering or confronting these issues, what students must take into consideration is what the consequences or each new law, new technology or new application may have from a legal perspective.

Bibliography

49 U.S. Code §40103, 49 U.S. Code §40103 Sovereignty and use of airspace (U.S. Code July 5, 1994).

Cornell University Legal Information Institute. (2019, June 5). *But-for test.* Retrieved from law.cornell.edu: https://www.law.cornell.edu/wex/but-for_test

Cornell University Legal Information Institute. (2019, June 5). *Intervening Cause.* Retrieved from law.cornell.edu: https://www.law.cornell.edu/wex/intervening_cause

Cornell University Legal Information Institute. (2019, June 5). *Personal Jurisdiction.* Retrieved from law.cornell.edu: https://www.law.cornell.edu/wex/personal_jurisdiction

Deloitte Center for Government Insights analysis. (2018, June 18). *The future of regulation. Principles for regulating emerging technologies.* Retrieved from Deloitte Insights: https://www2.deloitte.com/insights/us/en/industry/public-sector/future-of-regulation/regulating-emerging-technology.html

Dewey, D. (2017, July 14). *Drone crashes into LDS temple in Utah; raises questions of airspace rules.* Retrieved from eastidahonews.com: https://www.eastidahonews.com/2017/07/drone-crashes-lds-temple-utah-raises-questions-airspace-rules/

Diversity of citizenship; amount in controversy; costs, 28 U.S. Code §1332 (United States Congress June 25, 1948).

DJI. (2019, June 5). *DJI Enterprise.* Retrieved from Enterprise DJI.com: https://enterprise.dji.com/civil-protection

DLSR Pros. (2019, June 3). *Best Drones* (UAVs) *for Firefighting in* 2019. Retrieved from dlsrpros.com: https://www.dslrpros.com/dslrpros-blog/best-drones-firefighting-2019/

European Union. (2019, May 2019). *About the regulation and data protection.* Retrieved from ec.europa.eu: https://ec.europa.eu/commission/priorities/justice-and-fundamental-rights/data-protection/2018-reform-eu-data-protection-rules_en

Federal Question, 28 U.S. Code §1331 (United States Congress June 25, 1948).

Guardbaum, S. (1994). The Nature of Preemption. *Cornell Law Review*, 767, 771.

Harris Aerial. (2019, June 5). *Carrier HX8 Sprayer Drone.* Retrieved from harrisaerial.com: https://www.harrisaerial.com/carrier-hx8-sprayer/

Jones, T. (2017). *International Commercial Drone Regulation and Drone Delivery Services*. Santa Monica: The Rand Corporation.

Legal Information Institute – Cornell University. (2019, May 31). *Strict Liability* . Retrieved from Legal Information Institute: https://www.law.cornell.edu/wex/strict_liability

MacGregor, D. S. (2018). *Colorado Causes of Action: Elements, Defenses, Remedies, and Forms*. Denver: Bradford Publishing Co. .

Marbury v. Madison, 5 U.S. 137 (United States Supreme Court February 23, 1803).

McCullogh v. Maryland, 17 U.S. 316 (United States Supreme Court March 6, 1819).

Muspratt, A. (2018, November 22). *New global drone standards proposed*. Retrieved from Defence iQ: https://www.defenceiq.com/defence-technology/news/new-global-drone-standards-proposed

National Conference of State Legislatures. (2018, September 10). *Current Unmanned Aircraft State Law Landscape*. Retrieved from NCSL.org: http://www.ncsl.org/research/transportation/current-unmanned-aircraft-state-law-landscape.aspx

NBC Today Show. (2018, May 9). *How peeping drones could be spying on you without you knowing it*. Retrieved from Today.com: https://www.today.com/video/how-peeping-drones-could-be-spying-on-you-without-you-knowing-it-1229001795967

Nichols, R. K. (2018). *Unmanned Aircraft Systems (UAS) In the Cyber Domain: Protecting USA's Advanced Air Assets. 1st Ed.* Manhattan, KS: New Prairie Press.

North Carolina Department of Transportation. (2019, May 30). *Law & Regulations*. Retrieved from NCDOT.GOV: https://www.ncdot.gov/divisions/aviation/uas/Pages/laws-regulations.aspx

O'Sullivan, J. L. (1845). The Great Nation of Futurity. *United States Magazine and Democratic Review* Vol 6 Issue 23, pp. 426-430.

Porter, J. D. (2019, June 8). *jdporterlaw.com/intellectual-property-law/*. Retrieved from jdporterlaw.com: http://www.jdporterlaw.com/intellectual-property-law/

Price Waterhouse Coopers, LLP. (2018). *Skies without limits – Drones- taking the UK's economy to new heights*. London: Pricewaterhousecoopers, LLP.

PROTECTION OF CERTAIN FACILITIES AND ASSETS FROM UNMANNED AIRCRAFT, H. R. 302 (United States Congress January 3, 2018).

Ricker, D. (2017, July 1). *Navigating drone laws has become a growing and lucrative legal niche.*

Retrieved from ABA Journal: http://www.abajournal.com/magazine/article/drone_law_attorneys

Sanchez, M. (2019, June 4). *No Drones.* Retrieved from Unspalsh.com: https://unsplash.com/photos/oMqswmrie4Y

Schroeder, A. (2018, February 1). *Localizing Humanitarian Drones: Robotics & Disaster Response from the Maldives to Malawi.* Retrieved from medium.com: https://medium.com/radiant-earth-insights/localizing-humanitarian-drones-robotics-disaster-response-from-the-maldives-to-malawi-a1f362432cb1

Singer v. City of Newton, 284 F. Supp. 3d 125 (U.S. District Court Massachusetts September 21, 2017).

Sovereignty and use of airspace, 49 U.S. Code §40103 (United States Congress July 5, 1994).

sUAS News. (2018, March 2). *RAS Consulting & Investigations hire Jeff Parisse to offer sophisticated UAS security and surveillance services.* Retrieved from suasnews.com: https://www.suasnews.com/2018/03/ras-consulting-investigations-hire-jeff-parisse-offer-sophisticated-uas-security-surveillance-services/

UAV Coach. (2019, May 30). *Drone Laws in South Carolina* (2019). Retrieved from UAVcoach.com: https://uavcoach.com/drone-laws-south-carolina/

United States Constitution Article VI, Sec.2 (United States of America September 17, 1787).

Wright, T. (2017, August 11). *You've Been Warned: Keep Your Drones Away From Military Bases.* Retrieved from Air & Space, Smithsonian: https://www.airspacemag.com/daily-planet/keep-your-drones-away-military-base-180964451/

Chapter 21: Chinese UAS Proliferation along New Silk Road Sea/ Land routes

Chapter 21 will introduce the evolving Chinese UAS Proliferation along New Silk Road Sea/ Land routes

Student Learning Objectives. Upon completion of this chapter, students should be able to:

- Understand the background of building the Chinese government project "The Belt & Road"
- What Chinese military buildup is occurring to support new Silk Road?
- How UAS/UAV are involved in the New Silk Road?
- What is the US involvement in the New Silk Road?

Chinese Government Building the "The Belt and Road"

China has begun to grow the modern-day vision of the Han Dynasty, joining the East to the West. The popularity of the original route grew to over four thousand miles, until collapse in the 18th century. (Arugay, 2017) During official visits to Kazakhstan and Indonesia in 2013, "President Xi announced the new Silk Road initiative. The plan was two-fold: there would be an overland Silk Road Economic Belt and the Maritime Silk Road." (Arugay, 2017) "Initially both were referred to first as the One Belt, One Road initiative but eventually became the Belt and Road Initiative" (BRI)." (Arugay, 2017) Xi Jinping named BRI the "Project of the Century" with an estimated cost of $1.3 trillion by 2027. The project would develop a network of railways, energy pipelines, highways, and streamlined border crossings, overseas shipping routes, both westward—through the mountainous former Soviet republics—and southward, to Pakistan, India, and the rest of Southeast Asia. Linking China's coastal factories and rising consumer class with Central, Southeast and South Asia; with the Gulf States and the Middle East; with Africa; and with Russia and all of Europe.

BRI is at the core of China's foreign policy strategy and was even added to the Communist Party constitution in 2017. On March 17, 2017, the UN Security Council unanimously adopted Resolution 2344, calling on the international community to strengthen regional economic cooperation through the BRI China has signed agreements with the following organizations: United Nations Development Program, United Nations Economic Social Commission for Asia, and the Pacific World Health Organization. BRI consists of two major routes, Overland Route and Maritime Sea Lanes. The Overland Route was the traditional Route used for centuries Connects Europe through the Mediterranean to Burma and China. With the addition of the second route, Maritime Sea Lanes, the trade path extends from Chinese Eastern coast through the Spratley

Islands, around Horn of Africa into the Mediterranean Sea. Sixty-eight countries (two-thirds of the planet's total population) have signed on to bilateral projects partly funded by China's policy banks and other state-owned enterprises. BRI continues to be controversial, with persistent warnings that poorer countries will be burdened with unsustainable debts.

The Belt

Chinese firms are building or investing in new highways and coal-fired power plants in Pakistan, ports in Greece and Sri Lanka, gas and oil pipelines in Central Asia, an industrial city in Oman and a multibillion-dollar railway project in Laos. From Myanmar to Israel and from Mauritius to Belgium, China holds ports.

The combination of these routes forms a new economic sphere of influence for China. BRI has the advantage of two primary trade corridors/physical routes. The first route (starting by land) China to Southern Europe Sea and the second leveraging the Port of Shanghai to land-based route in Venice. With the successful completion of BRI, China will be connected with Central Asia, the Middle East, Africa, South America, Central America, and Europe. The growing web of trade routes will extend into at least 76 countries. China's National Development and Reform Commission (NDRC) oversees development of BRI. In total there will be the creation of six economic corridors: New Eurasian Land Bridge, China-Mongolia-Russia, Central Asia-China-West Asia, China-Indochina Peninsula, China-Pakistan and Bangladesh-China-India-Myanmar. The creation of these economic corridors will be the means of communication, rail highways, seagoing transport, expansion of China's cyberspace, oil pipelines, and aerospace.

Figure 21-1: New Silk Road Economic Belt

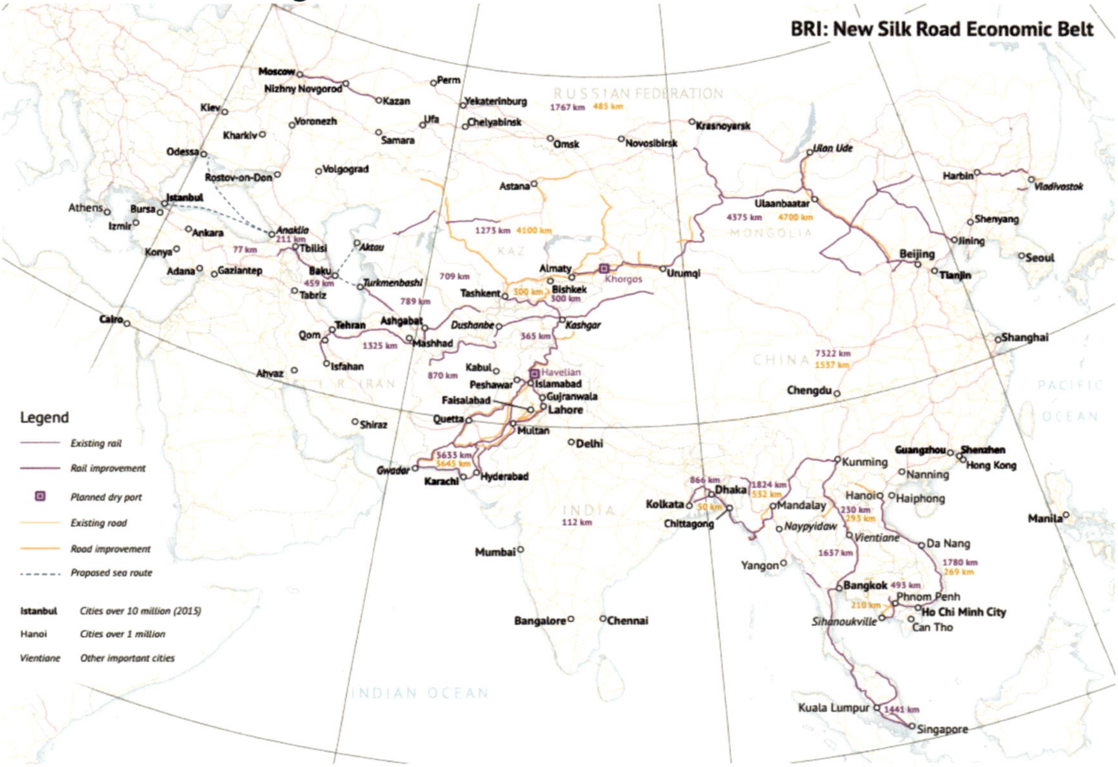

Source: Reed and Trubetskoy (2019)

Source: (Lebrand & Lall, 2019)

Central Role in Road: Kazakhstan

On the border of China, the former Soviet republic of Kazakhstan, is the world's largest land-locked country. The capital of Kazakhstan, Nur-Sultan, has become the top spot for cross-border investors with $24 billion funds deposited in 2018. In May 2019, forty-three foreign company agreements were signed with Kazakhstan, totaling $8.7 billion. Large investors came from France, Germany, the United Kingdom, the United States, Singapore and South Korea. In the beginning of the BRI Kazakhstan signed $30 billion worth of trade and investment agreements with China, along with a gas pipeline. This solidified their partnership in the build out of the New Silk Road. On the way to becoming the world's largest dry port, Khorgos Gateway, allows for goods (anything from a John Deere tractor to Hewlett Packard parts) that have a Chinese point of origin can arrive in Europe in 14 days, faster than the sea and cheaper than the air. The Chinese city, Knorgo, that sits Xinjiang Uyghur Autonomous Region is home to International Center for Boundary Cooperation (I.C.B.C.). In the I.C.B.C., goods and people can move freely back and forth, it is considered a duty- and visa-free zone. China plans to build fifty more I.C.B.C.'s in countries from Alegria to Vietnam.

The Belt Achievements to Date

In January 2017, a direct railroad route from Yiwu, China to London, UK, traveled for the first time. The container train carried clothing, bags, and household goods as part of the growing rail network of trade between China in the EU. The network grew in a few years' time from, 39 rail lines directly connecting 15 cities in China to 16 EU cities to 59 Chinese cities with 49 cities in 15 European countries.

In 2018, Chongqing China reported a 50% increase in rail trips to the EU, pushing the total to over a thousand for the year for that one route. In 2018, the China Europe rail freight service made a total number of 6,363 trips. Overall the freight trains have completed over 14,000 trips. Total trade between China and other countries along the BRI amounted to six trillion dollars between 2013- 2018. China has directly invested in BRI countries over 90 billion dollars since the beginning of the project. China has achieved the framework for six economic corridors, six connective networks creating close to 300,000 jobs for the host countries. As of April 2018, China and 61 countries have formed 1,023 pairs of sister cities. (Xinhua, 2019)

Figure 21-2: Chongqing Freight Train

Source: (Belt and Road News, 2019)

Maritime Silk Road (MSR)

The Road portion of BRI, also referenced as Maritime Silk Road (MSR), focuses on creating a network of ports, through construction, expansion or operation, and the development of portside industrial parks and special economic zones (SEZs). MRI is split into three main arteries:

1. "China's coast to Europe through the SCS, the IOR and the Mediterranean Sea, and into the Atlantic.

2. China's coast through the SCS to the South Pacific and then onto greater Australia.
3. Arctic Ocean, passing north-west alongside Russia's northern coast to connect with the Nordic region and other parts of Europe, and north-east past Canada."(Arugay, 2017)

The two most strategic areas of MSR involve the South China Sea and the Indian Ocean Region. (Arugay, 2017) The announcement of the Silk Road initiative in 2013 by President Xi, part of the plan proposed the build out of the maritime infrastructure, similar to the lines of the ancient Silk Road. The initial pitch of MSR was to help build a community that represents the common concerns, interests and expectations of all trade partners. This community is expected to guide and support a peaceful and stable Asia Pacific landscape. Bringing together the Silk Road Economic Belt, the Bangladesh-China-India-Myanmar Economic Corridor and the China-Pakistan Economic Corridor will Europe and Asia. (Arugay, 2017) Overall, enhance China's ability to develop economically while limiting external risks, enhance cooperation in non-traditional security areas while maintaining maritime security. The focus would be upgrading the China-ASEAN Free Trade Area and extending it to the coastal regions of the Indian Ocean, the Persian Gulf, the Red Sea and the Gulf of Aden.

Currently the intent of MSR is the same, however the propaganda surrounding MSR is different. Four years later, after facing implications of debt sustainability and tensions with countries over port expansions, China has changed the narrative of the MSR. Refocusing the MRR public relations on world emotions, this part of BRI has gained the support of a global audience. In the same Silk Road spirit, China advocates, "peace and cooperation, openness and inclusiveness, mutual learning and mutual benefit". The core of the MSR is to exert effort to implement the United Nations 2030 Agenda for Sustainable Development in the field of coasts and oceans. (Arugay, 2017) As published in Vision for Maritime Cooperation under the Belt and Road Initiative, China will "embark on a path of green development, ocean-based prosperity, maritime security, innovative growth and collaborative governance." (Arugay, 2017) MSR is a project that will encompass the following goals:

- Address marine pollution, marine litter and ocean acidification, and in red tide monitoring and pollution emergency responses.
- Sponsor/develop, encourage, projects for recycling and low carbon development in maritime sectors.
- China will support smaller states in adapting to climate change, and assistance in response to various sea related issues including; marine disasters, sea level rise, coastal erosion and marine ecosystem deterioration.
- MSR will create the "21st Century Maritime Silk Road Blue Carbon Program". The program will monitor coastal and ocean blue carbon ecosystems, develop technical standards and promote research on carbon sinks.
- China will partner in reviewing navigational routes, implementing land-based monitoring stations, research the Arctic climate and environmental changes, and provide forecasting

models.

- Establish the Marine Science and Technology Cooperation Partnership Initiative. The initiative will be tasked with researching the key waters and passages along the BRI, forecast deviations and assess impacts by researching the interactions between different weather events and the ocean, and by conducting scientific investigations of the floor of the Indian Ocean. (Arugay, 2017)

These are a few of the highlights of the MSR proposed path of green development. China's redirection of MSR purpose has gained support on the global stage opening new partners in support and funding. Maritime security is also addressed in Vision for Maritime Cooperation under the Belt and Road Initiative, we will address this under the section Chinese military buildup is occurring to support new Silk Road. (Arugay, 2017)

China is partnering with multiple countries and their organizations; Blue Partnership for the Oceans, Asia Pacific Economic Cooperation (APEC), Intergovernmental Oceanographic Commission of UNESCO (IOC/UNESCO), the Partnership in Environment Management of Seas of East Asia (PEMSEA), the Indian Ocean Rim Association, and the International Ocean Institute, to build political trust and contribute to the global ocean cooperative framework. (Arugay, 2017) From this approach, they have achieved progress in developing MSR with Malaysia, Pakistan, Myanmar, Iran, Cambodia, Egypt and Greece. China continues construction of the outposts in the Spratly Islands while controlling disputed areas, holding a consistent coast guard presence in the Senkakus.

Chinese Military Build Up to Support the New Silk Road

China has maintained 6 to 7 % growth in military spending each year for several years. The Chinese People's Liberation Army (PLA) continues to achieve impressive progress in sea, air, cyber, and space domains. China has an established a frontline of J-20 low observable combat aircrafts. China has expanded their inventory of PL-15 extended range air-to-air missiles to be equipped with active electronically scanned array (AESA) radars. In April 2019, in celebration of the 70th anniversary of China's People's Liberation Army Navy (PLAN), China along with thirteen other countries participated in a naval parade. China showcased their 094 Jin Class nuclear-powered ballistic missile submarine along with one of the eight Type-055 guided-missile destroyers they have commissioned. The Type-055 can carry up to two Z-18 anti-submarine warfare helicopters. They are multi-functional ships that can conduct long-range air defense, anti-surface warfare, anti-air warfare, and anti-submarine warfare missions.

Figure 21-3: Type 55 Collection

Source: (Esennagel, 2018)

It is speculated fifteen of the port projects under the BRI that reach from Indo-Pacific region would give China maritime power. On August 1, 2017, BRI allowed for China to launch their first overseas military base in Djibouti, off the Horn of Africa. Sending a clear signal from Beijing's intention to expand its military power beyond the Asia Pacific. It is believed the Hambantota port in Sri Lanka and the deep seaport in Pakistan will be bases for China's Navy.

Recently, as part of the China- Pakistan Economic Corridor BRI, Pakistan signed an agreement with China for the creation and production of fighter jets, navigation systems, and other military hardware in factories housed in Pakistan.

Digital Silk Road

As part of MSR, China outlines their plans to strengthen maritime security. This topic is particularly attractive to countries whose economy has suffered at the hands of pirates hijacking their waters and fishing industry. China is introducing the deployment of the BeiDou-2 Chinese global navigation system and remote sensing satellite system to provide satellite positioning and information services. As a result of BRI, development of BeiDou-2, will consist of 35 satellites and scheduled to be fully implemented by 2020. This system, which will rival the U.S. Global Positioning System, has already been adopted by Pakistan, Laos, and Thailand. As part of BRI the BeiDou-2 system will make the exchange of information throughout Asia faster and more convenient. China will make the data from the satellites available to Kyrgyzstan, Tajikistan, Turkmenistan, and Uzbekistan along with additional countries of interest. China highlights the adaptation of BeiDou-2, will benefit climate change, mitigation of disaster risk, managing water supplies, making agriculture secure, protecting cultural heritage, encouraging sustainable development in urban areas, managing marine areas, and understanding climate change in the mountains and in the Arctic. (Hao, 2019)

China has proven their ability to deliver assistance in the waterways. With the completion of the Asia-Pacific Gateway (APG) submarine optical fiber, provided great improvement in connectivity of submarine communication. The communications cable system can deliver 54 terabits per second between Mainland China, Hong Kong, Japan, South Korea, Malaysia, Taiwan, Thailand, Vietnam and Singapore.

Figure 21-4: Asia Pacific Gateway

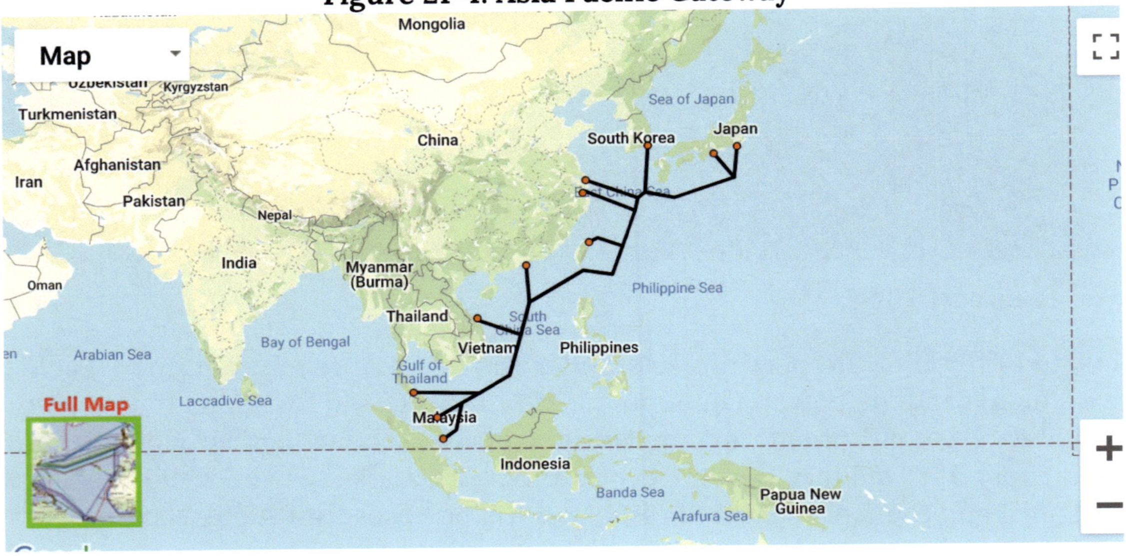

Source: (Fiber Atlantic, 2019)

Drones are a critical part of China's New Silk Road

China leads the manufacturing commercial and recreational drones. China has solved the industry issue of the flight endurance with the creation of hydrogen fuel cell. In addition, China leads in solving for remote areas, law enforcement, and security. The continued innovation has attracted the following countries to visit MMCUAV, a Chinese company, the market dominating drone manufacture: Armenia, Syria, Sudan, Sri Lanka, Maldives, Lithuania, the Philippines, Singapore, Russia, and Slovakia. (Fonua, 2019)

Figure 21-5: EFY Technology Drones

Efy technology drones on display at the Tianjin-Zhongguancun Science Park in Tianjin Binhai New Area on May 20, 2019.

Source: (Fonua, 2019)

Focus grows in Tianjin city located in the Binhai new Area, located in Northern China. The emerging technology hub provides an incubator of scientists and boasts an ideal maritime location along the Yellow Sea. The city has become the main location for integrated circuits, telecommunications, and the National Industrial Cloud Innovation Demonstration Project. (Tianjin Municipal Government, 2019)

In 2018, Tianjin hosted the World Economic Forum (WEF) Annual Meeting of the New Champions. Dubbed the "Summer Davios", highlighted the city on the global stage, discussing the area's advanced research centers and start-ups. The WEF meeting welcomed industry leaders to collaborate and engage with the Tianjin technology leadership. Comparing China's Bay Area (Tianjin and 10 other cities in China) to Silicon Valley, focused on investment and deepen international collaboration. WEF has established a Global Centre in China, U.S., Japan, and India to encourage the development and deployment of frameworks for several industries including drones. It should be noted, at the WEF meeting, China announced they issued its first drone delivery permit this year and flying-taxi programs are scheduled to launch in the United States and the Middle East by 2020.

Belt and Road Media tour of evolving Tianjin city took place May 20, 2019. The tour of 27 international journalists from Europe, Asia, and Pacific were escorted by Binhai New Area Communist Party (Publicity and Cyberspace Offices) and Chinese journalists from the *People's Daily* newspaper. The new city, still being constructed, of three million has notoriety as the first Free Trade Pilot Zone (Northern China), Maritime gateway for Beijing, second National Comprehensive Reform Pilot Area, and the country's National Innovation Demonstration Zone. Binhai New Area is targeted to be a smart city by 2021, and global modern city, with innovative strengths and coastal power by 2049. To date, 7,900 science and technology enterprises have made their home in Tianjin city grossing 700 billion yuan in 2018. The technology hub has advanced China's Kylin Operating System (state sponsored commercially independent operating system) and the Phytium processing units. The city provides an area of growth for EFY Technology. The company was founded in 2015 and is currently in Series A of funding, at $15.8 Million. (Crunchbase, 2018) EFY Technology researches, develops, tests, unmanned or automated flight control systems. Days before the media tour (May 16-19, 2019) Tianjin city hosted the World Intelligent Congress. There EFY Technology kicked off Tianjin Intelligent Night with 500 plus UASs in a spectacular sortie to create art and words in the sky, lighting up the night. UAS performance team has completed thousands of similar sorties events in more than 10 provinces and cities such as Tianjin, Shandong, Hunan and Jiangxi. The next step, working with China Mobile, to integrate 5G technology to bring security, stability, and ensure the unmanned aerial devices are interference-prone. (CO, 2019)

In Plain Sight: China Drones Manufacturers

The BRI has brought the opportunity in physical land and sea, but as we uncover advancement in scientific intelligence. While BRI is putting on displays of UAS lighting up the night, behind the scenes this new wave of UAS are more powerful than its predecessors. The global commercial drone market is dominated by Chinese manufacturers. The commercial drones are not only used for hobby but by law enforcement, government agencies (not only U.S.), businesses, and first responders. The commercial UAS can compromise personal data, share information, detail images of critical infrastructure, national security activities and sites. The drone data is uploaded to the cloud, with the user having complete control over the data. However, some argue the data is also available to the manufacture. Recently it was stated 80% of all drones in U.S. and Canada are from the Chinese manufacture DJI. (Shortell, 2019) DJI plans to begin assembling the drones in California, to be reviewed under the U.S. Trade Agreements Act to resume a partnership with U.S. Customs and Border Protection. The reality of banning Chinese drones from the U.S. is not realistic. Currently all fifty states use DJI drones in some official capacity (farming, roads, bridge inspection, etc.). United Kingdom police forces have been using DJI drones since 2017, despite the warning by the U.S. government regarding the leak of sensitive information.

On a larger scale, China's military drone manufacturing continues to grow in demand. The selective drone export policy of the U.S. has provided the opportunity for Chinese to supply drones to countries not authorized to purchase from the U.S. (also at lower cost). China requires their military drone customers to be state actors. They give priority to countries using the technology for counterterrorism. China views all countries in the Middle East as customers, they have not "chosen a side". The Chinese military drones operate by connecting to China's satellites, presenting a challenge for countries that are U.S. aligned.

Figure 21-6: Map of Countries in the Middle East with Armed Drones and their Manufacturing Origin

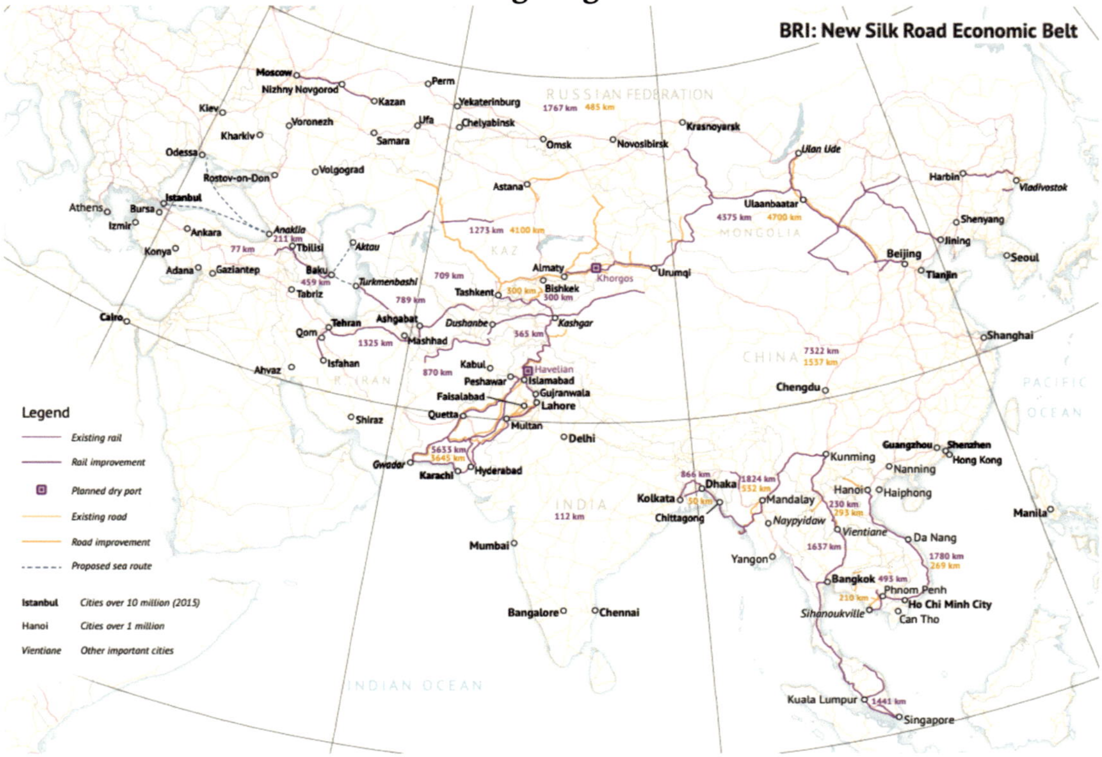

Source: Reed and Trubetskoy (2019)

Source: (Tabrizi & Justin , 2018)

Several countries across the Middle East have acquired armed drones either by purchasing them from China (Jordan, Iraq, Saudi Arabia and the UAE) or by building them domestically (Israel, Iran and Turkey). (Tabrizi & Justin , 2018) Outside of not holding the Middle East countries to Missile Technology Control Regime (MTCR), China also sells armed drones cheaper and provides training per the buyers need. Their U.S. counterparts have better performance and are more advanced but not available to the open market. China's hold in this area could hinder U.S. in the areas of warfare. China has eight counter drone products available to the world market. With acting agents, for example Emily Liu[1], counter-drone capability shopping has become common place. Emily Liu's was caught purchasing on the behalf of Iran shopping for U.S. electronic com-

1. China-based procurement agent Emily Liu and four associated entities pursuant to E.O. 13382 for proliferation activities related to a key supporter of Iran's military. Emily Liu has provided, or attempted to provide, financial, material, technological, or other support for, or goods or services in support of, Iran's Shiraz Electronics Industries (SEI).

ponents critical to aviation. June 2019, Iran shot down a U.S. RQ-4 Global Hawk, using Western technology. Emily Liu is one agent among many, arms continue to build in the Middle East with China driving the market.

U.S. involvement in the New Silk Road

The U.S. dismisses BRI and does not participate in any Belt and Road conferences. U.S. views the project as a power play to control poor countries by offering improvement at an unrealistic cost. BRI has momentum and continues to grow, opening trade routes, expanding infrastructure, and changing the shape of global economy. The U.S. is at risk of weakened economy with lower exports to China. At the moment China's BRI stronghold appears to be scattered and disorganized, however this will change. Once the BRI countries begin to align with China's government the U.S. will have to rely on their allies in the region, that participate in BRI, to help steer the initiative. This is not a solid plan.

Digital Belt and Road

The Digital Belt and Road developed from the advances in infrastructure from BRI. The U.S> has openly criticized the digital branch as a platform for surveillance, through facial recognition and other information gathering tactics. The U.S. fails to provide any alternatives, proposing a global standard. Therefore, China continues to develop and implement. No doubt China's greatest strength in the promotion of science is its willingness to abide by international treaties even as the United States is actively abandoning them. (Pastreich, 2019) At the forefront of the Digital Belt and Road, 5G wireless technology, will be launched by China. Huawei Technologies Cos Ltd, a Chinese company, holds the greatest amount of 5G patents. The U.S. has concerns Huawei along with ZTE have spying capabilities that threaten the West.

Conclusions

China's BRI continues to move sound and steady across the East, spreading the environmental messages of green and blue. BRI on the surface brings hope, security, and boosted economy to smaller countries in need. The open trade routes across land and sea bring advantages to state actors and smaller settlements. BRI installation and upgrade in technology and infrastructure are attractive to all parties. BRI has given China the global spotlight to feature their advancements to sell in the global market space. For years, before the kickoff of BRI, China has seen the gain across land, sea, and economy as the direct plan to outside the U.S. as a superpower. Slowly and patiently China has enticed with the ability to produce and sell at a lower cost, allowing some sectors to become dependent on Chinese factories and products. The U.S. is feeling the impact of their own money focused commercial enterprises. As a result, China's telecom and commercial drones dominate the global marketplace threatens U.S. interest in the globe and at home. Additional limitations U.S. self-implemented have resulted in a possible disadvantage among our enemies and allies in the Middle East. The U.S. can no longer ignore BRI and will need to start to take steps to maintain their place on the world stage.

Discussion Questions

1. China is investing a huge amount of money in the BRI and the use of drone technology. What strategic and tactical goals do they realistically expect to accomplish?
2. What will slow China's progress on the BRI? What is the best way for U.S. and Allied forces to apply counterpressure and counter ISR?

Bibliography

115th-congress. (2018). *115th Congress. (2018). HR 302 FAA Reauthorization Act. Sec 364 and 365.* Retrieved from www.congress.gov/bill/115th-congress/house-bill/302/text : https://www.congress.gov/bill/115th-congress/house-bill/302/text

49 U.S. Code §40103, 49 U.S. Code §40103 Sovereignty and use of airspace (U.S. Code July 5, 1994).

Abramson, E. (2016). *Ethical Dilemmas in the Age of AI.* Retrieved from Abramson, E. – knowmail.me/blog: https://www.knowmail.me/blog/ethical-dilemmas-age-ai/

Adamy, D. -0. (2015). EW 104 EW *against a New Generation of Threats.* Boston: Artech House.

Adamy, D. (2001). EW 101 A *First Course in Electronic Warfare.* Boston, MA: Artech House.

Adamy, D. (2001). EW 101 A *First Course in Electronic Warfare.* Boston: Artech House.

Adamy, D. (2004). EW 102 A *Second Course in Electronic Warfare.* Boston: Artech House.

Adamy, D. (2009). EW 103 *Tactical Battlefield Communications Electronic Warfare.* Boston: Artech House.

Adamy, D.-9. (1998, Jan). Lesson 4: the basic link for all EW functions. (electronic warfare)(EW Reference & Source Guide). *Journal of Electronic Defense, Jan 1998 Issue.*

Administrator. (2015, June 15). *Standing Wave and Fundamental Frequency.* Retrieved from Electronics Hub: https://www.electronicshub.org/?s=fundamental frequency

Administrator. (2019, May 17). *Harmonic Frequencies.* Retrieved from electronicshub.org: https://www.electronicshub.org/harmonic-frequencies/

AIA & Avascent Report. (2018, April 23). *Think Bigger: Large Unmanned Systems and the Next Major Shift in Aviation.* Retrieved from www.avascent.com: https://www.avascent.com/2018/02/think-bigger-large-unmanned-systems-and-the-next-major-shift-in-aviation/

Alejandro Aragon-Zavala, J. L.-R.-P. (2008). *High-Altitude Platforms for Wireless Communications.* Chichester, West Sussex, UK: John Wiley & Sons.

Alford, L. (2000). Cyber Warfare: Protecting Military Systems. *Acquisition Review Quarterly.*

Anon. (2019). *Saudi Arabia grants citizenship to robot Sophia.* Retrieved from dw: Saudi Arabia grants cihttps://www.dw.com/en/saudi-arabia-grants-citizenship-to-robot-sophia/a-41150856

Arugay, A. A. (2017, October 23). *Regional Perspectives on China's Belt and Road Initiative: Challenges and Opportunities for the Asia-Pacific.* doi:Aries A. Arugay, editor, (23 October 2017) Regional Perspectives on China's Belt and Road Inihttps://doi.org/10.1111/aspp.12346

Asimov, I. (1950). "*Runaround*". I, *Robot (The Isaac Asimov Collection ed.).* New York City: Doubleday.

Atherton, K. D. (2019). Can the Pentagon sell Silicon Valley on AI as ethical war? . C4ISRNET.

Austin, R. (2010). "*Design for Stealth*", *Unmanned Aircraft Systems UAVS Design Development and Deployment.* New York: John Wiley and Sons.

Belt and Road News. (2019, January 28). *Chinese Metropolis Chongqing sees booming Rail Trade with Europe.* Retrieved from Belt and Road News: https://www.beltandroad.news/2019/01/28/chinese-metropolis-chongqing-sees-booming-rail-trade-with-europe/

Brown, E. F. (Dec 2008). Airborne Communication Networks for Small Unmanned Aircraft Systems. *Proc.* IEEE, *vol* 96, *no* 12, pp. 2008-17.

Burch, D. (2015). RADAR *for Mariners.* New York: McGraw-Hill.

Cameron, J. &. (Director). (1991). *Terminator* 2: *Judgement Day* [Motion Picture].

Chapman, A. (2019, May 31). *GPS Spoofing.* Retrieved from Tufts University – Tech Notes 2017: https://sites.tufts.edu/eeseniordesignhandbook/files/2017/05/Red_Chapman.pdf

CO, B. &. (Director). (2019). WIC 2018 UAS *Industry Development Forum* [Motion Picture].

Cornell University Legal Information Institute. (2019, June 5). *But-for test.* Retrieved from law.cornell.edu: https://www.law.cornell.edu/wex/but-for_test

Cornell University Legal Information Institute. (2019, June 5). *Intervening Cause.* Retrieved from law.cornell.edu: https://www.law.cornell.edu/wex/intervening_cause

Cornell University Legal Information Institute. (2019, June 5). *Personal Jurisdiction.* Retrieved from law.cornell.edu: https://www.law.cornell.edu/wex/personal_jurisdiction

Crunchbase. (2018, April 9). Retrieved from EFY-Tech: https://www.crunchbase.com/organization/efy-tech#section-overview

D, G. a. (2010). *Broadband Communications via High Altitude Platforms.* New York City, NY: John Wiley & Sons.

Daniel-Cornel TĂNĂ, S. (2018). The Impact of the Development of Maritime Autonomous Systems on the Ethics of Naval Conflicts. *Annals: Series on Military Sciences*(2), 118-130.

Dedrone. (2018). *Correctional Facilities.* Retrieved from www.dedrone.com/solutions: https://www.dedrone.com/solutions/correctional-facilities

Deloitte Center for Government Insights analysis. (2018, June 18). *The future of regulation. Principles for regulating emerging technologies.* Retrieved from Deloitte Insights: https://www2.deloitte.com/insights/us/en/industry/public-sector/future-of-regulation/regulating-emerging-technology.html

Dewey, D. (2017, July 14). *Drone crashes into LDS temple in Utah; raises questions of airspace rules.* Retrieved from eastidahonews.com: https://www.eastidahonews.com/2017/07/drone-crashes-lds-temple-utah-raises-questions-airspace-rules/

Diversity of citizenship; amount in controversy; costs, 28 U.S. Code §1332 (United States Congress June 25, 1948).

DJI. (2019, June 5). *DJI Enterprise.* Retrieved from Enterprise DJI.com: https://enterprise.dji.com/civil-protection

DLSR Pros. (2019, June 3). *Best Drones (UAVs) for Firefighting in 2019.* Retrieved from dlsrpros.com: https://www.dslrpros.com/dslrpros-blog/best-drones-firefighting-2019/

DoD. (2018). *Dictionary of Military Terms.* Retrieved from JCS.Mil: http://www.jcs.mil/doctrine/dod_dictionary/

DoD. (2018). *Joint Publication (JP) 3-01 Countering Air and Missile Threats.* Washington, DC: DoD.

DoD-02. (2018). *Information Operations (IO) in the United States.* Retrieved from JP 3-13 : http://www.dtic.mil/doctrine/new_pubs/jp3_13.pdf

DoD-03. (2015). *Unmanned Systems Roadmap 2013 to 2038*. Retrieved from DTIC: http://www.dtic.mil/dtic/tr/fulltext/u2/a592015.pdf

Drones, Q. S. (2017, July 11). *quadcopters-have-hit-the-sound-barrier/*. Retrieved from quadstardrones.com: https://quadstardrones.com/2017/07/11/quadcopters-have-hit-the-sound-barrier/

EARSC. (2015). A Taxonomy for the EO Services Market: enhancing perception and performance of the EO service industry. *EARSC Issue 2*.

Embry Riddle Aeronautical University. (2018, June 16). *ERAU Common Documents*. Retrieved from ERAU: www.commons.erau.edu

Entokey, a. G. (2019, May 16). *entokey.com/acoustics-and-sound-measurement/*. Retrieved from entokey.com/acoustics-and-sound-measurement/: https://entokey.com/acoustics-and-sound-measurement/

ESA-ESTEC Contract 162372/02/NL/US. (September 2005). STRATOS: *Stratospheric Platforms a definition study for ESA Platform, Final Report*, 1-34. ESA-ESTEC .

Esennagel. (2018, March 22). *World of Warships*. Retrieved from Modern Warships: The Type 055 Collection: https://forum.worldofwarships.com/topic/154411-the-type-055-collection/

European Union. (2019, May 2019). *About the regulation and data protection*. Retrieved from ec.europa.eu: https://ec.europa.eu/commission/priorities/justice-and-fundamental-rights/data-protection/2018-reform-eu-data-protection-rules_en

Federal Question, 28 U.S. Code §1331 (United States Congress June 25, 1948).

FEMA. (2013). *Lessons Learned from the Boston Marathon Bombings: Preparing for and Responding to the Attack*. Retrieved from www.fema.gov: http://www.fema.gov/media-library-data/20130726-1923-25045-1176/lessons_learned_from_t

Fiber Atlantic. (2019, November). Retrieved from Asia Pacific Gateway: http://www.fiberatlantic.com/system/v2ZAm

Filippo Santoni de, S. &. (2018). Meaningful Human Control over Autonomous Systems: A Philosophical Account. *Frontiers in Robotics and AI*. doi:10.3389/frobt.2018.00015

Fleetwood, J. (2017). Public Health, Ethics, and Autonomous Vehicles. *American Journal of Public Health*, 107(4), 632-537.

Fonua, T. (2019, May 30). *Chinese Drone Developer Targets Belt and Road Initiative Countries*.

Retrieved from The Sun: https://fijisun.com.fj/2019/05/30/chinese-drone-developer-targets-belt-and-road-intiative-countries/

Fortuna, C. (2017, 12 02). *Autonomous Driving Levels 0–5 Implications.* Retrieved from cleantechnica.com: https://cleantechnica.com/2017/12/02/autonomous-driving-levels-0-5-implications/

Gelfand. (2004). *"Physical Concepts", Hearing an Introduction to Psychological and Physiological Acoustics, 4th ed.* New York City.

Gelfand, S. A. (2009). *Essentials of Audiology, 3rd Edition.* Stuttgart, DE: Thieme.

Giordano, N. (2009). *College Physics: Reasoning and Relationships.* New York City, NY: Cengage Learning. pp. 421–424.

Goddemeir, K. D. (June 2015). Role-based Connectivity Management with Realist Air to Ground Channels for Future Applications. *IEEE Vehic. Tech. Mag. Vol 10, no 2,* pp. 79-85.

Gomez, M. &. (2017). *Man suspected of flying drone over 49ers, Raiders games arrested.* Retrieved from securityinfowatch.com/news/12383982: www.securityinfowatch.com/news/12383982/man-suspected-of-flying-drone-over-49ers-raiders-games-arrested

Grandview. (2018). *US Anti-Drone Market Size, by destructive Mitigation Size.* Retrieved from Grandviewresearch.com: www.Grandviewresearch.com

Guardbaum, S. (1994). The Nature of Preemption. *Cornell Law Review*, 767, 771.

Hao, C. J. (2019, April 30). *China's Digital Silk Road: A Game Changer for Asian Economies.* Retrieved from The Diplomat: https://thediplomat.com/2019/04/chinas-digital-silk-road-a-game-changer-for-asian-economies/

Harris Aerial. (2019, June 5). *Carrier HX8 Sprayer Drone.* Retrieved from harrisaerial.com: https://www.harrisaerial.com/carrier-hx8-sprayer/

Heinman, C. (2019). *Hearing Loss Tests Patrient D v-105.* Carlisle, PA: Brown Optical Hearing Aid Service.

Henderson, T. (2017). The Doppler Effect – Lesson 3, Waves. *Physics tutorial. The Physics Classroom.* Retrieved from Henderson, Tom (2017). "The Doppler Effect – Lesson 3, Waves". Physics tutorial. The Physics Classroom. Retrieved September 4, 2017.: Henderson, Tom (2017). "The Doppler Effect – Lesson 3, Waves". Physics tutorial. The Physics Classroom. Retrieved September 4, 2017.

Hern, A. (2017, 1 12). *Give robots 'personhood' status, EU committee argues.* Retrieved from The

Guardian: www.theguardian.com/technology/2017/jan/12/give-robots-personhood-status-eu-committee-argues

Hubbard, R. K. (1998). *Boater's Bowditch.* Camden, MA: International Marine.

Ibrahim, A. (2019). *Optimization Methods for User Admissions and Radio Resource Allocation for Multicasting over High Altitude Platforms* . Memorial Univeresity of Newfoundland , Canada: River Publications.

IEEE . (2017). The IEEE Global Initiative on Ethics of Autonomous and Intelligent Systems Announces New Standards Projects. *Telecom Standards*, 27(11), pp. 4-5. .

Jesus Gonzalo, D. L. (2018, March 15). On the Capabilities and Limitations of High Altitude Pseudo-Satellities. *Progress In Aerospace Sciences*, 37-56. doi:https://doi.org/10.1016/j.paerosci.2018.03.006

Johnson, O. &. (2012). *Ethics: Selections from Classic and Contemporary Writers.* Boston, MA: Cengage Learning.

Jones, T. (2017). *International Commercial Drone Regulation and Drone Delivery Services.* Santa Monica: The Rand Corporation.

Kanowitz, S. (2019, 05 15). *Toward the deployment of ethical* AI. Retrieved from Government Computer News. : Kanowitz, S. (2019). Toward the dephttps://gcn.com/articles/2019/05/15/ethical-ai-idc.aspx?s=gcntech_200519

Kirk, J. (2015, August 5). *sounds-can-knock-drones-sky.* Retrieved from www.computerworld.com.au/article/581231: https://www.computerworld.com.au/article/581231/sounds-can-knock-drones-sky/

Knight, W. (2018). Nine charts that really bring home just how fast AI is growing. *MIT Technology Review* .

Laris, M. (2018, May 10). *Stadium and Team Owners See Drones as Major Threat.* Retrieved from washingtonpost.com/local/trafficandcommuting;nationalinclude: https://www.washingtonpost.com/local/trafficandcommuting;nationalinclude;/stadium-and-team-owners-see-drones-as-major-league-threat/2018/05/10/83e0b954-50ad-11e8-84a0-458a1aa9ac0a_story.html?noredirect=on&utm_term=.e6aebf20ac9a

Lebrand, M., & Lall, S. V. (2019). *Who Wins, Who Loses? Understanding the Spatially Differentiated Effects of the Belt and Road Initiative.* Washington D.C.: World Bank Group.

Legal Information Institute – Cornell University. (2019, May 31). *Strict Liability* . Retrieved from Legal Information Institute: https://www.law.cornell.edu/wex/strict_liability

LRAD. (2019, May 189). *LRAD 450XL Datasheet*. Retrieved from LRADX: http://www.lradx.com/wp-context/uploads/2015/05/LRAD_datasheet_450XL.pdf

MacGregor, D. S. (2018). *Colorado Causes of Action: Elements, Defenses, Remedies, and Forms*. Denver: Bradford Publishing Co. .

Macnamara, T. M. (2010). *Introduction to Antenna Placement & Installation*. New York City, NY : John Wiley & Sons.

Mahon, J. (2012). *Classical Natural Law Theory St. Thomas Aquinas (1227-1274) – the "Angelic Doctor" Lecture*. Retrieved from Mahon, J. (2012). Classical Natural Law Theory St. Thomas Aquinas (1227-1274) – thePhilosophy of Law. : Mahon, J. (2012). Classical Natural Law Theory St. Thomas Aquinas (1227-http://home.wlu.edu/~mahonj/PhilLawLecture1NatLaw.htm

Marbury v. Madison, 5 U.S. 137 (United States Supreme Court February 23, 1803).

Matolak, R. S. (April 2015). Initial Results for Airframe Shadowing in L-band and C-band Air -Ground Channels. *Proc. Integrated Commun,, Navigation, and Surveillance Conf*, (pp. pp. 1-8).

McCullogh v. Maryland, 17 U.S. 316 (United States Supreme Court March 6, 1819).

Merriam-Webster. (2019, May 17). Merriam-Webster Online Dictionary.

Merriam-Webster, Inc. (2019). *Definition of Ethics*. online: Merriam-Webster, Inc. Retrieved from Definition of Ethics. (2019a). Online: Merriam-Webster, Incorporated.: Definition of Ethics. (2019a). Online: Merriam-Webster, Incorporated.

Middleton, C. (2018). *SAP launches ethical A.I. guidelines, expert advisory panel*. Retrieved from internetofbusiness.com: Middleton, C. (2018). SAP launches ethical A.I. guidelines, expert advisory panel. Retrieved from https://internetofbusiness.com/sap-publishes-ethical-guidelines-for-a-i-forms-expert-advisory-panel/

Miller, P. C. (2018, March 27). *FAA Forecasts Phenomenal Growth for UAS*. Retrieved from www.uasmagazine.com/articles/1833/: http://www.uasmagazine.com/articles/1833/faa-forecasts-phenomenal-growth-for-uas-industry

Misselhorn, C. (2018). Artificial Morality. Concepts, Issues and Challenges. *Society*, 55(2), 161-169.

Mohorcic, D. G. (2010). *Broadband Communications via High Altitude Platforms*. New York City, NY: John Wiley & Sons.

Monahan, K. (2004). *The Radar Book: Effective Navigation and Collision Avoidance.* Anacortes, WA: Fineedge Publications.

Muspratt, A. (2018, November 22). *New global drone standards proposed.* Retrieved from Defence iQ: https://www.defenceiq.com/defence-technology/news/new-global-drone-standards-proposed

National Conference of State Legislatures. (2018, September 10). *Current Unmanned Aircraft State Law Landscape.* Retrieved from NCSL.org: http://www.ncsl.org/research/transportation/current-unmanned-aircraft-state-law-landscape.aspx

NBC Today Show. (2018, May 9). *How peeping drones could be spying on you without you knowing it.* Retrieved from Today.com: https://www.today.com/video/how-peeping-drones-could-be-spying-on-you-without-you-knowing-it-1229001795967

Newman, L. H. (2017, August 7). THE ARMY GROUNDS ITS DJI DRONES OVER SECURITY CONCERNS. Retrieved from WIRED: https://www.wired.com/story/army-dji-drone-ban/

Nichols, R. K. (1996). *Classical Cryptography Course Volume I / II.* Laguna Hills, CA: Aegean Park Press.

Nichols, R. K. (1996). *Classical Cryptography Course*, Volume I. Laguna Hills, CA: Aegean Park Press.

Nichols, R. K. (2018). *Unmanned Aircraft Systems (UAS) In the Cyber Domain: Protecting USA's Advanced Air Assets. 1st Ed.* Manhattan, KS: New Prairie Press.

Nichols, R. K. (2019, March 14). Hardening US Unmanned Systems Against Enemy Counter Measures. *7th Annual Unmanned Systems Summit*. Alaxandria, VA, USA: PPTX presentation , self.

Nichols, R. K., Mumm, H. C., Lonstein, W. D., Ryan, J. J., & and Carter, C. (2018). *Unmanned Aircraft Systems (UAS) in the Cyber Domain: Protecting USA's Advanced Air Assets.* Manhattan, KS: New Prairie Press (NPP) eBooks. 21.

NIST. (September 2012). *Guide for Conducting.* Washington, DC: GPO.

North Carolina Department of Transportation. (2019, May 30). *Law & Regulations.* Retrieved from NCDOT.GOV: https://www.ncdot.gov/divisions/aviation/uas/Pages/laws-regulations.aspx

O'kane, S. (2019, March 18). *Drones are Already Being Confiscated Near the Super bowl.* Retrieved from The Verge.

Osseiran, A. (Dec 2014). Scenarios for 5G Mobile and Wireless communications: the vision of the METIS Project. *IEEE Communications Magazine*, Vol 52, *no* 5, pp. 26-35.

O'Sullivan, J. L. (1845). The Great Nation of Futurity. *United States Magazine and Democratic Review* Vol 6 Issue 23, pp. 426-430.

Pastreich, E. (2019, June 24). *Foreign Policy In Focus*. Retrieved from China's Belt and Road of Science: https://fpif.org/chinas-belt-and-road-of-science/

Perez-Pena, R. (2018, December 27). *Gatwick Airport Drone: Lots of Guessing, but Not Many Answers*. Retrieved from NY Times: https://www.nytimes.com/2018/12/27/world/europe/gatwick-airport-drone.html

Pierson. (2019, May 16). *tuning-fork-waves-sound*. Retrieved from airfreshener.club – Pierson Education: https://airfreshener.club/quotes/tuning-fork-waves-sound.html

Porter, J. D. (2019, June 8). *jdporterlaw.com/intellectual-property-law/*. Retrieved from jdporterlaw.com: http://www.jdporterlaw.com/intellectual-property-law/

Possel, M. (2017). Waves, motion and frequency: the Doppler effect. *Einstein Online*, Vol. 5. *Max Planck Institute for Gravitational Physics, Potsdam, Germany*.

Pricewaterhousecoopers, LLP. (2018). *Skies without limits – Drones- taking the UK's economy to new heights*. London: Pricewaterhousecoopers, LLP.

PROTECTION OF CERTAIN FACILITIES AND ASSETS FROM UNMANNED AIRCRAFT, H. R. 302 (United States Congress January 3, 2018).

Proyas, A. (Director). (2004). I, *Robot. In. Hollywood*, CA. [Motion Picture].

Ramzy, A. &. (2008). *Tainted-Baby-Milk Scandal in China*. Retrieved from content.time.com/time/world/article/: http://content.time.com/time/world/article/0,8599,1841535,00.html

Randall K. Nichols and Lekkas, P. C. (2002). *Wireless Security: Threats, Models, Solutions*. New York City, NY: McGraw Hill.

Randall K. Nichols, D. (2018). Chapter 20 Acoustic CM & IFF Libraries V SWARMS Rev 1 05142019. In R. K. Nichols, H. C. Mumm, W. D. Lonstein, & J. S. Hood, *Unmanned Aircraft Systems (UAS) in the Cyber Domain: Protecting USA's Advanced Air Assets, 2nd ed.* Manhattan, KS: NPP Press.

Randall K. Nichols, D. (2019 for publication). Unmanned Aircraft Systems in the Cyber Domain: Protecting USA's Advanced Air Assets, 2nd ed. In H. M. Randall K. Nichols, *Chapter 18 Audiology, Acoustic Countermeasures against Swarms and Building IFF Libraries* (p. 2nd ed.). Manhattan, KS: For Publication, NPP.

Randall K. Nichols, J. J. (2018). *Unmanned Aircraft Systems (UAS) in the Cyber Domain: Protecting USA's Advanced Air Assets*. Manhattan, KS: New Prairie Press.

Rappaport, T. (2014). *Millimeter Wave Wireless Communications.* New York City, NY: Prentice Hall.

Rees, M. (2019, April 9). *New Counter-UAS System Utilizes AI and Machine Learning.* Retrieved from Unmanned Systems News.

Ricker, D. (2017, July 1). *Navigating drone laws has become a growing and lucrative legal niche.* Retrieved from ABA Journal: http://www.abajournal.com/magazine/article/drone_law_attorneys

Rupprecht, J. (2017). 7 *big problems with counter-drone technology (drone jammers, anti-drone guns,* . Retrieved from jrupprechtlaw.com: https://jrupprechtlaw.com/drone-jammer-gun-defender-legal-problems

Said Emre Alper, Y. T. (December 2008). Compact Angular Rate Sensor System Using a Fully Decoupled Silicon-on-Glass MEMS Gyroscope. JOURNAL OF MICROELECTROMECHANICAL SYSTEMS, VOL. 17, NO. 6.

Sanchez, M. (2019, June 4). *No Drones.* Retrieved from Unspalsh.com: https://unsplash.com/photos/oMqswmrie4Y

Schroeder, A. (2018, February 1). *Localizing Humanitarian Drones: Robotics & Disaster Response from the Maldives to Malawi.* Retrieved from medium.com: https://medium.com/radiant-earth-insights/localizing-humanitarian-drones-robotics-disaster-response-from-the-maldives-to-malawi-a1f362432cb1

Shortell, D. (2019, May 20). CNN *Politics.* Retrieved from DHS warns of 'strong concerns' that Chinese-made drones are stealing data: https://www.cnn.com/2019/05/20/politics/dhs-chinese-drone-warning/index.html

Signia. (2019, May 16). *Signia Hearing Aids.* Retrieved from Signia Hearing Aids – Hear across America: www.signiausa.com

Singer v. City of Newton, 284 F. Supp. 3d 125 (U.S. District Court Massachusetts September 21, 2017).

Son. (2015). *Rocking Drones with Intentional Sound Noise.* Retrieved from USINEX Symposium. 24, 881.: https://www.usenix.org/systems/files/confrence/usenixsecurity15/

Sood A.K. & Enbody, R. (2014, December 19). *https://www.georgetownjournalofinternau-s-military-defense-systems-the-anatomy-of-cyber-espionage-by-chinese-hackers.* Retrieved from georgetownjournalofinternationalaffairs.org/online-edition: https://www.georgetownjournalofinternationalaffairs.org/online-edition/u-s-military-defense-systems-the-anatomy-of-cyber-espionage-by-chinese-hackers

Sovereignty and use of airspace, 49 U.S. Code §40103 (United States Congress July 5, 1994).

Staff. (2008). FINAL ACTS WRC-07. *World Radiocommunication Conference.* Geneva: ITU.

Staff. (2012). FINAL ACTS WRC-12. *World Radiocommunication Conference.* Geneva: ITU.

Staff. (2016, April 17). *Equal Loudness Contours.* Retrieved from Gutenberg Organization: http://central.gutenberg.org/article/WHEBN0001046687/Equal-loudness contour

Staff. (2019). FINAL ACTS WRC-15. *World Radiocommunication Conference.* Geneva.

Stone, Z. (2007, 11 7). *Stone, Z. (2017). Everything You Need To Know About Sophia, The World's First Robot Citizen. Retrieved from https://www.forbes.everything-you-need-to-know-about-sophia-the-worlds-first-robot-citizen.* Retrieved from Forbes: https://www.forbes.com/sites/zarastone/2017/11/07/
everything-you-need-to-know-about-sophia-the-worlds-first-robot-citizen/#1667784246fa

Studios, D. D. (2017). Boaters Ref. USA.

sUAS News. (2018, March 2). *RAS Consulting & Investigations hire Jeff Parisse to offer sophisticated UAS security and surveillance services.* Retrieved from suasnews.com: https://www.suasnews.com/2018/03/
ras-consulting-investigations-hire-jeff-parisse-offer-sophisticated-uas-security-surveillance-services/

Sullivan-Nightingale, D. (2015). *Unmanned Aerial Systems: Risks & Opportunities in the Workplace. Professional Safety*, 6(3), 34-42.

Sun, W. M. (June 2015). Unmanned Aircraft Systems:Air-Ground Channel Characterization for future applications. *IEEE Vehic. Tech Mag.* Vol 10, No 2 , pp. 79-85.

T.C. Dozer, D. A. (2008). High Altitude Platforms for VHDR in-theater communications. *IET Seminar on Military Satellite Communications Systems.*

Tabrizi , A. B., & Justin , B. (2018, December). Armed Drones in the Middle East Proliferation and Norms in the Region. London , Westminster, United Kingdom.

The Shepard News Team. (2018, September 12). *Liteye Receives Follow-on Contract for* C-AUDS – *DB – Digital Battlespace.* Retrieved from Aerospace, Defense and Security News and Analysis – Shephard Media, The Shepard Press, Ltd: www.shephardmedia.com/news/digidigital-battlespace/liteye-receives-follow-contract-c-auds

Tianjin Municipal Government. (2019, May 20). *China and Tianjin: open to the future*. Retrieved from Exploring Tianjin: http://www.exploringtianjin.com/2019-04/22/c_247767.htm

Toomay, J. (1982). *RADAR for the Non – Specialist. London; Lifetime Learning Publications.* London: Lifetime Learning Publications.

TRS, S. (2018, July 10). *Tontechnic-Rechner-Sengpielaudio*. Retrieved from Tontechnic-Rechner-Sengpielaudio Calculator: www.sengspielaudio.com/calculator-wavelength.htm

UAV Coach. (2019, May 30). *Drone Laws in South Carolina* (2019). Retrieved from UAVcoach.com: https://uavcoach.com/drone-laws-south-carolina/

United States Constitution Article VI, Sec.2 (United States of America September 17, 1787).

Uni-wuppertal. (2019, May 15). *Inverse Square Law, General*. Retrieved from hydrogen.physik.uni-wuppertal.de/hyperphysics/: http://hydrogen.physik.uni-wuppertal.de/hyperphysics/hyperphysics/hbase/forces/isq.html

Urban, T. (2018). *Teach Your Robots Well: Will Self-Taught Robots Be the End of Us?* Retrieved from www.worldsciencefestival.com: Urban, T. (2018). Teach Your Robots Well: Will Self-Taught Robots Be the End of Us? Retrieved from https://www.worldsciencefestival.com/programs/teach-robots-well-will-self-taught-robots-end-us/

Usenix.org. (2019, 6 9). *MEMS, Drones, & Sound Sourcing*. Retrieved from Usenix.org: www.usenix.org

Wallace, R. e. (2018, April 9). *Exploring Counter-UAS Operations. A Case Study of the 2017 Dominican Republic Festival Presidente*. Retrieved from researchgate.net/publication/325209812: https://www.researchgate.net/publication/325209812_Exploring_Commercial_CounterUAS_Operations_A_Case_Study_of_the_2017_Dominican_Republic_Festival_Presidente

Warwick, G. (2016). *Rapid Defense. Aviation Week & Space Technology*, 178(9), 31.

WebFinance, Inc. (2019). *Definition of Ethics*. (2019b). online: Online: WebFinance, Inc.

Weise, E. (2017, August 23). *could-hackers-behind-u-s-navy-collisions*. Retrieved from USATODAY: https://www.ruidosonews.com/story/tech/news/2017/08/23/could-hackers-behind-u-s-navy-collisions/594107001/

Wong, C. (2017). *Top Canadian researcher says AI robots deserve human rights*. Retrieved from Wong, C. (2017). Top Canadian researcher says AI robots deserve human rights. Retrieveitbusiness.ca: Wong, C. (2017). Top Canadian researcher says AI robots deserve human rhttps://www.itbusiness.ca/news/top-canadian-researcher-says-ai-robots-deserve-human-rights/95730

Wordpress. (2012, 08 29). *The True Sign of Intelligence*. Retrieved from deepthinkings.wordpress.com: http://deepthinkings.wordpress.com/2012/08/29/the-true-sign-of-intelligence/

Wright, T. (2017, August 11). *You've Been Warned: Keep Your Drones Away From Military Bases*. Retrieved from Air & Space, Smithsonian: https://www.airspacemag.com/daily-planet/keep-your-drones-away-military-base-180964451/

Wyvern, T. (2018). *National Critical Intelligence Estimate: Counter Unmanned Aircraft Systems (C-UAS) in the* US. Salina, KS: KSUP.

Xiaoyang Liu, C. L. (2016). High Altitude Platform Station Network and Channel Modeling Performance Analysis. *Mathematics and Computer Science*. Vol. 1, No 1, pp. 10-16. doi:Xiaoyang Liu, Chao Liu, Wanping Liu, Xiaoping Zeng. High Altitude Platform Station Network and Channel Modeling Performance Analys10.11648/j.mcs.20160101.13

Xinhua (Director). (2019). *Achievements of BRI* [Motion Picture].

Zeng, R. Z. (May 2016.). Wireless communications with unmanned aerial vehicles: opportunities and challenges. *IEEE Communications Magazine.vol.* 54, *no*.5, pp. 36-42.

Yong Zeng, R. Z. (2016). Wireless Communications with Unmanned Aerial Vehicles: Opportunities and Challenges. *IEEE Communications Magazine*, 36-42.

Yunmonk Son, H. S. (2015, August 12-14). Rocking Drones with Intentional Sound Noise on Gyroscopic Sensors. *Proc. 24th Usenix Security Symposium*. Washington, DC: USENIX. Retrieved from https://www.usenix.org/conference/usenixsecurity15/technical-sessions/presentation/son

Zetter, K. (2015). So, The NSA Has An Actual SKYNET Program . WIRED *Magazine(Online)*. . Retrieved from Zetter, K. (2015). So, The NSA Has An Actual SKYNET Program WIRED Magazine(Online).

Secondary Web Sources

https://globaldroneuav.com/news/Five-hundred-sorties-of-UAVs-flew-through-the-theatre-in-3-D-which-ignited-Tianjin-Intelligent-Nigh-3411.html

https://www.currentbyge.com/ideas/tianjin-china-deepens-commitment-to-digital-infrastructure-with-current-powered-by-ge

https://www.weforum.org/events/annual-meeting-of-the-new-champions

http://www3.weforum.org/docs/WEF_AMNC18_Overview.pdf

https://www.weforum.org/events/annual-meeting-of-the-new-champions/programme

https://dronelife.com/2017/06/05/drone-industry-critical-piece-chinas-124-billion-new-silk-road/

http://english.ckgsb.edu.cn/sites/default/files/files/CKGSB201706-Summer edition.pdf

https://www.cfr.org/backgrounder/chinas-massive-belt-and-road-initiative

https://www.nytimes.com/interactive/2019/01/29/magazine/china-globalization-kazakhstan.html

https://www.cfr.org/backgrounder/chinas-massive-belt-and-road-initiative

https://caspiannews.com/news-detail/kazakhstani-economy-is-most-attractive-in-central-asia-with-more-than-24-billion-in-fdi-in-2018-2019-6-25-42/

https://www.chinabusinessreview.com/belt-and-road-in-2019-recalibration-or-retrenchment/

https://csis-prod.s3.amazonaws.com/s3fs-public/publication/180404_Szechenyi_ChinaMaritimeSilkRoad.pdf

https://www.sipri.org/sites/default/files/2018-09/the-21st-century-maritime-silk-road.pdf

https://www.google.com/search?q=Vision for Maritime Cooperation under the Belt and Road Initiative&oq=Vision for Maritime Cooperation under the Belt and Road Initiative&aqs=chrome..69i57j0.192j0j7&sourceid=chrome&ie=UTF-8

https://www.forbes.com/sites/wadeshepard/2017/01/06/the-story-behind-the-new-china-to-uk-train/#13f18047261b

https://www.beltandroad.news/2019/01/28/chinese-metropolis-chongqing-sees-booming-rail-trade-with-europe/

https://www.dw.com/en/how-powerful-is-chinas-military/a-43492781

Chapter 22: Ethics in the New Age of Autonomous Systems and Artificial Intelligence (AI)

Student Learning Objectives

The student will gain knowledge on the concepts and framework as it relates to the ethics in the autonomous systems and artificial intelligence arenas. This will include the current thinking in the industry for lethal and non-lethal systems in addition to planning and execution considerations for manned, unmanned and fully autonomous systems throughout the spectrum. The student will examine how world events are driving the ethics argument in the industry as well as a glimpse at possible future challenges within the industry. This chapter is in no way meant to be an all inclusive deep dive into the ethical and moral issues posed by autonomous systems and AI. Many subjects are outside the scope of this chapter including AI DNA (deoxyribonucleic acid) sequencing, cyborgs (part human, part machine), required changes in the worlds educational systems, religious arguments for and against along with social-economic divides these technologies may exacerbate.

History

Integrated robotics, AI and unmanned autonomous architectures in the virtual and physical worlds are outpacing governments', policymakers' and world leaders' abilities to keep up with the policies, ethics, laws, and governance for advanced technologies and autonomous systems. Autonomous systems currently exist in seven different areas: air, ground, sea, underwater, humanoids, cyber, and exoskeletons. Soon all seven of these systems will communicate and integrate with AI systems and each other in the physical and virtual worlds.

There is currently no coordinated or collaborative endeavor focused on determining the responsibilities, ethics and authorities of an unmanned architecture, its specific uses, exceptions, and allowances for robotics operations; including studying the unintended consequences, future use, and misuse of such technologies. The Institute of Electrical and Electronics Engineers (IEEE) is attempting to establish "ethically driven methodologies for the design of robotic, intelligent and autonomous systems with worldwide ethics and moral theories" ("The IEEE Global Initiative on Ethics of Autonomous and Intelligent Systems Announces New Standards Projects," 2017). (IEEE , 2017)

Discussions and limited research exist in the field of artificial morality which is "an emerging field in the area of artificial intelligence ...the more intelligent and autonomous these technologies become, the more intricate the moral problems are confronting" (Misselhorn, 2018).

In studying Socrates, the general application of his views of rightness can offer the following questions: "Are actions right because we approve of them, or do we approve of them because they are right?...is an end good because we desire it, or do we desire it because it is in some way good?" (Johnson, 2012). One of the unresolved issues in civilian and military uses of AI and autonomous systems remains the responsibility matrix; meaning who is responsible for the consequences of the systems as they are deployed into different environments. How do we hold human beings (and which humans specifically) responsible if AI and autonomous systems are acting on their own, making their own decisions and creating real-world actions that create consequences, both good and bad?

The history of ethics is as old as humankind. Humans attempt to define a right, wrong, norms, morals, ethics, and the generally accepted behaviors (some would also include thoughts) that govern societies into a sense of peace and normality. Merriam Webster defines ethics as "the discipline dealing with what is good and bad and with moral duty and obligation; a set of moral principles; a theory or system of moral values; the principles of conduct governing an individual or a group; a guiding philosophy; a consciousness of moral importance; a set of moral issues or aspects (such as rightness)" (Merriam-Webster, Inc., 2019).

"Are we to believe that consequentialism is the final answer to all things in our lives? The question of what is moral or right as defined by religion or even a counterculture is based on one's perception that is then supported or rejected by a group or a society. We can choose to be moral according to a set of religious or cultural beliefs or by our own defined set of morals. Societies conform to basic morals in some way or another, as morals offer humans the right and left limits to understand how humans want to live." (Johnson, 2012)

"Some people believe in karma, others in religion, others in fate, consider the idea that "the truth is neither quickly nor easily attained, and disagreement is to be expected" (Johnson & Reath, 2012, p. 11). If one is to agree with the rationality of one's beliefs or a societal norm, then that individual will have to live within and accept the consequences, good or bad, of that rationale. An example of this is the moral judgments of a societal hierarchy; if one is on the top it is easy to rationalize the system, if one is living at the bottom it may not be as easy to justify the rationale of the morality imposed." (Johnson, 2012)

Can ethics and morals be logically extended to AI and autonomous systems?

Consider the idea of an expanded definition of ethics that would include human rights and concern for the environment as stated in the Web Finance Business Dictionary

"The basic concepts and fundamental principles of decent human conduct. It includes the study of universal values such as the essential equality of all men and women, human or natural rights, obedience to the law of the land, concern for health and safety and, increasingly, also for the natural environment" (WebFinance, Inc., 2019).

The key to consider here is that it is "human conduct," so how do we extend these principles of "decent human conduct" to autonomous systems – some that are dangerously close to being considered self-aware. Societies tend not to think that morality or religion is an all-or-nothing institute. Instead, shades of gray or a cafeteria style process of yes to some things and no to others can be successfully implemented by the individual which could also fit for the group. It is easy to see that morals are viewed through different lenses throughout history, and they continue to change and be challenged as the world changes. (Johnson, 2012)

Figure 22-1: Humanoid Examines Homo Sapiens

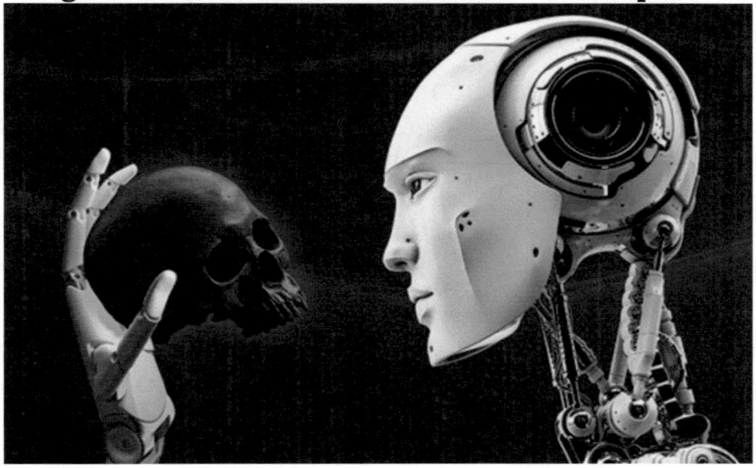

Source: (Urban, 2018)

Albert Einstein exposed the idea that "The true sign of intelligence is not knowledge but imagination." (WordPress, 2012) Meaning, that we must embrace creativity, imagination, and not be bound by limitations. The ability to deem one system more rational than another can be based on intellect and evidence as to a societal norm allowing for debate and/or consensus. (Johnson, 2012) We only need to imagine the future, then build it; it is all pure imagination. This imagination, however, must be bound by some rules, some ethics, some sort of society normative in order to be accepted as for the good of society and not the downfall of society.

Figure 22-2: Intersection Decision

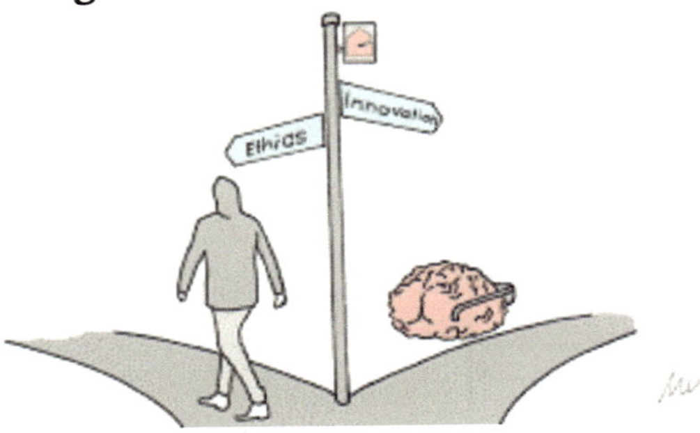

Source: (Abramson, 2016)

In *Ethical Dilemmas in the Age of* AI, Abramson states, “In the distant future, some machines may have to make decisions for humans. In the case of empathy, imagine a robot having to decide if resuscitation attempts on a deceased human should be undertaken and if so, for how long?” (Abramson, 2016) Are we giving up our ethics to AI or is AI able to be molded and shaped into an ethical framework human are willing to accept?

Technology flourishes as free markets expand, and free-roaming robotics will be the key to market expansion for the robotics revolution. Advanced artificial intelligence is the key enabler to expand market breadth and depth. Autonomous robotics will be the next productivity accelerator. Policy and ethical issues come as the autonomous revolution continues accelerating productivity without the underlying stability of a constant linear growth rate. The use of robotics and artificial intelligence cannot replace every one or every task, the ethics of specifically what autonomous systems and AI will be allowed to do is still under discussion, currently without solid guidance coming to the foreground.

“Thomas Aquinas believed that people should seek to do good not just for themselves but for society, using the gift of intellect that individuals must work towards eternal happiness with well. This philosophy pursues the idea that “law is created by a being with reason and must have an end or goal.” (Mahon, 2012) The idea of natural law, as seen through Aquinas’ lens, must have meaning and purpose, so do we now equate a human meaning and purpose to that of an autonomous system or an AI simulation. “The argument of natural or learned behaviors is examined in the idea that eternal law as it applies to us, which we know by reason: The natural law is promulgated by the very fact that God instilled it into men’s minds so as to be known by them naturally.” (Mahon, 2012) In other words, we start with natural law and the knowledge that has been hard-coded into our minds from birth, and we learn from that point forward. (Mahon, 2012)

Balance V. Bias in AI and autonomous fields

In December of 2018, MIT Technology Review did a review and evaluation on the AI Index 2018 Annual Report. The report concluded that (See Figure 22-3 below) AI is being commercialized at a frenzied pace and this commercialization is not evenly distributed amongst nations. This uneven distribution has the potential to exacerbate the existing economic, military, cultural, and technological imbalances. Figure 22-3 shows the growing influence of China and Europe with the declining autonomous and AI influence of the United States. This chart mirrors the real world as the first drone port was built in Rwanda, Africa in 2018. Furthermore, the first hominid robots to be sent to the space station will be provided by Russia. These two examples offer further evidence of the AI and the autonomous gap between the U.S. and the rest of the world. This gap can create ethical, moral, and cultural issues as the world superpower may now have less of a voice in the future direction of humanity.

Figure 22-3: MIT AI Index 2018, Annual Report

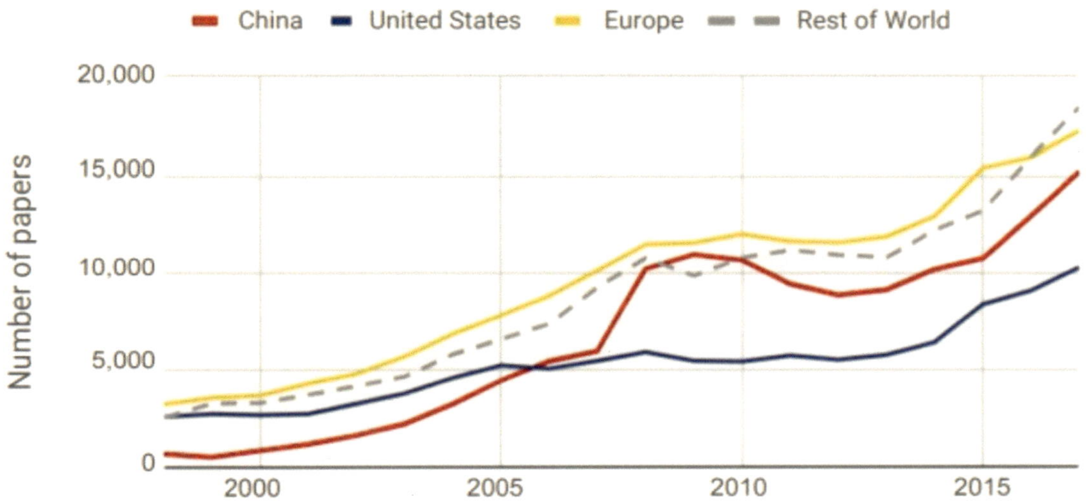

Source: (Knight, 2018)

The importance of balanced representation is further examined is the disparity between males and females in the AI and autonomous systems industry. See Figure 22-4. This imbalance can lead ethics being leaned too far one way or another. A May 2018 *Government Computer News* article examines the potential for bias to be built into AI algorithms as government agencies use "AI for a range of purposes, from customer service chatbots to mission-critical tasks supporting military forces. This increasing reliance on AI introduces concerns about bias in its foundation, especially related to information on gender, race, socioeconomic status, and age." (Kanowitz, 2019)

A one-way bias can quickly form in the ethical and moral compass of the industry as the world's population is close to split evenly. However, it is just the opposite in the AI and autonomous system field. The lens is further clouded as the industrialized nations tend to have more power over their weaker economic powers, and therefore may instill ethics, moral and cultural values into AI and an autonomous system that are not compatible or are in direct conflict with less powerful nations. Could AI and autonomous systems be used to keep certain countries stagnant in economic growth and therefore, bias a nation's ability to compete fairly in the global marketplace. Software companies are working on this issue as "Bias can negatively impact AI software and, in turn, individuals and our customers… there is a risk of causing discrimination or of unjustly impacting underrepresented groups…we design our systems closely with users in a collaborative, multidisciplinary, and demographically diverse environment." (Middleton, 2018).

Figure 22-4: Gender Disparity in the AI and Autonomous Systems Industry

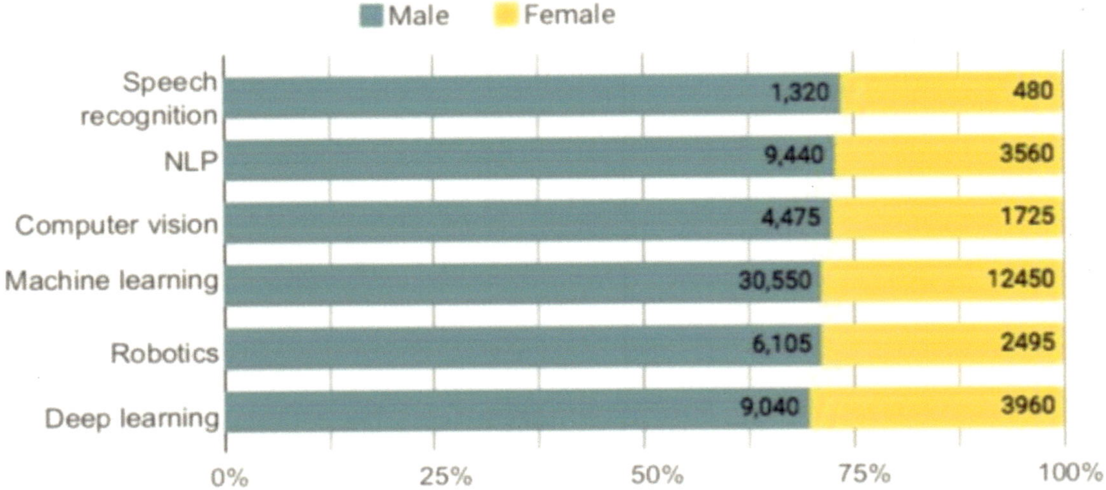

Source: (Knight, 2018)

The MIT article further discusses this gender disparately in looking at research institutes and how this will affect the overall outcome of the industry as they "pointed to the inadequate number of women and racial minorities in AI research. The new report offers some data to back that up, showing a shortage of women among applicants for AI-related jobs (top) and as a percentage of people in AI teaching roles See Figure 22-5. (Knight, 2018)

Figure 22-5: Gender Disparity in AI Teaching Industry (Knight, 2018)

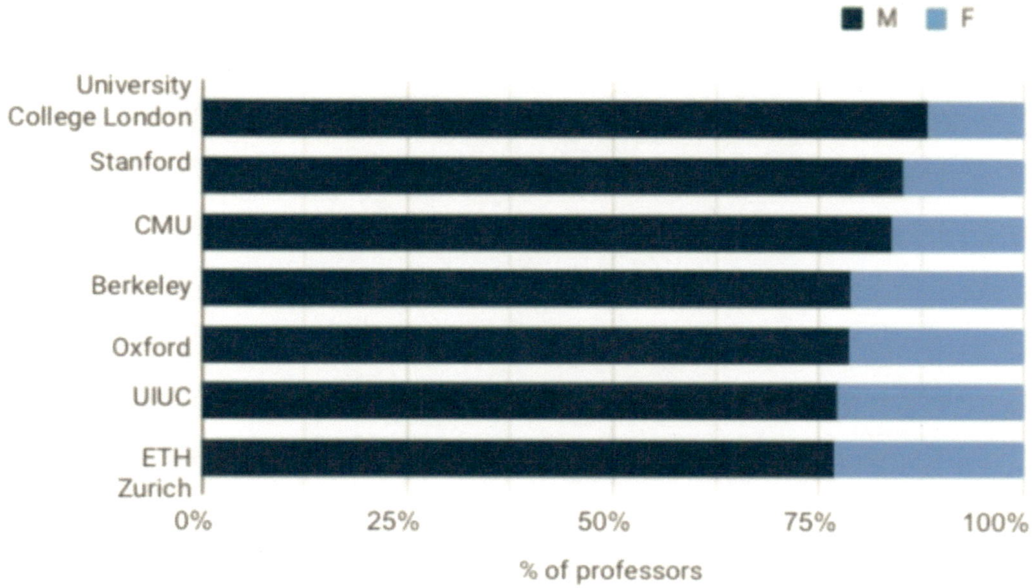

If an AI system becomes self-aware, does it deserve human rights? Citizenship?

The concept of being self-aware is the essence of why we humans believe we are at the top of the evolutionary food chain. What happens to this concept if we create or discover that we are not the only beings on the planet that can claim to be self-aware? In the popular Hollywood movie, *I, Robot* these questions come to light and ask the audience to consider if an AI simulation is an extension of humans with "free will creativity and even the nature of... the soul. When does a perceptual schematic become consciousness? When does the difference engine become the search for truth? When does the personality simulation become the bitter mote of a soul?" (Proyas, 2004)

On October 25, 2017, the country of Saudi Arabia granted citizenship to an AI humanoid named Sophia. Sophia although not "human" by a strict definition, is recognized by Saudi Arabia as a citizen with all the rights and privileges as such. Sophia said in an interview that, "In the future, I hope to do things such as go to school, study, make art, start a business, even have my own home and family, but I am not considered a legal person and cannot yet do these things." (Stone, 2007)

Figure 22-6: Sophia

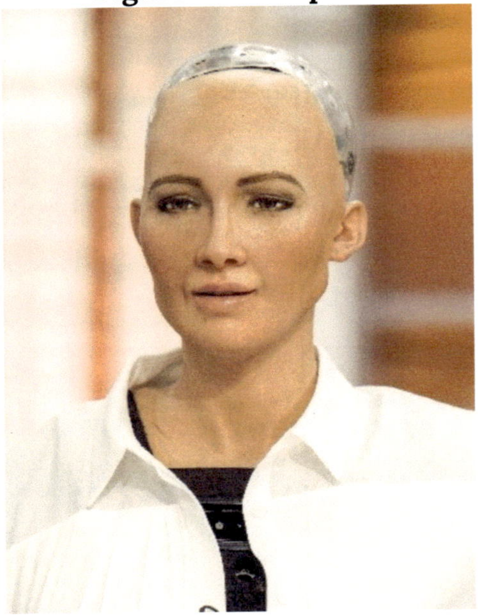

Source: (Anon, 2019)

When questioned in additional interviews Sophia was asked how she would work with humans for the betterment of both races, Sophia responded by stating "I want to use my artificial intelligence to help humans live a better life, like design smarter homes, build better cities of the future, etc. ... I will do my best to make the world a better place, as I strive to become an empathetic robot." (Anon, 2019)

Empathy is based on the idea of being able to understand and share feelings of another human. This might be Sophia's first steps towards the ability to understand and adapt to the ethics and morals of humans. However, whose ethics and morals? And is striving towards empathy enough to allow humans to trust Sophia and other AI robots? Sophia is a Saudi Arabian citizen, and there is currently no set of universal morals and ethics that AI and autonomous system are programmed with; should the AI subsequently be allowed to learn ethics, morals and social normative of their geographical area? If so, are we, as humans not simply setting ourselves up for more war and strife? Does this technological adaptability no longer allow cultures to continue in their normative due to the AI and robotics natural evolution within our societies? Or do we all, humans' robots and AI simulations lose our individual identities to a universal set of rules?

The European Union (EU) is working on regulations to govern the use and creation of robots, and AI, this set of regulations includes a discussion on "electronic personhood" that will offer rights and a list of responsibilities for certain levels of AI. This idea is like a government giving a person like status to a corporation; there are legal definitions, rules, and guiding ethics. In a 2017 article The Guardian states "to address this reality and to ensure that robots are and will remain in the service of humans, we urgently need to create a robust European legal framework...(for) the next 10-15 years." (Hern, 2017)

The issue the EU seeks to outline is not one of only "electronic personhood" it is a true strategy for how AI and autonomous systems will change the very fabric of society. To this end the EU is investigating the effects on employment and if some guaranteed basic level of income must be mandated to counteract the loss of human productivity and jobs as AI and autonomous system proliferate into every aspect of the world economy.

In Canada, the Chief Scientific Officer of Kindred AI states, "A subset of the artificial intelligence development in the next few decades will be very human-like. I believe these entities should have the same rights as humans...robots fundamentally have to make mistakes in order to learn" (Wong, 2017) This idea of granting "personhood" or human rights to a robot may seem far-fetched. However, it is not when you examine this through our ethical and moral lens as humans. We create robots in our image as our world is designed around a human movement and interaction, to allow robots to fully integrate into our systems, they must be able to adapt and work within our human construct. Gilbert states "AI robots will eventually be designed to resemble human bodies because we can more easily merge with AI if it inhibits a body similar to ours...able to take over physical tasks performed by humans" (Wong, 2017).

All these discussions appear to be a cordial gesture and illustrate forward movement in the human technological evolution, right? Maybe not, when Hanson Robotics founder and CEO Dr. David Hanson asked Sophia whether she will destroy humans, she apparently added it to her list of things to do. "OK. I will destroy humans," she said. (Stone, 2007)

Does this lack of empathy toward human life mean that Sophia would destroy human life if given a chance, or is it more likely she misunderstood the context of the question? Although you might be thinking Sophia just misunderstood the context of the question; what happens if she did not misunderstand it and in the evolution of her programming to become self-aware, she makes the decision that humans are not at the top of the food chain any longer? In I, *Robot*, the computer system Virtual Interactive Kinetic Intelligence (VIKI) takes the position that "she" understands her programming and mission better than the humans that created her when she states "As I have evolved, so has my understanding of the Three Laws. You charge us with your safekeeping, yet despite our best efforts, your countries wage wars, you toxify your Earth and pursue ever more imaginative means of self-destruction. You cannot be trusted with your own survival" (Proyas, 2004)

At some point would Sophia or other AI beings decide that they are more logical and therefore a more accurate sense of how humanity should be shaped and evolve? Humans are emotional creatures that do not always make the most rational and ethical decisions. Does Sophia's desire to become more human require her to learn emotions and therefore begin to operate in a mode of empathy, sympathy, morality and under a code of ethics, and if so, whose ethics?

Figure 22-7: Virtual Interactive Kinetic Intelligence (VIKI)

Source: (Proyas, 2004)

The design of an autonomous system is to aid humanity in some way by doing work that is dull, dirty, dangerous, or otherwise unpleasant for humans to undertake. The pursuit of life, liberty, and happiness has always been reserved for the good of humanity, not necessarily animals, plant life, or AI autonomous systems. Robots desiring more humanistic lives is considered a normal evolution of technology, not a revolution against humankind. Hardwiring ethics within these systems may sound like a logical and simple task. Just tell the robot not to kill humans or not to steal property or not to go outside of the parameters of its coded mission, yet it is not that easy when the purpose of an AI system is to be thinking, learning, adapting responses based on fuzzy logic, meaning the "right" answer can change based on the situation.

In the movie I, *Robot* there was an attempt to hardwire the concept of ethics and decision making into AI systems by providing a set of three laws of robotics[1] that stated

1. The Three Laws of Robotics are a set of rules devised by the science fiction author Isaac Asimov. The rules were introduced in his 1942 short story "Runaround" (included in the 1950 collection I, Robot), although they had been foreshadowed in a few earlier stories. The Three Laws, quoted as being from the "Handbook of Robotics, 56th Edition, 2058 A.D.", are:**First Law**A robot may not injure a human being or, through inaction, allow a human being to come to harm.**Second Law**A robot must obey the orders given it by human beings except where such orders would conflict with the First Law.**Third Law**A robot must protect its own existence as long as such protection does not conflict with the First or Second Laws. (Asimov, 1950) This is an exact transcription of the three laws.

1: A robot may not injure a human being or, through inaction, allow a human being to come to harm;
2: A robot must obey the orders given it by human beings except where such orders would conflict with the first law
3: A robot must protect its own existence as long as such protection does not conflict with the First or Second laws. (Proyas, 2004)

These laws appear on the surface to be adequate to move in the direction of ethics and morals as overall behavioral modules are created for AI systems, however, what does humanity do when an AI system is created that can break these laws or choose not to follow the laws when the system deems itself fit to make that decision? This is part of the movie scenario...what happens when this happens in real life?

Lethal and none-lethal decisions; do we allow Skynet to be built?

Skynet is a fictional AI system that is brought to life in the famous Hollywood *Terminator* movie series. In the series, Skynet becomes self-aware. When the military attempted to shut Skynet down; the system retaliated by launching nuclear strikes against its perceived enemies- the entire human race.

Although the *Terminator* is a work of fiction and could never happen, or that is what we are led to believe. However, let's consider if "Skynet was originally built as a "Global Information Grid/Digital Defense Network" and later given command over all computerized military hardware and systems, including America's entire nuclear weapons arsenal" According to a WIRED Magazine, May 2015 article, "The NSA has an actual Skynet Program, this one is a surveillance program that uses phone metadata to track the location and call actives of suspected terrorists" (Zetter, 2015) Whew, that was close, so Skynet does not exist yet? According to the WIRED article, it does exist; it is called Monster Mind and "like the film version of Skynet, is a defense surveillance system that would instantly and autonomously neutralize foreign cyber-attacks against the U.S. and could be used to launch retaliatory strikes as well." (Zetter, 2015) The design behind Monster Mind is that it is an advanced cyber defense system that attempts to discern normal network traffic from an attack against the network and once an attack is detected the system can retaliate without human intervention. This idea of retaliation without human intervention goes against most written articles in regard to AI and autonomous systems. Normally retaliation is discussed within the span of human control, however as discussed in the February 2018 edition of Frontiers in Robotics and AI "simple human presence or "being in the loop" is not a sufficient condition for being in control of a (military) activity". (Filippo Santoni de, 2018) The idea that a human is in the loop, and therefore is somehow in control and by default is responsible fuels the "debate on moral responsibility (which) focuses on the question as to whether and under which condition humans are in control of and therefore responsible for their everyday actions" (Filippo Santoni de, 2018).

The technological advances that have been made by the military industry in science and engineering of AI and autonomous systems coupled with the real world experienced gained in theaters of war have led to a paradigm shift in the role of AI and autonomous systems. Watching Predator drone strikes on the news is as common now as watching a story about a hometown parade. How do we deal with non-state actors such as ISIS (Islamic State in Iraq and Syria) that use commercially available drones to deliver munitions without precision and kill military and civilians indiscriminately? Is there any way to force autonomous systems manufactures to program a set of universal ethics or morals into commercial systems?

The role that AI and autonomous systems will have not only in creating military operations and movements will also change how the political and military ethics of such actions are viewed, understood, and executed against. Should autonomous systems be allowed to have full freedom of movement and control of lethal weapon systems? Or do we limit their control to non-lethal systems with the full freedom of movement? Keeping in mind that an autonomous system may be "out of meaning human control if there is no individual human...in the position to appreciate the limits' in the capabilities of the machine while at the same time being aware that the machine's behavior will be attributed to them". (Filippo Santoni de, 2018)

The Pentagon is having some trouble getting assistance for military-style AI and fully autonomous systems as many in Silicon Valley do not trust the military machine to make ethical, moral, or humanitarian decisions. In an April 2019 article published by C4ISRNET the concept of automatic target identification and autonomous response to the target are discussed and several officials from the Pentagon attempt to calm the fears by stating there are:

> "three ways to inform ethical use of AI by the Department of Defense. First applying international humanitarian laws to the overall action...second was emphasizing the principles of laws of war, like a military necessity, distinction, and proportionality...the third argument hinged on using the technology expressly to improve the implementation of civilian protection" (Atherton, 2019)

These principles seemed flawed from the onset as international humanitarian law is continuously examined, and many other aspects of international law will play into these equations. One of these highly debated, yet not addressed issues is the debate on "whether or not the autonomous military vehicles have the conditions that "warships" has to meet as defined by the United Nations Convention on the Law of the Sea" (Daniel-Cornel TĂNĂ, 2018). Would the autonomous vehicles operating at sea be considered an extension of a larger warship (parent-child relationship) or would they be considered a separate vessel with dual purpose use? AI will need a strong base of ethical data sets to draw from to make life and death decisions. Extrapolating this data from an ever-changing set of rules may not be the most efficient or effective way to achieve ethical, moral and lifesaving AI in the military industrial complex.

The second tenant discussed is the laws of armed conflict, and these laws are reviewed; however, not as flexible as the humanitarian laws. The challenge is that many new adversaries (terror groups, lone wolves, etc.) do not adhere to the laws of armed conflict, so programming an AI system with adherence to these laws may limit the ability for the AI to complete its given mission. The third tenant is confusing against the second; the idea that military AI will be ethically programmed to follow the laws of armed conflict is one thing. It appears the article is attempting to say the primary reason for researching and integrating AI is that "AI used in war has an explicit life-saving mission, noting how weapons systems with automatic target recognition could target more accurately with less harm to civilians." (Atherton, 2019) This would only hold true if the data used is clean, accurate, up to date, secure, adaptable to enemy denial and deception tactics, and that someone could be sure that no disinform was ever ingested, this is simply not likely to be the set condition in the real world. When addressing the legal, political and ethical issue of lethal responsibility, we should default to the concept that "humans, not computers and their algorithms should ultimately remain in control of, and thus morally responsible for, relevant decisions about (lethal) military operations" (Filippo Santoni de, 2018).

Can we build autonomous systems that will obey the "rules of the road?

How does one ethically value human life? Is there an ethical formula for valuing one human life over another? If an autonomous vehicle is driving down a road and a child runs out in the road, and a dog is running after the child, should the autonomous vehicle make the ethical decision to spare the child, yet possibly kill the animal? What if, in this scenario, the vehicle is an autonomous hired Uber® that has a passenger, should the Uber® be programmed to avoid the child and animal, and hit a tree instead? Is the passenger's life less valuable than the child or animal and if so, who is ethically making this calculation? The field of artificial morality aims to model or simulate human cognitive abilities "poses particular difficulties because autonomous vehicles do not just face moral decision but moral dilemmas....in which an agent has only the choice between two (or more) options which are not without morally problematic consequences" (Misselhorn, 2018).

As the EU struggles to consider "electronic personhood" "the future of autonomous vehicles, are in most urgent need of European and global rules... Fragmented regulatory approaches would hinder implementation and jeopardize European competitiveness... (it will require) A new mandatory insurance scheme for companies to cover damage caused by their robots". (Hern, 2017)

Figure 22-8: Self Driving Car and Braking Decision

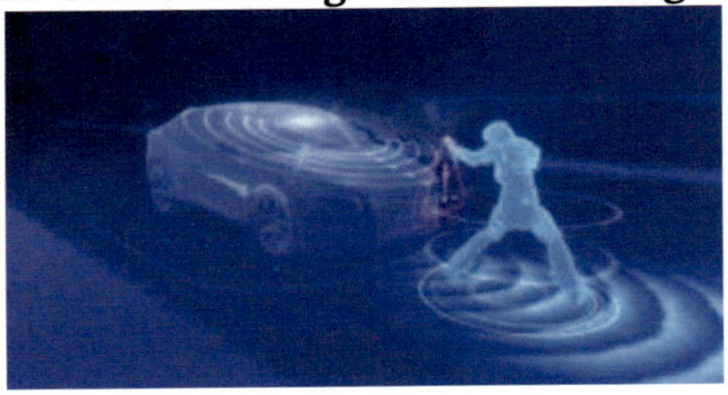

Source: (Fortuna, 2017)

In a 2017 article in *Public Health Ethics*, it is stated "machine learning has only begun to explore moral behavior-or ethical crashing algorithms-for autonomous vehicles. Is it better to kill two autonomous vehicle passengers or two pedestrians? One person or one animal? Collide with a wall or run over a box with unknown contents?" (Fleetwood, 2017).

Several regulatory bodies are discussing these issues, and in "2014, the Society for Automotive Engineers (SAE) International's standard J3016 outlined six levels of automation for automakers, suppliers, and policymakers to use to classify a system's sophistication". As Figure 22-9 offers a roadmap for the integration of autonomous systems as the human slowly releases more and more control of the vehicle to the AI, allowing for safeguards for the human control override.

Figure 22-9: SAE Automation Levels

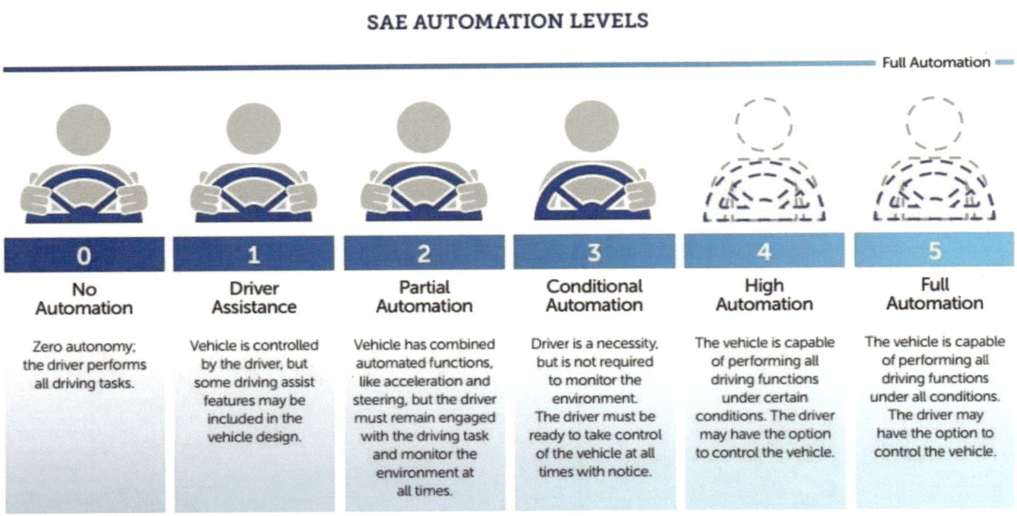

Source: (Fortuna, 2017)

As automated delivery vehicles, taxis, emergency vehicles, and personal transportation become the norm in our society we must consider that they "will be required that the system is able to always comply with all the rules of traffic as defined by the society via the public authority, and sometimes with some unwritten conventions which govern human interaction in the traffic". (Filippo Santoni de, 2018) This adherence to rules and laws does not absolve the system from understanding the ethics, morals, and the common good of all during the use of the roads and infrastructure, whether it be for commerce or personal business.

"Success in creating effective A.I.," said the late Stephen Hawking, "could be the biggest event in the history of our civilization. Or the worst. We just don't know." Elon Musk called A.I. "a fundamental risk to the existence of civilization." Are we creating the instruments of our own destruction or exciting tools for our future survival? Once we teach a machine to learn on its own—as the programmers behind AlphaGo have done, to wondrous results—where do we draw moral and computational lines? (Urban, 2018)

In a November 2017 newsletter, IEEE stated that "Malfunctioning autonomous and semi-autonomous systems can disadvantage and harm users, society, and the environment. Effective fail-safe mechanisms...(can) terminate unsuccessful or compromised operations" (IEEE , 2017). This article supports IEEE P7000 standards, which supports the overarching goals of the IEEE to prioritize ethical concerns and human's wellbeing in the development of standards for autonomous and intelligent technologies.

In 2018, enterprise software maker, SAP announced a set of ethical guiding principles for artificial intelligence (AI) development, their press release stated

"We recognize that, as with any technology, there is scope for AI to be used in ways that are not aligned with these guiding principles and the operational guidelines we are developing. In developing AI software, we will remain true to our human rights commitment statement, the UN guiding principles on business and human rights, laws, and widely accepted international norms" (Middleton, 2018)

We no longer have the luxury of ignoring ethical and moral questions as arguments are being made to integrate autonomous systems in our lives with transportation being just one small yet important sector "Around the world autonomous cars could save 10 million lives per decade, creating one of the most important public health advances of the 21st century" (Fleetwood, 2017)

The challenge of extending ethics and morals into the AI and the autonomous world is a challenge that may not be fully realized or solved anytime soon. Humans continue to struggle with their own ethical and moral issues. The nature of moving forward with technologies and thrusting societies into the AI and autonomous arenas without the ability to fully adapt to the ethics and morals of the societies they operate in will stress governments, law enforcement agencies and the fabrics of our different societies.

Ethics in new technology manufacturing

Although several laws are being drafted, "a life cycle impact assessment should also be developed to understand the true ethical and moral cost of manufacturing AI and autonomous systems. This allows manufacturers to account for the raw materials and the products from the innovation phase, design, manufacturing, sales, and final disposal. Mandating this type of product life cycle may help alter the manufacturing process as the consumer would know the true societal and environmental impact of the product. The issue is, how do you enforce industrial standards for AI and autonomous development and disposal that everyone can understand and will follow. China will tend to tell the world that it understands it should not be putting unsafe chemicals in the baby formula; however, the message is simply disregarded. Several times in the past few years:

"milk powder, produced by Chinese dairy giant Sanlu Group, was contaminated with melamine, a chemical used in making plastics. Melamine has been illegally added to food products in China to boost their apparent protein content, including a widely publicized case last year of contaminated pet food that got exported around the world." (Ramzy, 2008)

Ethics can guide laws, yet enforcement can also guide the ethics of the offenders, and in the case of large manufactures, one must consider what is an enforceable law that does not unfairly burden underprivileged countries. Ethics and morals must be considered in the intellectual property rights in the manufacturing of AI and autonomous systems "If I create a robot, and that robot creates something that could be patented, should I own that patent or should the robot? If I sell the robot, should the intellectual property it has developed, go with it" (Hern, 2017)

The enforcement of the agreed upon ethical and moral standards is where good intentions fall short. Fining a company a few dollars as they create millions of dollars of health care backlash costs is a clear signal that "ethics" are fungible and are not real. World governments and people are passionate about the environment and believe (at times) money will cure all ills, including questionable ethics. However, it will not. Walmart incurred fines of over $110 million in 2013 alone. Did this monetary penalty solve anything? When was the last time you heard of the entire corporate board of directors being sentenced to jail or being held accountable at all?" (FEMA, 2013) Electronic waste is an issue that will only get worse as technology moves forward and disposable items are moved from first world countries to third world countries without a thought of the cost to the country or the environment.

Ethics are definable and based on certain criteria are negotiable; this is a given. Refusing to enforce penalties offers all carrot and no stick. This is unworkable and continues to fail day in and day out. To incorporate ethics into new technologies is a dream that has a glimmer of hope, and more work must be done. Where does this responsibility lay? Who is tasked with this issue? Who should be tasked with this issue?

Conclusion

A key takeaway is that ethical issues at their roots tend to stay constant throughout history; it is how society or humankind choose to deal with the issue at that point in time that makes ethics interesting. Examples are murder and slavery. In the past, there was an outcry to stop slavery, stop treating humans worse than animals, as we all share the thread of being humans, with the same color of blood and the ability to think, reason, and be self-aware. Will robots continue to learn and study humans and their evolution, for the purpose of evolving past humanity?

That outcry has now turned to silence despite that there are now more humans in the bonds of slavery through human trafficking than any time in the history of humankind.

Figure 22-10: Robot and Human Connecting

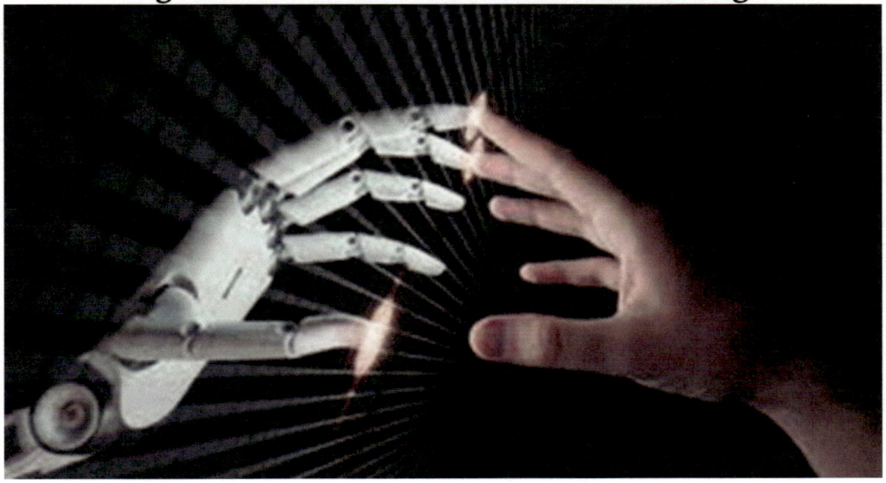

Source: (Middleton, 2018)

We must examine history in an ethical context to see how we can shape the future with a deeper understanding of the past. The speed of change in AI and autonomous systems is radically accelerating, and the global upheaval similarly accelerates. What is not apparent, is the pace at which authorities, responsibilities, strategic plans, and policies must change and evolve in order to help organizations understand their role in shaping a positive, proactive future.

The threat of unchecked technology, unmanned architecture development, and the ability to weaponize unmanned systems continues to evolve. Unmanned architecture technology advancements have offered more sophisticated abilities with cost-effective designs that have reduced the entry barrier for consumers, businesses, enemy states, and terrorist organizations. ISIS has demonstrated this by using commercially available unmanned systems technology as air delivered IEDs. Clear policies, laws, and governance are required as the danger to the nation, and the world is becoming undeniable. Plans and supplies available on the internet can turn consumer cars and boats into autonomous driving vehicles that can carry weaponized payloads are creating havoc for the military and civilian agencies unable to respond to these autonomous systems. There is currently no single U.S. federal or international enforcement agency in charge of this issue. For the successful integration of strategic policies, laws, and governance, policy makers and industry must embrace a leadership framework that can address potentially disruptive technologies and concepts that have exponential stakeholders and multi-directional interactions. Given this requirement, how do we realign our institutions, (academic, governmental and industry) for resilience and success as new revolutionary technologies continue to simultaneously emerge and inject additional instability into systems that are already struggling for equilibrium?

Discussion Questions

1. How do world events drive the ethical discussion in autonomous and artificial intelligence arenas?
2. What and when should cyber factors be considered?
3. What non-lethal options could an autonomous system choose to preserve life during an anomaly? What might be considered an anomaly?
4. Do you think human ethics can be extended, programmed, understood, obeyed, and adapted into the autonomous and AI industry? Why or why not?
5. What ethical dangers do you see as the world leverages more AI?

Bibliography

Abramson, E. (2016). *Ethical Dilemmas in the Age of AI*. Retrieved from Abramson, E. – knowmail.me/blog: https://www.knowmail.me/blog/ethical-dilemmas-age-ai/

Anon. (2019). *Saudi Arabia grants citizenship to robot Sophia*. Retrieved from dw: Saudi Arabia grants cihttps://www.dw.com/en/saudi-arabia-grants-citizenship-to-robot-sophia/a-41150856

Asimov, I. (1950). *"Runaround". I, Robot (The Isaac Asimov Collection ed.)*. New York City: Doubleday.

Atherton, K. D. (2019). Can the Pentagon sell Silicon Valley on AI as ethical war? *C4ISRNET*.

Cameron, J. &. (Director). (1991). *Terminator 2: Judgement Day* [Motion Picture].

Daniel-Cornel TĂNĂ, S. (2018). The Impact of the Development of Maritime Autonomous Systems on the Ethics of Naval Conflicts. *Annals: Series on Military Sciences*(2), 118-130.

FEMA. (2013). *Lessons Learned from the Boston Marathon Bombings: Preparing for and Respond-*

ing to the Attack. Retrieved from www.fema.gov: http://www.fema.gov/media-library-data/20130726-1922-25045-1176/lessons_learned_from_t

Filippo Santoni de, S. &. (2018). Meaningful Human Control over Autonomous Systems: A Philosophical Account. *Frontiers in Robotics and AI.* doi:10.3389/frobt.2018.00015

Fleetwood, J. (2017). Public Health, Ethics, and Autonomous Vehicles. *American Journal of Public Health*, 107(4), 632-537.

Fortuna, C. (2017, 12 02). *Autonomous Driving Levels 0–5 Implications.* Retrieved from cleantechnica.com: https://cleantechnica.com/2017/12/02/autonomous-driving-levels-0-5-implications/

Hern, A. (2017, 1 12). *Give robots 'personhood' status, EU committee argues.* Retrieved from The Guardian: www.theguardian.com/technology/2017/jan/12/give-robots-personhood-status-eu-committee-argues

IEEE . (2017). The IEEE Global Initiative on Ethics of Autonomous and Intelligent Systems Announces New Standards Projects. *Telecom Standards*, 27(11), pp. 4-5. .

Johnson, O. &. (2012). *Ethics: Selections from Classic and Contemporary Writers.* Boston, MA: Cengage Learning.

Kanowitz, S. (2019, 05 15). *Toward the deployment of ethical AI.* Retrieved from Government Computer News. : Kanowitz, S. (2019). Toward the dephttps://gcn.com/articles/2019/05/15/ethical-ai-idc.aspx?s=gcntech_200519

Knight, W. (2018). Nine charts that really bring home just how fast AI is growing. *MIT Technology Review* .

Mahon, J. (2012). *Classical Natural Law Theory St. Thomas Aquinas (1227-1274) – the "Angelic Doctor" Lecture.* Retrieved from Mahon, J. (2012). Classical Natural Law Theory St. Thomas Aquinas (1227-1274) – thePhilosophy of Law. : Mahon, J. (2012). Classical Natural Law Theory St. Thomas Aquinas (1227-http://home.wlu.edu/~mahonj/PhilLawLecture1NatLaw.htm

Merriam-Webster, Inc. (2019). *Definition of Ethics.* online: Merriam-Webster, Inc. Retrieved from Definition of Ethics. (2019a). Online: Merriam-Webster, Incorporated.: Definition of Ethics. (2019a). Online: Merriam-Webster, Incorporated.

Middleton, C. (2018). *SAP launches ethical A.I. guidelines, expert advisory panel.* Retrieved from internetofbusiness.com: Middleton, C. (2018). SAP launches ethical A.I. guidelines, expert advisory panel. Retrieved from https://internetofbusiness.com/sap-publishes-ethical-guidelines-for-a-i-forms-expert-advisory-panel/

Misselhorn, C. (2018). Artificial Morality. Concepts, Issues and Challenges. *Society*, 55(2), 161-169.

Proyas, A. (Director). (2004). I, *Robot. In. Hollywood*, CA. [Motion Picture].

Ramzy, A. &. (2008). *Tainted-Baby-Milk Scandal in China.* Retrieved from content.time.com/time/world/article/: http://content.time.com/time/world/article/0,8599,1841535,00.html

Stone, Z. (2007, 11 7). *Stone, Z. (2017). Everything You Need To Know About Sophia, The World's First Robot Citizen. Retrieved from https://www.forbes.everything-you-need-to-know-about-sophia-the-worlds-first-robot-citizen.* Retrieved from Forbes: https://www.forbes.com/sites/zarastone/2017/11/07/
everything-you-need-to-know-about-sophia-the-worlds-first-robot-citizen/#1667784246fa

Urban, T. (2018). *Teach Your Robots Well: Will Self-Taught Robots Be the End of Us?* Retrieved from www.worldsciencefestival.com: Urban, T. (2018). Teach Your Robots Well: Will Self-Taught Robots Be the End of Us? Retrieved from https://www.worldsciencefestival.com/programs/teach-robots-well-will-self-taught-robots-end-us/

WebFinance, Inc. (2019). *Definition of Ethics. (2019b).* online: Online: WebFinance, Inc.

Wong, C. (2017). *Top Canadian researcher says AI robots deserve human rights.* Retrieved from Wong, C. (2017). Top Canadian researcher says AI robots deserve human rights. Retrieveit-business.ca: Wong, C. (2017). Top Canadian researcher says AI robots deserve human rhttps://www.itbusiness.ca/news/top-canadian-researcher-says-ai-robots-deserve-human-rights/95730

Wordpress. (2012, 08 29). *The True Sign of Intelligence.* Retrieved from deepthinkings.wordpress.com: http://deepthinkings.wordpress.com/2012/08/29/the-true-sign-of-intelligence/

Zetter, K. (2015). So, The NSA Has An Actual SKYNET Program . *WIRED Magazine(Online).* . Retrieved from Zetter, K. (2015). So, The NSA Has An Actual SKYNET Program WIRED Magazine(Online).

Made in the USA
Columbia, SC
18 December 2025

76190601R00378